INTRODUCTION
TO
HUMAN
ANATOMY

ROBERTA BRUCK-KAN

Louisiana State University Medical Center

Harper & Row, Publishers
New York, Hagerstown, Philadelphia, San Francisco, London

**To my husband
MITCH KAN**

Sponsoring Editor: Bonnie K. Binkert/Kyle Wallace
Project Editor: H. Detgen
Senior Production Manager: Kewal K. Sharma
Compositor: Ruttle, Shaw & Wetherill, Inc.
Printer and Binder: Halliday Lithograph Corporation
Art Studio: Danmark & Michaels Inc.

Introduction to Human Anatomy

Library of Congress Cataloging in Publication Data

Bruck-Kan, Roberta.
    Introduction to human anatomy.

    Bibliography: p.
    Includes index.
    1. Anatomy, Human.   I. Title.
QM23.2.B78   611   78-5691
ISBN 0-06-041015-9

Contents

5

The Skeletal System
82

6

The Articular System
143

7

The Muscular System
171

8

The Cardiovascular System
256

13

The Endocrine System
454

14

The Reproductive System
491

15

The Nervous System
548

Preface

Today many college students are preparing for careers in allied health professions such as nursing, physical therapy, medical technology, dental hygiene, X-ray technology, and so on. Before taking professional-level courses in their specialty, these students generally take a course in human anatomy. Basic science programs for allied health students vary from one institution to another: In some, anatomy is combined with physiology; in others, anatomy is taught as a separate course. The instructor who teaches anatomy as a separate course is faced with a dilemma when choosing an appropriate textbook, as I was a number of years ago. Most college-level textbooks are designed for combined anatomy-physiology courses and are heavily weighted toward physiology. If the instructor chooses a combined anatomy-physiology textbook, the student has to sift through a large volume of physiology to glean a few anatomical facts. However, if the instructor chooses a textbook that deals solely with anatomy, the student may flounder in a morass of anatomical facts and be unable to relate structure to function.

After several years of teaching human anatomy to nursing students and experimenting with several different textbooks, I decided to create a textbook more suited to the needs of my students and others like them. My students have had a minimal exposure to biology and chemistry before taking human anatomy and take physiology as a separate but concurrent course. They have a heavy course load, with little time for reading, and are very goal-oriented. They want to relate what they are learning in their basic science courses to their future profession and do not want to waste valuable time in rote memorization of material they will never use. They are also very curious about disease processes and how disease disrupts normal structure and function.

Introduction to Human Anatomy is a textbook designed specifically for beginning students in the allied health professions who are taking anatomy as a separate course. In this book relatively sophisticated material is presented in what I believe is a well-organized and easily assimilated format. Much of the material is in list form, which reflects the book's evolution from an outline syllabus. Inasmuch as proper understanding and appreciation of a structure is impossible without knowing its function, I have included a section on function for nearly every organ and body part mentioned in the text. These functional descriptions are not intended to be as complete and detailed as they would be in a physiology textbook. Microscopic anatomy, blood supply, and innervation are included whenever they are relevant to understanding the function of a particular organ or body part. Because discussions of pathological anatomy attract the student's attention and reinforce his or her interest in learning normal human anatomy, I have included descriptions of some pertinent diseases and disorders at the end of each chapter or section. This material can be used or skipped over, depending on the individual instructor's needs and interests. To facilitate comprehension of the text, many of the illustrations were prepared by me and some were rendered by an art studio under my direction. The illustrations are semidiagrammatic in nature so that the student will not be confused by excessive detail and subtle shading effects.

I realize that it is impossible to write an "ideal" textbook, one that is suitable for every student and instructor of human anatomy at the undergraduate level. When one attempts to describe something as complex as the structure of the human body, it is extremely difficult to decide what material should be included and what should be omitted in an introductory textbook. Some students and instructors may feel that I have included too many details on microscopic structure and pathology for beginning students to handle; others may feel that I have not included enough details on gross structure. My choices were made on the basis of my own teaching experiences in anatomy at both the undergraduate and graduate levels, and I have included what I feel the allied health student needs to know.

For those students motivated to learn more, bibliographic references for further reading are included in the end-of-chapter "Suggested Readings" and in the end-of-text "References and Supplemental Readings." The readings that are of a more advanced level are preceded by a single asterisk, and those of a highly advanced level are preceded by two asterisks. To further facilitate study, the entries in the "References and Supplemental Readings" also are categorized by field.

I hope that *Introduction to Human Anatomy* proves to be a viable alternative to existing anatomy textbooks and that it will prove to both instructors and students alike as they engage in the study of the human body. I welcome all constructive criticisms and suggestions for improvement so that this book will come closer and closer to being the "ideal" textbook.

I would like to thank past and present colleagues who read and commented on segments of the manuscript during various stages of preparation: Drs. Jerry N. Bagwell, Ellen R. Batt, Anthony J. Castro, Robert F. Dyer, J. Ross McClung, Diane E. Smith, and Inia Hikawyj-Yevich. Other valued reviewers of the manuscript include: Lois L. Conrad, Jonathan Harrington, Eileen K. Howard, Sara E. Huggins, Patricia A. Lorenz, Patricia J. Morin, Henry D. Murphy, Richard L. Potter, and Lawrence M. Ross. I am indebted to an existing unpublished syllabus, prepared by Claudette Finley of the Department of Physical Therapy of the University of Florida, which gave me the idea of putting my own teaching materials into syllabus and then into book form. Thanks are also due to my students for their many valuable suggestions and criticisms of this text while it was in embryonic form. I am especially grateful to two fine typists, Eunice Schwartz Phillips and Mary Ann Wilde, who were able to interpret my handwriting and who patiently typed this material through two drafts and countless revisions. Finally, I would like to thank my husband for encouraging me when I was discouraged and for suffering through every deadline with me. Without his love and emotional support this book would never have come into being.

Roberta Bruck-Kan

1
The Language of Anatomy

Anatomy, like any other specialized subject, has its own terminology. The sooner you learn this terminology, the sooner you will be able to understand what you are reading.

Is your first reaction to think that anatomical terminology looks like Greek? You are partially right. Anatomical terminology is based on Greek and Latin words. The reason for this can be understood if we look back in history. The first anatomists were Greek and Roman doctors. During the Renaissance, when the science of anatomy was developing in the universities of Europe, Greek and Latin were still the common languages of educated men and women. It is not surprising that the early European anatomists used Greek and Latin words to describe the structures and relationships they encountered in the human body. A few Arabic words are also found in the language of anatomy; these are the contributions of the early anatomists from the Middle East. If you, like most present-day students of the health sciences, have not had instruction in Greek, Latin, or Arabic, you may not immediately recognize the derivations of anatomical terms. A Glossary is included at the end of this book to assist you in learning the language of anatomy. It is also suggested that you obtain a medical dictionary. The Glossary and a medical dictionary will give you the derivations and definitions of anatomical terms. You should make these terms part of your working vocabulary. Remember, the larger the working vocabulary you acquire, the more easily you can communicate with other members of the health professions.

The Scope of Anatomy

Because the science of anatomy is very old, anatomists have accumulated a vast amount of information about the human body. This information has been subdivided into anatomical *fields,* which are named according to the nature of the material studied or the means of studying the material.

Gross Anatomy.   Gross anatomy is concerned with the *macroscopic* structure of the human body, that is, structure that can be seen with the unaided eye. The oldest field of anatomy, gross anatomy, consists of observations made on living people (surface anatomy) and on carefully dissected human cadavers. It is mainly concerned with the relations of body structures to one another. Much of the anatomy discussed in this book is gross anatomy.

Microscopic Anatomy.   Microscopic anatomy is concerned with the structure of cells and tissues that can be seen with the light microscope. Within the last few decades our knowledge of human microscopic anatomy has been greatly expanded by information gained through use of phase and electron microscopy.

Developmental Anatomy.   Developmental anatomy is concerned with human growth and development from conception to maturity. *Embryology* is the subfield that deals with growth and development up to the time of birth.

Neuroanatomy.   Neuroanatomy is concerned with the gross and microscopic anatomy of the nervous system.

Positional and Directional Terms

The most fundamental words in the anatomical vocabulary are the positional and directional terms used to describe the spatial orientation of the body and the relations of its parts to one another. These are the first anatomical terms you should learn; if you fail to learn them, you will have great difficulty in understanding the rest of anatomy.

THE ANATOMICAL POSITION   For descriptive purposes, the human body is considered to be in an arbitrary fixed position called the *anatomical position.* In this position the body is regarded as standing erect, with the head and palms of the hands facing forward. That this position is neither natural nor comfortable is something you can discover for yourself by assuming it for a few minutes. However, the anatomical position is a useful frame of reference for terms of position and direction.

AXES OF ROTATION   The body has three cardinal axes of rotation, imaginary straight lines located at right angles to one another, which are named according to their placement in the body when it is in the anatomical position. These axes, which are largely related to the movable joints, are as follows:

1. *Longitudinal Axis.* The long vertical axis of the body or a body part.
2. *Anterior-Posterior Axis.* A horizontal axis that passes through the body or a body part from front to back.
3. *Transverse Axis.* A horizontal axis that passes through the body or a body part from one side to the other.

PLANES AND SECTIONS   The human body is a three-dimensional structure and so has three different planes of reference: the *sagittal, coronal,* and *transverse planes* (Fig. 1.1).

Sagittal Plane. A sagittal plane is any vertical plane that passes through the body from front to back, dividing it into right and left portions. A cut in the sagittal plane through the midline of the body produces a *midsagittal section,* while cuts parallel to the midline of of the body produce *parasagittal sections.*

Coronal Plane. A coronal plane (also termed *frontal plane*) is any vertical plane that passes through the body from one side to the other, dividing it into front and back portions. A cut through the body in this plane produces a *coronal* or *frontal section.*

Transverse Plane. A transverse plane (also termed *horizontal plane*) is any plane that passes through the body perpendicular to the sagittal and coronal planes, dividing it into upper and lower portions. It is usually designated by an anatomical reference point, for example, the transumbilical plane, which is a transverse plane through the umbilicus, or navel. A cut through the body in this plane produces a *transverse* or *horizontal section.*

DIRECTIONAL TERMS   Six pairs of directional terms enable the anatomist to locate body structures or describe the spatial relations of body parts to one another.

1. *Anterior-Posterior*
   a. *Anterior.* Nearer to the front of the body; this term replaces "front" and is used interchangeably with the term *ventral* in humans.
   b. *Posterior.* Nearer to the back of the body; this term replaces "back" and is used interchangeably with the term *dorsal* in humans.

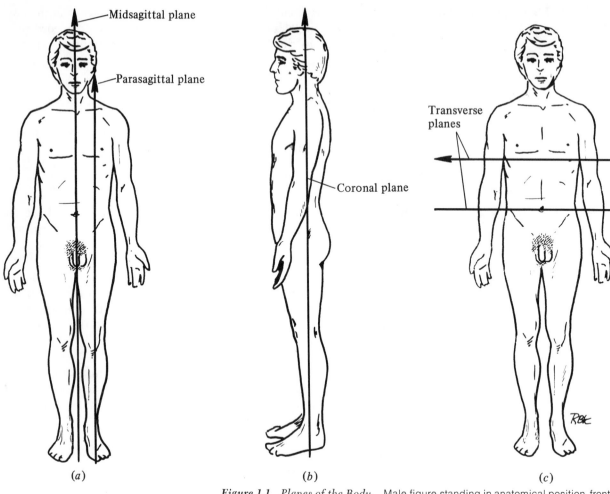

(a)   *(b)*   *(c)*

Figure 1.1   *Planes of the Body.*   Male figure standing in anatomical position, frontal and lateral views. (*a*) Midsagittal and parasagittal planes. (*b*) Coronal plane. (*c*) Two transverse planes.

2. *Superior-Inferior*
   a. *Superior.* Nearer to the crown of the head; in most instances this term replaces "upper" and is used interchangeably with the term *cranial.*
   b. *Inferior.* Nearer to the soles of the feet; in most instances this term replaces "lower" and is used interchangeably with the term *caudal.*
3. *Medial-Lateral*
   a. *Medial.* Nearer to the midline of the body.
   b. *Lateral.* Farther from the midline of the body.
4. *External-Internal*
   a. *External (Superficial).* Nearer to the surface of the body.
   b. *Internal (Deep).* Farther from the surface of the body.

5. *Central-Peripheral*
   a. *Central.* Nearer to the center of the body.
   b. *Peripheral.* Farther from the center of the body.
6. *Proximal-Distal*
   a. *Proximal.* Nearer to a point of attachment.
   b. *Distal.* Farther from a point of attachment.

Body Cavities

The human body has two major cavities that contain the internal organs, or viscera: the dorsal cavity and the ventral cavity (Fig. 1.2).

Dorsal Cavity. The dorsal cavity, which is nearer to the back, is the smaller of the two major cavities and contains the organs of the central nervous system. It is subdivided into the *cranial cavity,* which contains the brain, and the *vertebral cavity,* which contains the spinal cord.

Figure 1.2 *Body Cavities.* (a) Midsagittal section through human body showing dorsal and ventral body cavities in lateral view and contents of dorsal cavity. (b) Frontal section through human body showing dorsal and ventral body cavities in frontal view and contents of ventral cavity.

Ventral Cavity. The ventral cavity is the larger of the two major cavities and contains the organs of circulation, respiration, digestion, excretion, most hormone production, and reproduction. It is subdivided into two smaller cavities by a muscular partition called the dia-

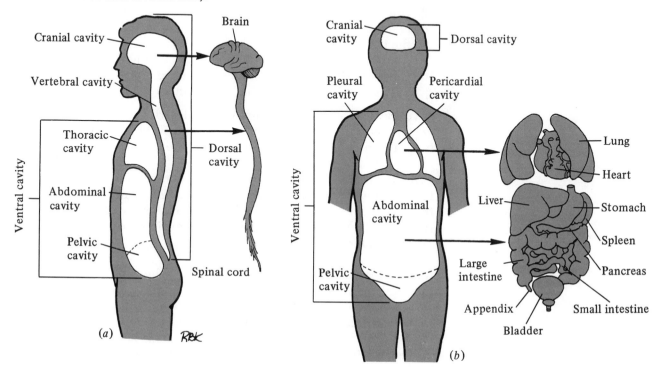

phragm. The *thoracic cavity* lies above the diaphragm; it contains the heart and great vessels, trachea, lungs, esophagus, and thymus gland. The abdominopelvic cavity lies below the diaphragm. Its abdominal portion contains the stomach, liver, gallbladder, pancreas, spleen, kidneys, adrenals, and intestines, while its pelvic portion contains the bladder, rectum, and internal reproductive organs.

It should be mentioned that the term "parietal" refers to the walls of the body cavities, while the term "visceral" refers to the internal organs.

Regions of the Human Body

Many of the geographical regions of the body's surface have special names. Some of the named regions most commonly used in anatomical descriptions are as follows:

1. *Axillary Region.* The depression between the arm and the trunk (armpit).
2. *Antecubital Region.* The depression at the front of the elbow.
3. *Inguinal Region.* The region and groove between the abdomen and the thigh (groin).
4. *Popliteal Region.* The depression at the back of the knee.
5. *Epigastric Region.* The region above the umbilicus (navel) on the front of the abdomen.
6. *Hypogastric Region.* The region below the umbilicus.
7. *Gluteal Region.* The region of the buttocks.
8. *Pudendal (Genital) Region.* The region around and including the external sex organs.
9. *Anal Region.* The area around and including the anus.
10. *Perineal Region.* The diamond-shaped region between the thighs that includes the anal and genital regions.

To localize a patient's symptoms or describe physical findings, a doctor may refer to certain surface regions of the abdomen. Two systems are in general use for subdividing the surface of the abdomen.

Four Quadrants. Under one system the abdomen is subdivided into four quadrants by drawing two lines, one vertical and one horizontal, through the umbilicus or navel (Fig. 1.3). For example, a surgeon refers to the appendix as being in the lower right quadrant and makes an incision for an appendectomy there.

Nine Regions. Under the other system, which provides for more precise localization, the abdomen is subdivided into nine regions by drawing two vertical lines, one through each collar bone, and two horizontal lines, one beneath the rib cage and the other through the crests of the hipbones (Fig. 1.4). For example, a person with an ulcer may complain to his or her doctor of pain in the epigastric region.

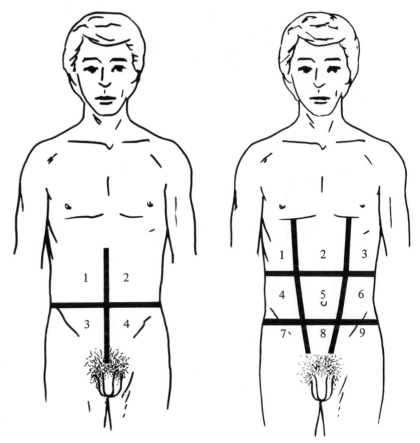

Figure 1.3 (Left) *The Four Quadrants of the Abdomen.* Quadrants: 1, upper right; 2, upper left; 3, lower right: 4, lower left.

Figure 1.4 (Right) *The Nine Regions of the Abdomen.* Regions: 1, right hypochondriac; 2, epigastric; 3, left hypochondriac; 4, right lateral; 5, umbilical; 6, left lateral; 7, right inguinal; 8, pubic; 9, left inguinal.

Parts of the Human Body

The human body has four major divisions: head, neck, trunk, and members (limbs). Each division contains a number of parts. There is a collection of terms for naming various parts of the body (Fig. 1.5). You probably know only the common or layperson's names for body parts. Now it is time to learn the anatomical names. Table 1.1 contains a list of body parts named according to both layperson's terminology and anatomical terminology.

It should be mentioned at this point that *member* (limb) is the appropriate anatomical term for an entire appendage — girdle plus "arm" or "leg." What you are accustomed to calling an "arm" or "leg" is termed an *extremity*. The upper extremity ("arm") has two major subdivisions: the (upper) arm, or brachium, and the forearm, or antebrachium. Likewise, the lower extremity ("leg") consists of the thigh, or femur, and the (lower) leg, or crus.

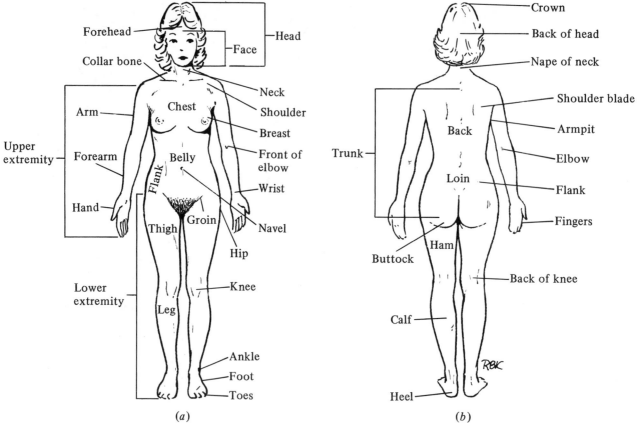

Figure 1.5 *Parts of the Human Body.* Female figure, front and back views. Major body parts and regions are indicated and named in layperson's terms.

Organization of the Human Body

There is a hierarchy of organization within the human body progressing from cells to tissues to organs to systems.

CELLS The fundamental unit of structure and function in the human body is the cell. All cells are not alike. Different cells are specialized to perform different functions; this is far more efficient than having each cell perform every function. The appearance of a cell is determined largely by the function it performs, so that cells specialized for different functions often differ markedly in appearance.

TISSUES Tissues represent the next degree of organization. The word "tissue" comes from a French word meaning "woven material," and tissues indeed form the fabric of the human body. A tissue consists of cells

Table 1.1   Parts of the Human Body

| Region | Layperson's term | ANATOMICAL TERM | |
		Noun	Adjective
Head (Caput)	Skull	Cranium	Cranial
	Forehead	Frons*	Frontal
	Back of head	Occiput	Occipital
	Face	Facies	Facial
	Nose	Nasus*	Nasal
Neck (Collum)	Neck	Cervix*	Cervical
	—	Nucha*	Nuchal
Trunk (Truncus)	Chest	Thorax	Thoracic
	—	Pectus*	Pectoral
	Breast	Mamma*	Mammary
	Belly	Abdomen	Abdominal
	Belly button (navel)	Umbilicus	Umbilical
	Flank (side)	Latus*	Lateral
	Loin (small of back)	Lumbus*	Lumbar
	Groin	Inguen*	Inguinal
	Back	Dorsum	Dorsal
	Backbone	Vertebrae	Vertebral
	Tailbone	Coccyx	Coccygeal
Upper Limb	Shoulder blade	Scapula	Scapular
(Membrum	Armpit	Axilla	Axillary
Superius)	Arm (upper)	Brachium	Brachial
	Elbow	Cubitus	Cubital
	Front of elbow	Antecubitus*	Antecubital
	Forearm	Antebrachium*	Antebrachial
	Wrist	Carpus	Carpal
	Hand	Manus	Manual
	Palm of hand	Vola*	Volar
	Back of hand	Dorsum	Dorsal
	Thumb	Pollex	Pollical
	Fingers	Digiti	Digital
Lower Limb	Hip	Coxa	—
(Membrum	Buttock	Gloutos*	Gluteal
Inferius)	Thigh	Femur	Femoral
	Knee	Genu	—
	Kneecap	Patella	Patellar
	Back of knee	Poples*	Popliteal
	Leg	Crus*	Crural
	Calf of leg	Sura*	Sural
	Ankle	Tarsus	Tarsal
	Heel	Calx*	Calcaneal
	Foot	Pes	Pedal
	Sole of foot	Planta*	Plantar
	Top of foot	Dorsum	Dorsal
	Big toe	Hallux	Hallucal
	Toes	Digiti	Digital

* Terms that are seldom used in contemporary anatomy.

of similar origin and appearance grouped together to perform a common function. Each tissue of the body fits into one of four major types: *epithelial tissue, connective tissue, muscular tissue,* or *nervous tissue.*

ORGANS   Organs represent a still higher degree of organization. An organ is a structure consisting of two or more tissue types that perform a common function or group of related functions. Many organs in the human body occur in pairs. If one organ of a pair is damaged or removed surgically, the remaining organ enlarges and performs all the functions formerly performed by both. For example, if one kidney is removed, the other kidney enlarges and works twice as hard.

SYSTEMS   Systems represent the highest degree of organization within the human body. A system consists of a group of organs that work together to perform a common function or group of related functions. A summary of the major systems of the human body is given next.

Integumentary System.   The integumentary system consists of the skin and its appendages. This system forms a protective barrier between the body and the external environment, receives sensory information, and helps regulate body temperature.

Skeletal System.   The skeletal system consists of all the bony, cartilaginous, and membranous structures that form the framework of the human body. It provides attachment sites for muscles, levers for muscular movement, protection for the viscera, sites for blood cell manufacture, and a storehouse for minerals.

Articular System.   The articular system consists of all the joints of the human body plus the ligaments and membranes associated with them. The joints provide for movement between the bones. The degree of movement at a given joint depends on the joint's structure.

Muscular System.   The muscular system consists of all the skeletal muscles of the body plus the protective sheaths and sacs associated with them. Skeletal muscles move the bones and certain other body parts.

Cardiovascular System.   The cardiovascular system consists of the heart, which functions as a pump, plus the vessels that distribute blood to the tissues and return it to the heart.

Lymphatic System.   The lymphatic system consists of vessels and organs that drain and filter excess fluid (lymph) from the tissue spaces and deliver it to the cardiovascular system. The lymphatic system also helps defend the body against foreign invasion, even when the invader is a potential lifesaving tissue or organ graft.

Urinary System. The urinary system consists of those organs that form, store, and expel urine from the body: the kidneys, ureters, bladder, and urethra.

Respiratory System. The respiratory system consists of all the organs that contribute to the exchange of respiratory gases between the air in the system and the blood: the nose, pharynx, larynx, trachea, bronchi, and lungs. It also participates in air conditioning, vocalization, and regulation of the pH of body fluids.

Digestive System. The digestive system consists of those organs that help ingest, digest, and absorb food and expel the waste products (feces) from the body: the mouth, tongue, teeth, pharynx, esophagus, stomach, intestines, rectum, anus, salivary glands, pancreas, liver, and gallbladder.

Endocrine System. The endocrine system consists of the ductless glands, whose secretions are of great importance in regulating body functions and maintaining homeostasis: the pituitary (hypophysis), ovaries, testes, placenta, adrenals, thyroid, parathyroids, pancreatic islets, and pineal body.

Reproductive System. The reproductive system consists of the organs that make human reproduction possible: the gonads (ovaries and testes), the reproductive ducts and glands, and the external genitalia.

Nervous System. The nervous system consists of the organs that receive and process information from inside and outside the body and send regulatory messages to the organs: the brain, spinal cord, peripheral nerves, and special sense organs. The two anatomical divisions of the nervous system are the central and peripheral nervous systems, while the two functional divisions are the voluntary (somatic) and involuntary (autonomic or visceral) nervous systems.

How the Human Body Is Studied

There are several ways of studying the human body; different approaches are used for different educational purposes and career goals.

Systemic Anatomy. One way of studying the human body is to consider it system by system. This method is called systemic anatomy. In systemic anatomy the student considers one system at a time throughout the entire body. Thus when the cardiovascular system is discussed, the student is presented with all the blood vessels from the head down to the big toe. The systemic approach is more suited to students whose biology backgrounds are not extensive and whose health career requirements are best fulfilled by a superficial exposure to anatomy. This book is based on the systemic approach to human anatomy.

Regional Anatomy. Another way of studying the human body is to break it down into a series of regions such as head and neck, thorax, abdomen and pelvis, upper extremity, and lower extremity. This method is called regional anatomy. In regional anatomy the student examines all the systems of a particular region — its muscles and bones, blood supply, lymphatic drainage, and innervation. A good deal of emphasis is placed on the spatial relations of the various organs and structures in the region under consideration. The regional approach is better suited to students who have had advanced biology courses, such as vertebrate anatomy and embryology, and whose health career requirements are best fulfilled by an in-depth exposure to anatomy.

Whatever way you study the human body, you should keep in mind that it is a single functional unit, not a series of isolated systems or parts. Many students make learning anatomy more difficult for themselves because they concentrate on the trees to the exclusion of the forest.

Diseases and Disorders of the Human Body

A *disorder* is a disturbance of body function, structure, or both. A partial list of the causes of disturbed function or structure includes hereditary factors, infectious organisms, harmful chemical and physical agents, environmental pollutants, aging, new growth (cancer), and the stresses of contemporary life. A *disease* is a disorder characterized by at least two of the following criteria: a recognized causative agent or agents, an identifiable group of signs (disturbances observed by a doctor) and symptoms (disturbances felt by the patient), or consistent structural alterations. The study of disturbed structure and function is called *pathology*. This book discusses those aspects of pathology that reinforce your knowledge of normal human structure and function.

Pathology also has a specialized vocabulary, and you are now asked to learn some basic pathological terms.

1. *Infection.* Invasion of the body by disease-producing organisms and the subsequent multiplication of these organisms accompanied by destruction of healthy cells and tissues.

2. *Immunity.* Resistance of the body to infectious agents or their products.
3. *Inflammation.* A local tissue response to injury characterized by pain, vascular congestion, and the outpouring of white blood cells and tissue fluid at the site of injury.
4. *Ischemia.* Decrease in blood supply to a tissue or organ.
5. *Infarction.* Loss of blood supply to a tissue or organ.
6. *Degeneration.* Regressive changes in cells or tissues in which structure is altered and certain functions are inhibited or lost.
7. *Necrosis.* Cell or tissue death.
8. *Atrophy.* Decrease in size of a normally developed cell, tissue, organ, or body part through disease or disuse.
9. *Hypertrophy.* Increase in size of a tissue or organ caused by an increase in size of existing cells.
10. *Hyperplasia.* Increase in size of a tissue or organ caused by an increase in number of cells.
11. *Metaplasia.* Reversible change of one adult cell or tissue type into another as the result of chronic irritation, altered nutrition, or altered function.
12. *Neoplasia.* Formation of abnormal and unchecked new growths (tumors).

Suggested Readings

Nomina Anatomica. 4th ed. Amsterdam: Exerpta Medica, 1977.

NYBAKKEN, OSCAR E. *Greek and Latin in Scientific Terminology.* Iowa State College Press, 1959.

SCHMIDT, J. E. *Visual Aids for Paramedical Vocabulary.* Springfield, Ill.: Thomas, 1973.

The Cell

The study of cells, *cytology*, is a highly specialized area of microscopic anatomy. Progress in relating cell structure with function has always been linked to technological advances in microscopy and new discoveries in the fields of physiology, biochemistry, and genetics.

The cell is the basic unit of structure and function in humans. All living organisms, except viruses, are composed of cells. Some organisms, such as bacteria, consist of a single cell. But most organisms, including humans, are composed of many cells.

The cells of the human body vary tremendously in size and shape. For example, a red blood cell is 7 microns in diameter, whereas an ovum is about 120 microns [a micron is 0.000001 meter or 0.000039 inch; the term micrometer (μm) is now preferred to micron (μ) by many scientific workers]. The shape of a cell depends upon its environment as well as its function. Blood cells that are freely suspended in plasma assume a spherical shape. The shape of cells in organs, partially determined by pressure from their neighbors, may be squamous (platelike), oval (egg-shaped), cuboidal (cube-shaped), pyramidal (pyramid-shaped), columnar (column-shaped), fusiform (spindle-shaped), or stellate (star-shaped).

Chemical Composition of Cells

The living substance of plant and animal cells, which has been called *protoplasm* since the 1830s, consists of the following major chemical components:

1. *Water.* An inorganic molecule that is the major component of protoplasm; all other components are suspended in water. Water also provides a medium for cellular chemical reactions.

2. *Salts.* Inorganic molecules such as sodium chloride. Salts help maintain acid-base balance and regulate osmotic pressure within the cell. Salts that separate into charged particles *(ions)* when dissolved in water are capable of conducting an electrical current. Such molecules are called *electrolytes.* The chief electrolytic ions in the cell and extracellular fluid are sodium, potassium, calcium, magnesium, and chloride.

3. *Proteins.* Organic molecules composed of small units called amino acids. Proteins serve as (a) *enzymes* (substances that speed up cellular chemical reactions) or (b) building materials for cellular structures.

4. *Carbohydrates.* Organic molecules that include sugars such as glucose and starches such as glycogen. Carbohydrates are the most abundant and readily available energy source in the cell and are used by many cells in preference to other energy sources.

5. *Lipids.* Organic molecules that include fats and fatlike substances. Lipids serve as a reservoir of stored energy in the cell. They also serve as building materials for cell membranes and as components of many *hormones* (substances that act as chemical messengers).

6. *Nucleic Acids.* Organic molecules such as deoxyribonucleic acid (DNA) and ribonucleic acid (RNA) that are formed from sugars, phosphoric acid, and nitrogen-containing bases. Nucleic acids constitute the genetic material of the cell.

Parts of the Cell

The cell (Fig. 2.1) is composed of two major parts: *cytoplasm* and *nucleus* (Table 2.1).

CYTOPLASM

Cytoplasm includes all the cell except the nucleus. Most of the cell's work is carried out in the cytoplasm, which functions as a miniature factory. The *cytoplasmic matrix* is separated from the external environment by the *cell membrane.* Embedded in the cytoplasmic matrix are *organelles,* permanent cell components that are the cell's machinery, and *inclusions,* usually temporary cell components that are the cell's raw materials and finished products.

Cell Membrane

The cell membrane *(plasma membrane* or *plasmalemma)* is a thin flexible membrane that encloses the cell. It is composed of lipids and proteins. The precise organization of its components is still not known, but many cell biologists now think that the cell membrane looks like the model shown in Fig. 2.2. In this lipid-globular protein mosaic model, the lipid molecules are arranged in a bilayer that

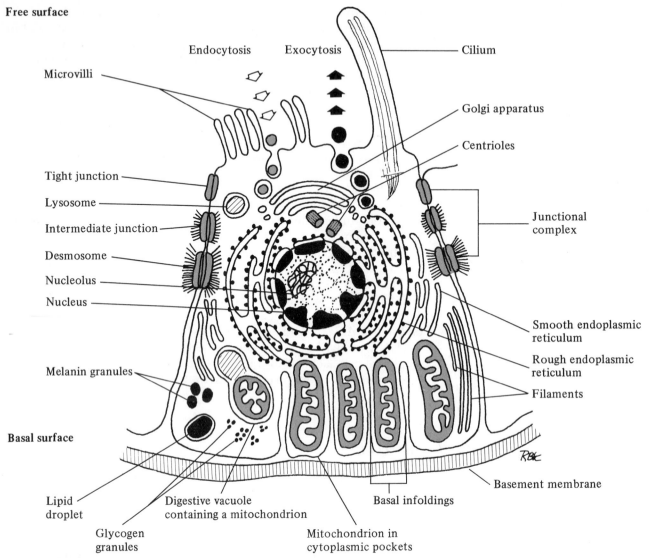

Free surface

Endocytosis   Exocytosis

Microvilli

Tight junction

Lysosome

Intermediate junction

Desmosome

Nucleolus

Nucleus

Basal surface

Melanin granules

Cilium

Golgi apparatus

Centrioles

Junctional complex

Smooth endoplasmic reticulum

Rough endoplasmic reticulum

Filaments

Basement membrane

Lipid droplet

Glycogen granules

Digestive vacuole containing a mitochondrion

Basal infoldings

Mitochondrion in cytoplasmic pockets

Figure 2.1 Generalized Cell. Organelles, inclusions, and cell surface specializations are shown as seen at electron microscopic level.

forms the framework of the membrane and provides support for the globular protein molecules that are embedded in it.

Electron microscopic examination of the cell membrane reveals a structure approximately 0.01 micron thick that consists of two dark layers surrounding a light layer. The dark layers, according to one interpretation, probably represent the proteins plus the hydrophilic (water-attracting) portions of the lipid molecules, while the light layer represents the hydrophobic (water-repellant) portions of the lipid molecules. The limiting membranes of cytoplasmic organelles have a similar appearance under the electron microscope, which suggests that all cellular membranes have a similar structure.

Table 2.1   Parts of the Cell

Boundary membranes	CYTOPLASM		NUCLEOPLASM	
	Organelles	Inclusions	Organelles	Inclusions
Cell membrane	Mitochondria	Pigment granules	Chromosomes	Messenger RNA
Nuclear envelope	Endoplasmic reticulum Golgi apparatus Lysosomes Filaments Microtubules Centrioles	Secretory granules Glycogen granules Lipid droplets	Nucleoli	Ribosomal RNA

The chief function of the cell membrane is to regulate the passage of substances into and out of the cell. The cell membrane has the ability to permit passage of certain substances and reject others on the basis of size, lipid solubility, and other molecular features

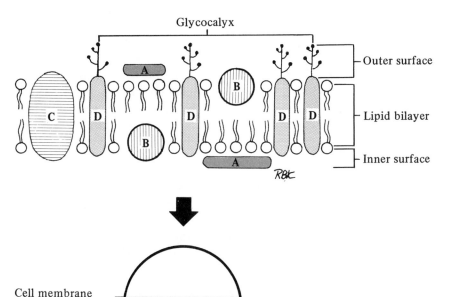

Figure 2.2 Cell Membrane. Top: model showing membrane to consist of lipid bilayer plus (A) extrinsic proteins, (B) intrinsic proteins partially embedded in lipid bilayer, and (C) intrinsic proteins completely embedded in lipid bilayer; glycoproteins (D) are intrinsic proteins whose carbohydrate branches form a fuzzy coat or glycocalyx on outer surface of some cell membranes. *Bottom:* trilaminar appearance of membrane as viewed in electron microscope.

such as electrical charge and affinity for carrier molecules in the membrane. This is called *selective permeability*.

Substances can pass through the cell membrane by passive or active means. If the energy or force for the transport of substances across the cell membrane is supplied by the extracellular environment, the process is referred to as *passive transport*. *Diffusion* is a passive transport process in which solute (molecules in solution) tends to achieve uniform distribution in a given space by passing from an area of higher concentration to an area of lower concentration. *Osmosis* is a special type of diffusion in which a solvent (water) passes through a semipermeable membrane (permeable to solvent but impermeable to solute) from an area of lower solute concentration to an area of higher solute concentration. *Filtration* is a form of passive transport in which solute and solvent are forced through a permeable membrane from an area of higher fluid pressure to an area of lower fluid pressure.

If the energy for the transport of substances across the cell membrane, especially against an electric or concentration gradient, is supplied by the cell, the process is referred to as *active transport*. Active transport is generally mediated by a carrier system. The solute molecules to be transported temporarily bind to carrier molecules that shuttle them across the cell membrane and release them at the appropriate places.

Substances present in the extracellular fluid may be taken into the cell *(endocytosis)* by other methods. Fluids are taken in by the process of *pinocytosis* (cell drinking); solids such as food particles, foreign bodies, or microorganisms are taken in by the process of *phagocytosis* (cell eating). The substances to be taken into the cell are engulfed by portions of the cell membrane that pinch off and break away from the surface to form either pinocytotic or phagocytic vacuoles. Materials may be expelled from the cell *(exocytosis)* by a reverse of the pinocytotic or phagocytotic processes. In such cases vacuoles fuse with the cell membrane and eject their contents from the cell. Waste products and secretory products such as hormones and neurotransmitters leave the cell in this way.

Cytoplasmic Matrix The cytoplasmic matrix, about which very little is known, is thought to consist mainly of large enzymatic and structural protein molecules suspended in water. These large molecules may be linked to carbohydrates, mineral salts, and other soluble substances. Whatever its composition, the cytoplasmic matrix is neither structurally nor biochemically homogeneous.

Organelles Organelles (little organs) are active living structures, most of which are separated from the cytoplasm by limiting membranes. Included in this category are mitochondria, ribosomes, endoplasmic reticulum, Golgi apparatus, lysosomes, filaments, microtubules, and centrioles.

Figure 2.3 Mitochondrion. Note that mitochondria contain nucleic acids (DNA and RNA) which are used in synthesis of some of their own respiratory enzymes. Drawing based on electron microscopic view.

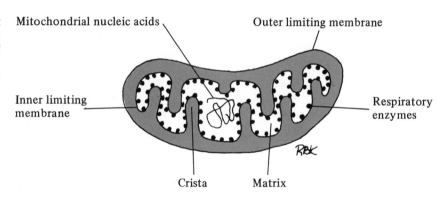

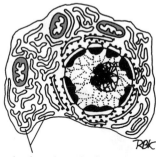

Smooth endoplasmic reticulum

Figure 2.4 Smooth Endoplasmic Reticulum in Adrenal Cortical Cell. Cells such as this, which produce steroid hormones for export, contain large amounts of smooth endoplasmic reticulum. Steroid biosynthesis takes place on smooth endoplasmic reticulum. Drawing based on electron microscopic view.

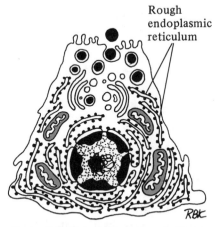

Figure 2.5 Rough Endoplasmic Reticulum in Pancreatic Acinar Cell. Cells such as this, which produce protein enzymes for export, contain large amounts of rough endoplasmic reticulum. Protein biosynthesis takes place on rough endoplasmic reticulum. Drawing based on electron microscopic view.

Mitochondria (Fig. 2.3) are small (0.2 to 5 microns) oval or rod-shaped bodies enclosed in a double membrane. The inner membrane of each mitochondrion is thrown into shelflike folds, called *cristae*, that project into its matrix (internal substance). The cristae increase the surface area of the inner mitochondrial membrane, which contains the enzymes necessary for cell respiration. With the aid of these respiratory enzymes, mitochondria break down the cell's nutrients into carbon dioxide and water. Since energy is derived from the breakdown of organic molecules during cell respiration, mitochondria are the power plants of the cell. The energy they supply is in the form of biologically useful, high-energy compounds such as ATP (adenosine triphosphate).

Ribosomes are tiny particles of protein and RNA that are the sites of cellular protein synthesis. They can be found free in the cytoplasm or attached to cytoplasmic membranes. Free ribosomes produce proteins for the cell's own use.

The *endoplasmic reticulum* is a vast network of membrane-bound cytoplasmic channels that store cell products and transport them to different regions of the cell. Since many of the cell's products are manufactured by this organelle, it also serves as an assembly line. *Smooth endoplasmic reticulum* (Fig. 2.4), which has a smooth outer membrane, synthesizes lipid products such as cholesterol or steroid hormones. *Rough endoplasmic reticulum* (Fig. 2.5), which has ribosomes attached to its outer membrane, is involved in the synthesis of protein products for export from the cell, such as enzymes or protein hormones.

The *Golgi apparatus* (Fig. 2.6) consists of stacks of flattened sacs plus a collection of small vesicles. This organelle is thought to regulate water movement within the cell. In secretory cells the chief function of the Golgi apparatus is to collect, concentrate, store, and package the cell products being produced for export. In cells that produce glycoproteins (molecules containing carbohydrates and proteins), the Golgi apparatus synthesizes the carbohydrate portion,

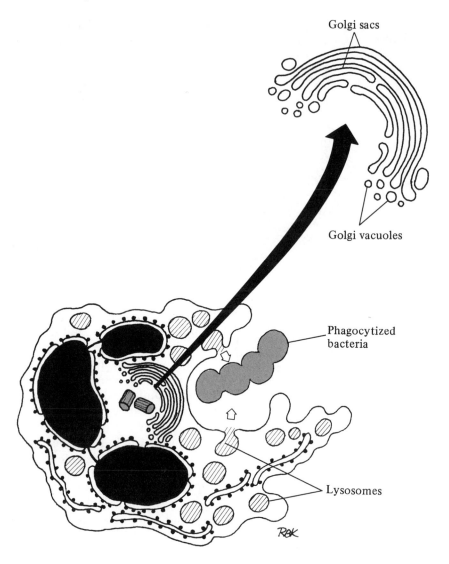

Golgi sacs

Golgi vacuoles

Phagocytized bacteria

Lysosomes

Figure 2.6 Golgi Apparatus and Lysosomes in Neutrophil. This scavenger white blood cell contains a prominent Golgi apparatus consisting of sacs and vesicles (enlargement, *top*). Golgi apparatus produces lysosomes, whose powerful digestive enzymes are about to attack a chain of bacteria the neutrophil has phagocytized. Drawing based on electron microscopic view.

which is then added to the protein portion it receives from the rough endoplasmic reticulum.

Lysosomes (Fig. 2.7) are spherical sacs containing powerful digestive enzymes in storage form. When these enzymes are activated, they break down particulate matter either inside or outside the cell. For example, when particulate matter is ingested by phagocytosis, the phagocytic vacuole fuses with a lysosome to form a digestive vacuole or secondary lysosome. Activated enzymes from the lysosome then attack and digest the particulate matter. When digestion is completed, the undigested residue is usually expelled from the cell, and

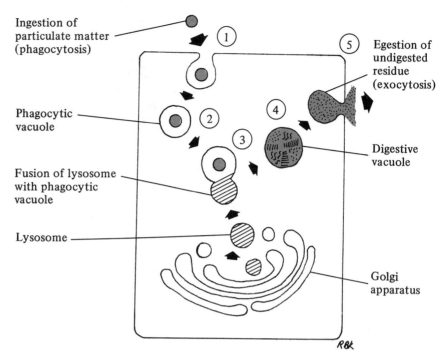

Ingestion of
particulate matter
(phagocytosis)

Egestion of
undigested
residue
(exocytosis)

Phagocytic
vacuole

Fusion of lysosome
with phagocytic
vacuole

Digestive
vacuole

Lysosome

Golgi
apparatus

Figure 2.7 Intracellular Digestion. Particulate material taken into cell by phagocytosis (1) is surrounded by cytoplasmic membrane pinched off from cell surface, which forms phagocytic vacuole (2). A lysosome fuses with phagocytic vacuole (3) to form digestive vacuole (4), in which particulate matter is digested by lysosomal enzymes. Undigested contents of digestive vacuole are released from cell by exocytosis (5).

the products of digestion are used as raw materials in the cell's synthetic processes. The cell also uses its lysosomes to dispose of useless organelles, such as worn-out mitochondria; this process is called *autophagy* (self-eating). Under certain conditions lysosomal membranes can rupture accidentally, releasing enzymes that destroy the entire cell. When a cell is digested by its own lysosomal enzymes, the process is called *autolysis* (self-dissolution). Lysosomes can also be extruded from the cell, in which case the lysosomal enzymes act in the extracellular fluid. This is what happens when bone tissue is broken down by certain types of bone cells.

Filaments and microtubules are cytoplasmic organelles that provide an internal framework for the cell and its processes. *Filaments* are solid threadlike structures that are involved in cellular contraction and protoplasmic movements as well as support. They are especially abundant in muscle, nerve, and epidermal cells. *Microtubules* are hollow tubular structures formed from protein subunits that can be assembled and disassembled with ease. They help the cell change its shape; they are responsible for contraction of motile cell processes; and, as part of the spindle apparatus, they participate in cell division.

Centrioles (Fig. 2.8) are short cylindrical organelles composed of microtubules. In the nondividing cell a pair of centrioles are found in a special region of the cytoplasm called the *centrosome*, which usually lies near the nucleus. Other cytoplasmic organelles orient

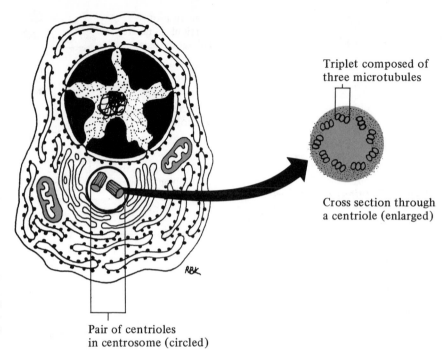

Triplet composed of
three microtubules

Cross section through
a centriole (enlarged)

Figure 2.8 Centrioles in Plasma Cell.
Centrosome (*circled*) containing two centrioles lying at right angles to one another.
Enlarged cross section of one centriole
(*right*) illustrates arrangement of microtubules within core of centriole. Drawings
based on electron microscopic views.

Pair of centrioles
in centrosome (circled)

themselves around the centrosome. Just before the cell divides, the
two centrioles move to opposite poles of the cell, from which points
they organize the spindle apparatus.

Inclusions Inclusions are passive, nonliving, often temporary inhibitants of the
cytoplasm. They are the raw materials and products of the cytoplasmic
factory.

Pigment granules occur in many nerve and epithelial cells. Some
granules contain a brown-black pigment called *melanin;* others contain *lipofuscin,* a golden-brown pigment formed from the residue of
intracellular digestion.

Cells that make products for export contain *secretory granules.*
These granules contain secretory products in transit from the Golgi
apparatus to the cell membrane.

Carbohydrate is stored in many cells in the form of *glycogen granules.* Liver and skeletal muscle cells often contain large amounts of
stored glycogen.

Lipid appears in the cytoplasm in the form of *lipid droplets.* Most
of the reserve lipid in the body is stored in fat cells. Other cells contain lipid droplets at one time or another.

NUCLEUS The nucleus (Fig. 2.9) is the control center of the cell. Executive orders regarding cell processes such as growth, differentiation, secre-

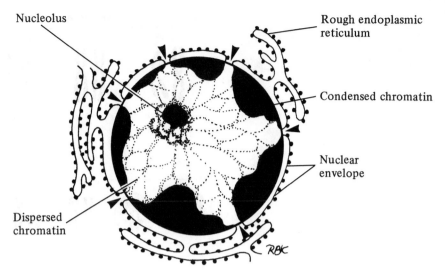

Nucleolus

Rough endoplasmic reticulum

Condensed chromatin

Nuclear envelope

Dispersed chromatin

Figure 2.9 *Interphase Nucleus and Its Constituents.* Nuclear envelope, which separates nucleus from cytoplasm, is studded with ribosomes and is in direct continuity with rough endoplasmic reticulum. *Arrows* indicate pores in nuclear envelope. Within nucleus lie nucleolus and chromatin, which appears in condensed and dispersed forms. Condensed chromatin is unavailable for RNA transcription. Drawing based on electron microscopic view.

tion, reproduction, and repair are issued by the nucleus and carried out by the cytoplasmic organelles. This highly specialized region of the cell is bounded by the *nuclear envelope* and contains *nucleoli* and *chromosomes* suspended in a matrix of *nucleoplasm.*

Nuclear Envelope

When the cell is not dividing, the nucleoplasm is separated from the cytoplasm by the nuclear envelope, a double-layered structure that communicates with the rough endoplasmic reticulum. Small pores exist in the nuclear envelope, presumably to permit passage of materials between the nucleoplasm and cytoplasm. Ribosomes are frequently attached to the outer membrane of the nuclear envelope.

Nucleoplasm

Nucleoplasm is the formless substance in which nuclear organelles, the nucleoli and chromosomes, are suspended. It is composed mainly of protein molecules (some of which have enzymatic activity), metabolites, ions, and water.

Nucleoli

Nucleoli are small rounded bodies present in the nuclei of non-dividing cells. There may be anywhere from one to four nucleoli present in each nucleus depending on cell type and function. Each nucleolus, part granular and part fibrillar in appearance, contains DNA, RNA, and proteins. It is the synthetic site of ribosomal RNA and protein. The newly synthesized ribonucleoprotein is assembled into ribosomal subunits, which are subsequently transferred to the cytoplasm.

Chromosomes

The most important organelles in the nucleus are the *chromosomes* (Fig. 2.10). There are 46 chromosomes in human *somatic* or body cells and 23 chromosomes in *gametes* or sex cells. During cell divi-

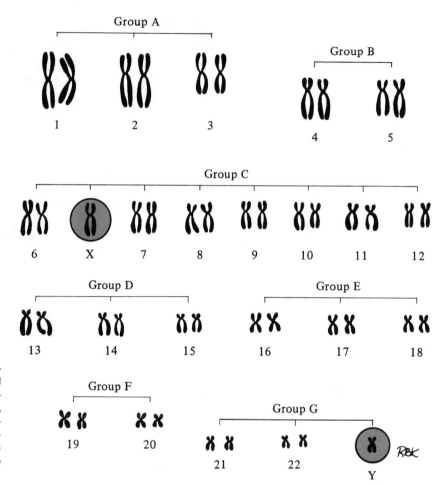

Figure 2.10 *Normal Human Male Karyotype.* The 46 chromosomes of normal human male have been arrested at metaphase, fixed, stained, photographed, paired, and arranged in seven groups (A–G) on basis of size and centromere position. Sex chromosomes of male (X and Y) are circled. Drawing based on karyotype made by Dr. Bruck-Kan.

sion chromosomes are visible as discrete bodies. This is because the *chromatin* strands that make up the chromosomes are tightly coiled. However, in the nondividing cell individual chromosomes cannot be identified because most of the chromatin strands are stretched out. Chromatin consists of DNA and protein. DNA is a nucleic acid that contains genetic messages. The genetic messages or *genes* carry the information for protein synthesis in code form. Each chromatid (longitudinal half of a chromosome) consists of a single DNA molecule that contains thousands of genes. Transcription of genes is the critical event in the transmission of orders from the nucleus to the cytoplasm. Genetic information is transcribed into strands of messenger RNA, another type of nucleic acid. Messenger RNA then enters the cytoplasm through pores in the nuclear envelope. The cytoplasmic ribosomes, with the aid of transfer RNA-amino acid complexes, translate these messages into proteins. Proteins, in turn, determine the cell's appearance, regulate its functions, and may even be exported.

Cell Surface Specializations

Many cells possess surface specializations. Surface specializations are modifications of the cell membrane and peripheral cytoplasm. They increase the cell's surface area, move materials over the cell's surface, and join cells to their neighbors.

INFOLDINGS

Infoldings of the cell membrane, deep enough to create cytoplasmic pockets, help increase the cell's surface area, They are present mainly on the basal surfaces of certain epithelial cells. For example, infoldings of the basal surface of kidney tubule cells (Fig. 2.11) increase the area available for transport of water and other substances across the cell membrane. Mitochondria contained in the cytoplasmic pockets provide the energy for active transport of substances such as sodium, potassium, and glucose.

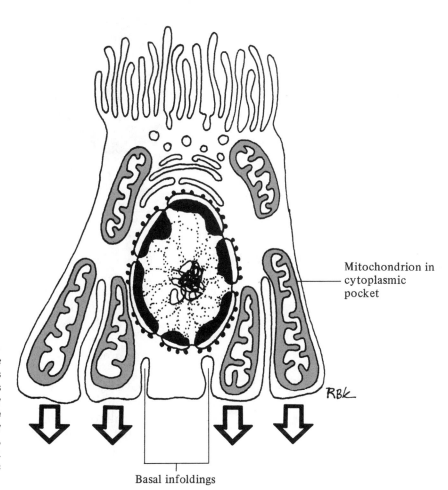

Mitochondrion in cytoplasmic pocket

Basal infoldings

Figure 2.11 Basal Infoldings and Associated Mitochondria in Kidney Tubule Absorptive Cell. Cell, which transports large volumes of fluid plus small molecules and ions across its basal surface, has many basal infoldings that increase its surface area. Cytoplasmic pockets, formed by basal infoldings, contain mitochondria, which supply energy for active transport. Drawing based on electron microscopic view.

Microvilli

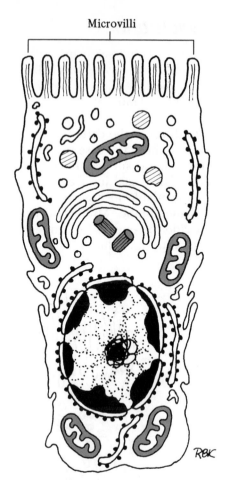

Figure 2.12 Microvilli in Intestinal Absorptive Cell. Prominent border of microvilli on cell's apical or free surface increases surface area available for absorption. Drawing based on electron microscopic view.

MICROVILLI   Microvilli (Fig. 2.12) are tiny nonmotile cytoplasmic projections from the free surfaces of many cells. The long slender microvilli of epithelial cells in the organs of taste, balance, and hearing are modified to function as sensory receptors. The dense array of microvilli on the free surface of intestinal absorptive cells and certain kidney tubule cells increases the surface area available for absorption and other exchange processes.

MOTILE PROCESSES   Some cells have motile cytoplasmic processes whose beating moves fluids or a mucous film over the cell surface. Long whiplike motile processes are called *flagella*. Sperm—male sex cells—are the only human cells that have flagella. Flagella help propel sperm through the fluids of the female reproductive tract to reach the ovum (female sex cell). Short motile processes are called *cilia* (Fig. 2.13). Most epithelial cells lining the respiratory tract bear numerous cilia; the beating of the cilia keeps a protective mucous sheet constantly moving over the free surface. Both cilia and flagella consist of a tapered

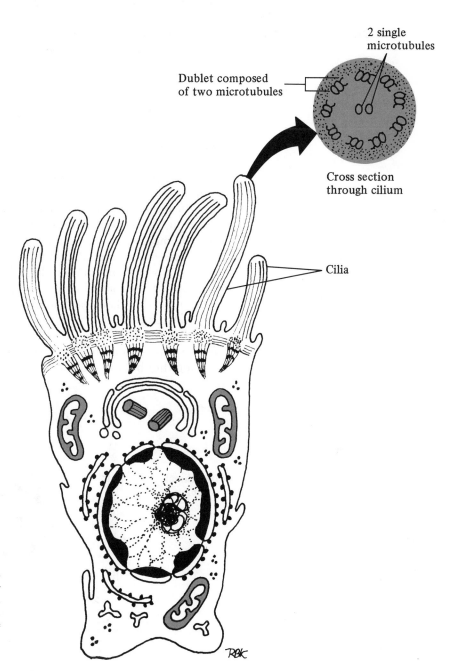

2 single
microtubules

Dublet composed
of two microtubules

Cross section
through cilium

Cilia

Figure 2.13 Cilia in Respiratory Epithelial Cell. Cilia move sheet of mucus over apical or free surface of cell. Cross section of cilium (enlarged) illustrates arrangement of microtubules within core of cilium. Drawings based on electron microscopic views.

cytoplasmic shaft supported by a core of microtubules whose contractions are responsible for their rhythmic beating movements. The microtubules extend from the tip of each process to a centriolelike structure anchored in the cytoplasm below the cell surface.

CELL ATTACHMENTS Epithelial, muscle, and nerve cells are attached to their neighbors by specializations of the cell membrane and peripheral cytoplasm. Many different attachment devices exist. The *junctional complex* (Fig. 2.14), consisting of tight junction, intermediate junction, and desmosome, is found on the lateral boundries of certain epithelial cells. The *tight junction*, found just below the free surface, is a zone in which the opposed cell membranes converge and appear to fuse. Tight junctions prevent substances from passing between cells instead of through them. The *intermediate junction*, found just below the tight junction, is a zone in which the opposed but unfused cell membranes are reinforced on their cytoplasmic side by fine filaments. The *desmosome*, found singly or in groups below the intermediate junction, is a cytoplasmic plaque in which the opposed but unfused cell membranes are reinforced by condensed cytoplasm and large filaments called tonofilaments. Desmosomes help prevent cells from pulling apart from one another. They are also present in the *intercalated disc*, an attachment device that maintains cohesion between the opposed ends of cardiac muscle cells. The *gap junction* or *nexus* is an attachment device that also mediates the flow of electrical current between cells, especially epithelial and smooth muscle cells. In this type of junction, the opposed cell membranes are very close to one another but do not fuse.

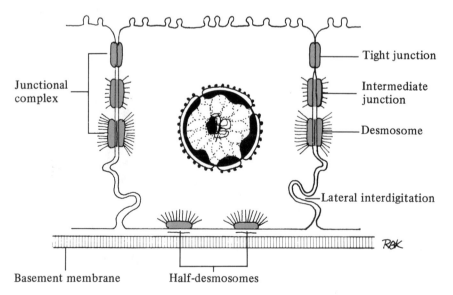

Figure 2.14 Attachments Between Two Epithelial Cells. Junctional complex, consisting of tight junction, intermediate junction, and desmosome, forms firm, watertight union between adjacent cells. Lateral interdigitations help prevent adjacent cells from pulling apart. Half-desmosomes attach basal surface of epithelial cell to basement membrane. Drawing based on electron microscopic view.

Junctional complex

Tight junction

Intermediate junction

Desmosome

Lateral interdigitation

Basement membrane

Half-desmosomes

The Life and Death of Cells

Cells are like people in some ways; they are born, grow up, go to work, reproduce themselves, and die.

CELL ORIGIN All cells come from preexisting cells. The trillions of cells in the human body have come from the fertilized ovum or *zygote*. Repeated divisions of the zygote give rise to the embryo, a mass of cells that initially look and act alike. Embryonic cells begin to specialize by separating into three *germ layers*. Cells from the germ layers in turn specialize to produce the different tissues of the body. The progressive specialization of cells to assume specific shapes and perform specific functions is called *differentiation*. Differentiation accounts for the division of labor among cells in a multicellular organism such as a human. It may also cause a cell to lose its ability to assume other shapes and functions. Once cells are fully differentiated, they usually do not divide again. However, even in the adult body there exist some undifferentiated cells, called *stem cells*, that have the potential ability to divide or differentiate if need arises.

When the adult form is reached and growth and differentiation have stopped, there is still a need for new cells to replace those that are worn out or that have been destroyed. New cells arise from existing cells, usually stem cells, by mitotic division. Many cells, such as epithelial cells, have a brief life span and are replaced. Some cells, such as highly specialized nerve cells, are not replaced if destroyed by disease or accident. Other cells, such as muscle cells, are replaced under certain conditions.

CELL CYCLE The life history of the cell is called the cell cycle. The cell cycle extends from one division to the next. The length of the cycle depends on the cell and tissue type. Cells in embryonic or rapidly renewing tissues usually complete their cycles in hours or days, while cells in other tissues may take weeks or months to complete their cycles. Some fully differentiated cells never complete their last cycles.

Interphase Interphase is the period between cell divisions. During interphase the cell is not resting but working hard. It carries out the metabolic activities needed for its own maintenance and growth, doubles its DNA content in preparation for its next division (if it is going to divide again), and synthesizes RNA and proteins.

DNA Replication All cells that are going to divide again replicate (make duplicate copies of) their DNA during interphase. DNA is a large ladderlike organic molecule made up of phosphoric acid, *deoxyribose* (a 5-carbon sugar), and nitrogen-containing bases called *purines* (adenine, guanine) and *pyrimidines* (thymine, cytosine). The sides of the ladder consist of two long strands composed of alternating units of phosphate and sugar from which the bases project (each base is attached to a sugar). The strands are held together by weak chemical bonds between complementary base pairs (the rungs of the ladder). In DNA

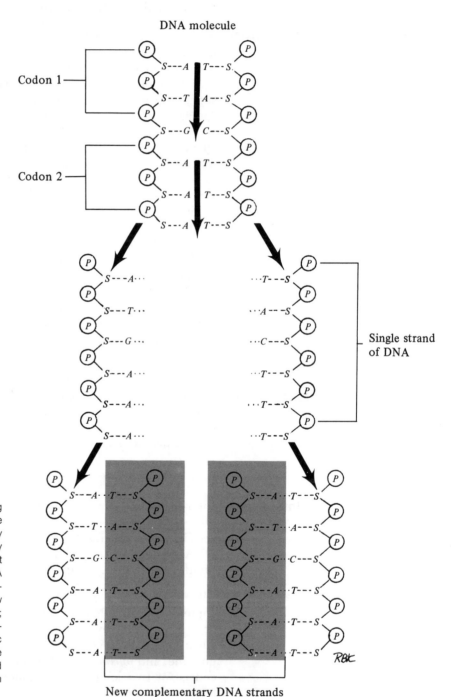

Figure 2.15 *DNA Replication.* Splitting of weak hydrogen bonds between base pairs in DNA molecule is followed by synthesis of two new complementary strands (*shaded*) on old strands, which act as templates. In this manner each DNA molecule gives rise to two molecules consisting of one old strand and one new strand. *A*, adenine base; *T*, thymidine base; *C*, cytosine base; *G*, guanine base; *S*, deoxyribose sugar molecule; *P*, phosphoric acid molecule. Note that each sequence of three DNA bases forms a code word (codon) that specifies one amino acid in a protein molecule.

adenine normally bonds only with thymine and guanine only with cytosine. The two strands are twisted around their long axes in a spiral configuration called a *double helix*. At the beginning of DNA replication (Fig. 2.15), the bonds between the base pairs are broken and the strands unwind. Each strand acts as a template for the synthesis of a complementary strand (its mirror image) with which it subsequently bonds. At the end of replication there are two new molecules of DNA for every parent molecule, each consisting of one old strand and one new strand, which are usually exact gene-for-gene copies of the original molecule. Thus DNA replication is the critical event in the transmission of genetic information from parent cells to daughter cells.

Protein Synthesis

Protein synthesis (Fig. 2.16) involves the translation of coded messages from the nucleus (in the form of messenger RNA molecules) by cytoplasmic ribosomes. The code words, sequences of three adjacent RNA bases, stand for amino acids, which are transferred to the ribosomes by transfer RNA molecules and linked into protein chains. Proteins are essential to the cell both as structural elements and as enzymes. Protein synthesis is restricted to those portions of interphase when the cell is not involved in DNA replication.

Cell Divisions

Cell division is the means by which cells reproduce themselves. The two types of cell division are *mitosis* and *meiosis*. Meiosis is restricted to the production of sex cells.

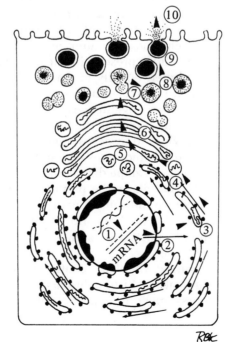

Figure 2.16 *Stages in Synthesis of Protein for Export.* (1) Formation of messenger RNA (mRNA) molecule on one strand of DNA molecule. (2) Passage of messenger RNA molecule out of nucleus through nuclear pore. (3) Translation of messenger RNA molecule on ribosomes of rough endoplasmic reticulum. (4) Entry of newly synthesized protein into rough endoplasmic reticulum. (5) Passage of protein molecules from rough endoplasmic reticulum to Golgi apparatus. (6) Packaging of protein into secretory droplets in Golgi apparatus. (7) Release of secretory droplets from Golgi apparatus and their subsequent fusion to form secretory granules (8) Condensation of proteins in secretory granules. (9) Migration of secretory granules to free surface of cell. (10) Release of protein molecules from free surface of cell.

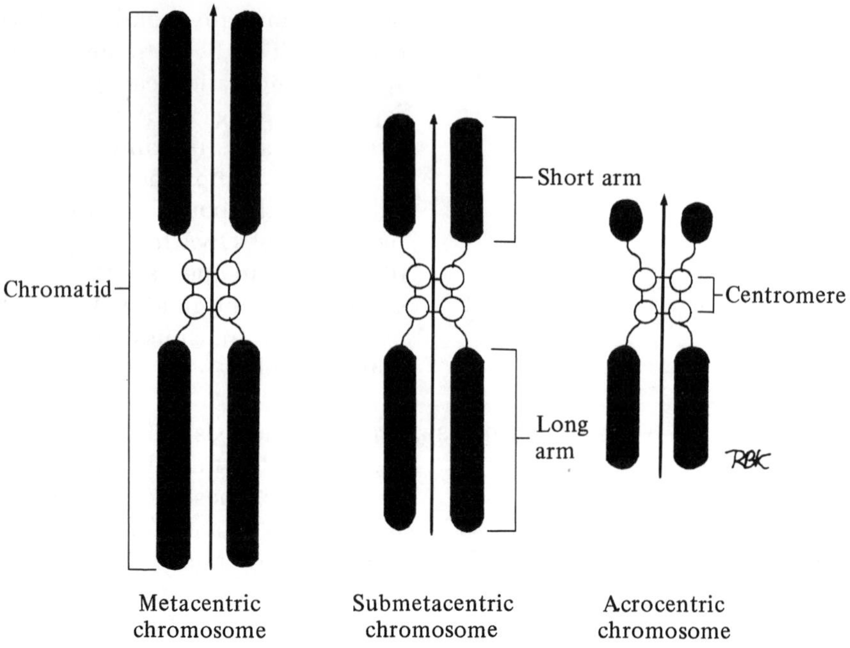

Figure 2.17 Morphology of Metaphase Chromosomes. Metaphase (premitotic) chromosomes consist of two longitudinal portions (sister chromatids) joined by centromere. During mitotic division centromere splits longitudinally (*arrow*) and sister chromatids migrate to opposite poles of cell. Metaphase chromosomes come in three morphological varieties: metacentric (median centromere), submetacentric (submedian centromere), and acrocentric (nearly terminal centromere). Submetacentric and acrocentric chromosomes have arms of unequal size.

Mitosis   In mitotic division a somatic cell gives rise to two daughter cells that are identical to it in chromosome number. During interphase, after DNA is replicated, each chromosome becomes a double structure consisting of two elongated structures called *chromatids*, which are held together by a *centromere* (Fig. 2.17). Each chromatid contains a single molecule of DNA plus various proteins and is equivalent to a chromosome after mitotic division. The 46 chromosomes (92 chromatids) present in the cell at the end of DNA replication are distributed equally to the two daughter cells during mitosis. Mitosis (Fig. 2.18) is a continuous process that has been subdivided into four phases for convenience of study.

During *prophase* the two centrioles migrate to opposite poles of the cell, where they act as organization centers for the spindle apparatus. As the chromosomes coil and become visible as separate bodies, the nuclear envelope disintegrates and nucleolar material disperses.

At *metaphase* the premitotic chromosomes line up on the mitotic spindle at the equator of the cell with their chromatids facing opposite poles. Each chromatid is attached to the spindle apparatus by a spindle fiber that extends from its centromere to the corresponding centriole.

During *anaphase* centromeres split longitudinally to release the chromatids. Sister chromatids, now known as postmitotic chromosomes, move to opposite poles of the cell; they appear to be pulled

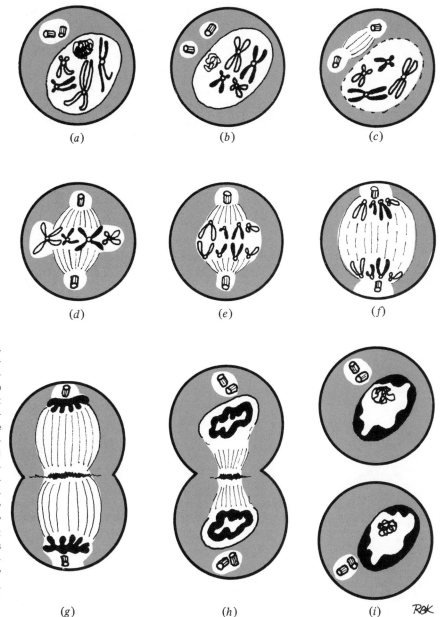

Figure 2.18 *Stages of Mitotic Division.* (a) Very early prophase: chromosomes begin to appear as separate bodies. (b) Mid-prophase: centrioles begin to migrate to opposite poles of cell. (c) Late prophase: nuclear envelope disintegrates and nucleolar material disperses; mitotic spindle begins to form. (d) Metaphase: chromosomes line up on mitotic spindle at cell equator. (e) Early anaphase: sister chromatids separate. (f) Late anaphase: chromatids (now postmitotic chromosomes) approach opposite poles of cell. (g) Early telophase: chromosomes lose their identity and cytokinesis begins. (h) Late telophase: nuclear envelopes begin to re-form as cytokinesis continues. (i) Interphase: completion of cytokinesis separates two new daughter cells from one another. *Black,* paternal chromosomes; *white,* maternal chromosomes.

by the spindle fibers. As the chromosomes separate, cytoplasmic construction (cytokinesis) begins at the equator of the dividing cell.

At *telophase* the chromosomes reach the opposite poles of the cell and begin to uncoil. A nuclear envelope forms around them. The spindle apparatus disappears, nucleoli reappear, and the cytoplasm constricts completely to separate the two daughter cells. The two daughter cells have identical chromosome complements and a share of the cytoplasm and cytoplasmic organelles of their parent.

Meiosis In meiosis the germ cells in the gonads (sex glands) give rise to the gametes (sex cells). To keep the chromosome number of the species constant from generation to generation, the sex cells must contain exactly half the number of chromosomes of somatic cells. Human somatic cells contain 46 chromosomes—22 pairs of *autosomes* plus 2 sex-determing chromosomes. Female somatic cells contain 44 autosomes plus 2 X chromosomes, while male somatic cells contain 44 autosomes plus X and Y chromosomes. After meiosis each ovum contains 22 autosomes plus one X chromosome while each sperm contains 22 autosomes plus one X or one Y chromosome. When the sperm

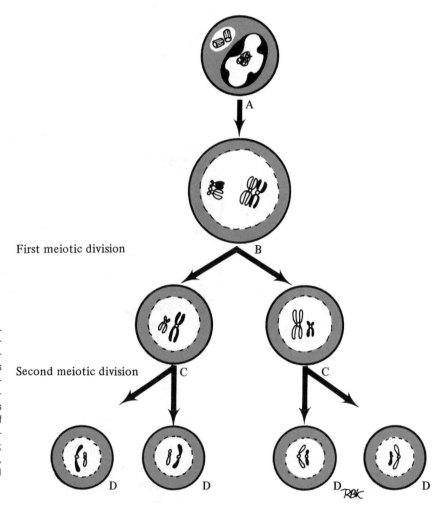

First meiotic division

Second meiotic division

Figure 2.19 Stages of Meiotic Division. (A) Spermatogonium or oogonium. (B) Primary spermatocyte or oocyte; like chromosomes pair. (C) Secondary spermatocytes or oocytes: products of first meiotic division contain half the number of chromosomes as their parent; like chromosomes have separated. (D) Sex cells: products of second meiotic division contain same number of chromosomes as their parents; sister chromatids have separated. *Black,* paternal chromosomes; *white,* maternal chromosomes.

and ovum unite at fertilization, the chromosome number is restored to 46 and the sex of the embryo is determined.

Reduction of chromosome number is accomplished by a series of two successive cell divisions called meiosis I and II. These two divisions are preceded by a single replication of chromosomal DNA. No DNA replication occurs between the two divisions. The stages of meiosis (Fig. 2.19) are summarized next.

Meiosis I is a unique type of division that features a long and complex prophase. During prophase, like chromosomes pair and may exchange genetic material. In anaphase, like chromosomes separate from one another and are distributed at random between the two daughter cells; each daughter cell receives 23 chromosomes. Since the chromosome number is reduced, meiosis I is a *reductional division.*

Meiosis II is similar to an ordinary mitotic division. Sister chromatids separate at anaphase so that each new daughter cell also receives 23 chromosomes. Since the chromosomes are distributed equally, meiosis II is an *equational division.*

CELL DEATH Cell death is a normal part of the growth of tissues and organs. In addition, a wide variety of physical, chemical, and infectious agents cause cell injury and death. If the injury is minor or of short duration, the cell usually recovers; if the injury is severe or prolonged, the cell dies.

Degeneration Degeneration is a reaction to injury that disrupts the cell's machinery. Morphological signs of degeneration include clumping of chromatin in the nucleus, coagulation of proteins in the cytoplasm, formation of cytoplasmic vacuoles, swelling of the endoplasmic reticulum, loss of ribosomes from rough endoplasmic reticulum, swelling of mitochondria, and accumulation of lipid in the cytoplasm.

Necrosis If the degenerative reaction is not reversed, cell death, or *necrosis,* occurs. As a cell is dying, various nuclear and cytoplasmic changes can be observed.

The nucleus usually becomes very small and dark-staining due to clumping of its chromatin. This nuclear reaction is called *pyknosis.* The pyknotic nucleus may then either break up into a series of smaller fragments or swell until it ruptures.

Necrotic changes in the cytoplasm often include swelling, which is caused by excessive water uptake. If lysosomal enzymes are released into the cytoplasm, the dying cell digests its own organelles. Eventually the swollen or autolysed cell bursts, and its contents are released into the general circulation. It is possible to detect cell necrosis by a rise in blood levels of certain enzymes. For example, when cardiac muscle cells die, they release myocardial enzymes such as SGOT (serum glutamic oxalic transaminase) into the bloodstream. A marked rise in these enzymes in a sample of a patient's blood indicates that this patient has probably suffered a heart attack.

Diseases and Disorders of the Cell

With each new discovery made in the basic medical sciences, we become more and more aware that disease begins at the cellular or subcellular level. Some causes of cell disorders or diseases are described here. For more detailed information you should consult recent textbooks on pathology, biochemistry, or genetics.

Inborn Errors of Metabolism.   Most inborn errors of metabolism are inherited defects of protein synthesis caused by mistakes made in gene copying during DNA replication. Essential proteins are either not synthesized or function wrongly or not at all if they are synthesized. Most enzyme defects as well as most structural protein defects arise in this manner. Metabolic storage diseases result from enzyme defects that prevent the breakdown or secretion of certain cell products.

Chromosomal Aberrations.   Chromosomal aberrations are abnormalities of chromosome number or structure caused by accidents usually occurring during meiosis or early mitotic divisions of the embryo. Chromosomal aberrations cause abnormalities of cell divisions or differentiation that produce some types of birth defects.

Infectious Agents.   Some infectious agents cause disease by invading and inhabiting cells; others cause disease with their by-products. For example, viruses are minute agents, composed of protein plus DNA or RNA, that invade the cell and take over its machinery for their own use. In reproducing themselves they usually destroy the cell. Bacteria are one-celled organisms that can poison cells with the toxic by-products of their metabolic activities.

Drugs and Metabolic Poisons.   Drugs and metabolic poisons can cause cell disease by interfering with vital cell functions. These agents may disrupt DNA replication, RNA transcription, protein synthesis, enzyme activity, or active transport, or they may have other effects.

Radiation.   X rays and other forms of radiant energy can produce severe cell damage, particularly if the cells are immature or dividing rapidly. Radiation effects on the nucleus range from changes in the genes (point mutation) to structural chromosome alterations. The greater the damage, the less effective the nucleus is as a control center. Severe chromosomal damage interferes with cell division, which may lead to cell death.

Suggested Readings

ALLFREY, VINCENT G., and ALFRED E. MIRSKY. "How cells make molecules," *Sci. Am.*, *205*, no. 3 (September 1961), pp. 74–82.

ALLISON, ANTHONY. "Lysosomes and disease," *Sci. Am.*, *217*, no. 5 (November 1967), pp. 62–72.

BEARN, A. G., and JAMES L. GERMAN, III. "Chromosomes and disease," *Sci. Am.*, *205*, no. 5 (November 1961), pp. 66–76.

CAPALDI, RODERICK A. "A dynamic model of cell membranes," *Sci. Am.*, *230*, no. 3 (March 1974), pp. 26–33.

DE DUVE, CHRISTIAN. "The lysosome," *Sci. Am.*, *208*, no. 5 (May 1963), pp. 64–72.

DEREUCK, A. V. S., and JULIE KNIGHT, eds. *Ciba Foundation Symposium: Cellular Injury*. Boston: Little, Brown, 1964.

FISCHBERG, MICHAIL, and ANTONIE W. BLACKLER. "How cells specialize," *Sci. Am.*, *205*, no. 3 (September 1961), pp. 124–140.

FOX, C. FRED. "The structure of cell membranes," *Sci. Am.*, *226*, no. 2 (February 1972), pp. 30–38.

GREEN, DAVID E. "The mitochondrion," *Sci. Am.*, *210*, no. 1 (January 1964), pp. 63–74.

HAYASHI, TERU. "How cells move," *Sci. Am.*, *205*, no. 3 (September 1961), pp. 184–204.

HAYFLICK, LEONARD. "Chromosomes and human disease," *Hosp. Pract.*, *2*, no. 2 (February 1967), pp. 54–63.

HOFFMAN, JOSEPH F. "Ionic transport across the plasma membrane," *Hosp. Pract.*, *9*, no. 10 (October 1974), pp. 119–127.

HOKIN, LOWELL E., and MABEL R. HOKIN. "The chemistry of cell membranes," *Sci. Am.*, *213*, no. 4 (October 1965), pp. 78–86.

HOLTER, HEINZE. "How things get into cells," *Sci. Am.*, *205*, no. 3 (September 1961), pp. 167–180.

HOLTZMAN, ERIC. "The biogenesis of organelles," *Hosp. Pract.*, *9*, no. 3 (March 1974), pp. 75–88.

HURWITZ, JERARD, and J. J. FURTH. "Messenger RNA," *Sci. Am.*, *206*, no. 2 (February 1962), pp. 32–49.

JAMIESON, JAMES D. "Membranes and secretion," *Hosp. Pract.*, *8*, no. 12 (December 1973), pp. 71–80.

KATZ, BERNHARD. "How cells communicate," *Sci. Am.*, *205*, no. 3 (September 1961), pp. 209–220.

KORNBERG, ARTHUR. "The synthesis of DNA," *Sci. Am.*, *219*, no. 4 (October 1968), pp. 64–78.

LEHNINGER, ALBERT L. "How cells transform energy," *Sci. Am.*, *205*, no. 3 (September 1961), pp. 62–73.

MARGULIS, LYNN. "Symbiosis and evolution," *Sci. Am.*, 225, no. 2 (August 1971), pp. 49–57.

MAZIA, DANIEL. "How cells divide," *Sci. Am.*, *205*, no. 3 (September 1961), pp. 100–120.

————. "The cell cycle," *Sci. Am.*, *230*, no. 1 (January 1974), pp. 54–64.

MILLER, O. L., JR. "The visualization of genes in action," *Sci. Am.*, *228*, no. 3 (March 1973), pp. 34–42.

MILLER, WILLIAM H., FLOYD RATLIFF, and H. K. HARTLINE. "How cells receive stimuli," *Sci. Am.*, *205*, no. 3 (September 1961), pp. 222–238.

MOSCONA, A. A. "How cells associate," *Sci. Am.*, *205*, no. 3 (September 1961), pp. 142–162.

NEUTRA, MARIAN, and C. P. LEBLOND. "The Golgi apparatus," *Sci. Am.*, *220*, no. 2 (February 1969), pp. 100–107.

NOMURA, MASAYASU. "Ribosomes," *Sci. Am.*, 221, no. 4 (October 1969), pp. 28–35.

PAPPAS, GEOGE D. "Junctions between cells," *Hosp. Pract.*, 8, no. 8 (August 1973), pp. 37–46.

ROBERTSON, J. DAVID. "The membranes of the living cell," *Sci. Am.*, *206*, no. 4 (April 1962), pp. 64–72.

RUSTAD, RONALD C. "Pinocytosis," *Sci. Am.*, *204*, no. 4 (April 1961), pp. 120–130.

SATIR, BIRGIT. "The final steps in secretion," *Sci. Am.*, *233*, no. 4 (October 1975), pp. 28–37.

SATIR, PETER. "Cilia," *Sci. Am.*, *204*, no. 2 (February 1961), pp. 108–116.

TRUMP, BENJAMIN F. "The network of intracellular membranes," *Hosp. Pract.*, 8, no. 10 (October 1973), pp. 111–121.

WESSELLS, NORMAN K. "How living cells change shape," *Sci. Am.*, *225*, no. 4 (October 1971), pp. 77–81.

WROBLEWSKI, FELIX. "Enzymes in medical diagnosis," *Sci. Am.*, *205*, no. 2 (August 1961), pp. 99–107.

Tissues

Cells of similar function (and frequently similar appearance) plus their intercellular products are assembled into functional groups called *tissues*. Histologists (scientists who study the microscopic anatomy of cells, tissues, and organs) place tissues into four primary groups on the basis of their appearance and function: *epithelial tissue, connective tissue, muscular tissue,* and *nervous tissue.* Each primary tissue is divided into several different subtypes (Table 3.1).

Table 3.1 The Four Primary Tissues

Epithelial tissue	Connective tissue	Muscular tissue	Nervous tissue
Epithelial Sheets	Loose Connective Tissue	Smooth Muscle	Neurons
1. Simple epithelia		Skeletal Muscle	1. Unipolar
2. Stratified epithelia	1. Areolar		2. Bipolar
	2. Adipose	Cardiac Muscle	3. Multipolar
Glands	3. Reticular		
			Neuroglia: CNS
1. Exocrine glands	Dense Connective Tissue		
a. simple			1. Astrocytes
b. compound	1. Dense irregular		2. Oligoden-
2. Endocrine glands	2. Dense regular		drocytes
a. follicular type			3. Microglia
b. cord type	Cartilage		
			Neuroglia: PNS
	1. Hyaline cartilage		
	2. Elastic cartilage		1. Schwann
	3. Fibrocartilage		cells
			2. Satellite cells
	Bone		
	1. Spongy		
	2. Compact		
	Blood-forming Tissue		
	1. Myeloid tissue		
	2. Lymphoid tissue		
	Blood		

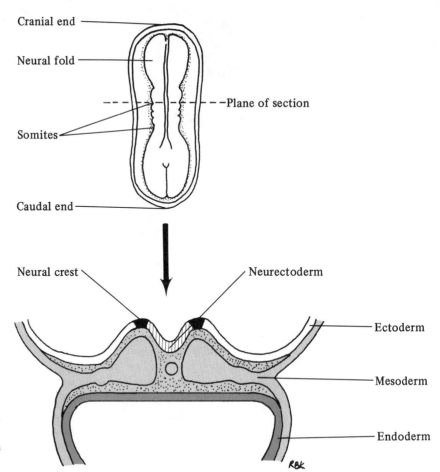

Figure 3.1 Three Embryonic Germ Layers. Top: early human embryo. *Bottom:* cross section through early human embryo showing germ layers from which human tissues derive. From ectoderm come epithelial tissue and nervous tissue (neurectoderm and neural crest); from mesoderm come epithelial tissue, connective tissue, and muscular tissue; from endoderm comes epithelial tissue.

The four primary tissues have their origins from the three *germ layers* of the embryo (Fig. 3.1). The three germ layers consist of *ectoderm* (outer skin), *mesoderm* (middle skin), and *endoderm* (inner skin). Generally speaking, we can say that ectoderm gives rise to nervous tissue, mesoderm gives rise to connective and muscular tissue, and all three germ layers give rise to epithelial tissue.

Epithelial Tissue

Epithelial tissue consists of cells in such close contact that there is little intercellular substance between them. Epithelial cells are firmly united to one another by a variety of attachment devices and are supported by a *basement membrane*. A basement membrane is a proteinaceous structure secreted by cells at interfaces between epithelial and connective tissue. Epithelial tissue contains no blood vessels; therefore epithelial cells receive oxygen and nutrients from connective tissue blood vessels by diffusion through the vessel walls and the basement membrane.

Epithelial tissue is specialized for protection, absorption, filtration, and secretion. It covers exposed body surfaces, lines body cavities and organs, and gives rise to glands. The two major subdivisions of epithelial tissue are *epithelial sheets* and *epithelial glands*.

EPITHELIAL SHEETS

Epithelial sheets range from one to several layers in thickness. They are classified according to the number of layers in the sheet and the shape of surface cells (Table 3.2).

Simple Epithelia

Simple epithelium consists of a single layer of cells supported by a basement membrane. The cells are either platelike, cuboidal, or columnar in shape.

Simple Squamous Epithelium

Simple squamous epithelium consists of a layer of thin platelike cells [Fig. 3.2(a)] that resemble old-fashioned bathroom floor tiles when seen from the surface. The thinness of these cells makes this type of epithelium suitable for diffusion and filtration, The simple squamous epithelium lining blood and lymph vessels is called *endothelium,* while that lining body cavities is called *mesothelium.* Both endothelium and mesothelium are of mesodermal origin.

Simple Cuboidal Epithelium

Simple cuboidal epithelium consists of a layer of cells as wide as they are high [Fig. 3.2(b)]. It is restricted to the ducts of certain glands and portions of the kidney tubules.

Table 3.2   General Classification of Epithelial Sheets

CELL SHAPE	CELL LAYERS	
	Simple	Stratified
Squamous	Simple squamous:* e.g., meso-thelial lining of body cavities	Stratified squamous:* e.g., epidermis (keratinized); lining of oral cavity (non-keratinized)
Cuboidal	Simple cuboidal:* e.g., kidney tubules	Stratified cuboidal: e.g., ducts of sweat glands
Columnar	Simple columnar:* e.g., lining of small intestine (nonciliated); lining of uterine tube (ciliated)	Stratified columnar: e.g., transition zones between simple epithelia and stratified squamous epithelium
	Pseudostratified columnar:* e.g., lining of most of respiratory tract (ciliated)	Transitional:* e.g., lining of most of urinary tract

* Common types of epithelia to be learned.

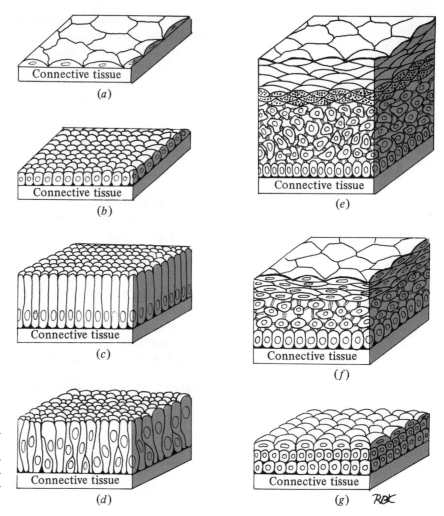

Figure 3.2 Types of Epithelia. (a) Simple squamous. (b) Simple cuboidal. (c) Simple columnar. (d) Pseudostratified columnar. (e) Stratified squamous (keratinized). (f) Stratified squamous (nonkeratinized). (g) Transitional.

Simple Columnar Epithelium

Simple columnar epithelium consists of a layer of cells that are higher than they are wide and whose nuclei all lie at the same level [Fig. 3.2(c)]. The cells in this type of epithelium are specialized for absorption or secretion and may bear cilia or microvilli on their free surface. A ciliated simple columnar epithelium lines the uterus and uterine tubes, while a simple columnar epithelium with a microvillous border lines the small intestine.

Pseudostratified Columnar Epithelium

Pseudostratified columnar epithelium consists of a single layer of columnar cells that all rest on the basement membrane but do not all reach the surface [Fig. 3.2(d)]. At first glance the sheet appears to be stratified because its cells vary in shape and their nuclei lie at different levels. Cells of this type of epithelium usually bear cilia or long microvilli on their free surface. Pseudostratified columnar epithelium lines portions of the respiratory and male reproductive tracts.

Stratified Epithelia

Stratified epithelium consists of two or more layers of cells; the basal layer rests on a basement membrane while the top layer has a free surface. Since several cell layers can withstand abrasion better than a single layer, stratified epithelia are located in areas subject to considerable wear and tear. Cells removed from the surface layer by abrasion are replaced by mitotic division of stem cells in the basal layer.

Stratified Squamous Epithelium

Stratified squamous epithelium, the most common stratified type, consists of three or more cell layers. The cells in the surface layer are platelike in shape while those in the basal layer are columnar. The epidermis [Fig. 3.2(e)], a stratified squamous epithelium exposed to the air, becomes *keratinized*—that is, its surface cells are converted into dead scalelike plates filled with keratin, a waterproofing material. The keratinized cells prevent the living cells below from drying out. Stratified squamous epithelia covering moist surfaces such as the oral cavity remain nonkeratinized [Fig. 3.2(f)].

Stratified Cuboidal and Columnar Epithelia

These types of stratified epithelia are relatively rare in the human body. Stratified cuboidal epithelium is restricted to the ducts of certain glands, while stratified columnar epithelium is found mainly at junctions between nonstratified epithelia and stratified squamous epithelium.

Transitional Epithelium

Transitional epithelium [Fig. 3.2(g)] is an unusual type of stratified epithelium that lines the urinary tract. This type of epithelium is capable of accommodating itself to a surface area that periodically expands and contracts. For example, when the bladder is distended with urine, the epithelium consists of two to three cell layers and the surface cells are flat; when the bladder is contracted, the epithelium consists of five to six cell layers and the surface cells are rounded.

EPITHELIAL GLANDS

Most glands are outgrowths from epithelial sheets that have invaded the underlying connective tissue.

Exocrine Glands

Exocrine glands are glands that have retained their connection to the epithelial sheet from which they originated. These connections or *ducts* carry secretions from the lumen (cavity) of the gland to the surface of the epithelial sheet.

Classification

Most exocrine glands are multicellular. The only unicellular exocrine gland present in the human body is the *goblet cell*. These cells

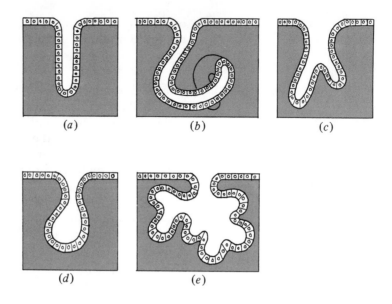

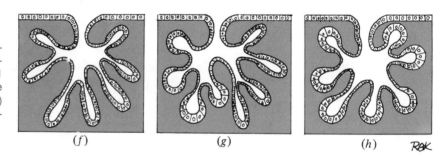

Figure 3.3 Types of Multicellular Exocrine Glands. (a) Simple tubular. (b) Simple coiled tubular. (c) Simple branched tubular. (d) Simple acinar. (e) Simple branched acinar. (f) Compound tubular. (g) Compound tuboacinar. (h) Compound acinar.

are columnar epithelial cells that secrete a lubricating fluid called *mucus*. Goblet cells are abundant in many moist epithelial sheets.

Multicellular exocrine glands are classified according to the shape of their secretory units and the complexity of their duct system (Fig. 3.3). Those with secretory units shaped like test tubes are classified as *tubular glands* while those with secretory units shaped like flasks are classified as *acinar glands*. *Simple exocrine glands* have ducts that generally do not branch and have few secretory units. *Compound exocrine glands* have branched, complex duct systems with multiple secretory units. Compound exocrine glands are sometimes classified according to the nature of the product they secrete. If they secrete a water fluid, they are called *serous glands;* if they secrete mucus, they are called *mucous glands.*

Method of Product Discharge   Cells in exocrine glands have several different methods of expelling their secretory products. These methods are illustrated in Fig. 3.4.

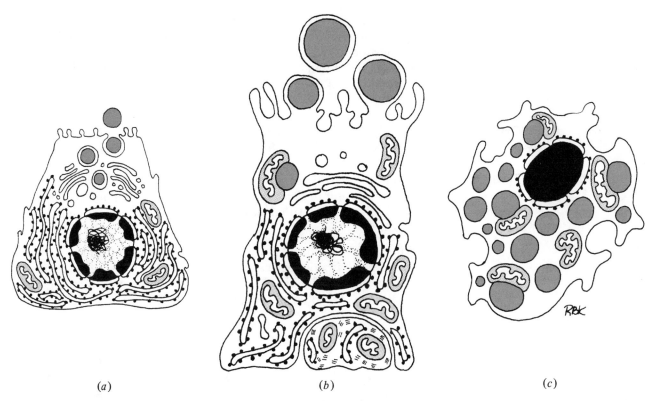

(a) *(b)* *(c)*

Figure 3.4 *Methods of Secretory Product Discharge.* (a) Merocrine secretion. Secretory product is released without loss of cytoplasm or cytoplasmic membrane, as in pancreatic acinar cell. (b) Apocrine secretion. Secretory product is released along with apical cytoplasm, as in mammary gland cell. (c) Holecrine secretion. Cell and its accumulated secretory product are discharged together, as in sebaceous gland cell.

Endocrine Glands Endocrine or ductless glands of epithelial origin have lost their connections to the epithelial sheets from which they originated. The secretory units of some endocrine glands, such as the thyroid gland, consist of follicles. A follicle is a hollow ball of cells surrounding a storage cavity. The secretory units of other endocrine glands, such as the pancreatic islets, consist of solid cords or clusters of cells. Endocrine glands contain extensive capillary networks, so secretory cells are able to discharge their products into nearby capillaries.

Connective Tissue

Connective tissue is the most widely distributed and useful tissue in the human body. It supports, protects, binds, and partitions almost all body parts. Connective tissue differentiates from *mesenchyme,* a loose primitive network of cells and scattered fibers derived from embryonic mesoderm. Mature connective tissue is characterized by a relatively small number of cells surrounded by a large amount of *matrix,* which consists of ground substance and fibers.

COMPONENTS OF CONNECTIVE TISSUE

In any type of connective tissue there are three components to be considered: the cells, the fibers, and the ground substance.

Connective Tissue Cells

Six different cell types are associated with connective tissue; not all types are present in dense and special connective tissues. Connective tissue cells are described in the following lists and are depicted in Fig. 3.5.

1. *Fibroblasts.* Fibroblasts are the most important and most numerous cells in connective tissue proper. They are responsible for producing the fibers and for secreting and maintaining the ground substance. Fibroblasts that appear to be inactive are called fibrocytes.

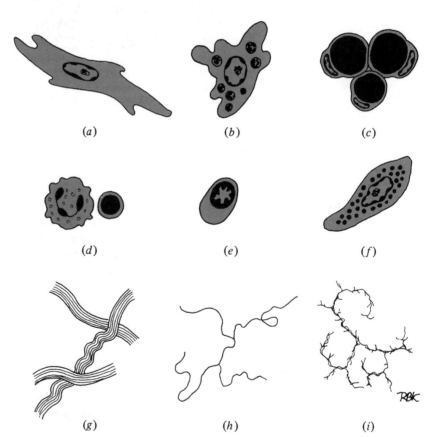

Figure 3.5 Connective Tissue Components. (a) Fibroblast. (b) Macrophage. (c) Fat cells (adipocytes). (d) White blood cells. (e) Plasma cell. (f) Mast cell. (g) Collagen fibers. (h) Elastic fibers. (i) Reticular fibers.

2. *Macrophages.*   Macrophages (histiocytes) are scavenger cells that move in amoeboid fashion through loose connective tissues to phagocytize dead cells, live bacteria, and particulate matter. Substances that are not digestible, like carbon particles, may remain in macrophages for a long time.

3. *Fat Cells.*   Fat cells (adipocytes) are fat-storing cells found in loose connective tissues. Immature fat cells, which contain little lipid, look like fibroblasts, while mature fat cells, which contain a single large lipid droplet, look like signet rings.

4. *White Blood Cells.*   White blood cells, to be described in Chapter 8, frequently are inhabitants of loose connective tissues. These cells normally migrate from the blood vessels into connective tissue, and their rate of migration is greatly increased during inflammation.

5. *Plasma Cells.*   Scattered throughout loose connective tissues are the plasma cells. Plasma cells synthesize proteins called *antibodies* (gamma globulin) that react against foreign substances called antigens.

6. *Mast Cells.*   Mast cells are generally found around small blood vessels in loose connective tissue. The cytoplasm of the mast cell is packed with secretory granules containing *heparin,* an anticoagulant, and *histamine,* an inflammatory substance that makes small blood vessels leaky.

Connective Tissue Fibers   Three types of protein fibers, secreted by fibroblasts or fibroblast-related cells, are associated with connective tissue. They may occur singly or in combination, depending on the tissue type.

1. *Collagen Fibers.*   Collagen fibers are found in nearly all types of connective tissue. These tough, white, coarse, sometimes branched fibers are composed of *collagen,* a proteinaceous substance that forms gelatin when boiled. Collagen fibers have great strength but little elasticity.

2. *Elastic Fibers.*   Elastic fibers are formed from the protein *elastin.* More slender than collagen fibers, elastic fibers appear as thin, wavy, yellow threads. Elasticity rather than strength is their chief characteristic.

3. *Reticular Fibers.*   Reticular fibers are thin, delicate, highly branched fibers also formed from collagen. Individually these fibers have little strength; collectively they form supportive frameworks for lymphoid tissues and organs.

Ground Substance   Connective tissue cells and fibers are embedded in a structureless intercellular material called ground substance. Ground substance is more complex and thicker than the ordinary intercellular material called *tissue fluid.* Chemically speaking, it is a colloidal suspension of several mucopolysaccharides (protein-sugar compounds) in water.

Ground substance varies in consistency from fluid to gel depending on the types of mucopolysaccharides present in the different connective tissues.

CONNECTIVE TISSUE PROPER

Connective tissue proper is found in all parts of the human body. Its appearance and functions are determined by the relative proportions of cells, fibers, and ground substance. Classification of connective tissue is based on the amount and arrangement of fibers and on the nature of ground substance. Loose connective tissues are characterized by a loose arrangement of cells and fibers in a relatively fluid ground substance. Dense connective tissues are characterized by a greater concentration of fibers, fewer cells, and less ground substance than the loose varieties.

Loose Connective Tissues
Areolar Connective Tissue

Areolar connective tissue (Fig. 3.6), which is more commonly called loose connective tissue, has a cobweblike appearance. It contains all the cell types and all the fibers previously described scattered in a relatively fluid ground substance. Areolar connective tissue is widely distributed throughout the body. It serves as packing material between muscles and internal organs, wraps blood vessels and nerves, binds muscles together, and separates tissue layers. It supports most epithelial membranes and forms the supporting framework for most

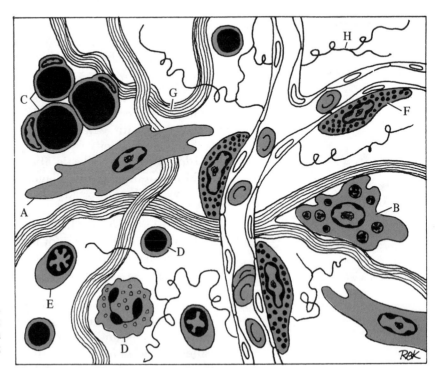

Figure 3.6 Areolar Connective Tissue. Components: A, fibroblast; B, macrophage; C, fat cells; D, white blood cells; E, plasma cell; F, mast cell (lying next to capillary); G, collagen fiber; H, elastic fiber.

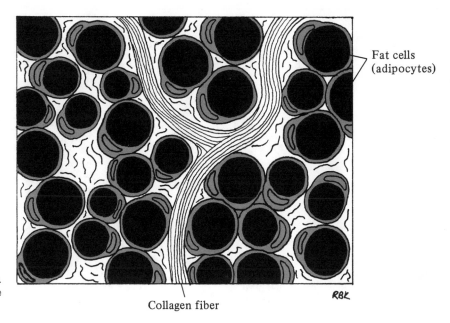

Fat cells (adipocytes)

Collagen fiber

Figure *3.7 Adipose Connective Tissue.* This is a type of loose connective tissue with a high concentration of fat cells.

organs. Because it contains many blood and lymph capillaries, areolar connective tissue also serves as a medium for the diffusion of nutrients and waste products between the vascular system and neighboring epithelial tissues.

Adipose Connective Tissue

Adipose connective tissue (Fig. 3.7) is a variant of areolar connective tissue in which fat cells predominate. The fat cells form clusters that are separated from one another by loose fiber meshworks. Adipose connective tissue is present in the superficial fascia, which lies beneath the dermis of the skin. It also forms protective pads around major joints and organs such as the heart and kidneys and forms fat deposits in the mesenteries. Functionally, adipose connective tissue serves as a shock absorber, as insulation, and a major food reserve.

Reticular Connective Tissue

Reticular connective tissue (Fig. 3.8) is a type of loose connective tissue in which reticular fibers and stellate reticular cells predominate. It forms the framework of lymphoid tissue, such as bone marrow, and lymphoid organs, such as the spleen and lymph nodes.

Dense Connective Tissues
Dense Irregular Connective Tissue

Dense irregular connective tissue (Fig. 3.9), commonly called fibrous connective tissue, is characterized by scattered fibroblasts plus large irregular bundles of collagen fibers arranged in a coarse meshwork. It occurs in the form of tough flexible sheets such as the dermis of the skin, various fascias, and the membranes that cover cartilage and bone. Many organs have protective capsules formed from dense irregular connective tissue.

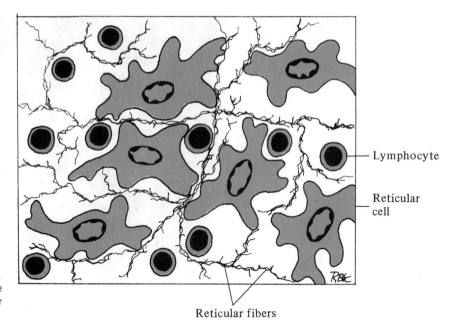

Figure 3.8 Reticular Connective Tissue. This is a type of loose connective tissue containing reticular cells plus irregular meshwork of reticular fibers.

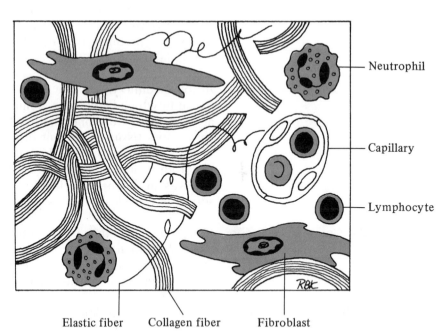

Figure 3.9 Dense Irregular Connective Tissue. This consists of irregularly arranged bundles of collagen fibers and elastic fibers plus scattered fibroblasts and white blood cells (neutrophils and lymphocytes).

Dense Regular Connective Tissue

Dense regular connective tissue (Fig. 3.10) is characterized by rows of fibrocytes lying between large parallel bundles of collagen and elastic fibers. This type of connective tissue has great strength be-

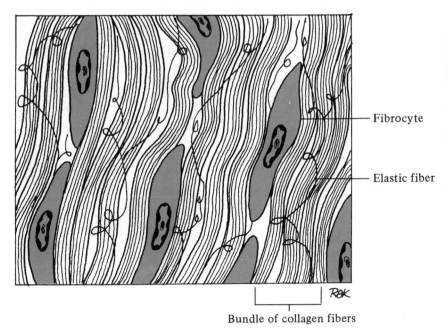

Fibrocyte

Elastic fiber

Bundle of collagen fibers

Figure 3.10 Dense Regular Connective Tissue. This consists of large parallel bundles of collagen fibers plus fibrocytes (mature fibroblasts), which lie between fiber bundles.

cause the fiber bundles are oriented along the line of stress. Dense regular connective tissue forms *tendons,* fibrous cords that join muscle to bone; *ligaments,* fibrous bands that join bone to bone; and *aponeuroses,* fibrous sheets that generally join muscle to muscle.

SPECIAL CONNECTIVE TISSUES

At first glance the special connective tissues do not look like connective tissue because of the unusual nature of their matrices. Yet on closer inspection, all special connective tissues contain fibroblast-related cells and fibers of some sort. Like connective tissue proper, they are derived from mesenchyme.

Cartilage

Cartilage is a special connective tissue in which the matrix is a firm gellike substance containing a dense network of fibers. Cartilage cells, or *chondrocytes,* lie in *lacunae* (small cavities) in the matrix. Cartilage is enclosed by *perichondrium,* a fibrous connective tissue membrane containing blood vessels. Since blood vessels are generally absent from cartilage, chondrocytes receive oxygen and nutrients from perichondrial vessels by diffusion through the semisolid matrix. For this reason cartilage has a low metabolic rate and is slow to regenerate when damaged. If the matrix becomes calcified, diffusion stops and the chondrocytes die.

Hyaline Cartilage

Hyaline cartilage (Fig. 3.11), commonly called gristle, is the most widely distributed cartilage in the human body. It has a translucent,

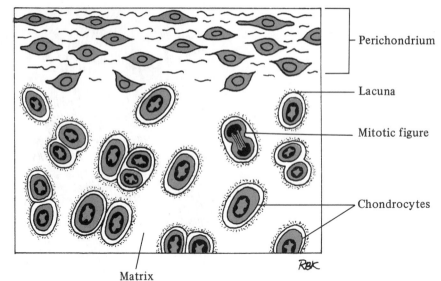

Figure 3.11 Hyaline Cartilage. Chondrocytes in their lacunae are scattered throughout homogenous-appearing matrix. New chondrocytes arise from fibroblastlike cells in perichondrium and from mitotic division of existing chondrocytes.

bluish white, homogenous matrix that contains a fine meshwork of collagen and elastic fibers. Hyaline cartilage is found in the embryo, where it acts as the precursor of one type of bone. In the adult it covers the articular surfaces of bones in freely movable joints, unites the ribs to the sternum, and forms most of the supporting cartilages of the respiratory tract.

Elastic Cartilage

Elastic cartilage (Fig. 3.12) has a more flexible and opaque matrix than hyaline cartilage. The matrix contains many yellow elastic fibers,

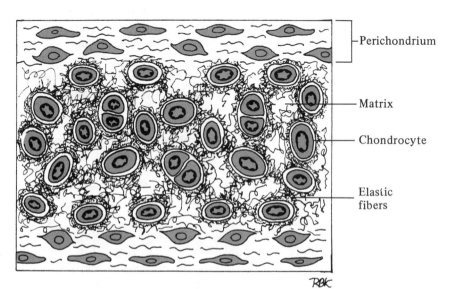

Figure 3.12 Elastic Cartilage. This contains many elastic fibers in its matrix.

which give it a yellowish color. Elastic cartilage is found in structures that require great flexibility, such as the cartilages of the larynx and the pinna of the external ear.

Fibrocartilage

Fibrocartilage (Fig. 3.13) is an unusual type of cartilage that closely resembles dense regular connective tissue. It is characterized by rows of chondrocytes, surrounded by a small amount of ground substance, that lie between thick parallel bundles of collagen fibers. Fibrocartilage is less firm but stronger than other types of cartilage; therefore it can act as a shock absorber. It forms the discs between adjacent vertebral bodies, the pubic symphysis, and the discs in the wrist, knee, and jaw joints.

Bone

Bone is a type of special connective tissue in which the matrix is very hard. Bone matrix is arranged in layers or lamellae that contain parallel bundles of collagen fibers embedded in a small amount of ground substance. It acquires its great rigidity when calcium and other mineral salts are deposited on the collagen fibers. Calcification enables bone to bear great weights without bending or breaking. Since the high density of the calcified matrix inhibits diffusion, bone cells, or *osteocytes,* cannot be supplied with oxygen and nutrients from blood vessels in the *periosteum* (the fibrous connective tissue membrane that encloses bone). Instead they are supplied by blood vessels that pass through tunnels in the matrix. The osteocytes in their lacunae maintain contact with one another and with their blood vessels through microscopic channels called *canaliculi.*

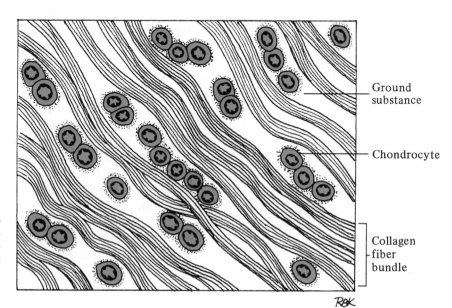

Figure 3.13 *Fibrocartilage.* Unusual type of cartilage in which chondrocytes are embedded in scanty amount of ground substance and lie in rows between thick compact bundles of collagen fibers. Collagen fibers give fibrocartilage its strength and toughness.

Ground substance

Chondrocyte

Collagen fiber bundle

RBK

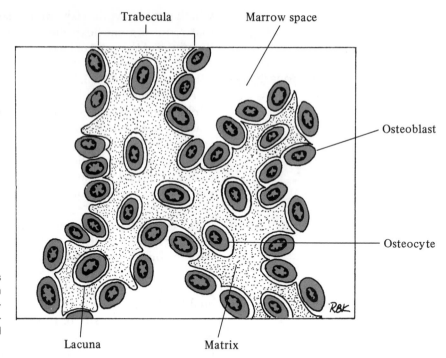

Trabecula   Marrow space

Osteoblast

Osteocyte

Figure 3.14 Spongy Bone. Trabecula is functional unit of spongy bone. Within trabecula, osteocytes lie in lacunae; outside trabecula, osteoblasts lay down additional bone matrix. Trabecula is surrounded by bone marrow spaces.

Lacuna   Matrix

Spongy Bone

Spongy bone (Fig. 3.14) is a type of bone that looks spongy on gross examination because its lamellae are organized into slender bars, or *trabeculae.* The trabeculae form a latticework whose spaces are filled with bone marrow. The resulting bone is strong but light in weight. Spongy bone is found in newly forming bone, in embryonic bone, and around the marrow spaces of adult bone.

Compact Bone

Compact bone (Fig. 3.15) is a type of bone that looks solid when examined grossly because its lamellae are organized into Haversian systems, or *osteons.* Each osteon consists of a Haversian canal surrounded by concentric lamellae of osseous tissue whose cells draw their oxygen and nutrients from vessels in the canal. Compact bone is very strong but heavy. It forms the protective outer shell of all bones.

Blood-forming Tissue and Blood
Blood-forming Tissue

Blood-forming (hematopoietic) tissue is a highly cellular type of special connective tissue. It comes in two different varieties: *myeloid tissue* and *lymphoid tissue.* Myeloid tissue, which produces red blood cells and most white blood cells, is found in the marrow cavities of long bones. Lymphoid tissue, which contains white blood cells called lymphocytes and cells that derive from them, is found in lymph nodes and other lymphoid organs.

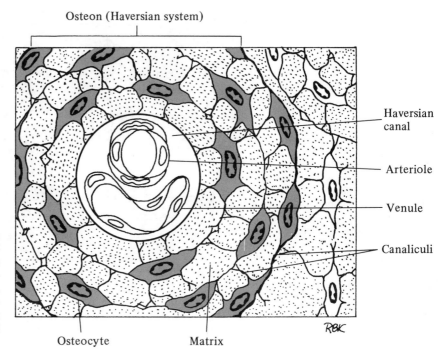

Osteon (Haversian system)

Haversian canal

Arteriole

Venule

Canaliculi

Osteocyte   Matrix

Figure 3.15 Compact Bone. Osteon (Haversian system) is functional unit of compact bone. Osteocytes form concentric rings around Haversian canal, the longitudinal vascular axis of Haversian system. Osteocytes maintain contact with Haversian canal and with one another through canaliculi, minute channels within calcified bone matrix.

Blood   Blood, which is described in Chapter 8, is a type of special connective tissue in which the cells are suspended in a fluid matrix called plasma. When blood clots, fibers form in the fluid matrix to convert it into a semisolid mass.

Muscular Tissue

Muscular tissue is specialized for contraction. The contraction of muscular tissue is responsible for movement of the bones, pumping of the blood, and the motility of visceral organs. Muscular tissue consists primarily of cells, called *muscle fibers,* that are specialized for contraction. It also contains connective tissue elements, which form protective wrappings around individual and groups of muscle cells, blood vessels, and nerve fibers.

Most muscle cells are derived from mesenchyme. Myoblasts (embryonic muscle cells) grow along their axis of contraction to form the highly elongated cells we call muscle fibers. Three types of muscular tissue are recognized on the basis of muscle fiber structure and function.

SMOOTH MUSCLE   Smooth muscle (Fig. 3.16) consists of long tapering muscle fibers that have single centrally located nuclei. Filaments of the contractile

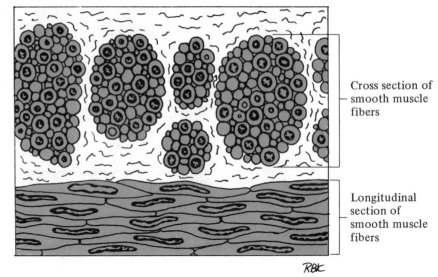

Cross section of smooth muscle fibers

Longitudinal section of smooth muscle fibers

Figure 3.16 Smooth Muscle. Long tapered smooth muscle fibers contain centrally located nuclei. Because smooth muscle fibers have staggered arrangement in muscular sheets, all do not have same diameter in cross section; this explains why nucleus does not appear in all cross-sectional profiles of smooth muscle fibers.

proteins actin and myosin are present in smooth muscle fibers, but they are not highly organized and do not preserve well in routine tissue preparation. This is why smooth muscle fibers do not have the distinct banding pattern characteristic of striated muscle fibers.

Smooth muscle fibers are frequently organized into thin sheets. Sheets of smooth muscle are present in the walls of the hollow muscular organs of the digestive, respiratory, urinary, and reproductive systems and in the walls of blood and large lymphatic vessels. Smooth muscle is also present in the dermis of the skin and in the iris and ciliary body of the eye. Within smooth muscle sheets all the fibers run in the same direction but are staggered so that the broad central region of one fiber lies next to the tapered end of an adjacent fiber.

Smooth muscle, which can contract spontaneously, is regulated by the autonomic, or visceral, nervous system and is said to be *involuntary*. It does not respond readily to our will.

SKELETAL MUSCLE

Skeletal muscle (Fig. 3.17) consists of extremely large fusiform fibers that possess many peripherally located nuclei. Each mature skeletal muscle fiber represents the fusion of many immature muscle cells, which explains the presence of the large number of nuclei. Filaments of actin and myosin present in the cytoplasm of skeletal muscle fibers are organized into bundles called *myofibrils*. The high degree of organization of myofibrils in skeletal muscle fibers creates a pattern of striations that appears to extend across the width of the fiber. For this reason skeletal muscle is also called *striated muscle*.

The bulk of muscular tissue in the human body is skeletal muscle. Skeletal muscles, which are used primarily to move the bones, are the organs of the muscular system (Chapter 7).

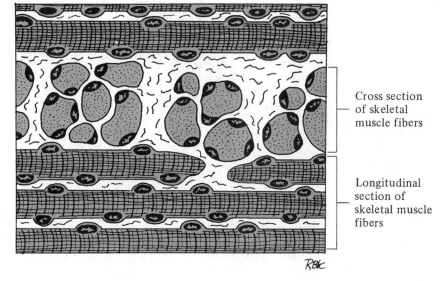

Figure 3.17 Skeletal Muscle. Large tapered skeletal fibers contain many peripherally located nuclei. In longitudinal section the many nuclei and cross striations are quite obvious. In cross section the peripheral location of nuclei is readily apparent.

Cross section of skeletal muscle fibers

Longitudinal section of skeletal muscle fibers

Skeletal muscle is under control of the somatic motor nervous system and is said to be *voluntary*. This type of muscle responds readily to our will.

CARDIAC MUSCLE   Cardiac muscle (Fig. 3.18) consists of large branched fibers with single centrally located nuclei. Like skeletal muscle fibers, cardiac muscle fibers contain myofibrils whose organization creates a striated pattern.

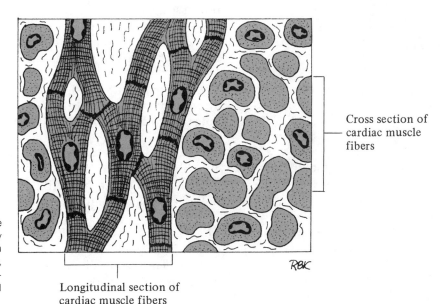

Figure 3.18 Cardiac Muscle. These branched muscle fibers contain centrally located nuclei. Intercalated discs, which unite cardiac muscle cells to one another, plus cross striations are seen in longitudinal section. In cross section central location of nucleus is apparent.

Cross section of cardiac muscle fibers

Longitudinal section of cardiac muscle fibers

Cardiac muscle fibers also have the ability to contract on their own, independent of innervation. However, the intrinsic rhythmicity of cardiac muscle, like that of smooth muscle, is regulated by the autonomic nervous system.

Nervous Tissue

Nervous tissue is specialized to receive stimuli and to conduct electrical impulses. It enables the body to react to environmental changes and coordinate the functions of its organs.

Nervous tissue consists of nerve cells, or *neurons,* and supporting cells, or *neuroglia.* Neurons receive stimuli and conduct impulses, while the neuroglia provide them with support and protection. All neurons and most neuroglia are derived from embryonic ectoderm.

NEURONS

Neurons are the most highly specialized cells in the human body. They are extremely sensitive to lack of oxygen and changes in blood glucose concentration. A serious consequence of their high degree of specialization is that they are unable to replace themselves by mitotic division. Destruction of neurons following oxygen deprivation, for example, can lead to permanent loss of nervous tissue function.

Neurons usually occur in groups. A *ganglion* is a group of nerve cell bodies in the peripheral nervous system, while a *nucleus* is a group of nerve cell bodies in the central nervous system. Each neuron consists of a cell body and one or more processes. Neurons have two kinds of processes, axons and dendrites. *Axons* carry electrical impulses *away* from the cell body, while *dendrites* carry electrical impulses *toward* the cell body. Structural classification of neurons is based on the number of processes they possess.

Unipolar Neurons

The unipolar neuron [Fig. 3.19(*a*)] has a single axonlike process that divides into peripheral and central branches near the cell body. Unipolar neurons are found in the ganglia of all spinal and certain cranial nerves.

Bipolar Neurons

The bipolar neuron [Fig. 3.19(*b*)] has two processes, an axon and a dendrite, which are located at opposite poles of the cell body. Bipolar neurons are found in the organs of special senses.

Multipolar Neurons

The multipolar neuron [Fig. 3.19(*c*)] has one axon and many dendrites. Multipolar neurons are found throughout the central nervous system. The axons of multipolar neurons contribute to peripheral nerves as well as to fiber tracts of the central nervous system.

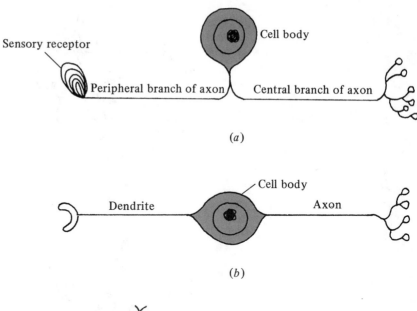

(a)

(b)

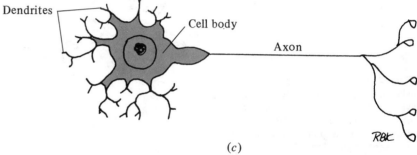

(c)

Figure 3.19 *Neurons.* (a) Unipolar neuron has one process divided into peripheral and central branches. Cell body sits to side of process. (b) Bipolar neuron has two processes, axon and dendrite, which arise from opposite poles of cell body. (c) Multipolar neuron has one axon and many dendrites.

NEUROGLIA   Neuroglia are cells that do not conduct electrical impulses. They are specialized to provide physical and metabolic support for the neurons. Neuroglia, which are ten times as numerous as neurons, fall into two categories: those located in the central nervous system (CNS) and those located in the peripheral nervous system (PNS).

Neuroglia: CNS   There are three major types of neuroglia in the central nervous system: *astrocytes*, *oligodendrocytes*, and *microglia*.

Astrocytes   Astrocytes are large stellate cells with many radiating cell processes. The protoplasmic astrocyte [Fig. 3.20(a)] has short thick processes, while the fibrillar astrocyte [Fig. 3.20(b)] has long slender processes containing many fibrils. Astrocytes provide metabolic support for CNS neurons.

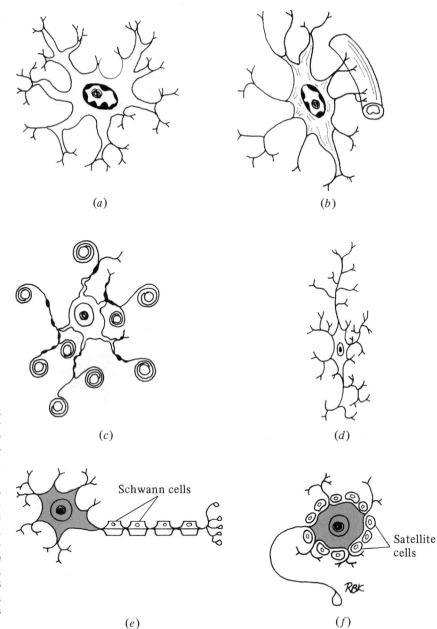

Figure 3.20 *Neuroglia.* (a) Protoplasmic astrocyte, a CNS glial cell with short, thick cytoplasmic processes, is located in gray matter of brain and spinal cord. (b) Fibrillar astrocyte, a CNS glial cell with slender processes containing fibrils, is located mainly in white matter of brain and spinal cord. Fibrillar astrocytes transmit nutrients from CNS capillaries to neurons. (c) Oligodendrocyte, a CNS glial cell, forms and maintains myelin sheaths of CNS neurons. (d) Microgliocyte, a small cell of mesodermal origin, acts as scavenger in the CNS. (e) Schwann cells, PNS glial cells, form and maintain myelin sheaths of PNS neurons. (f) Satellite cells, PNS glial cells, surround and protect cell bodies of neurons in peripheral ganglia.

Schwann cells

Satellite cells

(a)   (b)   (c)   (d)   (e)   (f)

Oligodendrocytes   The oligodendrocyte [Fig. 3.20(c)] is a stellate cell, smaller than the astrocyte, with relatively few radiating cell processes. Oligodendrocytes form and preserve the myelin sheaths that insulate nerve fibers in the central nervous system.

Microglia	The microgliocyte [Fig. 3.20(*d*)] is a small fibroblastlike cell of mesodermal origin that has invaded the central nervous system. Microglia are phagocytic cells that rid the central nervous system of foreign organisms and cell debris.
Neuroglia: PNS	There are two types of neuroglia in the peripheral nervous system: *Schwann cells* and *satellite cells.*
Schwann Cells	Schwann cells [Fig. 3.20(*e*)] are small supporting cells that form and preserve the myelin sheaths that insulate nerve fibers in the peripheral nervous system.
Satellite Cells	Satellite cells [Fig. 3.20(*f*)] are small supporting cells that surround the cell bodies of neurons in peripheral ganglia.

Tissue Combinations

The four primary tissues combine in various ways to form membranes and organs.

MEMBRANES Membranes consist of thin layers of tissue—some fragile, some tough—that line or cover body structures.

Epithelial Membranes Epithelial membranes line hollow muscular organs and body cavities. An epithelial membrane consists of an epithelial sheet with a connective tissue backing. The connective tissue layer strengthens the membrane and contains blood vessels that provide oxygen and nutrients for the epithelial sheet. There are two distinct types of epithelial membranes: *mucous membranes* and *serous membranes.*

Mucous Membranes The mucous membrane, or *mucosa* (Fig. 3.21), contains an epithelial sheet, one to several layers thick, that secrets *mucus,* a thick sticky substance that forms a lubricating film on the surface of the epithelial sheet. Mucous membranes line the hollow muscular organs and pas-

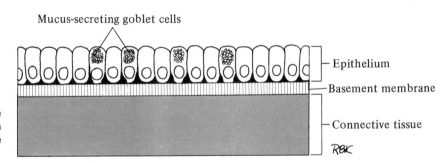

Figure 3.21 Mucosa. This membrane consists of a sheet of epithelial cells, which secrete mucus, lying on connective tissue substrate.

sageways belonging to the digestive, respiratory, urinary, and reproductive systems.

Serous Membranes

The serous membrane, or *serosa* (Fig. 3.22), contains a sheet of mesothelium that secretes a thin watery fluid. Serous membranes line body cavities and cover organs within body cavities. The slippery surface film provides for frictionless movement of organs within their respective cavities. Serous membranes form the *pericardium,* a sac that contains the heart, the *pleurae,* sacs that contain the lungs, and the *peritoneum,* a sac that contains most of the abdominopelvic organs.

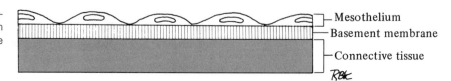

Figure 3.22 Serosa. This membrane consists of a sheet of mesothelial cells, which secrete serous fluid, lying on connective tissue substrate.

Mesothelium
Basement membrane
Connective tissue

Connective Tissue Membranes

Connective tissue membranes are composed entirely of connective tissue elements. They provide support and protection for various body structures.

Supportive Membranes

The various fascias consist of connective tissue sheets and tubes. There are two distinct fascial systems: *external* and *internal.* The external fascia is subdivided into superficial and deep varieties. *Superficial fascia* is an extensive adipose membrane that lies beneath the skin and binds it to deeper structures. *Deep fascia* consists of tough fibrous membranes that invest individual muscles, serve as partitions between muscle masses, provide protective sheaths for blood vessels and nerves, and form restraining bands (retinacula) for tendons. Internal fascia consists of fibrous connective tissue membranes that lie deep to the serosas of the thoracic and abdominopelvic cavities. Most organ capsules are fibrous membranes that protect and provide support for the soft tissues of the organs they enclose. *Perichondrium* and *periosteum* are fibrous membranes that cover the outer surface of cartilage and bone, respectively.

Secretory Membranes

Secretory membranes, which are associated with the articular and muscular systems, secrete a lubricating fluid that reduces friction between moving parts. *Synovial membranes* line the joint cavities of freely movable joints except at the articular cartilage surfaces. The slippery synovial fluid secreted by these membranes reduces friction between moving bones. *Bursae* are fluid-filled sacs of secretory tissue found between muscles or between muscles and bone in areas of friction. *Tendon sheaths* are secretory membranes that enclose and lubricate tendons subject to friction.

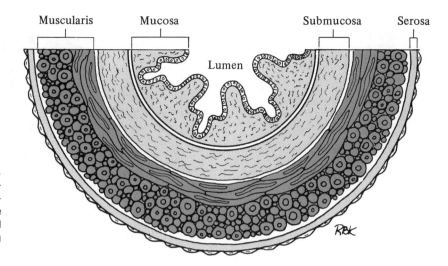

Muscularis   Mucosa   Submucosa   Serosa

Lumen

Figure 3.23 Hollow Muscular Organ. Cross section (half) shows mucosa, epithelial sheet plus underlying connective tissue; submucosa, fibrous connective tissue layer; muscularis, smooth muscle coat; and serosa, mesothelial sheet plus underlying connective tissue.

ORGANS   Organs are structures composed of two or more primary tissues. They occur in two basic forms: the *hollow muscular organ* and the *solid glandular organ.*

Hollow Muscular Organs   The hollow muscular organ (Fig. 3.23) is usually a tube or a modified tube that has a lumen (central cavity) surrrounded by a muscular wall. Organs of this type, such as the stomach, intestines, bladder, or ureters, have the following components:

1. *Mucosa.* A mucosa lines the hollow muscular organ. It consists of an epithelial sheet plus a highly vascular connective tissue layer (lamina propria) that supports and nourishes the epithelial sheet. A thin layer of smooth muscle fibers may be present.
2. *Submucosa.*   The submucosa is a layer of fibrous tissue that underlies the mucosa. This layer may be absent in some hollow muscular organs. When present it provides additional support for the mucosa.
3. *Muscularis.*   The muscularis is a muscular coat consisting of one or more layers of smooth muscle fibers. The muscularis permits alteration of the diameter of the organ. The alternate contraction and relaxation of smooth muscle fibers in a hollow muscular tube is called *peristalsis.*
4. *Adventitia/Serosa.*   The adventitia is a protective fibrous coat that covers the hollow muscular organ and blends in with surrounding connective tissue. Organs suspended in a body cavity are covered with a serosa instead of an adventitia.

Solid Glandular Organs   The solid glandular organ (Fig. 3.24) is generally an organ with a

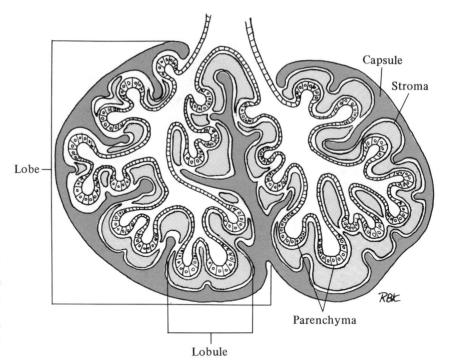

Figure 3.24 *Solid Glandular Organ.* Longitudinal section shows parenchyma, soft tissue containing functional cells of organ; stroma, loose connective tissue framework that supports functional cells; and capsule, fibrous connective tissue membrane that surrounds and protects organ.

secretory function. Solid glandular organs, such as the liver, pancreas, adrenals, kidneys, or spleen, have the following components:

1. *Soft Tissue.* The soft tissue, or *parenchyma,* is frequently an epithelial tissue that is specialized for secretion.
2. *Stroma.* The stroma is the connective tissue framework that supports the parenchyma.
3. *Capsule.* The organ capsule is the connective tissue membrane that surrounds the solid glandular organ. Inward projections from the capsule may partition the parenchyma into smaller units called lobes and lobules.

Diseases and Disorders of Tissues

Aplasia and Hypoplasia. Failure of a tissue or organ to form is termed aplasia if complete and hypoplasia if partial.

Atrophy. A decrease in the size of a tissue or organ after formation has taken place is termed atrophy. The decrease may be due either to loss of cells or to decrease in cell size. Some of the causes of atrophy include aging, disuse, decreased blood supply, denervation, pressure, and lack of endocrine stimulation. For example, when muscle cells are not used at all, they decrease in size.

Hypertrophy.   The increase in the size of a tissue or organ due to an increase in cell size is termed hypertrophy. Hypertrophy is a response to increased work load or injury in tissues or organs whose cells have lost their ability to divide. For example, an increase in the heart's work load will cause hypertrophy of cardiac muscle fibers.

Hyperplasia.   An increase in the size of a tissue or organ due to an increase in cell number is termed hyperplasia. Hyperplasia is a response to increased work load or injury in tissues or organs whose cells still retain the ability to divide. For example, an increase in a gland's work load will often cause hyperplasia of secretory cells.

Metaplasia.   A reversible change of one adult tissue type into another related but different type is termed metaplasia. Metaplasia is most common in epithelial and connective tissues. Squamous metaplasia refers to the transformation of an absorptive or secretory epithelium into stratified squamous epithelium. For example, squamous metaplasia occurs in the respiratory tract of chronic cigarette smokers as a response to the irritation of smoke inhalation.

Dysplasia.   A reversible alteration in the size, shape, and organization of adult cells, especially in epithelial tissue, is termed dysplasia. Dysplasia is usually the result of chronic irritation, and dysplastic tissue is more likely to become cancerous than normal tissue.

Neoplasia.   The proliferation of cells to form a mass whose growth exceeds and is not coordinated with that of the tissue in which it arises is termed neoplasia. The mass so formed is termed a *neoplasm* (new growth) or *tumor*. Tumor formation is a response to stimuli such as chronic irritation, x-irradiation, or exposure to cancer-causing chemicals. Though tumors arise from all four primary tissues, epithelial and connective tissue tumors are more common than other types. Perhaps it is because these two tissues have the greatest ability to renew and repair themselves.

Tumors are classified as *benign* or *malignant* depending primarily on their ability to kill their host; this distinction, however, is not always clear. As a rule, a benign tumor is well differentiated, slow growing, encapsulated (which facilitates surgical removal), and does not kill its host. Some benign tumors can cause considerable damage and even death by the pressure they exert on vital structures. A malignant tumor, also termed a *cancer*, generally is poorly differentiated, rapidly growing, unencapsulated, and kills its host. *Metastasis* (spread of a malignant tumor to a distant site) occurs by dispersion of cancer cells through the cardiovascular or lymphatic systems. It is often impossible to eradicate cancers by surgery alone due to their lack of encapsulation and tendency to metastasize.

Tumors are generally named according to the tissue of origin and to

Table 3.3 Classification of Tumors

Tissue of origin	Benign form	Malignant form
Epithelial Tissue		
1. Epithelial sheets	Papilloma	Carcinoma
2. Glandular epithelium	Adenoma	Adenocarcinoma
Connective Tissue		
1. Adipose tissue	Lipoma	Liposarcoma
2. Fibrous tissue	Fibroma	Fibrosarcoma
3. Cartilage	Chondroma	Chondrosarcoma
4. Bone	Osteoma	Osteosarcoma
5. Blood-forming tissue		
a. myeloid tissue	—	Myelogenous leukemia
b. lymphoid tissue	—	Lymphatic leukemia
Muscle Tissue		
1. Smooth muscle	Leiomyoma	Leiomyosarcoma
2. Skeletal and cardiac muscle	Rhabdomyoma	Rhabdomyosarcoma
Nervous Tissue		
1. CNS neurons	—	Neuroblastoma, neurocytoma
2. CNS neuroglia	—	Astrocytoma, oligodendroglioma
3. PNS neurons	Ganglioneuroma	Neuroblastoma
4. PNS neuroglia	Neurofibroma	Neurofibrosarcoma
5. Melanocytes (pigment cells)	Melanoma	Malignant melanoma

whether or not they are benign or malignant (Table 3.3). The naming of tumors is variable; however, some generalizations can be made. The names for tumors end in "-oma" ("tumor of"). Malignant tumors of epithelial cell origin are called *carcinomas* ("crablike"). Malignant tumors of mesenchymal cell origin are called *sarcomas* ("fleshy"). When both epithelial and mesenchymal tissues are involved, the name of the tumor reflects this fact, for example, carcinosarcoma.

Suggested Readings

CAIRNS, JOHN. "The cancer problem," *Sci. Am., 233*, no. 5 (November 1975), pp. 64–78.

GROSS, JEROME. "Collagen," *Sci. Am., 204*, no. 5 (May 1961), pp. 120–130.

ROSS, RUSSELL. "Wound healing," *Sci. Am., 220*, no. 6 (June 1969), pp. 40–50.

———, and PAUL BORNSEIN. "Elastic fibers in the body," *Sci. Am., 224*, no. 6 (June 1971), pp. 44–52.

VERZÁR, FREDERIC. "The aging of collagen," *Sci. Am. 208*, no. 4 (April 1963), pp. 104–114.

4

The Integumentary System

The skin, or integument, is a vital organ that covers the human body. It defends the body from invasion by harmful microorganisms, functions as a sensory organ, prevents the body from drying out, helps regulate body temperature, and protects the body from the damaging effects of ultraviolet radiation.

The integumentary system is the first organ system we will study. In this system we will bring together previously discussed material about cells and tissues.

Anatomy of the Skin

The skin (Fig. 4.1) consists of two interacting tissue sheets, the epidermis and the dermis. It rests on the superficial or subcutaneous fascia, an adipose connective tissue membrane that is an important part of the integumentary system.

EPIDERMIS The epidermis is a superficial sheet of keratinized stratified squamous epithelium that varies in thickness depending on its location. It is thickest on friction areas of the skin, such as the palms of the hands and the soles of the feet, and thinnest on the lips and eyelids.

In humans the epidermis has a wavy surface of ridges and grooves. This is because the underlying dermis is thrown into a series of projections called *dermal papillae*. The epidermis lying on top of these papillae is thrown into ridges. The patterns of epidermal ridges on the palms and soles, which appear as loops, arches, and whorls, are

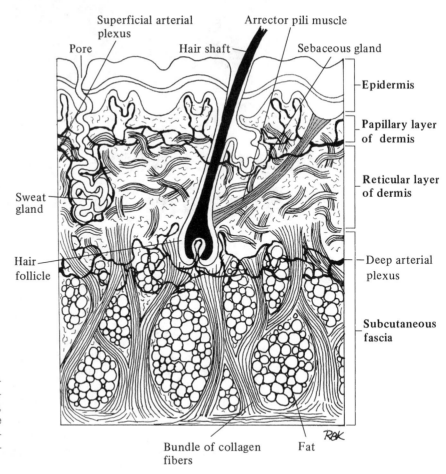

Figure 4.1 Integument. Section cut perpendicular to surface shows the three divisions of integumentary system: epidermis, dermis, and subcutaneous fascia. Note structural organization of dermis and subcutaneous fascia and location of integumentary appendages and blood vessels.

largely inherited and unique to each individual. In the form of fingerprints and footprints, they serve as a means of identification.

Like other epithelial sheets the epidermis is constantly being worn away from the surface and renewed by mitotic division of basal cells. Newly formed cells mature, become keratinized, and die as they are pushed toward the surface.

From base to surface the epidermis is subdivided into four layers, or strata, that are depicted in Fig. 4.2 and described next.

Stratum Germinativum

The stratum germinativum consists of several rows of cells ranging from low columnar in the bottom row to polyhedral (many-sided) in the top row. A basement membrane separates the stratum germinativum from the underlying dermis. The cells in this layer still retain the capacity to divide; hence the name germinativum. Cells of the stratum germinativum are united by desmosomes, attachment devices that extend across tissue fluid spaces between adjacent cells. The

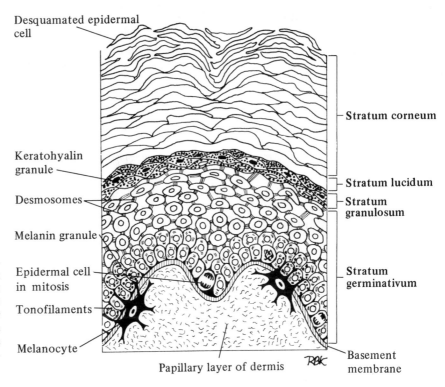

Desquamated epidermal cell

Keratohyalin granule

Desmosomes

Melanin granule

Epidermal cell in mitosis

Tonofilaments

Melanocyte

Papillary layer of dermis

Stratum corneum

Stratum lucidum

Stratum granulosum

Stratum germinativum

Basement membrane

Figure 4.2 Epidermis. Section cut perpendicular to surface shows the four strata of epidermis: stratum germinativum, stratum granulosum, stratum lucidum, and stratum corneum. Note mitotic figures in basal row of germinativum cells plus desmosomes that unite germinativum cells. Melanocytes at dermo-epidermal junction transfer melanin granules to epidermal cells.

presence of many desmosomes gives these cells a prickly appearance in light microscopic preparations. Stratum germinativum cells actively synthesize *tonofilaments*, components of the desmosomes and precursors of the waterproofing protein known as *keratin*. Keratin is a hard fibrous protein with a high sulfur content.

Stratum Granulosum   The stratum granulosum consists of one or two rows of fusiform cells that produce *keratohyalin*, another precursor of keratin. The presence of keratohyalin granules in these cells gives the name granulosum to this layer. Keratohyalin combines with the tonofilaments to form keratin in the upper two layers of the epidermis. Mitotic division does not occur in the stratum granulosum. Also, as their cytoplasm fills with keratohyalin granules, the nucleus and cytoplasmic organelles of the granulosum cells begin to degenerate.

Stratum Lucidum   The stratum lucidum consists of several rows of flattened, closely compacted cells that are filled with keratin and lack nuclei. In light microscopic preparations it appears as a clear wavy band that separates the stratum granulosum from the stratum corneum. The stratum lucidum is more prominent in thick skin than in thin skin and may even be absent.

Stratum Corneum   The stratum corneum consists of several to many rows of scalelike keratinized cells lacking nuclei and cytoplasmic organelles. These cells are cemented together with intercellular substance to form a tough waterproof barrier that protects the living cells below from dehydration. No living cell could stand the constant exposure to air, water, and temperature changes that these dead cells face. Ultimately, the dead cells are sloughed off from the surface of the epidermis as dry flaky scales.

DERMIS   The dermis, also called the *corium* or *cutis,* is a sheet of connective tissue that supports and maintains the overlying epidermis. It consists of two poorly demarcated layers that are described here.

Papillary Layer   The papillary or superficial layer of the dermis consists of a layer of areolar connective tissue lying directly beneath the epidermis. This layer includes the dermal papillae, with their networks of blood and lymphatic vessels, as well as numerous sensory receptors. Since the epidermis has no blood vessels of its own, oxygen and nutrients reach the epidermis by diffusion from papillary blood vessels.

Reticular Layer   The reticular or deep layer of the dermis consists of a layer of dense irregular connective tissue lying below the papillary layer. Large irregularly oriented bundles of collagen and elastic fibers are interwoven to form a tough flexible meshwork that, if treated with tanning chemicals, becomes leather. The reticular layer of the dermis possesses great powers of elasticity and distensibility, as is seen in pregnant women and extremely obese people. However, it can be overstretched, which causes tearing of the dermis. Tears in the dermis are visible as white streaks called stretch marks, or *linea albicantes.* Linea albicantes are frequently found on the hips, thighs, breasts, and abdomen.

SUBCUTANEOUS FASCIA   The superficial fascia (tela subcutanea or hypodermis) is a sheet of adipose connective tissue lying between the dermis and the deep fascia covering the muscles. Its collagen and elastic fibers are continuous with those of the dermis and form fiber bundles that mainly run parallel to the long axis of the skin. These fiber bundles are thickest and most numerous on the palms and soles, where the superficial fascia is firmly attached to underlying structures. The number of fat cells in the superficial fascia varies with sex, age, nutritional state of the individual, and region of the body. When properly distributed, the fat content of the superficial fascia gives pleasing contours to the body. The superficial fascia is pierced by many large blood vessels, which, with the high content of fat cells, permits the superficial fascia to act as a fat reservoir, an insulator, and a temperature-regulating device.

Skin Color

NORMAL COLOR Normal skin color is produced by the interaction of several pigments present in the skin. These pigments include *melanin, carotene,* and *hemoglobin.*

Melanin Melanin is a brown-black pigment formed in *melanocytes,* pigment cells derived from a type of embryonic ectoderm called neural crest. Melanocytes are located at the dermo-epidermal junction and transfer melanin-producing granules into young epidermal cells. Melanin helps protect epidermal cells from the damaging effect of ultraviolet rays in sunlight. For example, people develop a tan when exposed to strong sunlight, and human populations living close to the equator, where the sunlight is intense, have darker pigmentation than those who live nearer to the poles. The number of melanocytes in the skin is approximately the same in all races; it is the type of melanin, the amount of melanin produced, and the degree of granule aggregation that determines whether an individual's color is black, brown, red, tan, or white.

Carotene Carotene is a yellow pigment, normally found in epidermal cells, that gives the skin its yellowish tones. This same pigment is responsible for the color of carrots and other yellow vegetables. Excessive ingestion of carrots causes a person's skin to develop a pronounced yellow color.

Hemoglobin Hemoglobin is an oxygen-binding pigment found in red blood cells. When combined with oxygen, it colors the blood bright red. Oxygenated blood flowing through superficial dermal vessels gives the skin its pinkish tones.

ABNORMAL COLOR Certain physical conditions are characterized by abnormal skin coloration. Some of these conditions are described next.

Cyanosis Cyanosis is a condition in which the skin takes on a dark bluish color from the deoxygenated blood flowing through superficial dermal vessels. Cyanosis appears in people with certain cardiovascular or respiratory diseases. Also, when breathing is interrupted, people become cyanotic.

Jaundice Jaundice is a condition in which the skin takes on a yellowish or greenish color from excess bile pigments in the bloodstream. People with liver diseases frequently become jaundiced.

Hemochromatosis Hemochromatosis is a condition in which the skin takes on a bronze color when the body mishandles iron pigment metabolism. This

condition often occurs in people receiving multiple blood transfusions or intravenous iron therapy.

Skin Appendages

There are three classes of skin appendages in humans: glands, hair,
and nails. All are derivatives of the epidermis.

GLANDS Integumentary glands are exocrine glands whose secretory units are
located in the dermis or subcutaneous fascia. Excretory ducts connect them either to the free surface of the epidermis or to a hair follicle.

Sweat Glands Sweat, or sudoriferous, glands (Fig. 4.3) occur in two types: *eccrine
sweat glands* and *apocrine sweat glands.*

Eccrine Sweat Glands Approximately 2 million eccrine sweat glands are present in the skin
of the average adult. These glands are widely distributed over all the
hairless areas of the body, except for the eardrum and the glans
penis, and are most numerous on the palms and soles. Eccrine sweat

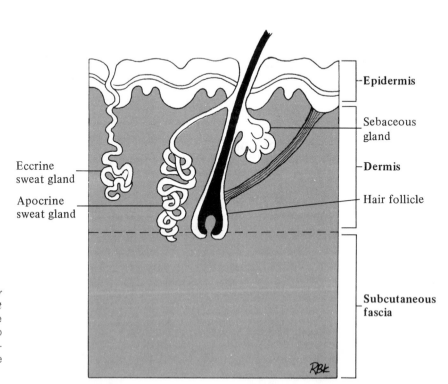

*Figure 4.3 Relationship Among Hair
Follicle, Sebaceous Gland, and Sweat
Glands.* Sebaceous gland and apocrine
sweat gland empty their secretions into
hair follicle; eccrine sweat gland has independent opening or pore on free surface
of epidermis.

glands are simple tubular glands whose coiled secretory units lie in the dermis; their excretory ducts spiral through the epidermis to terminate in minute pores on the free surface. This type of gland produces a colorless, odorless, watery secretion called *sweat*. Sweat plays an important role in temperature regulation; its evaporation from the surface of the skin rapidly lowers body temperature.

Apocrine Sweat Glands
Apocrine sweat glands are highly coiled simple tubular glands located primarily in the armpits and around the anus. These glands are associated with hair follicles, into which they release their secretions. Apocrine sweat glands are basically scent glands that produce a thick milky secretion in response to stress or sexual stimulation. This secretion is odorless until degraded by bacteria on the skin. Negative public reaction to the secretions of apocrine sweat glands causes perfume and deodorant industries to flourish.

Ceruminous Glands
Ceruminous glands are modified sweat glands located in the external auditory canal of the ear. They secrete cerumen, or earwax, a substance that protects the eardrum. However, excessive accumulation of earwax in the external ear canal can cause a partial hearing loss.

Mammary Glands
Mammary glands (breasts) are specialized integumentary organs containing modified sweat glands that secrete milk. The breasts of the human female, which are discussed in Chapter 14, reach their greatest development during her childbearing years under the stimulus of pituitary and ovarian hormones.

Sebaceous Glands
Sebaceous glands are simple branched acinar glands derived from developing hair follicles. Most sebaceous glands remain connected to hair follicles into which they release their secretion. Only on the lips, eyelids, and external genital organs are sebaceous glands found independent from hair follicles. The oily secretion of the sebaceous gland, called *sebum*, lubricates the skin and helps waterproof it. *Acne vulgaris*, an inflammation of the sebaceous glands commonly seen in adolescents, is due to sebaceous gland hyperactivity, which is probably stimulated by rising levels of sex hormones. Some forms of acne leave permanent pits and scars in the skin. Comedones, or *blackheads*, are plugs of keratin and dirty dried-up sebum that block the openings of hair follicles.

HAIR
Hair in humans is now largely an ornamental rather than a protective appendage. It is distributed in varying amounts over the surface of the body; the amount and patterns of distribution depend on race, sex, and age. Hair is most abundant on the scalp, in the armpits, and in the pubic region. It is totally absent from the palms and soles.

The *hair follicle* (Fig. 4.4), which produces and supports the hair,

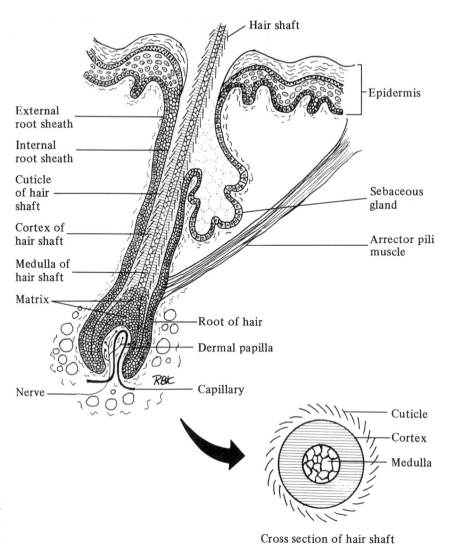

External root sheath

Internal root sheath

Cuticle of hair shaft

Cortex of hair shaft

Medulla of hair shaft

Matrix

Nerve

Hair shaft

Epidermis

Sebaceous gland

Arrector pili muscle

Root of hair

Dermal papilla

Capillary

Cuticle

Cortex

Medulla

Cross section of hair shaft

Figure 4.4 Hair Follicle. Note location of sebaceous gland; contraction of arrector pili muscle squeezes sebaceous gland, which discharges sebum into hair follicle.

is an epithelial tube extending deep into the dermis. At its deep end the follicle expands into the hair bulb. The bulb surrounds a connective tissue papilla that contains nerve endings and blood vessels.

Hair formation occurs in the matrix, the layer of epithelial cells on top of the papilla, which is comparable to the stratum germinativum of the epidermis. The matrix produces the hair root, a column of growing and dividing epithelial cells that become keratinized as they move upward. The hair shaft is a cylinder of keratinized cells, continuous with the root, whose free end protrudes above the surface of the skin. The shaft consists of the medulla (an inner layer of loosely packed cells and air spaces), the cortex (a middle layer of tightly packed, heavily keratinized cells), and the cuticle (an outer layer of cells whose overlapping shinglelike arrangement prevents erupted hairs from matting). The growing hair receives melanin granules from melanocytes located over the tip of the dermal papilla. Hair color is determined by the amount and aggregation of melanin granules in the cortex and the amount of air spaces in the medulla.

Hair follicles lie diagonal to the surface of the skin. The *arrector pili* muscle, a bundle of smooth muscle fibers attached to the hair follicle, runs obliquely from the greater angle of the follicle to the papillary layer of the dermis. Contraction of this muscle in response to cold, fear, or anger straightens the hair and elevates the skin surrounding the hair. This is how the so-called goose bumps are produced in humans. Also, contraction of the arrector pili muscle causes the release of sebum into the hair follicle by squeezing the sebaceous gland between the muscle and follicle wall.

NAILS Fingernails and toenails of the human body are protective keratinized appendages, closely related to the hooves and claws of animals, that develop on the dorsal surface of the terminal phalanges of fingers and toes. Nails (Fig. 4.5) are formed in folliclelike structures made up of the nail bed and fold. The nail bed consists of the skin lying beneath the nail, while the nail fold consists of a fold of skin surrounding the proximal surface and lateral borders of the nail. The visible portion of the nail is the body of the nail plate, and the hidden portion the root.

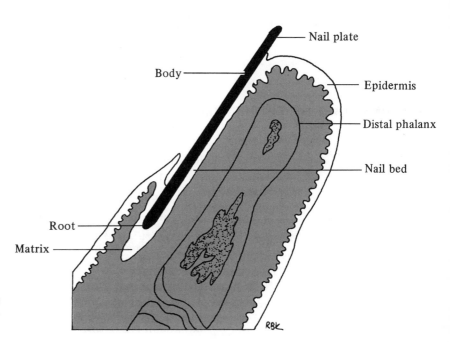

Figure 4.5 Nail Follicle. Fingernails and toenails develop in folliclelike structures similar to hair follicle.

The matrix is the portion of the nail follicle that gives rise to the nail root. As new cells are formed by the matrix, the nail plate moves forward over the nail bed and older cells become keratinized. Nail growth rate is dependent on the health and nutritional state of the individual. When a nail is injured, it will often die and drop off. If the nail matrix is still intact, a new nail will form under the dead one before it drops off.

Blood Supply of the Skin

The skin contains extensive networks of blood vessels. Integumentary arterioles form large plexuses in the subcutaneous fascia and in the deep and superficial layers of the dermis. Branches from the superficial plexus form extensive capillary networks around sweat glands, hair follicles, and sebaceous glands. The superficial plexus also gives off capillary loops to the dermal papillae that supply oxygen and nutrients to the deeper cells of the epidermis. Veins draining the skin form networks that run roughly parallel to the arteriolar plexuses. A complex system of arteriovenous shunts is present in the skin so that blood can be shunted from the superficial dermal arterioles directly into deeper venous vessels. Operation of this shunt mechanism reduces heat loss from the skin when the body attempts to conserve heat.

Innervation of the Skin

The integumentary system receives many stimuli from the external environment. Accordingly, it is richly supplied with a wide variety of receptors and effectors.

INTEGUMENTARY RECEPTORS   Integumentary receptors (Fig. 4.6) consist of endings of sensory nerve fibers that have been modified to receive stimuli from the external environment. Primary skin sensations include pain, temperature, touch, and pressure. These sensations are received by the following nerve endings and transmitted to the central nervous system:

1. *Pain.*   Pain receptors are unencapsulated, or free, nerve endings located in the deep layers of the epidermis.
2. *Temperature.*   In the past the receptors thought to be responsible for the perception of heat and cold were the encapsulated receptors of Ruffini and Krause, respectively. This view has been challenged recently by neurophysiologists and is probably no longer valid. Heat receptors, are now thought to be free nerve endings similar to pain receptors, while cold receptors are thought to be encapsulated nerve endings (possibly those of Krause).

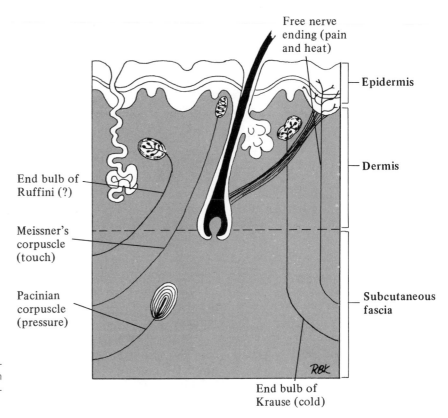

Free nerve ending (pain and heat)

Epidermis

End bulb of Ruffini (?)

Dermis

Meissner's corpuscle (touch)

Pacinian corpuscle (pressure)

Subcutaneous fascia

End bulb of Krause (cold)

Figure 4.6 Integumentary Sensory Receptors. Major receptors are shown in relation to region of integument they normally occupy.

3. *Touch.* Touch receptors include Meissner's corpuscles, encapsulated nerve endings located in the dermal papillae, tactile discs (of Merkel), discoid nerve endings located in the epidermis, and the free nerve endings that surround hair follicles.

4. *Pressure.* Pressure receptors consist primarily of pacinian corpuscles, encapsulated nerve endings located in the superficial fascia. In longitudinal section this corpuscle resembles an onion bulb.

INTEGUMENTARY EFFECTORS Integumentary effectors consist of the muscles and glands that respond to messages transmitted from the central nervous system to the skin by certain autonomic nerve fibers. These fibers, which belong to the sympathetic division of the autonomic nervous system, can be classified as follows:

1. *Vasomotor Fibers.* Sympathetic fibers that innervate smooth muscle in integumentary blood vessels. Impulses from these nerves cause constriction of integumentary vessels.

2. *Sudomotor Fibers.* Sympathetic fibers that innervate sweat

glands. Impulses from these nerves cause increased secretion of sweat.

3. *Pilomotor Fibers.* Sympathetic fibers that innervate arrector pili muscles. Impulses from these nerves cause muscle contraction with subsequent erection of hairs and secretion of sebum.

Functions of the Skin

Functions of the skin include protection, temperature regulation, fluid regulation, and excretion.

PROTECTION

Protection from environmental assault is a major function of the skin. It acts as a barrier between the deeper structures of the body and the external environment. Where subject to excessive friction, the skin thickens and may even form protective calluses. The skin is also the body's first line of defense against invasion by disease-producing microorganisms. In addition, the melanin pigment in the skin helps protect deeper body structures from the harmful effects of ultraviolet radiation.

TEMPERATURE REGULATION

The skin plays an important role in the regulation of body temperature. Temperature regulation is the result of coordinated activity between autonomic nerve fibers, blood vessels, and sweat glands in the skin. When the outside temperature is high or when the body is overheated, as after strenuous exercise, superficial arterioles dilate. More blood flows through the dilated arterioles, which increases the amount of heat lost through radiation into the air. At the same time, sweat glands increase sweat secretion; the evaporation of sweat from the surface of the skin has a major cooling effect. Conversely, when the outside temperature is low or when body temperature drops, superficial arterioles constrict and the bulk of the blood is shunted into deep integumentary vessels. This action reduces heat loss from radiation. Sweat secretion is also diminished, which reduces heat loss from evaporation.

FLUID REGULATION

The skin serves as a barrier to massive loss of tissue fluid into the external environment. Fluid loss normally occurs through the sweat glands and is regulated by the autonomic nervous system. Otherwise, people who live in hot dry climates would shrivel up like prunes. Only when the skin is severely injured or destroyed by burns are excessive amounts of tissue fluid lost into the external environment. The rapid loss of large quantities of tissue fluid following extensive burns can upset the body's water balance and lead to cardiovascular collapse and death. The skin is also virtually impermeable to water from the external environment. The epidermis contains keratin plus lipid substances that act as water repellants. Otherwise, people would swell up like sponges when they took baths or swam in fresh water.

EXCRETION The skin also serves as an excretory organ. Excretion of water, sodium, chloride, and potassium ions, lactic acid, and minute amounts of urea (a nitrogenous waste product) occurs through the sweat glands. However, compared to the kidneys, the skin is a minor excretory organ.

Diseases and Disorders of the Integumentary System

Because of its location, the skin is subject to a wide variety of assaults and insults that produce skin lesions. A summary of terms describing skin lesions appears in Table 4.1. Some of the many disorders and diseases of the skin are described here:

Albinism. Albinism is an inherited pigment disorder characterized by complete absence of melanin production in skin, hair, and eyes. Albinos have white skin, yellow hair, and pink eyes regardless of their race. They also have no protection from the damaging effects of sunlight.

Table 4.1 Definitions of Integumentary Lesions

Lesion	Definition
Macula	A flat red or brown spot on the skin.
Papule	A small, raised, red skin lesion; often pointed.
Wheal	A firm, raised, plateaulike skin lesion caused by allergic reactions.
Vesicle	A small blister; often seen in clusters on the skin.
Bulla	A large blister.
Pustule	A raised, pus-filled lesion.
Cyst	A fluid-filled sac; contains no pus.
Nodule	A small solid node that is irregular in shape.
Keratosis	A raised, often scaly, wartlike lesion.
Verruca	A wart.
Polyp	A stalked or nonstalked raised growth on the surface of the skin.
Fissure	A cleft or groove in the skin.
Ulcer	An erosion of the skin extending into the dermis or subcutaneous fascia.

Bedsores.   A bedsore, or *decubitus ulcer,* is an area of skin ulceration over a bony prominence such as the shoulder or sacrum. Bedsores usually occur in patients who are bedridden for long periods of time and are caused by prolonged pressure upon bony prominences. Compression of the skin against bone in these areas decreases blood circulation, which leads to skin necrosis and ulceration. Frequent turning of bedridden patients, general cleanliness, and massage of pressure areas to improve circulation helps prevent the occurrence of bedsores. Recently, water beds have been installed in some hospitals for use by patients likely to develop bedsores. The shifting surface of the waterbed helps decrease skin pressure.

Burns.   A burn is a skin injury caused by contact with dry heat, hot liquids, steam, chemicals, electricity, or ultraviolet radiation. Burns are classified by degrees according to the amount of damage done to the skin:

1. *First-Degree Burn.*   A burn that only involves the epidermis; it is characterized by reddening of the skin.
2. *Second-Degree Burn.*   A burn that involves both the epidermis and dermis; it is characterized by blistering and destruction of the epidermis plus some damage to the dermis.
3. *Third-Degree Burn.*   A burn that involves the epidermis, dermis, and subcutaneous fascia; it is characterized by severe damage to the dermis such that epidermal regeneration from glands and hair follicles cannot take place. Skin grafts are frequently used to repair the sites of third-degree burns.

   The size of a burn is often more important than its degree of severity in determining the extent to which a particular skin function is impaired. For example, a third-degree burn that covers a large area of skin is more serious than a third-degree burn that covers a small area in terms of fluid loss.

Dermatrophic Viral Diseases.   Dermatrophic viral diseases are viral infections that cause skin eruptions. Some of the most common ones are listed here:

1. *Variola.*   Smallpox.
2. *Varicella.*   Chicken pox.
3. *Rubeola.*   Regular measles.
4. *Rubella.*   German measles.
5. *Herpes Simplex.*   Cold sores.
6. *Verruca Vulgaris.*   Warts.

Keratoses.   Keratoses are brownish warty growths that occur in older people and in people who have been exposed to the sun for long periods of time. These lesions, which are sometimes precancerous, represent overgrowths of the epidermis.

Malignant Skin Lesions.   Three types of malignant skin lesions arise from the epidermis; they are as follows:

1. *Basal Cell Carcinoma.*   An epidermal cancer of low-grade malignancy that arises from the basal portion of the stratum germinativum. It frequently occurs in people who have had prolonged exposure to intense sunlight.
2. *Squamous Cell Carcinoma.*   An epidermal cancer of medium-grade malignancy that arises from the upper portion of the stratum germinativum. It occurs in sites of chronic skin irritation or in preexisting skin lesions such as keratoses.
3. *Malignant Melanoma.*   An epidermal cancer of high-grade malignancy that arises from pigment cells. It occurs most commonly in blue black moles on the face, neck, and extremities.

Suggested Readings

BICKERS, DAVID R., and ATTALLAH KAPPAS. "Metabolic and pharmacological properties of the skin," *Hosp. Pract.*, *9*, no. 5 (May 1974), pp. 97–106.

DANIELS, FARRINGTON, JR., JAN C. VANDER LEUN, and BRIAN E. JOHNSON. "Sunburn," *Sci. Am.*, *219*, no. 1 (July 1968), pp. 38–46.

FRASER, R. D. B. "Keratins," *Sci. Am.*, *221*, no. 2 (August 1969), pp. 86–96.

JARRETT, A., R. I. C. SPEARMAN, and P. A. RILEY. *Functional Dermatology.* Philadelphia: Lippincott, 1966.

LERNER, AARON B. "Hormones and skin color," *Sci. Am.*, *205*, no. 1 (July 1961), pp. 98–108.

MARPLES, MARY J. "Life on the human skin," *Sci. Am.*, *220*, no. 1 (January 1969), pp. 108–115.

MONTAGNA, WILLIAM. "The skin," *Sci. Am.*, *212*, no. 2 (February 1965), pp. 56–66.

————, and PAUL F. PARAKKAL. *The Structure and Function of Skin.* 3d ed. New York: Academic Press, 1974.

PENROSE, L. S. "Dermatoglyphics," *Sci. Am.*, *221*, no. 6 (December 1969), pp. 72–84.

5

The Skeletal System

In general, the human skeletal system consists of 206 bones. The skeletal system, which forms the rigid framework of the human body, has five major functions: support, protection, muscle attachment, blood cell production, and mineral storage. This chapter is divided into two parts: Part A deals with the structure, function, and formation of bones, while Part B deals with the bones of the human body.

Part A. Bones as Organs

Gross Anatomy

Bones are organs that consist of bone tissue, bone membranes, and bone marrow. They are classified as long, short, sesamoid, flat, or irregular, largely on the basis of shape.

LONG BONES Long bones are bones that are longer than they are wide. A typical long bone consists of a shaft, or *diaphysis,* which is hollow, and two or more ends, the *epiphyses* (Fig. 5.1). The epiphyses do not form a bony union with the diaphysis until longitudinal bone growth is completed. The hollow interior of the diaphysis, called the *medullary* (marrow) *cavity,* is filled with bone marrow. Spongy bone forms the core of the epiphyses and surrounds the medullary cavity, while compact bone forms the thin outer shell of the epiphyses and the bulk of the diaphysis. Long bones include all the bones of the upper and lower extremities except for the wrist and ankle bones.

SHORT BONES Short bones are bones that are as high as they are wide. The short bone consists of a core of spongy bone surrounded by a shell of com-

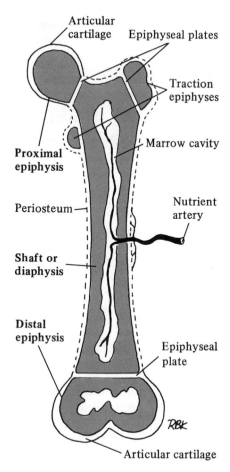

Articular
cartilage

Epiphyseal plates

Traction
epiphyses

Marrow cavity

**Proximal
epiphysis**

Periosteum

Nutrient
artery

**Shaft or
diaphysis**

**Distal
epiphysis**

Epiphyseal
plate

RBK

Articular cartilage

Figure 5.1 Long Bone. Coronal section shows major parts.

pact bone. It does not contain a medullary cavity. Short bones include the wrist and ankle bones.

SESAMOID BONES Sesamoid bones resemble sesame seeds in shape and short bones in structure. In general, they are small, variable in number, and develop in tendons that cross bony prominences or joints. The largest and most important sesamoid bone in the human body is the patella (kneecap), which develops in the tendon of the quadriceps femoris muscle.

FLAT BONES Flat bones are bones that are wider than they are high. They consist of two layers of compact bone separated by a layer of spongy bone. Flat bones include the ribs and the bones of the skull vault. In the flat bones of the skull (Fig. 5.2), the layers of compact bone are called the inner and outer tables, while the layer of spongy bone is called the diploë.

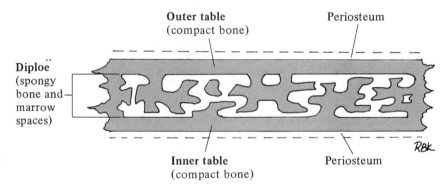

Figure 5.2 Flat Bone of Skull. Coronal section shows major parts.

IRREGULAR BONES   Irregular bones are bones whose shapes do not fit into any of the previous categories. Irregular bones include the vertebrae, shoulder blades, hipbones, facial bones, and bones of the skull floor.

Microscopic Anatomy

Bone, as described in Chapter 3, consists of cells plus a calcified matrix.

BONE CELLS   Three different kinds of cells, which represent different functional states of the same cell type, occur in bone: osteoblasts, osteocytes, and osteoclasts. The *osteoblast* [Fig. 5.3(*a*)], a relative of the fibroblast, is the bone-building cell. Osteoblasts, which are found on the growing surfaces of bone, secrete bone matrix, or osteoid. When surrounded by calcified matrix, the osteoblast becomes an *osteocyte*, or bone-maintaining cell [Fig. 5.3(*b*)], whose metabolic influence is exerted on the surrounding matrix. Its processes extend through microscopic canals or canaliculi in the matrix to contact and communicate with other osteocytes. The canaliculi also provide for the exchange of metabolites between the blood stream and the tissue fluid bathing the osteocytes. The *osteoclast* [Fig. 5.3(*c*)] is the bone-destroying cell. It is a large multinucleate cell probably formed by the fusion of several osteoblasts. Osteoclasts, which are the agents of bone resorption, are found in shallow cavities on the surface of partially demineralized bone. It is thought that they secrete powerful lysosomal enzymes that break down organic matrix and dissolve the mineral matter, whose ions (namely, calcium and phosphate) are released into the bloodstream.

INTERCELLULAR SUBSTANCE   The intercellular substance of bone consists of an organic matrix and inorganic salts. Bone matrix contains bundles of collagen fibers ce-

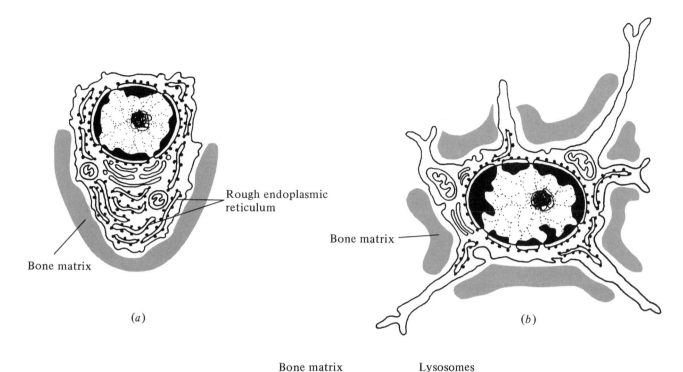

Rough endoplasmic
reticulum

Bone matrix

(a)

(b)

Bone matrix

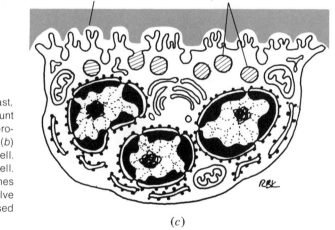

Bone matrix   Lysosomes

(c)

Figure 5.3 Bone Cells. (a) Osteoblast, the bone-building cell. Note large amount of rough endoplasmic reticulum for production of bone matrix components. (b) Osteocyte, the bone-maintaining cell. (c) Osteoclast, the bone-destroying cell. Note lysosomes, whose digestive enzymes are released extracellularly to dissolve bone matrix components. Drawings based on electron microscopic views.

mented together and arranged in layers, or *lamellae*, 3 to 7 microns thick. The inorganic salts, which consist primarily of calcium phosphate crystals, are deposited on the surface of the collagen fibers. The organic matrix gives bone its toughness and flexibility, while the inorganic salts give it its strength.

ORGANIZATION OF BONE   In Chapter 3 bone tissue was described as occurring in two forms: *spongy bone* and *compact bone*. Both types occur together in most bones of the human body (Fig. 5.4).

Spongy Bone   The structural units of spongy bone consist of *trabeculae,* slender bars of bone tissue containing varying numbers of lamellae. The trabeculae form a three-dimensional latticework whose pattern is determined by mechanical stresses on the bone and whose spaces are occupied by bone marrow. Since the osteocytes in trabeculae can receive oxygen and nutrients from nearby bone marrow capillaries, the trabeculae usually do not contain blood vessels.

Compact Bone   The structural units of compact bone consist of Haversian systems, or *osteons.* The osteon (Fig. 5.5) is organized around a Haversian canal, a longitudinal channel containing blood vessels and nerves. Up to half a dozen ring-shaped concentric lamellae surround the Haversian canal. A radiating system of canaliculi connect the osteocytes in the osteon with one another and with the Haversian canal from which they draw their oxygen and nutrients. The spaces between individual osteons are filled by interstitial lamellae, which are the remnants of osteons partially destroyed in the bone remodeling process. In addition, there are circumferential lamellae, which run parallel to the internal and external surfaces of compact bone.

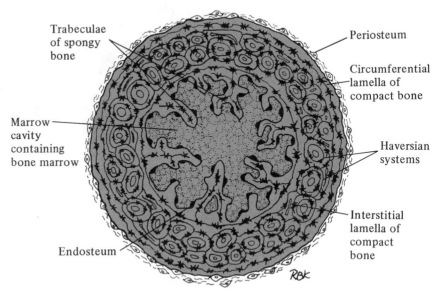

Trabeculae of spongy bone

Periosteum

Circumferential lamella of compact bone

Marrow cavity containing bone marrow

Haversian systems

Interstitial lamella of compact bone

Endosteum

Figure 5.4   Shaft of Long Bone. Cross section shows spongy or medullary bone surrounding marrow cavity. Compact or cortical bone surrounds spongy bone and contains Haversian systems plus interstitial and circumferential lamellae. Endosteum lines marrow cavity; periosteum covers cortical bone.

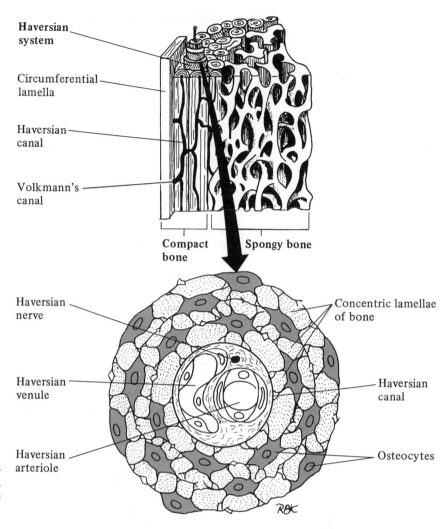

Haversian
system

Circumferential
lamella

Haversian
canal

Volkmann's
canal

Compact
bone

Spongy bone

Haversian
nerve

Concentric lamellae
of bone

Haversian
venule

Haversian
canal

Haversian
arteriole

Osteocytes

Figure 5.5 *Haversian System.* *Top:* section of long bone shows Haversian system. *Bottom:* cross section through Haversian system.

BONE MEMBRANES   Two types of connective tissue membranes are associated with bone: *endosteum* and *periosteum*.

Endosteum   Endosteum is a thin membrane of reticular connective tissue that lines the marrow cavity and the channels within compact bone. It has the ability to produce both new osteoblasts and new bone marrow cells.

Periosteum   Periosteum is a thick membrane of fibrous connective tissue that covers a bone except for its articular surfaces. Under the proper stimulus, the fibroblastlike cells of the inner portion of the periosteum differentiate into new osteoblasts that can participate in bone growth or repair.

BONE MARROW

Bone marrow, or myeloid tissue, is a type of special connective tissue found in the marrow cavities and spaces of bones. It consists of blood-forming cells and fat cells supported by a reticular fiber stroma. In the fetus all bone marrow is red; in the adult, bone marrow occurs in red and yellow varieties. *Red bone marrow* gives rise to red blood cells, most white blood cells, and platelets. It is red because it contains many blood vessels and red blood cells. In the adult, red bone marrow is found in the flat bones of the skull, the proximal epiphyses of the arm and thighbones, the ribs and breastbone, the bodies of the vertebrae, and the crests of the hipbones. The breastbone, or sternum, is the site doctors frequently use to obtain bone marrow samples for diagnostic purposes. *Yellow bone marrow* is yellow because it has a high content of fat cells. Except for the preceding locations, all red marrow is replaced by yellow marrow during childhood and adolescence.

Blood Supply

Bones have an intrinsic blood supply derived from systemic arteries. The largest artery to a bone, particularly a long bone, is called the *nutrient artery*. The nutrient artery, which gives off periosteal branches, enters the bone through a hole in the diaphysis called the nutrient foramen. It then divides into several *medullary arteries*, which supply the bone marrow. The epiphyses receive a separate supply from the *epiphyseal arteries*. According to recent studies, the capillary network within compact bone receives blood from medullary arteries and is drained by periosteal veins. Blood vessels travel through compact bone by way of horizontally oriented Volkmann canals and the vertically oriented Haversian canals.

Innervation

Peripheral nerve fibers innervate the periosteum and travel through the vascular channels to innervate compact bone. The periosteum, which is well supplied with pain receptors, is highly sensitive to pain; bone itself, however, is very insensitive to pain.

Function

Support. Bones provide support for the soft tissues of the body. Without the rigid framework provided by the skeleton, humans would be as shapeless as amoebae.

Protection. The skeleton provides protection for vital organs by enclosing them in bony containers. The skull forms a protective case for the brain. Likewise, the vertebral column surrounds and protects the spinal cord. The thoracic cage forms an inverted bony basket for the heart and lungs, while the pelvis forms a bony basin for the bladder, rectum, and internal reproductive organs.

Muscle Attachment. Bones provide attachment sites for skeletal muscles. They act as levers in pulley systems, formed by skeletal muscles, that make body movements possible.

Blood Cell Production. Before birth the bone marrow takes over the job of producing blood cells. These cells are formed in red marrow and are released into the bloodstream by way of the medullary vessels when they are mature.

Mineral Storage. The skeleton serves as the principal storage depot for calcium and phosphate ions and as a minor storage depot for sodium and magnesium ions. Bone is the chief source of exchangeable calcium in the body. Calcium may be drawn out of bone in large quantities during pregnancy, prolonged bed rest, or hyperactivity of the parathyroid glands.

Bone Formation, Growth, and Remodeling

BONE FORMATION AND GROWTH

Bone formation begins in the fetus about the eighth week of prenatal life. The first bone that forms is a loosely woven bone called *immature bone.* Immature bone, which is similar to the bone that takes part in fracture repair, has relatively more cells and fibers and less organic cement substance and minerals than mature bone. Almost all immature bone formed during fetal development is replaced by mature lamellar bone.

Bones develop from mesenchymal connective tissue. In the human fetus there are two types of bone formation: *intramembranous,* or direct ossification, and *endochondral,* or indirect ossification.

Intramembranous Ossification

Bones that form directly in mesenchymal membranes are known as *membrane bones;* the process by which they form is called intramembranous ossification. The flat bones of the cranial vault and face as well as the the clavicles (collarbones) and shafts of long bones form by this process.

Intramembranous ossification (Fig. 5.6) begins in a fibrous mesenchymal membrane that forms at the site of the future bone. Mesenchymal cells at one or more ossification centers in the membrane differentiate into osteoblasts. The osteoblasts then lay down irregular islands of bone called spicules. The spicules merge to form thin radiating trabeculae of spongy bone. From these ossification centers spongy bone formation spreads peripherally to the edges of the membrane.

Later, some of the spongy bone is replaced by compact bone, which forms the inner and outer tables. The spongy bone that remains between these plates of compact bone becomes the diploë. Connective tissue surrounding the growing bone mass gives rise to its periosteum.

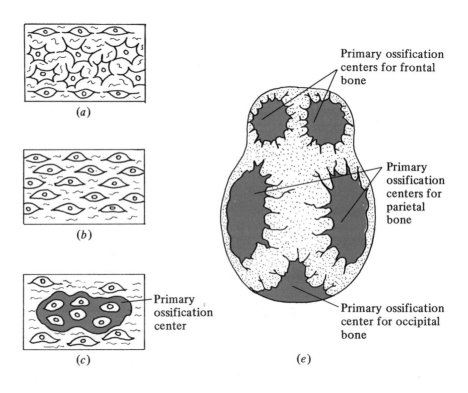

Primary ossification centers for frontal bone

Primary ossification centers for parietal bone

Primary ossification center for occipital bone

Primary ossification center

(a)

(b)

(c)

(e)

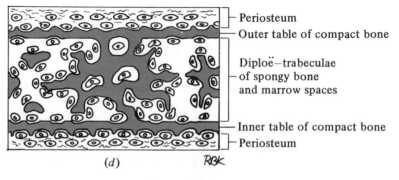

Periosteum

Outer table of compact bone

Diploë—trabeculae of spongy bone and marrow spaces

Inner table of compact bone

Periosteum

(d)

Figure 5.6 Stages in Intramembranous Ossification of Skull Bone. (a) Mesenchyme. (b) Fibrous mesenchymal membrane. (c) Initiation of ossification in primary ossification center in fibrous mesenchymal membrane. (d) Section through developing membrane bone showing compact and spongy bone. (e) Roof of fetal skull showing primary ossification centers of skull bones.

Endochondral Ossification

Bones that form by replacement of a temporary cartilaginous model are known as *endochondral bones;* the process by which they form is called endochondral ossification. Most bones of the human body form by this process.

Endochondral ossification (Fig. 5.7) begins in a hyaline cartilage model of the future bone. During bone development and growth, the cartilage is gradually replaced by bone except at the joint surfaces. If the future bone is a long bone, bone formation begins within a band-shaped area of the connective tissue membrane that surrounds the central portion of the diaphysis. Here the connective tissue cells

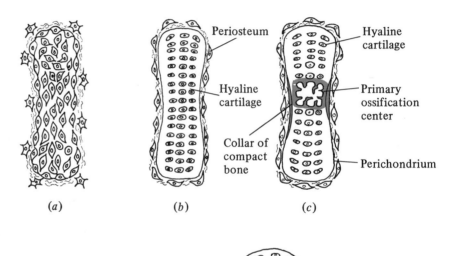

(a)   *(b)*   *(c)*

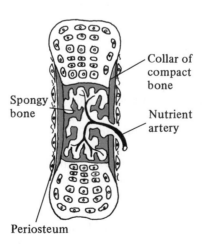

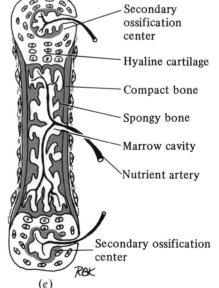

Figure **5.7** *Stages in Endochondral Ossification of Long Bone.* (a) Mesenchymal model. (*b*) Hyaline cartilage model. (c) Initiation of ossification in diaphysis. (*d*) Invasion of primary ossification center by nutrient artery. (*e*) Establishment of secondary ossification centers in epiphyses.

(d)   *(e)*

in contact with the cartilage differentiate into osteoblasts. These osteoblasts begin to form intramembranous bone, which constitutes the periosteal collar.

While the periosteal collar is forming, a primary ossification center develops in the center of the diaphysis. In this site the cartilage cells hypertrophy and the surrounding matrix calcifies. This is followed by the death of the hypertrophied chondrocytes and destruction of some of the calcified matrix. Next, osteogenic cells arrive at the primary ossification center along with the invading nutrient artery. These osteogenic cells begin to lay down spongy bone on the remnants of calcified cartilage. Osteoclastic resorption of some of the newly formed spongy bone creates the marrow cavity.

The zone of endochondral ossification extends toward both ends of the diaphysis by a series of changes similar to those that occurred in the primary ossification center. At the same time, the periosteal bone collar thickens and also extends to both ends of the diaphysis. It serves to strengthen the diaphysis of the developing bone while cartilage is replaced by spongy bone and the marrow cavity is being formed.

At the time of birth and for several years thereafter, secondary ossification centers appear in the epiphyses of developing long bones. The sequence of events in spongy bone formation and resorption in these epiphyseal centers is identical to that observed in the diaphyseal center. Spongy bone formation spreads out from both epiphyseal and diaphyseal centers until only two plates of hyaline cartilage are left. These *epiphyseal plates* are located at each end of the bone between the epiphyses and diaphysis. They are the source of new cartilage cells for longitudinal bone growth and persist until bone growth is completed.

BONE GROWTH, REMODELING, AND REPAIR

Most bones continue to grow long after birth. Bones of the skull vault are not completed until the second year of life. Some of the long bones of the extremities continue to grow throughout childhood and adolescence. Bone growth is usually completed at 18 years of age in the human female and 20 years in the human male.

Longitudinal Growth

The long bone continues to grow in length (Fig. 5.8) by endochondral ossification. Its epiphyseal plates form new cartilage at the surfaces facing the diaphysis, which moves the plates further away from one another and increases the length of the diaphysis. This new cartilage undergoes the same sequence of changes that occurred in the primary and secondary ossification centers. The newly formed spongy bone is then incorporated into the diaphysis. When a long bone attains its mature length, cartilage proliferation stops and the epiphyseal plates are replaced by compact bone. This event, called closure of the epiphyseal plates, prevents further longitudinal bone growth. The

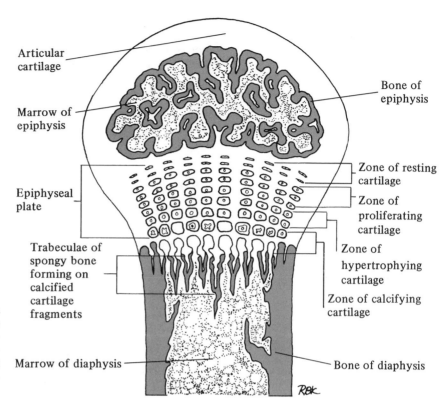

Articular cartilage

Marrow of epiphysis

Epiphyseal plate

Trabeculae of spongy bone forming on calcified cartilage fragments

Marrow of diaphysis

Bone of epiphysis

Zone of resting cartilage

Zone of proliferating cartilage

Zone of hypertrophying cartilage

Zone of calcifying cartilage

Bone of diaphysis

Figure 5.8 Longitudinal Growth in Long Bone. Epiphyseal plate is source of cartilage cells that proliferate, hypertrophy, and die; they leave behind fragments of calcified cartilage matrix that serves as framework for spongy bone formation.

bony union between an epiphysis and the diaphysis is visible in a mature bone as the *epiphyseal line.*

Circumferential Growth

The long bone continues to grow in width by intramembranous ossification. Periosteal osteoblasts deposit new bone on the periosteal surface of the diaphysis while osteoclasts resorb old bone from the endosteal surface. Bone resorption from the endosteal surface enlarges the marrow cavity and maintains diaphyseal bone at optimal thickness. Bone deposition and resorption continue until circumferential bone growth is completed.

Remodeling

Mature bones are not static, although their appearance leads students to assume that they never change. Bone remoldeling occurs both before and after bone growth stops and is necessary to help bones adjust to the demands of different stages of life. Before growth stops, a good deal of spongy bone is converted to compact bone for greater strength. After bone growth is completed, there is constant remodeling of compact bone in response to mechanical stresses, changing levels of calcium-regulating hormones, and systemic disease. Old osteons are destroyed and new ones are laid down.

Repair Following a bone fracture, blood from torn blood vessels forms a clot in the fracture site. The clot is then invaded by capillaries and connective tissue cells which first convert the clot into fibrous connective tissue and then into a mass of cartilage, the *temporary callus*. Osteogenic cells derived from the periosteum and endosteum invade the temporary callus and achieve bony union of the fracture by replacing the cartilage with spongy bone. This *bony callus* is subsequently remodeled into compact bone.

Bone Markings

The surface of a bone tells a story about the bone's function. Every bone has a collection of projections, depressions, cavities, or holes created by articulations, muscle attachments, and the passage of nerves and blood vessels. These are called bone marking.

A list of some of the many types of bone markings follows. You should be familiar with these markings before proceeding to the study of individual bones.

Articular Projections

1. *Head.* A rounded articular projection separated from the shaft or body of a bone by a neck.
2. *Condyle.* A large articular projection directly attached to the shaft of a bone.

Nonarticular Projections

1. *Process.* A projection generally used for muscle attachment.
    a. *Trochanter.* A very large blunt process.
    b. *Tuberosity.* A large blunt or rounded process.
    c. *Tubercle.* A small rounded process.
    d. *Epicondyle.* A small rounded process above a condyle.
2. *Spine.* A pointed projection.
3. *Trochlea.* A pulley-shaped projection.
4. *Hamulus.* A hook-shaped projection.
5. *Crest.* A prominent ridge.
6. *Line.* A low ridge.

Depressions, Cavities, and Holes

1. *Fossa.* A shallow depression in a bone.
2. *Sulcus.* A deep depression or groove in a bone.
3. *Sinus.* An air-filled cavity in the interior of a bone.
4. *Antrum.* A large sinus.
5. *Foramen.* A hole in a bone for the passage of nerves and blood vessels.
6. *Fissure.* A narrow cleft.
7. *Meatus.* A short wide canal.

Part B. **Bones of the Human Body**

The average human skeleton (Fig. 5.9) contains 206 bones, whose distribution is indicated in Table 5.1. Some skeletons contain more or less than 206 bones. There are a variable number of sesamoid bones associated with tendons of the hands and feet. A bone may fail to ossify; also, parts of a bone with several ossification centers may fail to fuse at the proper time.

To facilitate study we divide the human skeleton into two major divisions: the axial skeleton and the appendicular skeleton. The *axial skeleton* consists of the skull, the vertebral column, and the thoracic

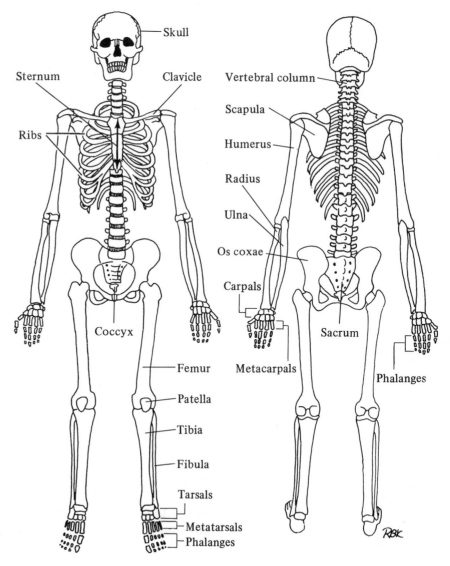

Figure 5.9 Human Skeleton. Anterior and posterior views.

Table 5.1 Distribution of Bones in Human Skeleton

Division	Subdivision	Number of bones
AXIAL SKELETON		80
	Skull	
	Cranium	8
	Face	14
	Ear ossicles	6
	Hyoid bone	1
	Vertebral Column	
	Cervical vertebrae	7
	Thoracic vertebrae	12
	Lumbar vertebrae	5
	Sacrum	1
	Coccyx	1
	Sternum	1
	Ribs	24
APPENDICULAR SKELETON		126
	Shoulder Girdle	
	Scapula	2
	Clavicle	2
	Upper Extremity	
	Humerus	2
	Ulna	2
	Radius	2
	Carpals	16
	Metacarpals	10
	Phalanges	28
	Pelvic Girdle	
	Os coxae	2
	Lower Extremity	
	Femur	2
	Patella	2
	Tibia	2
	Fibula	2
	Tarsals	14
	Metatarsals	10
	Phalanges	28

cage, These structures comprise the longitudinal axis of the body. The *appendicular skeleton* consists of the appendages of the body (limbs), the upper and lower extremities plus the girdles that connect them to the axial skeleton.

The Axial Skeleton

THE SKULL The skull (Figs. 5.10–5.13) consists of 29 flat and irregular bones. For study purposes we will consider them in the following order: cranial bones, facial bones, and hyoid bone. The three pairs of ossicles (ear

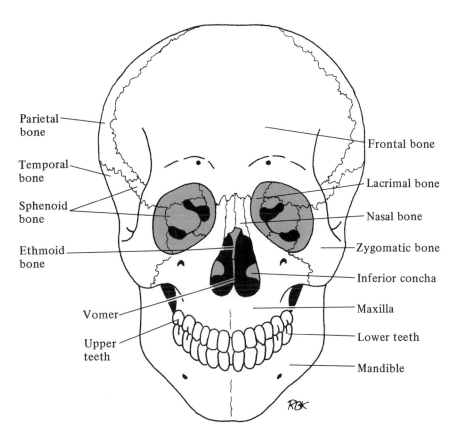

Figure 5.10 *Frontal View of Skull.*

Parietal bone
Temporal bone
Sphenoid bone
Ethmoid bone
Vomer
Upper teeth
Frontal bone
Lacrimal bone
Nasal bone
Zygomatic bone
Inferior concha
Maxilla
Lower teeth
Mandible

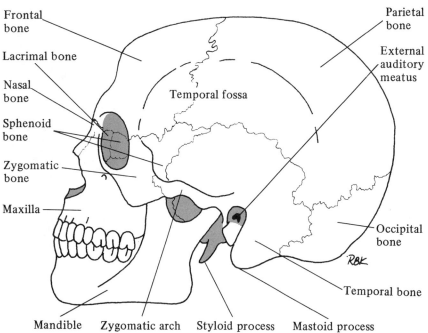

Figure 5.11 *Lateral View of Skull: Left Side.*

Frontal bone
Lacrimal bone
Nasal bone
Sphenoid bone
Zygomatic bone
Maxilla
Temporal fossa
Parietal bone
External auditory meatus
Occipital bone
Temporal bone
Mandible
Zygomatic arch
Styloid process
Mastoid process

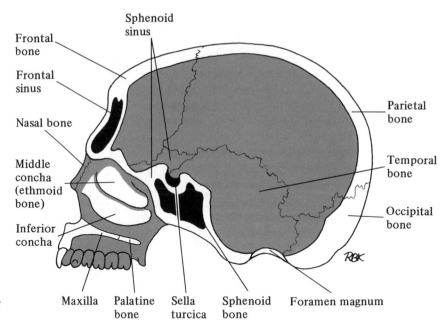

Figure 5.12 *Interior View of Skull: Right Side.*

Frontal bone

Frontal sinus

Nasal bone

Middle concha (ethmoid bone)

Inferior concha

Sphenoid sinus

Parietal bone

Temporal bone

Occipital bone

Maxilla   Palatine bone   Sella turcica   Sphenoid bone   Foramen magnum

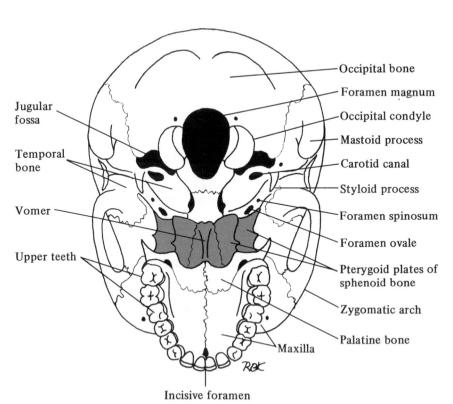

Figure 5.13 *Inferior View of Skull.*

Jugular fossa

Temporal bone

Vomer

Upper teeth

Occipital bone

Foramen magnum

Occipital condyle

Mastoid process

Carotid canal

Styloid process

Foramen spinosum

Foramen ovale

Pterygoid plates of sphenoid bone

Zygomatic arch

Palatine bone

Maxilla

Incisive foramen

bones) will be considered with the ear in Chapter 15. With the exception of the lower jawbone, the ear ossicles, and the hyoid bone, the bones of the skull are united by immovable joints called *sutures.*

Cranial Bones

The cranium consists of eight bones that form a case for the brain: the frontal, parietal (two), occipital, temporal (two), sphenoid, and ethmoid bones.

The *frontal bone* (Fig. 5.14) is a large curved bone consisting of a flat vertical portion (squama) and a horizontal portion. The squama forms the forehead and front of the cranial vault, while the horizontal portion forms the upper part of the eye sockets (orbits) and part of the roof of the nasal cavity. The *superciliary arches* are prominent bony ridges above the orbits; they can be palpated (felt through the skin) beneath the eyebrows. A pair of air-filled cavities, the *frontal sinuses,* occur above the orbits in the interior of the frontal bone.

The *parietal bones* are a pair of curved rectangular bones that form the top and part of the sides of the cranial vault. They are typical flat bones with no special markings.

The *occipital bone* (Fig. 5.15) is a trapezoid-shaped bone consisting of a squama and a basal portion. The squama forms the back of the cranial vault, while the basal portion forms part of the base of the skull. The occipital bone contains a large hole, the *foramen magnum,* that transmits the spinal cord. On either side of the foramen magnum are the *occipital condyles,* two rounded projections that articulate with the uppermost vertebra. A prominent rounded projection from the squama, the *external occipital protuberance,* can be felt through the scalp at the back of the head.

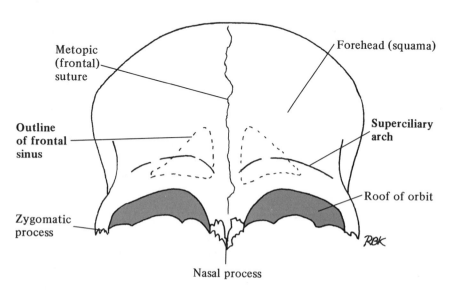

Figure 5.14 *Frontal Bone: Frontal View.*

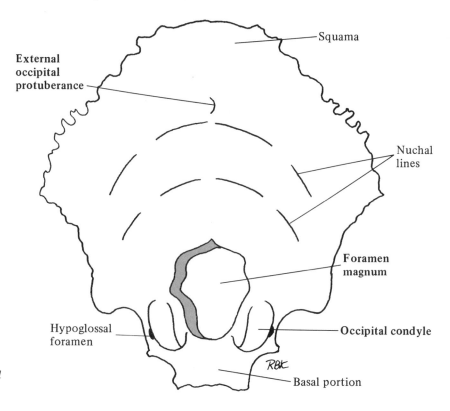

Figure 5.15 *Occipital Bone: External View.*

The *temporal bones* are a pair of irregular bones that form part of the base of the skull and the sides of the cranial vault. Each temporal bone (Fig. 5.16) consists of squamous, mastoid, tympanic, and petrous portions. The squamous portion (squama) is a thin plate of flat bone that forms part of the cranial vault. A curved projection, the *zygomatic process*, arches forward from the squama to articulate with the temporal process of the zygomatic (cheek) bone. Beneath the root of the zygomatic process is the *mandibular fossa*, a depression that receives the articular condyle of the lower jawbone, or mandible. The mastoid portion is located behind and below the squama. The *mastoid process*, a large rounded process created by the pull of a powerful neck muscle, projects downward from this portion of the temporal bone. It contains a collection of air cells called the *mastoid sinuses*. The tympanic portion of the temporal bone consists of a curved plate of bone located below the squama and in front of the mastoid portion. It forms part of the wall of the *external auditory meatus*, the bony canal of the external ear. The petrous portion is the rock-hard, wedge-shaped part of the temporal bone that contributes to the base of the skull. It contains the tympanic cavity (middle ear), the ear ossicles, and the organs of balance and hearing. The *internal auditory meatus*, a short bony canal, can be seen on the medial surface of the internal

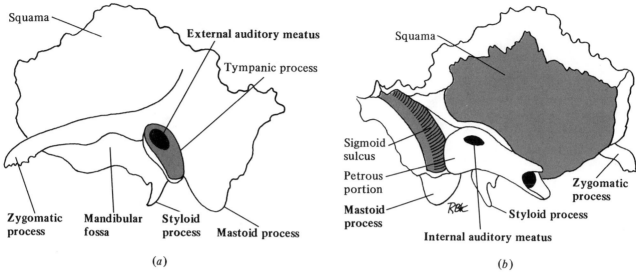

Squama

External auditory meatus

Tympanic process

Zygomatic
process Mandibular
fossa Styloid
process Mastoid process

(*a*)

Squama

Sigmoid
sulcus

Petrous
portion

Mastoid
process

Internal auditory meatus

Zygomatic
process

Styloid process

(*b*)

Figure 5.16 Left Temporal Bone.
(a) External view.   (b) Internal view.

aspect of the petrous portion. The *styloid process* is a sharp spine for muscle and ligament attachment that projects downward from its external aspect.

The *sphenoid bone* (Fig. 5.17), which is the central bone of the base of the skull, resembles a bat with outstretched wings. It consists of a central *body*, paired *greater* and *lesser wings*, which extend outward from the body, and two *pterygoid processes*, which extend downward from the body. A pair of air-filled sinuses, the *sphenoid sinuses*, are located within the body of the sphenoid bone. On the upper surface of the body is a depression named the *sella turcica* (for its resemblance to a Turkish saddle). In life the sella turcica contains the pituitary gland. The sphenoid bone also contains several important foramina, which are listed later in this chapter in Table 5.2.

The *ethmoid bone* (Fig. 5.18) is a small delicate bone located in the floor of the cranial cavity between the orbits. It consists of a cribriform plate, a perpendicular plate, and two lateral masses. The *cribriform plate* is a horizontal sheet of bone perforated by many small foramina. Its upper surface forms part of the floor of the cranial cavity while its lower surface forms part of the roof of the nasal cavity. The *crista galli* (cock's comb) projects upward from the cribriform plate into the cranial cavity, where it acts as the attachment site for the outer covering of the brain. The *perpendicular plate* projects downward from the cribriform plate into the nasal cavity, where it forms the upper part of the bony *nasal septum*. The lateral masses contribute to formation of the orbits and the lateral walls of the nasal cavity. Each lateral mass bears two scroll-like bony shelves, the *superior* and *middle conchae* (turbinates), which project medially into the nasal cavity.

Figure 5.17 *Sphenoid Bone.* (a) Superior view. (b) Anterior view.

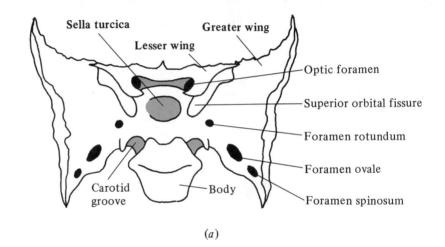

Sella turcica

Lesser wing

Greater wing

Optic foramen

Superior orbital fissure

Foramen rotundum

Foramen ovale

Carotid groove

Body

Foramen spinosum

(a)

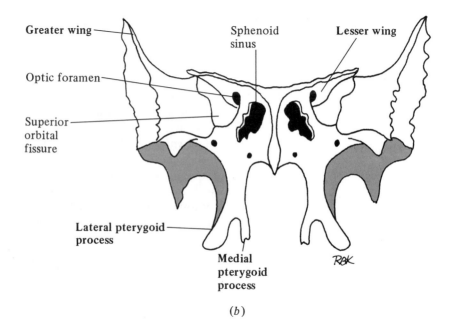

Greater wing

Sphenoid sinus

Lesser wing

Optic foramen

Superior orbital fissure

Lateral pterygoid process

Medial pterygoid process

(b)

Figure 5.18 *Ethmoid Bone.* (a) Superior view. (b) Lateral view. (c) Anterior view.

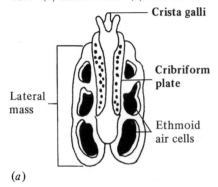

Crista galli

Cribriform plate

Lateral mass

Ethmoid air cells

(a)

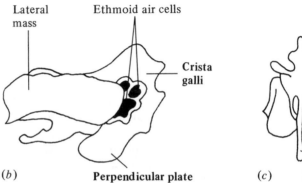

Lateral mass

Ethmoid air cells

Crista galli

Perpendicular plate

(b)

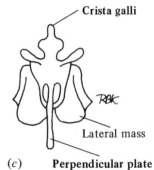

Crista galli

Lateral mass

Perpendicular plate

(c)

Facial Bones   The facial skeleton consists of 14 irregularly shaped bones that can be divided into external and internal groups for the purpose of study.

External Facial Bones   The external facial bones can be seen on the exterior of the skull; they include the maxillae (two), mandible, zygomatic (two), nasal (two), and lacrimal (two) bones.

The *maxillae* are paired facial bones that form the upper jaw, most of the hard palate (roof of the mouth), and part of the orbits and the nasal cavity. Each maxilla (Fig. 5.19) consists of a body that bears four processes. The *body* is a pyramidal structure that contains a large air-filled cavity called the *maxillary sinus* or *antrum* (of Highmore). The frontal process articulates with the frontal bone, while the zygomatic process articulates with the zygomatic bone; they both contribute to formation of the orbits and the nasal cavity. The *palatine process* articulates with its counterpart from the opposite maxilla to form the anterior part of the hard palate and part of the floor of the nasal cavity. The *alveolar process* on the underside of the body contains sockets for the upper teeth.

The *mandible* (Fig. 5.20) is a large unpaired bone that forms the lower jaw; it is the only movable bone of the face. Shaped like the letter U, the mandible consists of a horizontal *body* and two vertical *rami* that extend upward at right angles to the body. Where the ramus joins the body, there is a rounded projection called the *angle* of the jaw. The *alveolar process* on the upper surface of the body contains sockets for the lower teeth. A forward projection from the body forms the *mentum* (chin). Each ramus bears two processes on its upper

Figure 5.19 Left Maxilla. (a) External view. (b) Internal view.

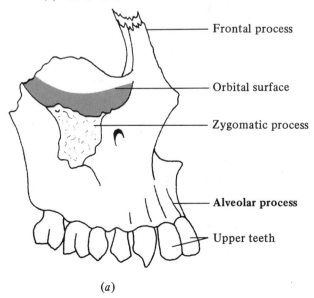

- Frontal process
- Orbital surface
- Zygomatic process
- **Alveolar process**
- Upper teeth

(a)

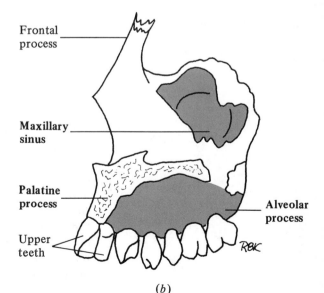

- Frontal process
- **Maxillary sinus**
- **Palatine process**
- Upper teeth
- **Alveolar process**

(b)

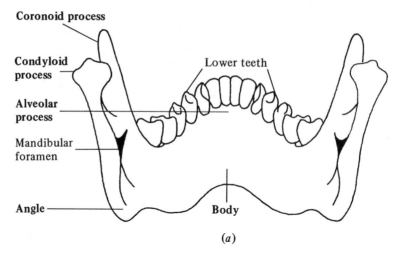

Coronoid process

Condyloid process

Alveolar process

Mandibular foramen

Angle

Lower teeth

Body

(*a*)

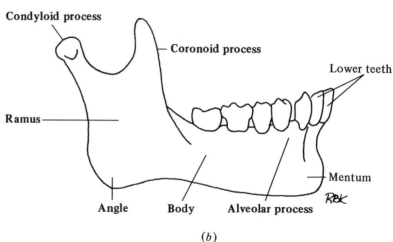

Condyloid process

Coronoid process

Lower teeth

Ramus

Mentum

Angle　　Body　　Alveolar process

(*b*)

Figure 5.20 *Mandible.* (a) Posterior view. (*b*) Lateral view.

surface: the coronoid and condyloid processes. The anterior, or *coronoid,* process is a sharp projection created by the pull of one of the chewing muscles. The posterior, or *condyloid,* process articulates with the mandibular foramen of the temporal bone to form the *temporomandibular joint.*

The *zygomatic (malar) bones* are paired quadrangular bones that form the cheeks. Each zygomatic bone (Fig. 5.21) consists of a *body,* which forms the prominence of the cheek, and four processes. The *frontal, maxillary,* and *orbital processes* contribute to formation of the orbit; the *temporal process* articulates with the zygomatic process of the temporal bone to form the *zygomatic arch.* The zygomatic arch is a prominent curved structure that can be palpated through the skin on the side of the face.

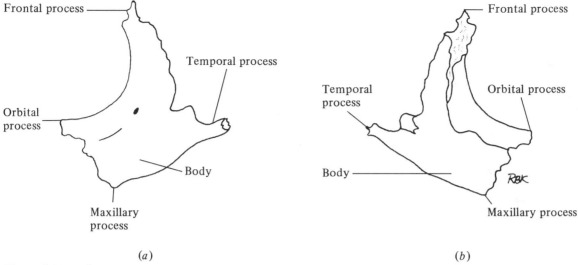

Frontal process

Temporal process

Orbital process

Body

Maxillary process

Frontal process

Temporal process

Orbital process

Body

Maxillary process

(a)

(b)

Figure 5.21 *Left Zygomatic Bone.* (a) External view. (b) Internal view.

The *nasal bones* are two small rectangular bones that lie side by side, forming the bridge of the nose.

The *lacrimal bones* are the smallest bones in the face. Each rectangular lacrimal bone lies in the medial wall of the orbit. It has a groove that contains the tear duct, a channel for the drainage of tears that extends from the orbit to the nasal cavity.

Internal Facial Bones

The internal facial bones are found in the interior of the skull; they are difficult to see unless the skull is sectioned sagittally. Internal facial bones include the vomer, the palatine bones (two), and the inferior conchae (two).

The *vomer* [Fig. 5.22(*a*)] is a thin flat bone shaped like an old-fashioned plowshare. It is located in the nasal cavity, where it forms the back and lower part of the nasal septum.

The *palatine bones* are paired L-shaped bones that lie behind the maxillae. Each palatine bone [Fig. 5.22(*b*)] consists of a horizontal plate and a vertical plate. The horizontal plate unites with its counterpart from the opposite palatine bone to form the posterior part of the hard palate and part of the floor of the nasal cavity. The vertical plate contributes to formation of the lateral wall of the nasal cavity.

The *inferior conchae* (turbinates) are paired bones that lie directly beneath the lateral masses of the ethmoid bone. Each inferior concha [Fig. 5.22(*c*)] is a scroll-like bone that projects medially into the nasal cavity.

The Hyoid Bone

The *hyoid bone* (Fig. 5.23) is a U-shaped bone that lies in the anterior portion of the neck between the angle of the jaw and the larynx (voice box). It consists of a central *body* and two pairs of curved lateral pro-

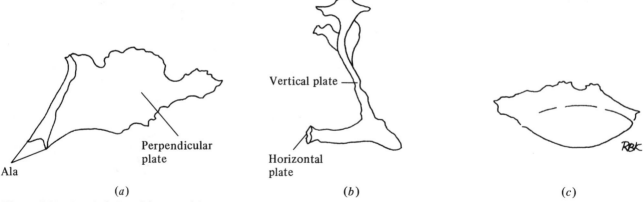

Ala

Perpendicular
plate

(a)

Vertical plate

Horizontal
plate

(b)

(c)

Figure 5.22 *Internal Facial Bones.* (a) Vomer, lateral view. (b) Left palatine bone, frontal view. (c) Right inferior concha, lateral view.

jections, the *greater* and *lesser cornua* (horns). The hyoid bone does not articulate with the skull; it is suspended from the styloid processes of the temporal bones by the *stylohyoid ligaments.* This bone, which is sometimes called the tongue bone, is the attachment site for some of the muscles that move the tongue and aid in speaking and swallowing.

Special Features of the Skull The skull is a very complex structure; certain areas require additional description at this time: the sinuses, the cranial vault, and the floor of the cranial cavity. The nasal cavity will be discussed with the respiratory system (Chapter 11), the oral cavity with the digestive system (Chapter 12), and the orbits with the eye (Chapter 15).

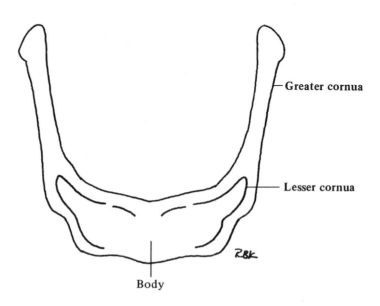

Greater cornua

Lesser cornua

Body

Figure 5.23 *Hyoid Bone: Frontal View.*

Sinuses

Paranasal Sinuses. The paranasal sinuses (Fig. 5.24) are air-filled cavities located in skull bones surrounding the nasal cavity. Named for the bones in which they are located, paranasal sinuses include the frontal sinuses, the maxillary sinuses, the sphenoid sinuses, and the ethmoid air cells. Because they are connected to the nasal cavity, the paranasal sinuses are frequently invaded by infections originating in the nasal cavity. Inflammation of the mucous membrane lining the paranasal sinuses is known as *sinusitis.*

Mastoid Sinuses. The mastoid sinuses are air cells located in the mastoid process of the temporal bone. They communicate with each other and with the middle ear. Infections can easily spread from the middle ear into the mastoid air cells; the resulting inflammation of their mucosal lining is called *mastoiditis.* In the days before antibiotics, mastoiditis was quite a serious disease and was difficult to cure; severe cases often spread to the brain as meningitis or encephalitis and were usually fatal.

Cranial Vault

The cranial vault, which forms the roof of the cranial cavity, consists of the two parietal bones plus the squamous portions of the frontal, temporal, and occipital bones.

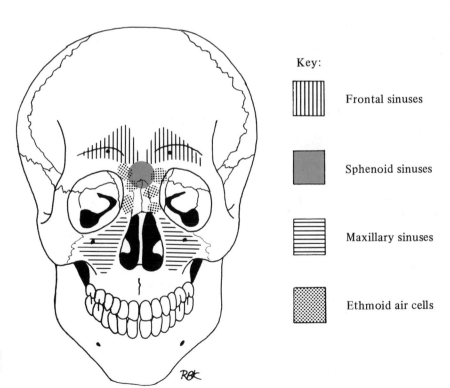

Figure 5.24 *Paranasal Sinuses.* Frontal view of skull with positions of paranasal sinuses indicated.

Sutures. The following immovable joints, called sutures, occur between the bones of the adult cranial vault (Fig. 5.25): the *sagittal suture*, a longitudinal suture between the parietal bones; the *coronal suture*, a transverse suture between the frontal and parietal bones; the *lambdoidal suture*, a suture shaped like the Greek letter lambda (λ) between the parietal bones and the occipital bone; and two *squamosal sutures*, curved sutures between the parietal bones and the squamae of the temporal bones. In the cranial vault of the newborn infant, a suture is present between the right and left halves of the frontal bone. It usually disappears by the sixth year; if it persists in the adult, it is called the *metopic suture.*

Fontanelles. In the newborn infant (Fig. 5.26) the flat bones of the cranial vault have not completed ossification. At the angles of the parietal bones there are six wide areas of unossified fibrous connective tissue membrane called fontanelles, or "soft spots." The *anterior fontanelle* is a large diamond-shaped area between the parietal bones and the two halves of the frontal bone; it usually closes (becomes ossified) by the middle of the second year. The *posterior fontanelle* is a small triangular area between the parietal bones and the occipital bone; it closes a few months after birth. The *lateral fontanelles* are four small irregular areas at the lateral angles of the parietal bones which close a month or two after birth. The fontanelles are clinically

Figure 5.25 Sutures of Adult Cranial Vault. (a) Superior view. (b) Lateral view.

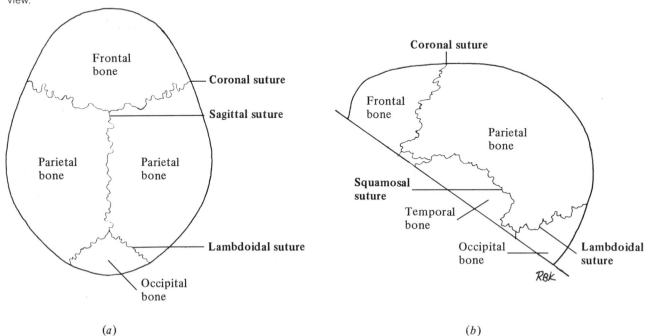

(a)

(b)

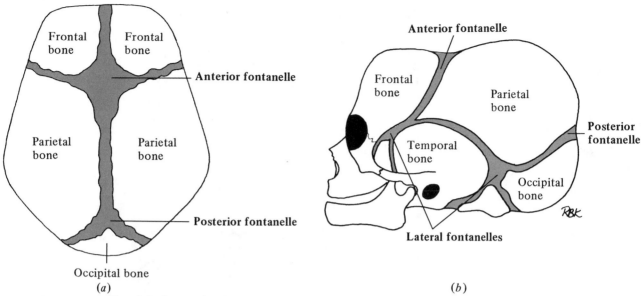

Figure 5.26 Fontanelles of Fetal Cranial Vault. (a) Superior view. (b) Lateral view.

important since they permit overriding and molding of the bones of the cranial vault as the baby's head passes through the birth canal.

Floor of the Cranial Cavity

The floor of the cranial cavity (Fig. 5.27), which contains various parts of the brain (Chapter 15), consists of a series of three depressions, or *fossae;* in descending order from front to back, they are the anterior, middle, and posterior cranial fossae.

Anterior Cranial Fossa. The anterior cranial fossa is a shallow depression formed by the orbital plates of the frontal bone and the cribriform plate of the ethmoid bone. Its posterior boundary is formed by the lesser wings of the sphenoid bone. In life it contains the frontal lobes and olfactory bulbs of the brain.

Middle Cranial Fossa. The middle cranial fossa is a butterfly-shaped depression formed by the body and greater wings of the sphenoid bone plus the squamous portions of the temporal bone. Its posterior boundary is a curved line drawn through the crests of the petrous portions of the temporal bones. In life the middle cranial fossa contains the brain stem, the pituitary gland, and the temporal lobes of the brain.

Posterior Cranial Fossa. The posterior cranial fossa is a large deep depression formed by the petrous portions of the temporal bones and the base of the occipital bone. In life the posterior cranial fossa contains the cerebellum and medulla oblongata.

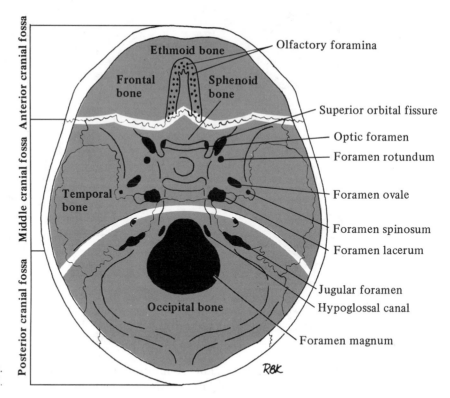

Figure 5.27 Floor of Cranial Cavity. Foramina and cranial fossae are indicated.

Foramina. A number of important openings or foramina are found in the floor of the cranial cavity for the transmission of cranial nerves and blood vessels. These are depicted in Fig. 5.27 and listed in Table 5.2.

THE VERTEBRAL COLUMN

The vertebral column, or spine, is a flexible bony column that encloses the spinal cord and forms the longitudinal axis of the body. It articulates with the skull superiorly and the hipbones inferiorly. The vertebral column is composed of bony units called *vertebrae*. An infant's vertebral column consists of 33 vertebrae separated by cartilaginous intervertebral discs. Fusion of some of the vertebrae and their discs reduces the number of separate units to 26 in the adult vertebral column (Fig. 5.28). The vertebrae are numbered from the superior end of the column downward and are subdivided into the following five groups: (a) 7 *cervical* vertebrae in the neck, (b) 12 *thoracic* vertebrae in the upper back, (c) 5 *lumbar* vertebrae in the lower back, (d) 5 *sacral* vertebrae (fused) in the sacrum, and (e) 4 *coccygeal* vertebrae (fused) in the coccyx.

Table 5.2 Foramina in Floor of Cranial Cavity

Foramen/ Passageway	Cranial fossa	Location	Structures transmitted
GROUP A			
Olfactory foramina	Anterior	Cribriform plate of the ethmoid bone	Olfactory nerves
Optic foramen	Anterior	Lesser wing of the sphenoid bone	Optic nerve
Superior orbital fissure	Anterior	Between the greater and lesser wings of sphenoid bone	Oculomotor nerve, trochlear nerve, oph-thalmic division of the trigeminal nerve, and abducens nerve
GROUP B			
Foramen ro-tundum	Middle	Greater wing of the sphenoid bone	Maxillary division of the trigeminal nerve
Foramen ovale	Middle	Greater wing of the sphenoid bone	Mandibular division of the trigeminal nerve
Foramen spinosum	Middle	Greater wing of the sphenoid bone	Middle meningeal artery
GROUP C			
Internal auditory meatus	Posterior	Petrous portion of tem-poral bone	Facial nerve and vestibulocochlear nerve
Jugular foramen	Posterior	Between the temporal and occipital bones	Glossopharyngeal nerve, vagus nerve, and spinal accessory nerve; internal jugu-lar vein
Foramen magnum	Posterior	Occipital bone	Spinal cord and spinal accessory nerve
Hypoglossal canal	Posterior	Occipital bone	Hypoglossal nerve

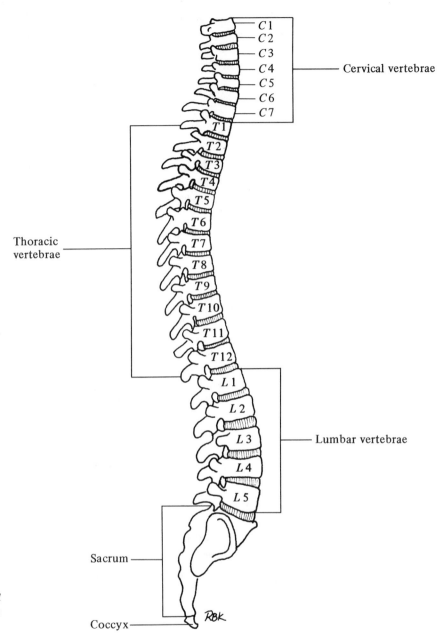

C 1
C 2
C 3
C 4
C 5
C 6
C 7

Cervical vertebrae

T 1
T 2
T 3
T 4
T 5
T 6
T 7
T 8
T 9
T 10
T 11
T 12

Thoracic
vertebrae

L 1
L 2
L 3
L 4
L 5

Lumbar vertebrae

Sacrum

Coccyx

Figure 5.28 Vertebral Column: Lateral View. Individual vertebrae are numbered; vertebral regions are indicated.

Structural Features of Vertebrae With the exception of the first and second cervical vertebrae and the fused vertebrae of the sacrum and coccyx, all vertebrae have the following basic structural features (Fig. 5.29):

1. *Body.* The body is a solid block of bone that functions as the weight-bearing portion of the vertebra.

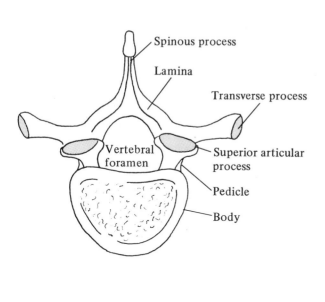

(a)

(b)

Figure 5.29 *Structural Features of Vertebrae.* (a) Typical vertebra, superior view. (b) Two articulated vertebrae, lateral view.

2. *Vertebral Arch.* The vertebral arch is a ring of bone that encloses the spinal cord and is attached to the posterior aspect of the body. It consists of two *pedicles,* which form the roots of the arch, and two *laminae,* which fuse in the midline to complete the arch.

3. *Processes.* Seven processes arise from each vertebral arch: two *superior articular processes* for articulation with the vertebra above, two *inferior articular processes* for articulation with the vertebra below, two *transverse processes* for either rib articulation or muscle attachment, and a *spinous process* for muscle attachment.

4. *Vertebral Foramen.* The vertebral foramen is the large hole in the center of the arch that transmits the spinal cord.

5. *Intervertebral Foramina.* The superior and inferior vertebral notches on adjacent vertebrae form a pair of intervertebral foramina for the transmission of spinal nerves.

Regional Variations of Vertebrae

Vertebrae from different parts of the vertebral column show regional modifications of the basic structural features, which correlate with their different functions.

Cervical Vertebrae

The seven cervical vertebrae, which are located in the neck, are the smallest vertebrae in the vertebral column. Most cervical vertebrae [Fig. 5.30(c)] have small bodies, bifid (cleft) spinous processes, triangular vertebral foramina, and foramina for transmission of the vertebral arteries in their transverse processes.

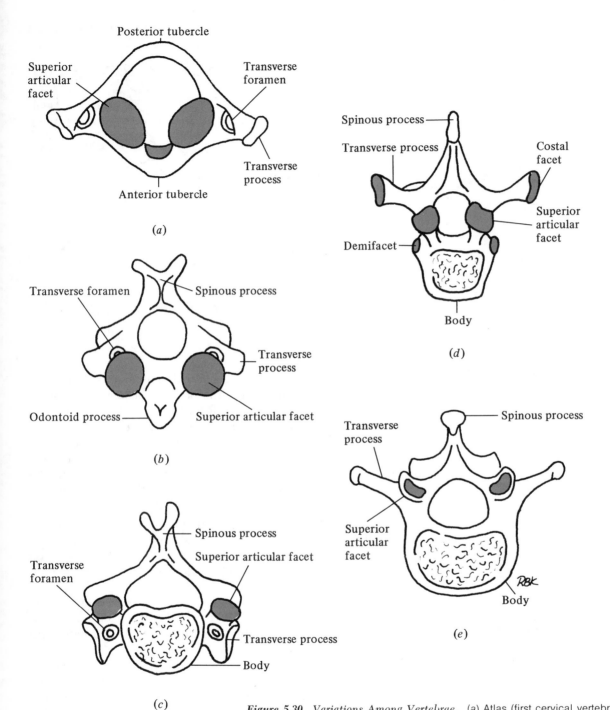

Figure 5.30 *Variations Among Vertebrae.* (a) Atlas (first cervical vertebra), superior view. (b) Axis (second cervical vertebra), superior view. (c) Typical cervical vertebra, superior view. (d) Typical thoracic vertebra, superior view. (e) Typical lumbar vertebra, superior view.

The first, second, and seventh cervical vertebrae have some unusual features that require explanation. The first cervical vertebra [Fig. 5.30(a)], called the *atlas,* has no body and no spinous process. It consists of a ring of bone whose superior surface bears cup-shaped articular processes. These processes articulate with the occipital condyles of the skull to form a joint that permits nodding movements of the head to take place. The second cervical vertebra [Fig. 5.30(b)], called the *axis,* has a large toothlike *odontoid process* projecting superiorly from its body. The odontoid process forms a pivot on which the atlas can rotate. This pivot joint permits head shaking and other rotary head movements to take place. The seventh cervical vertebra, called the *vertebra prominens,* has a prominent, nonbifid spinous process that can be palpated at the nape of the neck.

Thoracic Vertebrae

The 12 thoracic vertebrae, which are located in the upper back, are larger than the cervical vertebrae and smaller than the lumbar vertebrae; their bodies gradually increase in size from T1 to T12. All thoracic vertebrae [Fig. 5.30(d)] have long pointed spinous processes, that are directed obliquely downward, and small round vertebral foramina. Since thoracic vertebrae articulate with ribs, all bodies have *facets* or *demifacets* for articulation with rib heads; the transverse processes of vertebrae T1 to T10 have *costal facets* for articulation with rib tubercles.

Lumbar Vertebrae

The five lumbar vertebrae, which are located in the lower back, are the largest unfused vertebrae in the vertebral column. Lumbar vertebrae [Fig. 5.30(e)] have bean-shaped bodies whose large size is correlated with the amount of weight they bear. They also have short blunt spinous processes, long slender transverse processes, and triangular vertebral foramina. Their articular processes have a different orientation from those of other vertebrae; superior articular processes are directed medially while inferior articular processes are directed laterally.

The Sacrum

The sacrum [Fig. 5.31(a)] is a large triangular bone formed by fusion of the five sacral vertebrae and the cartilaginous discs between them. It articulates with the fifth lumbar vertebra superiorly, with the hipbones laterally, and with the coccyx inferiorly.

The smooth concave anterior surface of the sacrum consists of a central bar of bone, which represents the bodies of the sacral vertebrae. It is crossed by four transverse ridges that represent ossified intervertebral discs. Four pair of *anterior foramina,* which transmit the anterior divisions of sacral spinal nerves, lie opposite the transverse ridges. The *sacral promontory,* the prominent anterior lip on the upper surface of the first sacral vertebra, is an important anatomical landmark for the pelvis. The rough convex posterior surface of the sacrum contains a *median crest,* which represents the spinous

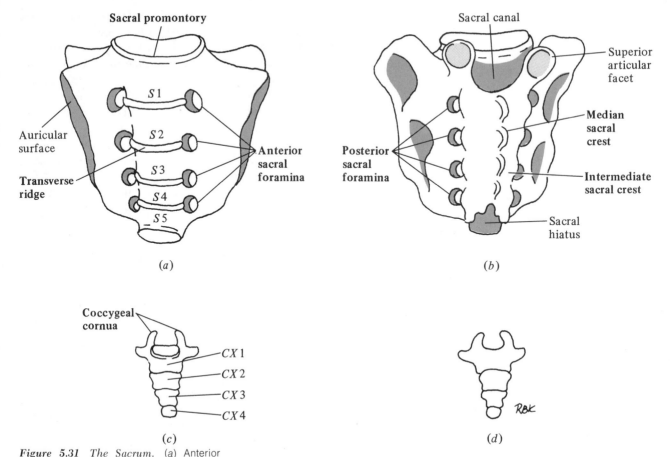

Sacral promontory

S 1
S 2
S 3
S 4
S 5

Auricular surface

Transverse ridge

Anterior sacral foramina

Sacral canal

Superior articular facet

Median sacral crest

Posterior sacral foramina

Intermediate sacral crest

Sacral hiatus

(a)

(b)

Coccygeal cornua

CX 1
CX 2
CX 3
CX 4

RBK

(c)

(d)

Figure 5.31 *The Sacrum.* (a) Anterior surface. (b) Posterior surface.
   The Coccyx. (c) Anterior surface. (d) Posterior surface.

processes of the sacral vertebrae. The medial crest is flanked by a pair of *intermediate crests*, which represent the articular processes of the sacral vertebrae. Four pairs of *posterior foramina* transmit the posterior divisions of sacral spinal nerves. The sides of the sacrum contain ear-shaped *auricular surfaces* that articulate with the hipbones.

The Coccyx   The coccyx, or tailbone [Fig. 5.31(b)] represents the vestigial remains of a tail in humans. It is a small triangular bone, formed by fusion of the four rudimentary coccygeal vertebrae, that articulates with the sacrum superiorly.

The Vertebral Column as a Functional Unit   The adult vertebral column is a structure about 27 inches long. The vertebrae are assembled in such a way that the vertebral bodies face anteriorly and the spinous processes face posteriorly.

Except in the sacrum and coccyx, individual vertebral bodies are separated by cartilaginous intervertebral discs, which act as shock

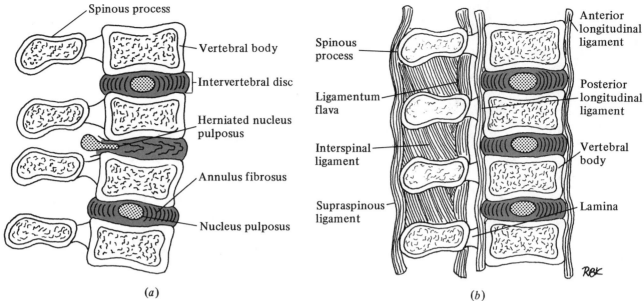

Figure 5.32 *Sagittal Section of Vertebral Column.* (a) Middle intervertebral disc has ruptured and nucleus pulposus has herniated posteriorly. (b) Ligaments of vertebral column.

absorbers and permit a high degree of flexibility. Each intervertebral disc consists of an outer wrapping of fibrocartilage, called the *annulus fibrosus,* and a soft pulpy center, called the *nucleus pulposus.* If the annulus fibrosus tears or degenerates, the nucleus pulposus usually herniates out into the spinal canal [Fig. 5.32(a)]. This condition, called *herniated disc,* can be quite painful when the nucleus pulposus presses on the spinal cord or the root of a spinal nerve.

The vertebrae are lashed together and held in position by several tough flexible ligaments [Fig. 5.32(b)]. *Anterior* and *posterior longitudinal ligaments* extend down the vertebral bodies from the axis to the sacrum to keep them in position. The tips of the spinous processes are united by the *ligamentum nuchae,* which extends from the external occipital protuberence to the seventh cervical vertebra, and by the *supraspinous ligament,* which extends from the seventh cervical vertebra to the sacrum. *Ligamenta flava,* elastic ligaments, unite adjacent laminae; *interspinous ligaments* unite adjacent spinous processes.

Curvature of the Vertebral Column

When viewed from the side, as in Fig. 5.33, the vertebral column shows certain concave and convex curves that cannot be appreciated in other views.

Normal Curves

During fetal life the vertebral column assumes the shape of the letter C. Portions of this concave primary curve are retained as the *thoracic* and *sacral curves* of the adult vertebral column. Secondary curves are acquired after birth. The convex *cervical curve* develops

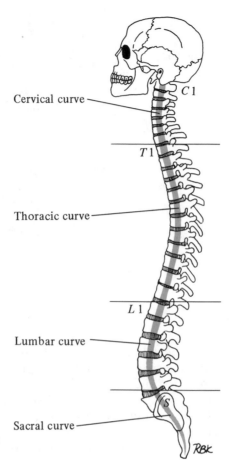

Cervical curve

C 1

T 1

Thoracic curve

L 1

Lumbar curve

S

Sacral curve

Figure 5.33 *Normal Curvature of Adult Vertebral Column.*

about three months after birth when the infant begins to hold its head erect. When the infant begins to stand erect and walk, the convex *lumbar curve* develops.

Abnormal Curves   Abnormal curvature of the vertebral column results from injury, disease, or poor posture. *Kyphosis*, or hunchback, is a condition in which the thoracic curve is exaggerated. Kyphosis most commonly results from vertebral tuberculosis or other inflammatory diseases that attack the vertebrae. An exaggeration of the lumbar curve is called *lordosis*, or swayback; it is usually due to poor posture. *Scoliosis*, an acquired lateral curvature of the vertebral column, is often caused by a collapsed vertebra or disease of the back muscles.

THE THORACIC CAGE   The thoracic cage (Fig. 5.34) is an inverted basket formed by the bodies of the 12 thoracic vertebrae, the sternum, and the 12 pairs of ribs plus their costal cartilages. Shaped like a flattened cone,

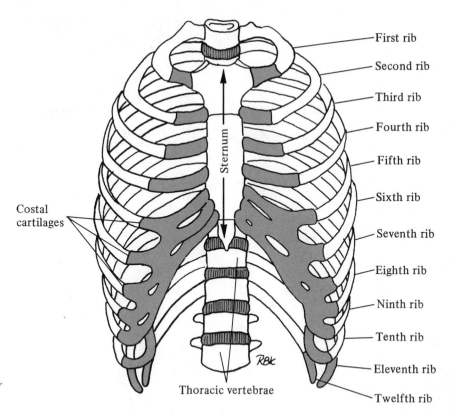

First rib
Second rib
Third rib
Fourth rib
Fifth rib
Sixth rib
Seventh rib
Eighth rib
Ninth rib
Tenth rib
Eleventh rib
Twelfth rib

Sternum

Costal cartilages

Thoracic vertebrae

Figure 5.34 Thoracic Cage: Anterior View.

the thoracic cage is narrow at its blunt apex, broad at its base, and flattened anteroposteriorly. It encloses the heart and lungs and provides shelter for the liver and spleen. It also plays a vital role in the process of breathing.

The Sternum The sternum, or breastbone (Fig. 5.35), is a flat sword-shaped bone that lies in the midline of the anterior thoracic wall. It consists of three parts: the *manubrium* above, the elongated *gladiolus* (body) in the middle, and the pointed *xiphoid process* below. A deep groove, the *jugular notch,* is present on the upper border of the manubrium; it can be palpated between the sternal ends of the collar bones. The manubrium joins the body at an angle, the *sternal angle* (of Louis), which is an important anatomical landmark for thoracic structures. The second pair of ribs articulate with the sternum at this point. The collarbones (clavicles) articulate with the upper end of the manubrium, while the upper seven pairs of ribs articulate with the sides of the manubrium and body. The xiphoid process acts as an attachment site for several thoracic and abdominal muscles.

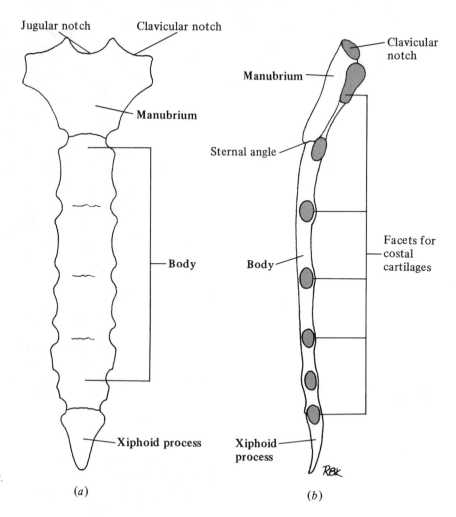

Figure 5.35 *Sternum.* (a) Anterior view. (b) Lateral view.

(a)

(b)

The Ribs The greater part of the thoracic cage is formed by the 12 pairs of ribs and their costal cartilages. The ribs (Fig. 5.36) are slender, flattened, curved bones that articulate with thoracic vertebrae posteriorly. They are classified according to their anterior attachments. The upper 7 pairs of ribs are classified as *vertebrosternal* because they articulate directly with the sternum by way of their costal cartilages. Pairs eight through ten are classified as *vertebrochondral* because they articulate with the costal cartilages of the ribs directly above them. Pairs eleven and twelve are classified as *vertebral* or *floating ribs* because their anterior ends are unattached.

Structural Features of Ribs Each rib consists of a vertebral end, a body or shaft, and a sternal end:

1. *Vertebral End.* The vertebral end of a typical rib contains a head, neck, and tubercle. The head articulates with the body of a thoracic

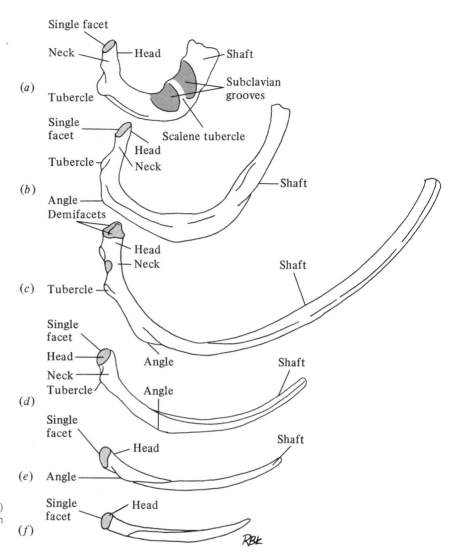

Figure 5.36 Ribs. (a) First rib. (b) Second rib. (c) Central rib. (d) Tenth rib. (e) Eleventh rib. (f) Twelfth rib.

vertebra. A narrow neck separates the head from the shaft; it bears a tubercle that articulates with the transverse process of a thoracic vertebra.

2. *Body.* The body or shaft of a typical rib is a thin curved piece of bone with inner and outer surfaces and upper and lower borders. The lower border is marked by a groove that transmits the intercostal vein, artery, and nerve. The angle of the rib is the region where the shaft begins to curve sharply.

3. *Sternal End.* The sternal end of a typical rib bears a costal cartilage, a bar of hyaline cartilage that articulates directly or indirectly with the sternum.

Variations of Ribs The first, second, tenth, eleventh, and twelfth ribs are atypical; they have some unusual features, which are described here.

The first rib, located at the narrow end of the thoracic cage, is the shortest rib and has the greatest curvature. Unlike the other ribs, it is flattened in the horizontal plane. Its shaft has no angle; the upper surface of its shaft is grooved by the passage of subclavian blood vessels.

The second rib is intermediate in size and shape between the first rib and a typical rib. Its shaft is flattened in the vertical plane and has a slight angle.

The tenth, eleventh, and twelfth ribs, located at the broad end of the thoracic cage, have large heads and pointed anterior ends. The anterior ends of the eleventh and twelfth ribs end blindly in the musculature of the lateral body wall. Neither necks nor tubercles are present on these ribs.

The Appendicular Skeleton

The appendicular skeleton is the other major subdivision of the skeletal system. It consists of the upper and lower limbs—the shoulder girdle and upper extremities plus the pelvic girdle and lower extremities.

THE SHOULDER GIRDLE The shoulder girdle consists of two scapulae (posterior) plus two clavicles (anterior), which attach the upper extremities to the axial skeleton. The human body's erect posture has freed the upper extremity from weight bearing and locomotion; consequently it has become highly mobile and adapted for performing precision tasks. The mobility of the upper extremity is due to the lightness and flexibility of the shoulder girdle. Its only firm attachment to the axial skeleton occurs at the articulations of the clavicles with the sternum.

Scapula The scapula, or shoulder blade (Fig. 5.37), is a large flat bone, shaped like an inverted triangle, that lies on the posterior wall of the thorax over the second to seventh ribs. Though it is buried in the musculature of the upper back, it is superficial enough to be palpated in its entirety.

The scapula has superior, lateral (axillary), and medial (vertebral) borders plus superior, lateral, and inferior angles. The lateral angle contains a small shallow depression, the *glenoid fossa*, which forms a socket for the head of the humerus. The inferior angle of the scapula can be palpated quite easily when the scapula is rotated up and outward.

The scapula has two surfaces. The concave costal surface, which faces the thoracic wall, contains a large shallow depression called

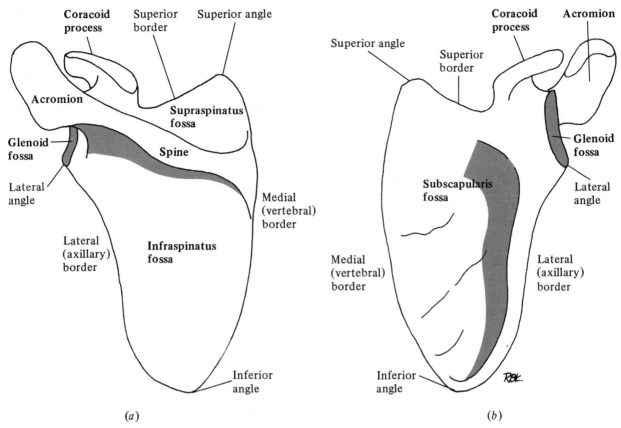

Coracoid process Superior border Superior angle

Acromion

Glenoid fossa

Lateral angle

Lateral (axillary) border

Supraspinatus fossa

Spine

Infraspinatus fossa

Medial (vertebral) border

Inferior angle

(a)

Superior angle

Coracoid process Acromion

Superior border

Glenoid fossa

Subscapularis fossa

Lateral angle

Medial (vertebral) border

Lateral (axillary) border

Inferior angle

(b)

Figure 5.37 *Left Scapula.* (a) Posterior view. (b) Anterior view.

the *subscapular fossa.* The convex posterior surface is divided into two unequal parts by the scapular spine. Above the spine is a small shallow depression called the *supraspinatus fossa.* Below the spine is a large deep depression called the *infraspinatus fossa.* All three of these scapular fossae are sites of muscle attachment.

Three processes arise from the scapula. The *scapular spine,* created by the pull of the trapezius muscle, runs diagonally across the posterior surface of the scapula. The *coracoid process* (crow's beak) is a hooked process arising from the superior border of the scapula; it is the attachment site of three muscles. The *acromion* (tip of the scapular spine) forms the point of the shoulder. It is the portion of the scapula that articulates with the clavicle.

Clavicle The clavicle, or collarbone (Fig. 5.38), is a double curved bone that lies in the root of the neck. It consists of a medial end that articulates with the sternum, a flattened curved shaft, and a lateral end that articulates with the acromion. The clavicle forms a strut or prop for the shoulder; without its support the shoulder and arm would sag.

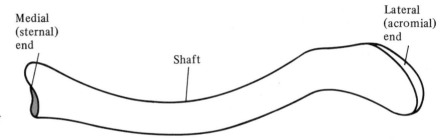

Medial
(sternal)
end

Shaft

Lateral
(acromial)
end

Figure 5.38 Left Clavicle: Superior View.

Because it is superficial in location, the clavicle is very vulnerable to blows on the shoulder and is the most frequently broken bone in the whole body.

THE UPPER EXTREMITY

The upper extremity is divided into arm, forearm, wrist, hand and fingers. Bones of the upper extremity consist of the humerus (arm), radius and ulna (forearm), carpals (wrist), metacarpals (hand), and phalanges (fingers).

Humerus

The arm (brachium) contains a single bone, the humerus (Fig. 5.39), which is the largest and longest bone in the upper extremity.

The proximal end of the humerus consists of the *head,* a rounded ball that articulates with the glenoid fossa of the scapula. The head is separated from the shaft by a slight constriction called the *anatomical neck.* Below the anatomical neck are two projections that act as sites of muscle attachment: the *greater tubercle* (on the lateral border) and the *lesser tubercle* (on the anterior surface). The *bicipital groove,* located on the anterior surface between the two tubercles, was created by passage of one of the tendons of the biceps brachii muscle. Just below the two tubercles is the *surgical neck,* which is one of the most common humeral fracture sites.

The long twisted humeral shaft has both anterior and posterior surfaces. The *deltoid tuberosity,* the attachment site of the deltoid muscle, is a small projection from the lateral border of the humerus midway between the two ends of the shaft. The *radial groove,* created by passage of the radial nerve, spirals diagonally down the posterior surface of the humeral shaft.

The distal end of the humerus is divided into two articular condyles. The pulley-shaped medial condyle, called the *trochlea,* articulates with the ulna, while the ball-shaped lateral condyle, called the *capitulum* (little head), articulates with the radius. The *medial* and *lateral epicondyles* are muscle attachment sites located on the humeral borders above the condyles. The *radial* and *coronoid fossae* are depressions on the anterior surface above the condyles; they receive the head of the radius and the coronoid process when the elbow is bent.

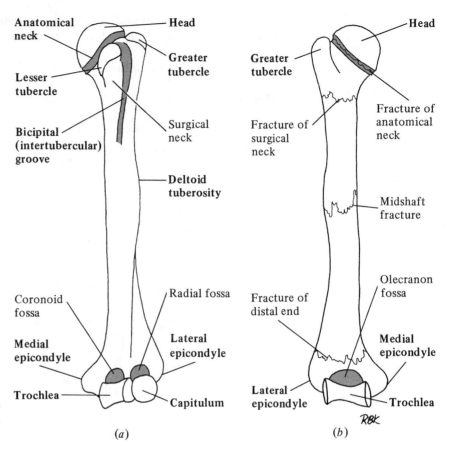

Anatomical neck — Head
Lesser tubercle — Greater tubercle
Bicipital (intertubercular) groove — Surgical neck
Deltoid tuberosity
Coronoid fossa — Radial fossa
Medial epicondyle — Lateral epicondyle
Trochlea — Capitulum
Head
Greater tubercle
Fracture of surgical neck — Fracture of anatomical neck
Midshaft fracture
Fracture of distal end — Olecranon fossa
Medial epicondyle
Lateral epicondyle — Trochlea
RBK

Figure 5.39 *Left Humerus.* (a) Anterior view. (b) Posterior view. Common fracture sites are indicated.

(a)   (b)

The *olecranon fossa* is a depression on the posterior surface above the trochlea; it receives the olecranon process of the ulna when the elbow is straightened.

Ulna The ulna, or elbow bone (Fig. 5.40), is located on the medial (little finger) side of the forearm. It is a long slender bone that is larger at its proximal end.

The proximal end of the ulna, which is shaped like the head of a wrench, consists of *olecranon* and *coronoid processes* separated by a groove called the *trochlear notch*. It articulates with the trochlea of the humerus. The *radial notch*, a small depression that articulates with the head of the radius, lies lateral to the coronoid process.

The shaft of the ulna, which decreases in diameter as it descends toward the wrist, has three surfaces and three borders. Its sharp interosseous border, which faces the radius, provides attachment for the interosseous membrane that extends between the two forearm bones.

The distal end of the ulna consists of a small *head,* which articulates with the ulnar notch of the radius but does not articulate with the wrist bones. The ulnar *styloid process* projects downward from the medial side of the head and can be palpated at the wrist.

Radius The radius (Fig. 5.40) is located on the lateral (thumb) side of the forearm. It is a long slender bone that is larger at its distal end.

The proximal end of the radius consists of a flat disc-shaped *head* that articulates with the capitulum of the humerus and the radial notch of the ulna. A cylindrical *neck* separates the radial head from the shaft. Just below the neck on the medial border of the radius is a projection called the *radial tuberosity,* an attachment site for the biceps brachii muscle.

The shaft of the radius, which increases in diameter as it descends toward the wrist, also has three surfaces and three borders; the sharp

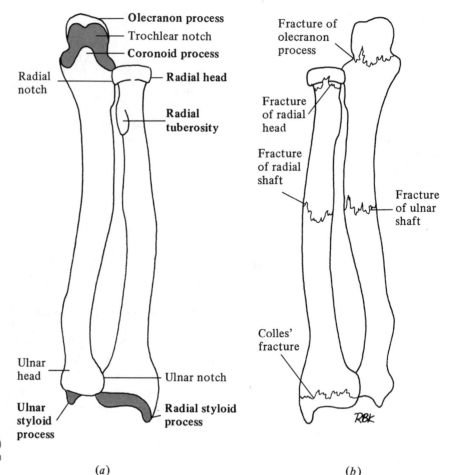

Figure 5.40 *Left Radius and Ulna.* (a) Anterior view. (b) Posterior view. Common fracture sites are indicated.

(a)

(b)

interosseous border, which faces the ulna, is the other attachment site for the interosseous membrane.

The club-shaped distal end of the radius has a concave inferior surface that articulates with two of the wrist bones. Its medial surface contains a depression, the *ulnar notch*, that articulates with the head of the ulna. The radial *styloid process*, which can be palpated at the wrist, projects downward from its lateral surface. Fracture of the distal third of the radius, called a *Colles' fracture*, is a very common fracture that results from a fall on the palm of the outstretched hand.

Due to the pivotal articulations between the proximal and distal ends of the radius and ulna, it is possible for the radius either to lie parallel to the ulna with the palm of the hand up (supination) or to lie across the ulna with the palm of the hand down (pronation), as is shown in Fig. 5.41.

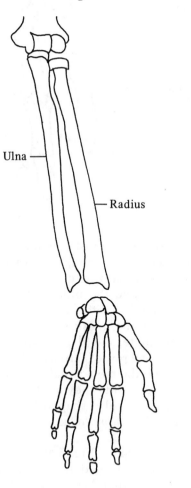

Figure 5.41 *Bones of the Forearm and Hand.* (a) In supination. (b) In pronation.

(a)

(b)

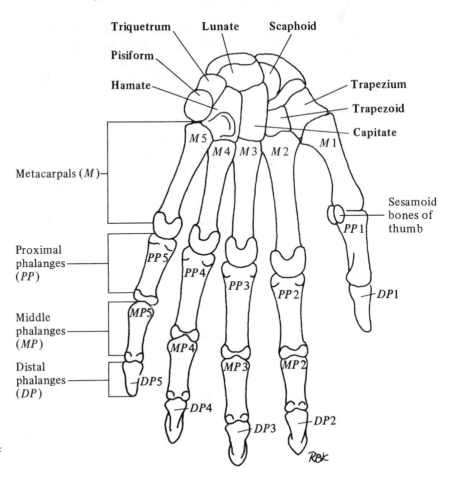

Figure 5.42 *Bones of the Left Hand: Palmar Surface.*

Carpals The wrist, or carpus, contains eight small bones called carpals (Fig. 5.42), which are arranged in two rows of four.

    The proximal row of carpals articulates with the carpal surface of the radius and the ulnar articular disc above. From lateral (thumb) side to medial (little finger) side they are the (a) *scaphoid,* shaped like a boat; (b) *lunate,* shaped like a half-moon; (c) *triquetrum,* shaped like a triangle; and (d) *pisiform,* shaped like a pea.

    The distal row of carpals articulates with the proximal row of carpals above and the metacarpals below. From lateral (thumb) side to medial (little finger) side they are the (a) *trapezium,* shaped like a trapezoid; (b) *trapezoid,* also shaped like a trapezoid; (c) *capitate,* has a small head; and (d) *hamate,* has a small hook (hamulus).

Metacarpals The metacarpals (Fig. 5.42) are five miniature long bones, numbered from lateral to medial, that form the bony framework of the hand. Their squared-off bases articulate with the distal row of carpals above, while their rounded heads articulate with the proximal phalanges below. The heads of the metacarpals form the knuckles when one makes a fist. The articular surfaces of both the first metacarpal and the trapezium are reciprocally saddle-shaped, which permits the thumb to have a wide range of motions including opposition. The human's ability to oppose the thumb to the other fingers is a marked evolutionary advance.

Phalanges The phalanges (Fig. 5.42) are fourteen miniature long bones that form the bony core of the thumb and fingers. The thumb contains only proximal and distal phalanges, while each finger contains proximal, middle, and distal phalanges. A pair of sesamoid bones is commonly found over the metacarpophalangeal joint of the thumb.

THE PELVIC GIRDLE The pelvic girdle, or pelvis, consists of the two hipbones plus the sacrum and coccyx. The hipbones articulate with the sacrum posteriorly and with each other anteriorly. The pelvic girdle serves to attach the lower extremities to the axial skeleton. The human's erect posture has limited the lower extremities to weight bearing and locomotion, which has increased the size and weight of the pelvic girdle. The firm union of the pelvic girdle with the axial skeleton provides a rigid support for our erect posture and bipedal gait.

Hipbone Each hipbone, or os coxae (Fig. 5.43), is a large irregular bone shaped like two fan blades radiating out from a central hub. The hipbone is formed from three separate bones that fuse after birth: the ilium, ischium, and pubis. All three contribute to formation of the *acetabulum*, a deep cup-shaped depression that forms a socket for the head of the femur and acts as the hub of the hipbone.

The *ilium* extends upward from the acetabulum. It consists of a lower portion or *body*, which makes up two-fifths of the acetabulum, and a flattened flaired upper portion, or *ala*. The upper border of the ilium is a curved bony rim called the *iliac crest*, which can be palpated just below the waist. The iliac crest terminates in the four iliac spines. The prominent *anterior superior iliac spine*, which is the attachment site of the inguinal ligament and several thigh muscles, is an important anatomical landmark. The *greater sciatic notch* is a deep indentation on the posterior border of the ilium that transmits the sciatic nerve. The auricular surface of the ilium articulates with the sacrum to form the *sacroiliac joint*.

The *ischium* extends downward and backward from the acetabulum. It consists of a *body*, which forms two-fifths of the acetabulum, and the *ischial ramus*. The posterior border of the ischium gives rise to the *ischial spine*, a sharp projection that is an important anatomical landmark, and the *lesser sciatic notch*, a shallow depression that transmits the internal pudendal nerve and blood vessels. The body of the ischium terminates in the *ischial tuberosity*, a large rough

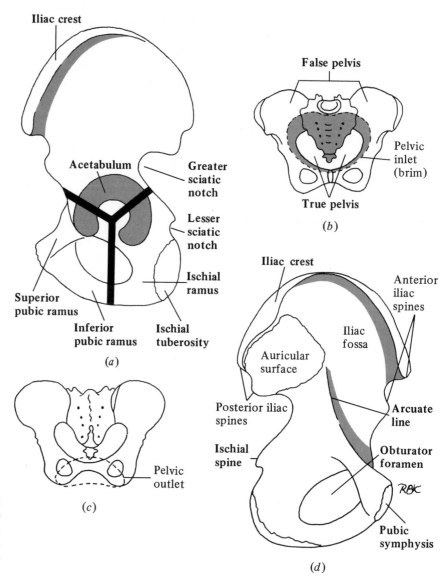

Figure 5.43 Hipbone and Pelvis. (a)
Left hipbone external view. Lines indicate
three major divisions: *top,* ilium; *bottom left,*
pubis; *bottom right,* ischium. (b) Pelvis,
anterior view. (c) Pelvis, posterior view.
(d) Left hipbone, internal view.

projection that bears the weight of the body when one sits down. The
ischial ramus is a bar of bone that curves forward from the body to
unite with the inferior ramus of the pubis.

The *pubis* extends downward and forward from the acetabulum.
It consists of a *body*, which forms one-fifth of the acetabulum, and
two rami. The body of the pubis articulates anteriorly with its counterpart from the opposite hipbone to form a joint called the *pubic
symphysis.* The *superior pubic ramus* ascends posteriorly to unite
with the ilium, while the *inferior pubic ramus* descends posteriorly
to unite with the ischial ramus. Together the inferior rami form the

pubic arch. The pubic and ischial rami form the borders of the *obturator foramen,* a large hole that transmits the obturator nerve and blood vessels.

The Pelvis as a Functional Unit

The pelvis (Fig. 5.43) is a bony basin formed by the two hipbones anteriorly and laterally and by the sacrum and coccyx posteriorly. It provides protection for the urinary bladder, some of the reproductive organs, and the distal portion of the large intestine.

The pelvis is divided into two parts by the *pelvic brim,* a curved line that passes through the sacral promontory, the arcuate lines of the ilia, and the pectineal lines and crests of the pubes. The *false pelvis,* which contains abdominal rather than pelvic organs, lies above the pelvic brim. It is bounded by the lumbar vertebrae posteriorly, by the flaired portions of the ilia laterally, and by abdominal wall muscles anteriorly. The pelvic brim, or *pelvic inlet,* marks the superior boundary or entrance to the true pelvis. The *true pelvis* lies below the pelvic brim. It is a bony cavity bounded by the sacrum and coccyx posteriorly, the ischia and lower portions of the ilia laterally,

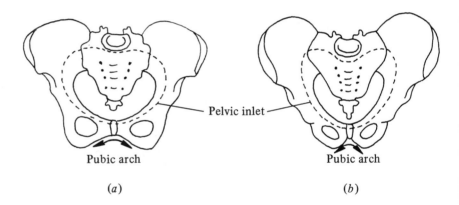

Pelvic inlet

Pubic arch

Pubic arch

(a)

(b)

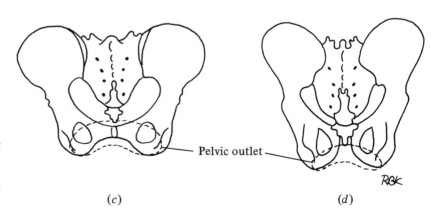

Pelvic outlet

Figure 5.44 Comparison of Female and Male Pelvises. (a) Female pelvis, anterior view. (b) Male pelvis, anterior view. (c) Female pelvis, posterior view. (d) Male pelvis, posterior view.

(c)

(d)

and the pubic rami anteriorly. In life the urinary bladder is located behind the pubic symphysis, while the rectum is located in front of the sacrum and coccyx. The vagina and uterus lie between the bladder and rectum in the female. The exit from the true pelvis, called the *pelvic outlet*, is bounded by the tip of the coccyx posteriorly, the ischial tuberosities laterally, and the pubic arch anteriorly.

Because the female true pelvis also functions as the birth canal, there are structural differences between the male and female pelvises (Fig. 5.44):

1. *Overall Construction.* The male pelvis is large and heavy; the female pelvis is light and delicate.
2. *Pelvic Inlet.* The male pelvic inlet is heart-shaped; the female pelvic inlet is round or broadened into an oval.
3. *Pelvic Cavity.* The male true pelvis is narrow and deep; the female true pelvis is wide and shallow.
4. *Pelvic Outlet.* The male pelvic outlet is narrow because the angle of the pubic arch is less than 90°; the female pelvic outlet is wide because the angle of the pubic arch is greater than 90°.

THE LOWER EXTREMITY

The lower extremity is divided into thigh, leg, ankle, foot, and toes. Bones of the lower extremity consist of the femur (thigh), tibia and fibula (leg), tarsals (ankle), metatarsals (foot), and phalanges (toes).

Femur

The thigh contains a single long bone, the femur, or thighbone (Fig. 5.45), which corresponds to the humerus and is the largest and longest bone in the human body.

The proximal end of the femur consists of a large ball-shaped *head* that articulates with the acetabulum of the hipbone. The head is separated from the femoral shaft by a long thick *neck* that joins the shaft at an angle of 125°. The length and angularity of the neck make it a frequent site of fracture, especially in older people whose bones are brittle. Fracture of the femoral neck is commonly referred to as a broken hip. Below the neck are two large projections that correspond to the tubercles of the humerus: the *greater trochanter* (on the lateral border) and the *lesser trochanter* (on the posterior surface).

The long massive femoral shaft has both anterior and posterior surfaces. The shaft is smooth except for the *linea aspera*, a low ridge for muscle attachment on the posterior surface.

The distal end of the femur is divided into two large articular condyles: the *medial femoral condyle*, which articulates with the medial tibial condyle, and the *lateral femoral condyle*, which articulates with the lateral tibial condyle. There is a deep depression, the *intercondylar fossa*, between the condyles on the posterior surface. The *medial* and *lateral epicondyles* are muscle attachment sites on the femoral borders above the condyles. The *adductor tubercle*, a

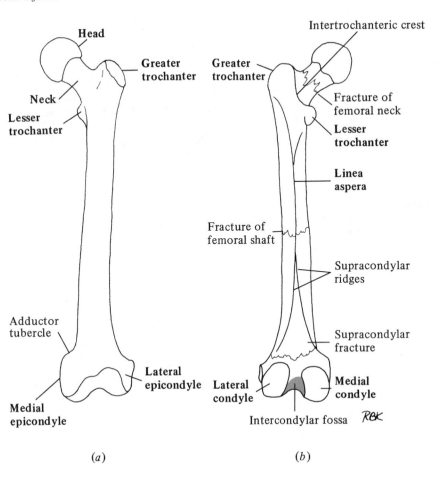

Figure 5.45 Left Femur. (a) Anterior view. (b) Posterior view. Common fracture sites are indicated.

(a)   *(b)*

projection from the medial epicondyle, marks the attachment site of the adductor magnus muscle.

Patella   The patella, or kneecap (Fig. 5.46), is a large sesamoid bone that develops in the tendon of the quadriceps femoris muscle where it crosses the knee joint. Shaped like an inverted triangle, the patella has a convex anterior surface and a concave posterior surface. The posterior surface rides on the femoral condyles when the leg is straightened. Because it is superficial and located in front of the knee joint, the patella is frequently fractured.

Tibia   The tibia, or shinbone (Fig. 5.47), is located on the medial (big toe) side of the leg. It is a long heavy bone that corresponds to the radius.

The broad proximal end of the tibia consists of two articular condyles separated by a ridge called the *intercondylar eminence*. The *medial* (oval) and *lateral* (round) *tibial condyles* articulate with the

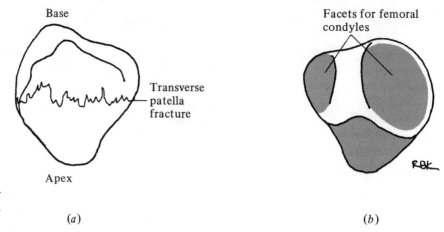

Base

Transverse
patella
fracture

Apex

Facets for femoral
condyles

Figure 5.46 *Left Patella.* (*a*) Anterior
view. Common fracture site is indicated.
(*b*) Posterior view.

(*a*)

(*b*)

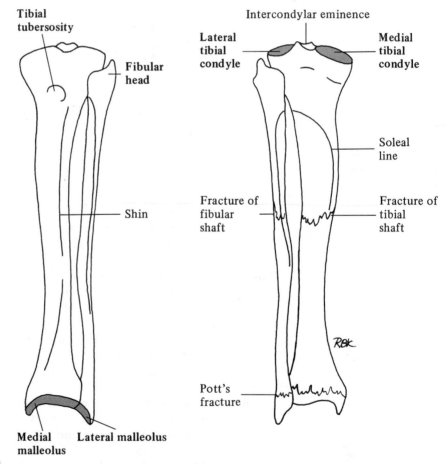

Tibial
tubersosity

Fibular
head

Shin

Medial
malleolus

Lateral malleolus

Intercondylar eminence

Lateral
tibial
condyle

Medial
tibial
condyle

Soleal
line

Fracture of
fibular
shaft

Fracture of
tibial
shaft

Pott's
fracture

Figure 5.47 *Left Tibia and Fibula.* (*a*)
Anterior view. (*b*) Posterior view. Common fracture sites are indicated.

(*a*)

(*b*)

corresponding femoral condyles. On the border of the lateral condyle is a facet that articulates with the head of the fibula. The *tibial tuberosity* is a small projection on the anterior surface just below the tibial condyles; it is the attachment site of the patellar ligament.

The triangular shaft of the tibia has three borders and three surfaces. The anterior border of the tibia is both sharp and superficial; the skin overlying the tibia at this site is easily bruised when the shin is bumped. The sharp interosseous border, which faces the fibula, provides attachment for the interosseous membrane that extends between the two leg bones. The horseshoe-shaped *soleal line* is a muscle attachment site on the posterior surface of the tibial shaft.

The distal end of the tibia is quadrilateral in shape. Its inferior surface articulates with the talus, an ankle bone. The *medial malleolus* (inner ankle prominence) projects downward from the medial side; it corresponds to the radial styloid process. On the lateral side is a facet that articulates with the distal end of the fibula.

Fibula
The fibula (Fig. 5.47) is a very slender bone located on the lateral (little toe) side of the leg. The fibula, which corresponds to the ulna, does not articulate with the femur and is not a weight-bearing bone. However, it does participate in forming the ankle joint.

The proximal end of the fibula consists of a blunt pyramidal *head*, which acts as a site of muscle attachment. The fibular head bears a facet on its medial surface for articulation with the lateral femoral condyle.

The fibular shaft is thin, twisted, and triangular in shape. Like the tibia, it has three borders and three surfaces. The sharp interosseous border, which faces the tibia, is the other attachment site for the interosseous membrane.

The pyramidal distal end of the fibula bears a facet on its medial surface for articulation with the distal end of the tibia. The *lateral malleolus* (outer ankle prominence) projects downward from the lateral side; it corresponds to the ulnar styloid process. The medial surface of the lateral malleolus also articulates with the talus; both malleoli participate in forming the ankle joint. Fracture of the distal end of the fibula and the medial malleolus is known as a *Pott's fracture.* This type of fracture, common in skiing accidents, results from a sudden sharp twist of the ankle when the foot is immobile.

Tarsals
The ankle, or tarsus, contains seven bones called tarsals (Fig. 5.48), which are divided into anterior and posterior groups. The shapes of the tarsals differ considerably from those of the carpals, their counterparts in the wrist.

The posterior group of tarsals consists of two large bones that are named as follows:

1. *Talus.* An L-shaped bone that participates in the formation of the ankle joint and bears the weight of the tibia.

2. *Calcaneus.* An oblong bone that forms the prominence of the heel and supports the talus. It is also known as the os calcis.

The anterior group of tarsals consists of five small bones that fit together like pieces of a jigsaw puzzle; they are named as follows:

1. *Navicular.* A boat-shaped bone on the medial (big toe) side of the foot that articulates with the anterior end of the talus.
2. *Cuboid.* A cube-shaped bone on the lateral (little toe) side of the foot that articulates with the anterior end of the calcaneus.
3. *Cuneiforms.* Three wedge-shaped bones that articulate with the anterior end of the navicular.

Metatarsals The metatarsals (Fig. 5.48) are five miniature long bones, numbered from medial to lateral, that form the bony framework of the foot. Their

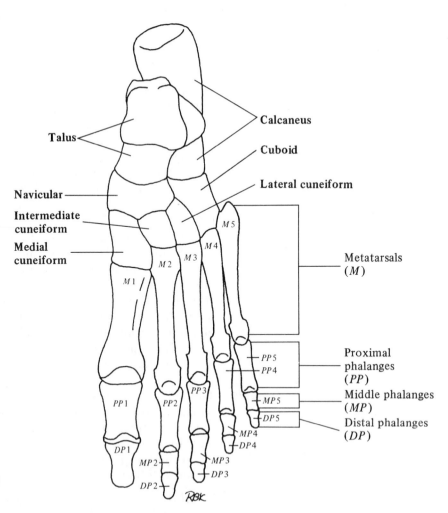

Figure 5.48 *Bones of Left Foot: Dorsal Surface.*

squared-off bases articulate with the cuboid and cuneiforms posteriorly, while their rounded heads articulate with the proximal phalanges anteriorly. The heads of the metatarsals form the ball of the foot. The metatarsal of the big toe is modified for weight bearing and has a very limited degree of mobility.

Phalanges

The phalanges (Fig. 5.48) are 14 miniature long bones that form the bony core of the toes. The big toe contains only proximal and distal phalanges, while all the other toes contain proximal, middle, and distal phalanges. A pair of sesamoid bones is commonly found over the metatarsophalangeal joint of the big toe; they are similar to those found in the thumb.

Arches of the Foot

The foot is a complex structure designed to support body weight. The tarsals and metatarsal normally do not rest flat on the ground but are arranged in the form of arches that help distribute body weight to different parts of the foot. These arches are bound together by ligaments and supported by muscle tendons. The *longitudinal arch* [Fig. 5.49(a)] extends from the calcaneus to the heads of the metatarsals, with the talus acting as keystone and summit of the arch. Anterior

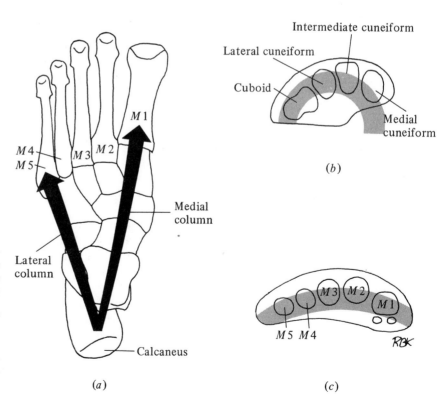

Figure 5.49 *Arches of Foot.* (a) Longitudinal arch from calcaneus posteriorly to heads of matatarsals anteriorly. *Arrows* show its division into medial and lateral columns. (b) Transverse arch (coronal section shown) extends across width of foot through cuboid, navicular (not shown), and cuneiform bones. (c) Metatarsal arch (coronal section shown) extends across width of foot through heads of the five metatarsals.

to the talus the longitudinal arch splits into medial and lateral columns. The *transverse arch* [Fig. 5.49(*b*)] extends across the width of the foot from the navicular to the cuboid. The *metatarsal arch* [Fig. 5.49(*c*)] extends across the heads of the metatarsals. *Flat foot* (pes planus) is a foot disorder due either to skeletal deformity, such as a tilted talus, or to laxity of the ligaments that support the longitudinal arch.

COMPARISON OF THE UPPER AND LOWER EXTREMITIES

By now it should be apparent to you that there are many similarities between the upper and lower extremities. This is not accidental. Both the upper and lower extremities develop according to a basic plan inherited from our primitive tetrapod (four-legged) ancestors. Initially, all four extremities performed the same function (paddling in water and walking on dry land) and possessed the same pentadactyl (five-digit) structure. This is why bones of the lower extremity correspond to those of the upper extremity. The differences we observe between the bones of the upper and lower extremities are due mainly to their adaptation to different functions.

You may still be confused by the fact that in the anatomical position the thumb side of the upper extremity is lateral while the big toe side of the lower extremity is medial. If these extremities correspond to one another, why aren't the thumb and the big toe on the same side? In the embryo both the thumb side of arm buds and the big toe side of leg buds originally are directed upward. Later in the course of prenatal development, and continuing postnatally, the upper extremity undergoes adduction so that it almost parallels the trunk. The lower extremity, however, undergoes adduction, extension, and almost 180° medial rotations. This causes the knee to point forward and the big toe to lie medially, which is just the reverse of the situation in the upper extremity, where the elbow points backward and the thumb lies laterally.

Diseases and Disorders of the Skeletal System

CONGENITAL MALFORMATIONS

Congenital malformations of the skeletal system are structural defects present at birth. Some are not nearly as hopeless as they once were, as many successful new treatments have been instituted for them.

Craniostenosis. Craniostenosis is a condition in which the skull sutures close prematurely. The abnormally small skull prevents the brain from growing to its proper size. This in turn causes mental retardation. Surgically created artificial sutures in the stenotic skull allow the brain to grow normally, which prevents the tragedy of mental retardation.

Cleft Palate. Partial or complete failure of the palatine processes of the maxillae to fuse is known as cleft palate. The infant with a cleft palate is unable to nurse effectively because the opening between his or her mouth and nose prevents suction. The infant must be fed with a syringe and tube, a cup, or a spoon. Cleft palate is usually corrected by surgical reconstruction at about 18 months of age, before the child learns to talk.

Spina Bifida. Spina bifida is a vertebral column malformation in which the arches of one or more vertebrae fail to form; the spinal cord is exposed in the region of the defect. Most cases of spina bifida are asymptomatic and are detected only upon x-raying the spine (spina bifida occulta). If the defect is extensive enough, there may be herniation of the spinal cord and its coverings through the skin (spina bifida cystica). Spina bifida cystica is considerably more serious since it is associated with some degree of neurological damage and the risk of meningitis.

Phocomelia. Phocomelia (seal flipper limb) is a relatively rare condition in which the intermediate portion of an extremity is absent. In some cases of phocomelia, a well-formed hand or foot arises directly from the limb girdle. During the early 1960s the drug *thalidomide*, taken by women in early pregnancy, was responsible for many occurrences of phocomelia.

Congenital Hip Dislocation. Congenital hip dislocation, caused by underdevelopment of the hip joint, is a malformation more common in female than male infants. In this disorder the acetabulum is shallow and its upper lip does not form a large enough shelf to hold the femoral head in place. Also, the greater trochanter is smaller than normal, allowing the femur to slip out of the acetabulum with ease. With early detection and simple physical therapy, this condition is often corrected without surgery.

METABOLIC AND ENDOCRINE DISORDERS

Rickets and Osteomalacia. Rickets (in children) and osteomalacia (in adults) are skeletal diseases characterized by softening of the bones (excess of osteoid tissue) due to a failure of osteoid mineralization. Rickets is further characterized by enlargement of the epiphysis and costochondral junctions, scoliosis, pelvic deformity, bowed legs, and enlargement of the skull. A major cause of these two related diseases is vitamin D deficiency. Vitamin D promotes the absorption of calcium from the digestive tract and regulates the metabolism of calcium and phosphorus so that these two minerals are available to the bones. Simple vitamin D deficiency is rather uncommon in the United States today, due to the widespread use of vitamin-D-fortified food products. However, the deficiency may be secondary to hepatobiliary, kidney, and parathyroid disease or to the use of certain drugs. Also, dietary deficiency of calcium or phosphate may lead to changes identical to those seen in vitamin-D-deficient rickets or osteomalacia.

Scurvy. Scurvy is a connective tissue disease caused by vitamin C deficiency. The major skeletal effect of scurvy is the stagnation of bone growth from the epiphyseal plates through decreased productions of collagen and matrix. In severe cases of scurvy, hemorrhage and necrosis occur in epiphyseal growth zones; this can lead to epiphyseal separation. Scurvy can be prevented by good nutrition and reversed by vitamin C therapy.

Osteoporosis. Osteoporosis is a type of bone atrophy characterized by reduction in bone mass. There is a thinning of cortical bone and an enlargement of the marrow cavity at the expense of spongy bone. Some of the causes of osteoporosis include prolonged immobility, malnutrition, estrogen deficiency, prolonged corticosteroid therapy, hyperparathyroidism, and tumors of the skeletal system. Postmenopausal women frequently develop osteoporosis; they are usually treated with estrogenic compounds, which appears to reverse the disease process.

FRACTURES A fracture is a break in the continuity of bony tissue. Most fractures are caused by trauma due to a fall or an automobile accident. Fractures can also result from stress or fatigue in the absence of any trauma. Pathological conditions, such as bone infections or bone cancer, weaken bone and predispose it to fracture.

Fractures are classified as *simple*, or *closed*, if the skin over the fracture site is unbroken and *compound*, or *open*, if the skin over the fracture site is broken. Compound fractures are more serious than simple fractures since the danger of infection is increased. Different types of fractures are illustrated in Fig. 5.50.

Fractures cause varying amounts of disability depending on the type of fracture, its location, and its subsequent treatment. Soft tissue may be injured by fractured bones; a fractured rib can puncture a lung, and a muscle can be displaced or torn by a bone fragment. Jagged bone fragments may even sever nerves and blood vessels. For additional information on fractures, the student is referred to surgical and orthopedic textbooks.

INFLAMMATORY DISEASE Osteomyelitis is an inflammatory disease of bone and bone marrow caused by pyogenic (pus-forming) microorganisms. The microorganisms can enter a bone through an open fracture site or arrive by way of the bloodstream. The infection generally begins in the marrow cavity of a long bone and spreads to cortical bone by way of vascular channels. The inflammatory response to the infection causes bone necrosis and

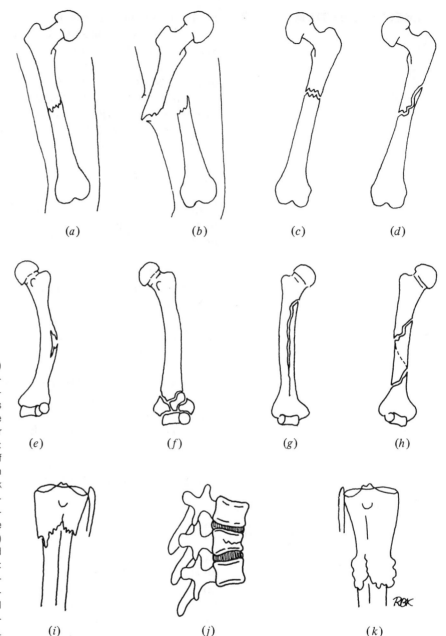

(a) *(b)* *(c)* *(d)*

(e) *(f)* *(g)* *(h)*

(i) *(j)* *(k)*

Figure 5.50 *Types of Fractures.* (a) Simple or closed fracture: skin over fracture site is unbroken. (b) Open or compound fracture: bone fragment protrudes through break in skin. (c) Transverse fracture: fracture line runs perpendicular to long axis of bone. (d) Oblique fracture: fracture line runs diagonal to long axis of bone. (e) Greenstick fracture: type seen in children in which bone does not break through completely. (f) Comminuted fracture: bone is broken into three or more fragments. (g) Linear fracture: fracture line runs parallel to long axis of bone. (h) Spiral fracture: fracture line spirals around long axis of bone. (i) Impacted fracture: bone fragments are crushed into one another. (j) Compression fracture: bone fragments are compressed. (k) Pathological fracture: fracture occurring in bone previously weakened or damaged by disease.

abscess formation. Before the advent of antibiotics, the chances that a patient would recover from osteomyelitis without severe bone damage were very poor. Today, with antibiotic therapy and drainage of existing abscesses, osteomyelitis can be arrested before it does much damage.

SKELETAL SYSTEM TUMORS

Skeletal system tumors may originate as primary growths in cartilage, bone, and bone marrow or they may metastasize to the skeletal system from other organs.

Chondrogenic Tumors. *Chondromas* are benign cartilaginous tumors that arise from the small bones of the hands and feet. *Chondrosarcomas* are malignant tumors arising from adult cartilage. Though chondrosarcomas of the trunk, shoulders, and hips are quite common, they grow slowly and metastasize late.

Osteogenic Tumors. Benign tumors of osteoid tissue, called *osteomas*, are relatively rare. *Osteosarcomas*, which arise from the long bones of the extremities, are even rarer than osteomas but are highly malignant and usually fatal. Current treatment consists of amputation of the affected extremity followed by administration of anticancer drugs.

Multiple Myeloma. Multiple myeloma is a highly malignant tumor of bone marrow elements. It more often affects the skull, ribs, sternum, and vertebral bodies, all of which are red marrow sites. This tumor, which arises in many locations at the same time, destroys bones from within by expansion.

Metastatic Tumors. Metastatic tumors are more common in the skeletal system than primary tumors. Tumors of the prostate gland, breast, lung, thyroid gland, and kidney frequently metastasize to bone. Tumor cells reach the bone marrow either through the bloodstream or the lymphatic system. Metastatic tumors also destroy bones by expansion from within the marrow cavity, which causes considerable pain and often pathological fractures.

Suggested Readings

BASSETT, ANDREW L. "Electrical effects in bone," *Sci. Am., 213*, no. 4 (October 1965), pp. 18–25.

KOLODNY, A. LEWIS, and ANDREW R. KLIPPER. "Bone and joint diseases in the elderly," *Hosp. Pract., 11*, no. 11 (November 1976), pp. 91–101.

° LICHTENSTEIN, LEWIS. *Diseases of Bone and Joints*. St. Louis: Mosby, 1970.

SOKOLOFF, LEON, and JOHN H. BLAND. *The Musculoskeletal System*. Baltimore: Williams & Wilkins, 1975.

TAUSSIG, HELEN B. "The thalidomide syndrome," *Sci. Am., 207*, no. 2 (August 1962), pp. 29–35.

° Advanced level.

6

The Articular System

The articular system consists of the functional connections between the bones of the human body. An articulation, or *joint,* is a union between two or more bones that generally permits some degree of movement. Human joints range from immovable to freely movable. To provide you with a good working knowledge of joints, the development, anatomy, and function of joints is presented in Part A of this chapter. Part B provides you with an overview of the clinically important human joints plus their diseases and disorders.

Part A. Joints as Organs

Joint Development

Joints develop from the mesenchymal interzone between the ends of developing bones. This region of undifferentiated mesenchyme is called the *primitive joint plate.* The subsequent fate of the primitive joint plate (Fig. 6.1) determines the type of joint that is formed and the degree of movement that it permits. The three different developmental types of joints can be classified as follows:

1. *Fibrous Joints.* In some joints the mesenchyme develops into fibrous connective tissue. These joints generally become synarthroses, or immovable joints.
2. *Cartilaginous Joints.* In other joints the mesenchyme develops into cartilage. Some of these joints become amphiarthroses, or slightly movable joints.
3. *Synovial Joints.* In still other joints the central portion of the mesenchyme is resorbed during development, leaving behind

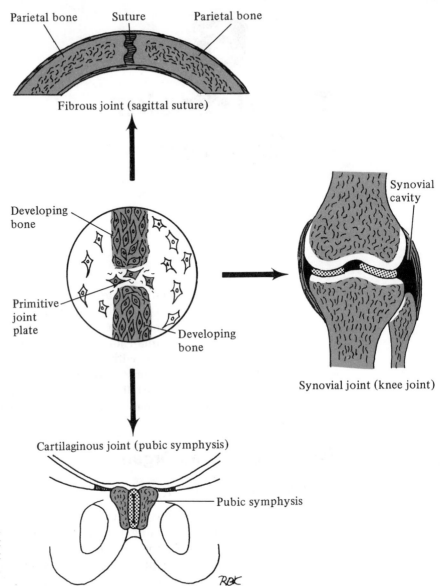

Parietal bone   Suture   Parietal bone

Fibrous joint (sagittal suture)

Developing bone

Primitive joint plate

Developing bone

Synovial cavity

Synovial joint (knee joint)

Cartilaginous joint (pubic symphysis)

Pubic symphysis

Figure 6.1 *Fate of Primitive Joint Plate.*
Subsequent development of primitive joint
plate (*middle*) determines whether joint
will be fibrous (*top*), cartilaginous (*bottom*),
or synovial (*right*).

a cavity lined by a synovial membrane. These joints become
diarthroses, or freely movable joints.

Joint Anatomy

One way of classifying the joints of the human body is to consider
whether or not a true synovial cavity is present between the artic-

ulating bones. If no cavity is present and the ends of the articulating bones are united by fibrous, cartilaginous, or bony tissue, the joint is called a *synarthrosis*. If a small nonsynovial cavity is present, as in a certain type of cartilaginous joint, the joint is called an *amphiarthrosis*. However, if a true synovial cavity is present, the joint is called a *diarthrosis*.

Another way of classifying human joints is to consider the type of tissue that unites the ends of articulating bones: fibrous, cartilaginous, or synovial tissue.

FIBROUS JOINTS

Fibrous joints are those in which the ends of the participating bones are united by fibrous connective tissue. Most fibrous joints are completely immovable due to the shapes of the articular surfaces and the shortness of the connective tissue fibers uniting them. Two important fibrous joints are the *suture* and the *syndesmosis*.

Suture

A suture is a fibrous union between skull bones. Only in fetal life and early infancy are the skull bones separated by a sufficient amount of fibrous tissue to be somewhat movable. Eventually, the fibrous membrane ossifies; the resulting bony joint, which is called a *synostosis*, is completely immovable. In a typical suture, such as the lambdoidal suture [Fig. 6.2(a)], the serrated articular surfaces of the participating bones interlock.

Syndesmosis

A syndesmosis is a type of fibrous union in which the connective tissue fibers are long enough to permit a slight amount of side-to-side movement while still holding the participating bones in close approximation. The interosseous membrane between the radius and ulna is a good example of the syndesmosis [Fig. 6.2(b)].

CARTILAGINOUS JOINTS

Cartilaginous joints are those in which the ends of the participating bones are united by cartilage. Cartilaginous joints are slightly movable due to the flexibility of the intervening cartilage. There are two types of cartilaginous joints: the *synchondrosis* and the *symphysis*.

Synchondrosis

A synchondrosis [Fig. 6.2(c)] is a joint in which the participating bones are united by hyaline cartilage. A common type of synchondrosis, seen in children, is the *epiphyseal plate*. The epiphyseal plate, described in Chapter 5, is a strip of hyaline cartilage that unites the epiphysis of a long bone to its diaphysis. Usually the epiphyseal plate is immovable, but a strong pulling force can cause some movement and even epiphyseal separation. When the skeleton is fully mature, most epiphyseal plates have ossified and become synostoses.

Symphysis

A symphysis [Fig. 6.2(d)] is a joint in which the participating bones are united by fibrocartilage. Symphyses permit a small amount of

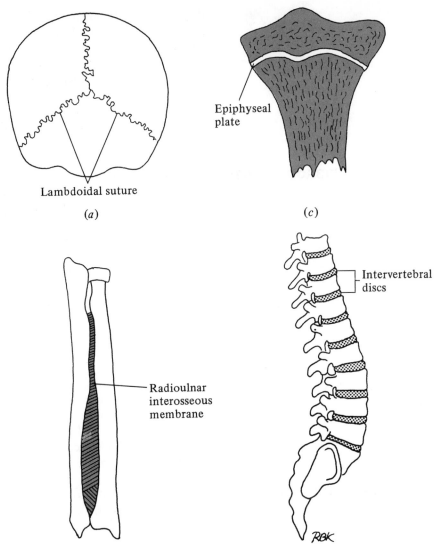

Lambdoidal suture

(a)

Epiphyseal
plate

(c)

Radioulnar
interosseous
membrane

Intervertebral
discs

(b)

(d)

Figure 6.2 Fibrous Joints. (a) Suture: lambdoidal suture of skull is a typical one. (b) Syndesmosis: radioulnar interosseous membrane is a syndesmosis that permits considerable movement between participating bones.

Cartilaginous Joints. (c) Synchondrosis: epiphyseal plate is a type of synchondrosis present in long bones of children. (d) Symphysis: intervertebral discs between adjacent vertebral bodies are classified as symphyses.

movement due to the flexibility and compressibility of the fibrocartilage. The *pubic symphysis* is a disc of fibrocartilage containing a small nonsynovial cavity that unites the pubic bones anteriorly. The intervertebral discs, which unite the bodies of adjacent vertebrae, are also symphyses. Symphyses can also be classified as amphiarthroses.

SYNOVIAL JOINTS   Synovial joints (Fig. 6.3) are highly specialized and flexible joints that have the following characteristics:

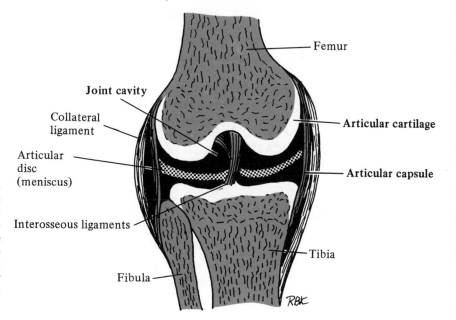

Femur

Joint cavity

Collateral ligament

Articular cartilage

Articular disc (meniscus)

Articular capsule

Interosseous ligaments

Tibia

Fibula

RBK

Figure 6.3 Synovial Joint (Knee Joint). Structural features (coronal section seen here) include joint cavity, articular (hyaline) cartilage covering articular surfaces of participating bones, and articular capsule uniting ends of participating bones. Note interosseous ligaments, which assist in uniting participating bones, and articular discs, menisci, which improve fit between articular surfaces.

1. A synovial cavity.
2. A synovial membrane that secretes synovial fluid.
3. An articular capsule.
4. Articular cartilages.

The articular surfaces of the participating bones are covered with *articular cartilages,* thin sheets of hyaline cartilage that reduce friction and protect the articular surfaces from pressure. The ends of the participating bones are united by a sleevelike *articular capsule* of fibrous connective tissue. The articular capsule is usually reinforced by external thickenings called capsular ligaments. The inner surface of the articular capsule is lined by a *synovial membrane,* which is reflected (turned back) over the surfaces of the participating bones up to their articular cartilages. It forms a closed sac around the *synovial* or *joint cavity.* The synovial membrane secretes a highly viscous fluid, called *synovial fluid,* which has the consistency and appearance of raw egg white. Synovial fluid lubricates the articular cartilages so they glide on one another without friction. The cavity of a normal joint is a slitlike potential space that contains only a thin film of synovial fluid, but the cavity of an injured or inflamed joint may expand to contain a considerable volume of fluid.

Some of the larger joint cavities contain projections from the synovial membrane in the form of folds, fringes, and fat pads. These projections fill in clefts and crevices between the bones and act as cushions. *Articular discs,* composed of fibrocartilage and covered by synovial membrane, occur in some joint cavities. In some instances they promote better fit between the articular surfaces; in other instances they subdivide the joint cavity, which permits the joint to perform several different movements.

Joint Function

CLASSIFICATION OF SYNOVIAL JOINTS

Synovial joints, being freely movable, are classified according to the amount and type(s) of movement their structure permits. The different classes of synovial joints are described next.

Uniaxial Joints

The uniaxial joint has one axis of rotation (see Chapter 1 for definition) and permits movement in only one plane. Uniaxial joints are divided into the following subgroups:

1. *Hinge Joint.* In the hinge, or ginglymus, joint [Fig. 6.4(*a*)], the axis of rotation (transverse axis) is at right angles to the long axis of the participating bones. The only movements permitted are back-and-forth movements called flexion and extension, which use the transverse axis of rotation. Examples of the hinge joint include the elbow, knee, and interphalangeal joints.
2. *Pivot Joint.* In the pivot, or trochoid, joint [Fig. 6.4(*b*)], the axis of rotation (vertical axis) is parallel to the long axis of the participating bones. A bone either rotates around the long axis of another bone or it rotates around its own long axis. The movements permitted at the pivot joint are medial and lateral rotation, which only use the vertical axis of rotation. Examples of the pivot joint include the proximal and distal radioulnar joints and the joint between the atlas and odontoid process of the axis.

Biaxial Joints

Biaxial joints possess two axes of rotation (transverse and anterior-posterior axes) at right angles to one another and permit movements in two planes. Biaxial joints are divided into the following subgroups:

1. *Condyloid Joint.* The condyloid, or ellipsoid, joint [Fig. 6.4(*c*)] is a joint in which a convex condyle fits into a concave ellipse. Movements permitted at the condyloid joint are flexion and extension plus side-to-side movements called abduction and adduction, which use the anterior-posterior axis of rotation. The radiocarpal (wrist) and metacarpophalangeal (knuckle) joints are good examples of the condyloid joint.
2. *Saddle Joint.* The saddle, or sellaris, joint [Fig. 6.4(*d*)] is a joint in which the two articular surfaces are reciprocally concave-convex. Movements permitted by the saddle joint are flexion, extension,

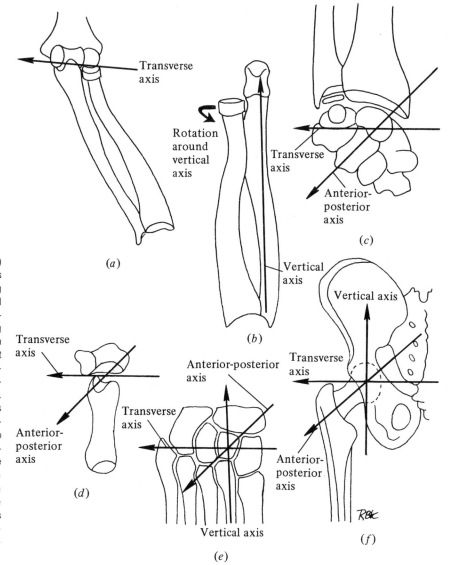

Figure 6.4 *Types of Synovial Joints.* (a)
Hinge joint, such as elbow. Note that axis
of rotation is at right angles to participating
bones. (*b*) Pivot joint, such as proximal
radioulnar joint. Note that rotational movement of radial head occurs around long
axis of the ulna. (*c*) Condyloid joint, such
as wrist joint. Two axes of rotation permit
back-and-forth plus side-to-side movements to occur between radius and articular disc and proximal row of carpals.
(*d*) Saddle joint, such as thumb. Two axes
of rotation permit back-and-forth, side-to-
side and some rotational movement to
occur between trapezium and first metacarpal. This accounts for thumb's wide
range of mobility. (*e*) Gliding joints, such
as intertarsal joints. Three axes of rotation
permit short gliding movements in three
planes. (*f*) Ball-and-socket joint, such as
hip joint. Three axes of rotation permit back-
and-forth, side-to-side, and rotational movements to occur.

abduction, adduction, plus a limited amount of rotation. Because
the saddle joint permits rotational movement, some anatomists
classify it as a multiaxial joint. The best example of a saddle joint
is the carpometacarpal joint of the thumb.

Multiaxial Joints   Multiaxial joints possess three axes of rotation (transverse, anterior-
posterior, and vertical) and permit movements in three planes. Multi-
axial joints can be divided into the following subgroups:

1. *Gliding Joints.* In the gliding, or plane, joint [Fig. 6.4(e)], the articular surfaces are almost flat. Gliding joints permit short gliding movements in all three planes. Some examples of gliding joints are the joints between the carpals of the wrist and between the tarsals of the ankle.

2. *Ball-and-Socket Joints.* In the ball-and-socket joint, or enarthrosis [Fig. 6.4(f)], one articular surface is a rounded ball while the other is a cup-shaped socket. Movements permitted by the ball-and-socket joint, which is the most freely movable type of joint, include back-and-forth, side-to-side, and rotational movements. The shoulder and hip joints are both ball-and-socket joints.

Movements Occurring at Synovial Joints

Various types of movements are possible at synovial joints; usually more than one type of movement can occur at a given joint. These movements are described next.

Angular Movements

Angular movements are movements that increase or decrease the angle between bones in a given joint. The different types of angular movements are as follows:

1. *Flexion.* Flexion [Fig. 6.5(a)] is an angular movement that decreases the angle between bones or brings bones closer together, such as bending the upper extremity at the elbow or the lower extremity at the knee.

2. *Extension.* Extension [Fig. 6.5(b)] is an angular movement that increases the angle between bones or moves bones further apart, such as straightening the upper extremity at the elbow or the lower extremity at the knee. *Hyperextension* is a movement that increases the angle between bones to greater than 90°.

3. *Abduction.* Abduction [Fig. 6.5(c)] is an angular movement away from the midline of the body, such as moving the arms away from the sides of the body.

4. *Adduction.* Adduction [Fig. 6.5(d)] is an angular movement toward the midline of the body, such as moving the arms back toward the sides of the body.

5. *Circumduction.* Circumduction [Fig. 6.5(e)] consists of successive movements of flexion, abduction, extension, and adduction, such as performed at the shoulder or hip joints, in which the proximal or fixed end of an extremity acts as a pivot while the distal or free end moves in a circle. The pitcher's wind up and the tennis serve are examples of circumduction from the sports world.

Rotational Movements

Rotational movements are movements in which a bone moves around a vertical axis. Rotational movements are as follows:

1. *Medial Rotation.* Medial, or internal, rotation [Fig. 6.5(f)] consists of rotating a part of the body around its long axis so that it

turns inward toward the midline of the body. In medial rotation of the leg, the toes point inward.

2. *Lateral Rotation.* Lateral, or external, rotation [Fig. 6.5(g)] consists of rotating a part of the body around its long axis so that it turns outward from the midline of the body. In lateral rotation of the leg, the toes point outward.

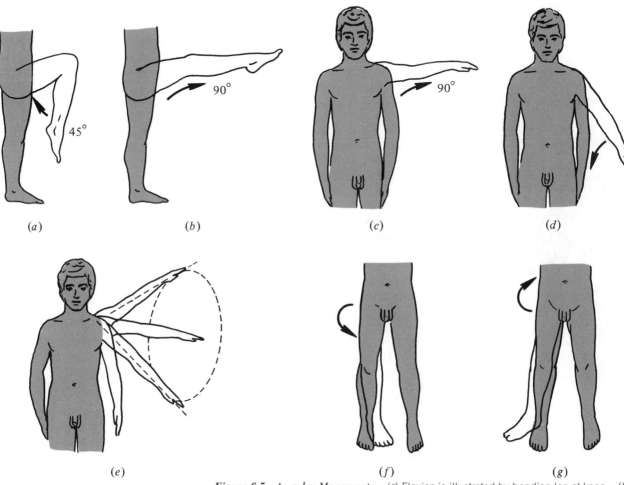

(a)   (b)   (c)   (d)

(e)   (f)   (g)

Figure 6.5 *Angular Movements.* (a) Flexion is illustrated by bending leg at knee. (b) Extension is illustrated by straightening leg at knee. (c) Abduction is illustrated by moving arm away from side of body. (d) Adduction is illustrated by moving arm toward side of body. (e) Circumduction is illustrated by swinging arm in circular fashion so that it describes a cone in space. Circumduction is a combination of angular movements in which flexion, abduction, extension, and adduction succeed one another.

Rotational Movements. (f) Medial rotation is illustrated by rotating leg toward midline of body so that toes turn inward. (g) Lateral rotation is illustrated by rotating leg away from midline of body so that toes turn outward.

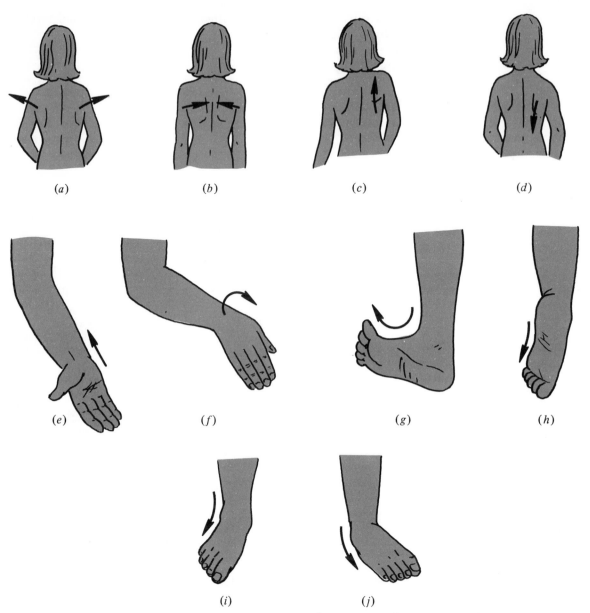

Figure 6.6 *Special Movements.* (*a*) Protraction is illustrated by moving scapulae forward and away from vertebral column in horizontal plane. (*b*) Retraction is illustrated by moving scapulae backward and toward vertebral column in horizontal plane. (*c*) Elevation is illustrated by raising scapula in vertical plane. (*d*) Depression is illustrated by lowering scapula in vertical plane. (*e*) Supination is illustrated by moving forearm so that palm of hand faces upward. (*f*) Pronation is illustrated by moving forearm so that palm of hand faces downward. (*g*) Dorsiflexion is illustrated by moving foot at ankle so that dorsum of foot is brought closer to leg. (*h*) Plantarflexion is illustrated by moving foot at ankle so that sole of foot is brought closer to leg. (*i*) Eversion is illustrated by moving foot at subtalar joint so that sole of foot is turned outward. (*j*) Inversion is illustrated by moving foot at subtalar joint so that sole of foot is turned inward.

Special Movements

In addition to angular and rotational movements, there are special movements performed by only a few body parts. These special movements are as follows:

1. *Protraction.* Protraction [Fig. 6.6(*a*)] is a movement of part of the body forward in the horizontal plane, such as moving the scapulae forward and away from the vertebral column.
2. *Retraction.* Retraction [Fig. 6.6(*b*)] is a movement of part of the body backward in the horizontal plane, such as in moving the scapulae backward and toward the vertebral column.
3. *Elevation.* Elevation [Fig 6.6(*c*)] is a raising of part of the body, such as raising the scapulae during the act of shrugging the shoulders.
4. *Depression.* Depression [Fig. 6.6(*d*)] is a lowering of part of the body, such as lowering the scapulae when the shoulders droop.
5. *Supination.* Supination [Fig. 6.6(*e*)] is a movement of the forearm at the radioulnar joints such that the radius lies parallel to the ulna and the palm of the hand faces up or forward, as in carrying soup.
6. *Pronation.* Pronation [Fig. 6.6(*f*)] is a movement of the forearm at the radioulnar joints such that the radius crosses over the ulna and the palm of the hand faces down or backward.
7. *Dorsiflexion.* Dorsiflexion [Fig. 6.6(*g*)] is a movement of the foot at the ankle joint such that the dorsum of the foot is brought closer to the leg.
8. *Plantarflexion.* Plantarflexion [Fig. 6.6(*h*)] is a movement of the foot at the ankle joint such that the sole (plantar surface) of the foot is brought closer to the leg.
9. *Eversion.* Eversion [Fig. 6.6(*i*)] is a movement of the foot at the subtalar joint such that the sole of the foot is turned outward.
10. *Inversion.* Inversion [Fig. 6.6(*j*)] is a movement of the foot at the subtalar joint such that the sole of the foot is turned inward.

Part B. **Joints of the Human Body**

Joints of the human body are subdivided into joints of the axial skeleton and joints of the appendicular skeleton. The following sections present an overview of the joints of the human body, with emphasis placed on the clinically more important joints. To assist you in familiarizing yourself with the joints, Table 6.1 contains a synopsis of most of the joints of the human body, including their names, locations, classifications, and actions. You should familiarize yourself with the joints and their movements before proceeding to study the muscular system.

Table 6.1 Synopsis of Joints of Human Body

Joint	Location	Type	Movement(s)
SKULL			
Majority of joints	Between bones of the skull	Suture	None after early infancy
Temporoman-dibular	Between mandibular condyle and mandibular fossa of temporal bone	Hinge and gliding	Elevation, depression, protraction, retraction, side-to-side
VERTEBRAL COLUMN			
Intervertebral discs	Between adjacent vertebral bodies	Symphysis	Flexion, extension, hyperextension, rotation
Vertebral arches	Between superior and inferior articular processes	Gliding	Same as above
Atlantooccipital	Between occipital condyles of the skull and atlas	Condyloid	Flexion, extension
Atlantoaxial	Between odontoid process of axis and atlas	Pivot	Rotation
THORACIC CAGE			
Costovertebral	Between rib heads and vertebral bodies plus their discs	Gliding	Elevation, depression
	Between rib tubercles and transverse processes of vertebrae	Gliding	Same as above
Costochondral	Between rib shafts and costal cartilages	Synchondrosis	Slight
Sternochondral	Between costal cartilages and sternum	Gliding	Elevation, depression
PECTORAL GIRDLE			
Sternoclavicular	Between manubrium of sternum and clavicle	Hinge and gliding	Elevation, depression, protraction, retraction; upward and downward rotation
Acromioclavicular	Between acromion of scapula and clavicle	Gliding	Slight
Coracoclavicular	Between coracoid process of scapula and clavicle	Syndesmosis	Slight
UPPER EXTREMITY			
Glenohumeral	Shoulder—between glenoid fossa of scapula and head of humerus	Ball-and-socket	Elevation, depression, adduction, abduction, medial and lateral rotation, circumduction
Humeroulnar	Elbow—between humerus and ulna	Hinge	Flexion, extension

Table 6.1 *(Continued)*

Joint	Location	Type	Movement(s)
Humeroradial	Elbow—between humerus and radius	Hinge	Flexion, extension
Superior radioulnar	Forearm—between radius and ulna	Pivot	Rotation; pronation, supination
Intermediate radioulnar	Forearm—between shafts of radius and ulna	Syndesmosis	Pronation, supination
Inferior radioulnar	Forearm—between radius and ulna	Pivot	Rotation; pronation, supination
Radiocarpal	Wrist—between radius and proximal row of carpals	Condyloid	Flexion, extension, abduction, adduction
Intercarpal	Wrist—between adjacent carpals	Gliding	Slight
Carpometacarpal (thumb)	Hand—between trapezium and first metacarpal	Saddle	Flexion, extension, abduction, rotation, circumduction
Carpometacarpal (fingers)	Hand—between distal row of carpals and metacarpals	Gliding	Slight
Metacarpo-phalangeal	Hand—between heads of metacarpals and proximal phalanges	Condyloid	Flexion, extension, adduction, abduction
Interphalangeal	Hand—between phalanges	Hinge	Flexion, extension

PELVIC GIRDLE

Joint	Location	Type	Movement(s)
Sacroiliac	Between auricular processes of sacrum and ilium	Symphysis and syndesmosis	None except during childbirth
Pubic symphysis	Between pubic bones	Symphysis	None except during childbirth

LOWER EXTREMITY

Joint	Location	Type	Movement(s)
Acetabulofemoral	Hip—between acetabulum of hipbone and head of femur	Ball-and-socket	Flexion, extension, adduction, abduction, medial and lateral rotation, circumduction
Tibiofemoral	Knee—between femoral and tibial condyles	Hinge	Flexion, extension, hyperextension, rotation (in certain positions)
Femoropatellar	Knee—between femoral condyles and patella	Gliding	Gliding of patella on femoral condyles
Superior tibio-fibular	Leg—between tibia and fibula	Gliding	Slight
Intermediate tibiofibular	Leg—between shafts of tibia and fibula	Syndesmosis	Slight
Inferior tibio-fibular	Leg—between tibia and fibula	Syndesmosis	Slight
Talocrural	Ankle—between talus, tibia, and fibula	Hinge	Dorsiflexion, plantarflexion

Table 6.1 (Continued)

Joint	Location	Type	Movement(s)
Subtalar	Foot—between talus and calcaneus	Gliding	Inversion, eversion
Intertarsal	Foot—between adjacent tarsals	Gliding	Slight
Tarsometatarsal (big toe)	Foot—between first cuneiform and first metatarsal	Saddle	Flexion, extension, adduction, abduction
Tarsometatarsal (toes)	Foot—between cuboid plus second and third cuneiforms and second to fifth metatarsals	Gliding	Slight
Metatarso-phalangeal	Foot—between heads of metatarsals and proximal phalanges	Condyloid	Flexion, extension, adduction, abduction
Interphalangeal	Foot—between phalanges	Hinge	Flexion, extension

Joints of the Axial Skeleton

Joints of the axial skeleton include the joints of the skull, vertebral column, and thoracic cage.

JOINTS OF THE SKULL   All but one of the joints between skull bones are sutures. The only freely movable skull joint is the joint between the temporal bone and the condyloid process of the mandible (Fig. 6.7). The cavity of the *temporomandibular joint* (jaw joint) is subdivided by an articular disc that permits this joint to function as both a hinge and a gliding joint. Since the articular capsule is very loose, the jaw joint is stabilized by surrounding muscles. Movements possible at this joint include elevation (closing the mouth) and depression (opening the mouth) plus protraction, retraction, and side-to-side gliding movements used in biting and chewing. The jaw joint may be subluxated (partially dislocated) when the teeth do not occlude properly.

JOINTS OF THE VERTEBRAL COLUMN   The most important joints of the vertebral column are the joints between the first cervical vertebra and the skull and between the first and second cervical vertebrae. The *atlantooccipital joint* [Fig. 6.8(a)] consists of a pair of condyloid joints between the atlas and the occipital bone. Movements at this joint consist mainly of flexion and extension, such as seen when nodding the head "yes." The *atlantoaxial joint* [Fig. 6.8(b)] is a pivot joint between the atlas and axis. The atlas rotates around the odontoid process of the axis to produce the head-shaking "no" movement.

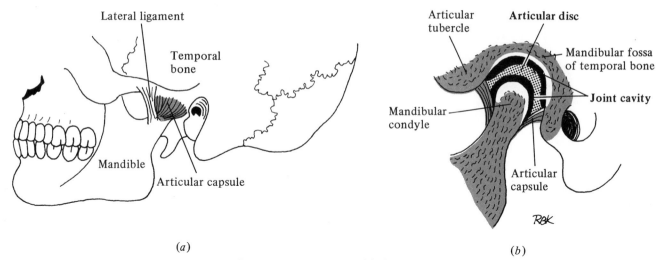

(a)

(b)

Figure 6.7 *Temporomandibular Joint.* (a) Lateral view; joint shown enclosed in its articular capsule. (b) Sagittal section, showing articular disc that subdivides joint into upper and lower cavities.

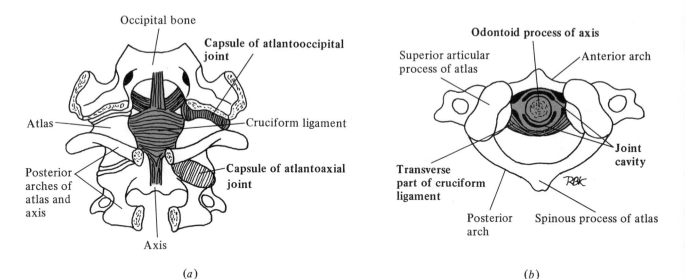

(a)

(b)

Figure 6.8 *Joints of Vertebral Column.* (a) Atlantooccipital and atlantoaxial joints, posterior view. Back of skull plus atlas and axis have been cut away to expose atlantooccipital and lateral atlantoaxial joints. Median atlantoaxial joint is hidden beneath cruciform ligament. (b) Median atlantoaxial joint, superior view. Odontoid process of axis is shown articulating with anterior arch of atlas. It is held in position by transverse portion of cruciate ligament.

The joints between adjacent vertebral bodies consist of the fibro-cartilaginous intervertebral discs described in Chapter 5, which are classified as symphyses. The joints between the superior and inferior articular processes of adjacent vertebral arches are synovial joints of the gliding variety. Both types of joints permit flexion, lateral flexion (bending to the side) extension, hyperextension, and rotation of the vertebral column. While there is limited movement at each joint, the simultaneous movement of many joints throughout the vertebral column allows for considerable flexibility, especially in the cervical and lumbar regions.

JOINTS OF THE THORACIC CAGE

The ribs articulate with thoracic vertebrae by means of gliding joints. Rib heads articulate with vertebral bodies, while rib tubercles articulate with transverse processes [Fig. 6.9(a)]. The costal cartilages of the first to tenth pairs of ribs are united to the sternum or to each other by small gliding joints. Each rib is capable of its own range of movements, primarily elevation and depression. Due to its attachments to the vertebral column posteriorly and to the sternum or costal cartilages anteriorly, each rib swings up or down like a bucket handle [Fig. 6.9(b)]. The movements of all the ribs are combined in the respiratory movements of the thorax. During inspiration the ribs move up and out, which increases the anterior-posterior diameter of the thoracic cage; during expiration the ribs move down and in, which decreases the anterior-posterior diameter of the thoracic cage.

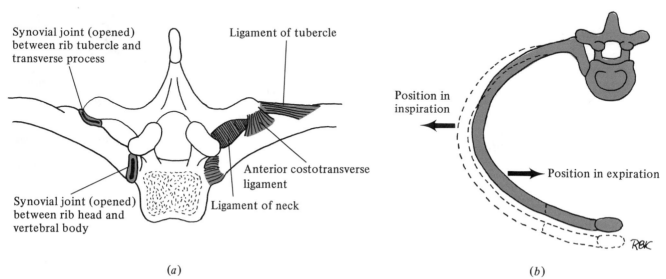

Synovial joint (opened) between rib tubercle and transverse process

Ligament of tubercle

Anterior costotransverse ligament

Ligament of neck

Synovial joint (opened) between rib head and vertebral body

Position in inspiration

Position in expiration

(a)

(b)

Figure 6.9 Joints of Thoracic Cage. (a) Costovertebral joints, superior view. These include synovial joints between rib head and vertebral body and between rib tubercle and transverse process. (b) Movements of vertebrosternal rib, superior view. Position of a true rib is shown in inspiration *(dashed)* and expiration *(solid)*.

Joints of the Appendicular Skeleton

Joints of the appendicular skeleton include the joints of the shoulder girdle and upper extremity plus the joints of the pelvic girdle and lower extremity.

JOINTS OF THE SHOULDER GIRDLE

The joints of the shoulder girdle provide the upper extremity with a high degree of mobility. The *sternoclavicular joint* (Fig. 6.10) is the joint between the sternum and clavicle that attaches the shoulder girdle to the axial skeleton. Its joint cavity is subdivided by an articular disc that permits the sternoclavicular joint to function like a ball-and-socket joint. Movements possible at the sternoclavicular joint include elevation, depression, protraction, retraction, upward rotation, and downward rotation. Strong extrinsic ligaments, such as the sternoclavicular and interclavicular ligaments, support the joint and prevent dislocation of the clavicle. The *acromioclavicular joint* is a relatively unimportant gliding joint that attaches the scapula to the clavicle. Actually, the strongest attachment between the clavicle and scapula is formed by the trapezoid and conoid ligaments. Traumatic rupture of the acromioclavicular joint, which frequently occurs during sports events, is referred to as a shoulder separation.

JOINTS OF THE UPPER EXTREMITY
Shoulder Joint

The shoulder joint [Fig. 6.11(*a*)] is a ball-and-socket joint between the glenoid fossa of the scapula and the head of the humerus. It permits flexion, extension, abduction, adduction, and circumduction plus medial and lateral rotation of the arm. The shoulder joint is the most mobile and least stable joint in the entire body. The glenoid fossa has a smaller articular surface than the head of the humerus; it is deepened to some extent by the *glenoid labrum,* a fibrocartilaginous

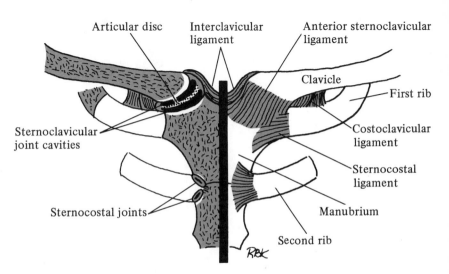

Figure 6.10 Sternoclavicular and Sternocostal Joints. Anterior view and coronal section. Note that sternoclavicular joint is subdivided by articular disc. First rib is united to sternum by immovable joint, while second rib is united to sternum by pair of synovial joints.

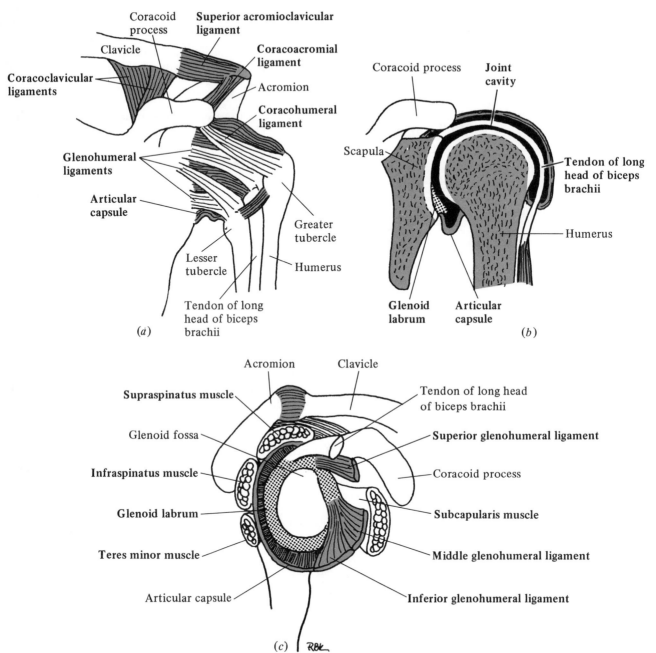

Figure 6.11 *Shoulder Joints.* (a) Left shoulder and acromioclavicular joints, anterior view. Lax articular capsule of shoulder joint is strengthened anteriorly by glenohumeral ligaments and superiorly by coracohumeral ligament. Ligaments attach clavicle to acromion and coracoid process of scapula. (b) Left shoulder joint, coronal section, showing joint cavity. Glenoid labrum helps promote better fit between humeral head and glenoid cavity. Note that tendon of long head of biceps muscle passes through joint cavity before entering bicipital groove of humerus. (c) Interior of right shoulder joint, opened to expose glenoid labrum, articular capsule (cut), biceps tendon, and muscles of rotator cuff (supraspinatus, infraspinatus, teres minor, and subscapularis). Muscles of rotator cuff provide only effective support for shoulder joint.

ring applied to the rim of the socket. The loose articular capsule is provided with *glenohumeral* and *coracohumeral ligaments,* which contribute little to the support of the joint [Fig. 6.11(*b*)]. The main support for this joint comes from the tendon of the long head of the biceps brachii muscle and the muscles of the *rotator cuff* (supraspinatus, subscapularis, infraspinatus, and teres minor muscles), which cross the shoulder joint to insert on the humerus [Fig. 6.11(*c*)]. These structures hold the head of the humerus in the glenoid fossa and support the joint from above, in front, and behind. The *coracoacromial arch* forms a secondary socket that helps prevent upward dislocation of the shoulder. Several bursae, fluid-filled sacs, are associated with the shoulder joint; the most important of these are the *subacromial bursa,* which reduces friction between the coracoacromial arch and the supraspinatus tendon, and the *subscapularis bursa,* which reduces friction between the subscapularis tendon and the joint capsule.

Because it is structurally weak, the shoulder joint is subject to sprains and dislocation; recurrent dislocation of the shoulder is a common problem, especially in younger persons. The rotator cuff degenerates with age, and tears of the rotator cuff are common. A

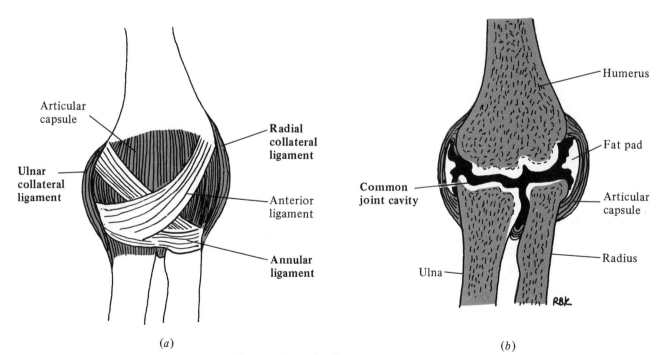

(a)

(b)

Figure 6.12 Left Elbow Joint. (a) Anterior view, showing joint enclosed in loose articular capsule that is supported by radial and ulnar collateral ligaments plus anterior and annular ligaments. (b) Coronal section. Note that humeroulnar, humeroradial, and superior radioulnar joints share a common joint cavity.

torn rotator cuff reduces a person's ability to abduct his or her arm. Inflammation of the subacromial bursa (bursitis) is a painful condition that also greatly reduces mobility of the arm.

Elbow Joint

The elbow joint (Fig. 6.12) is a hinge joint that unites the arm to the forearm. Movements permitted at the elbow joint include flexion and extension; hyperextension is prevented by strong ligaments and not by the olecranon process of the humerus, as you might expect. The elbow joint consists of the humeroulnar and humeroradial articulations described in Chapter 5. These articulations are enclosed in a loose articular capsule supported by *radial* and *ulnar collateral ligaments*. Partial dislocation of the radial head frequently occurs in young children when impatient adults drag them by the forearm. Traumatic dislocation of the distal end of the humerus can result from a fall on the outstretched palm.

Radioulnar Joints

Three joints (Fig. 6.13) between the radius and ulna permit pronation and supination to occur:

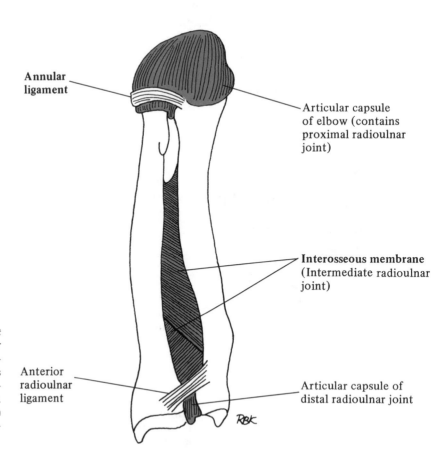

Annular ligament

Articular capsule of elbow (contains proximal radioulnar joint)

Interosseous membrane (Intermediate radioulnar joint)

Anterior radioulnar ligament

Articular capsule of distal radioulnar joint

Figure 6.13 Radioulnar Joints of Right Forearm. Anterior view shows superior radioulnar joint enclosed in articular capsule of elbow joint; annular ligament keeps radial head in contact with ulna. Intermediate radioulnar joint is a syndesmosis. Distal radioulnar joint is enclosed in its own capsule, which is strengthened by a ligament.

1. *Superior Radioulnar Joint.* A pivot joint between the head of the radius and the radial notch of the ulna.
2. *Intermediate Radioulnar Joint.* A syndesmosis formed by the interosseous membrane of the forearm.
3. *Inferior Radioulnar Joint.* A pivot joint formed by the head of the ulna and the ulnar notch of the radius.

Wrist Joint   The wrist joint (Fig. 6.14), which attaches the hand to the forearm, is a condyloid joint formed by the articulation of the distal end of the radius and ulnar articular disc with the proximal row of carpals. Movements permitted at this joint include flexion, extension, abduction (radial deviation), and adduction (ulnar deviation). Due to its exposed position and frequent use as a defensive shield, the wrist is subject to considerable trauma, which can cause sprains and dislocations of the wrist bones.

JOINTS OF THE PELVIC GIRDLE   The *sacroiliac joint* [Fig. 6.15(a)] is a combined symphysis and syndesmosis that occurs between the auricular surfaces of the sacrum and ilium. It attaches the pelvic girdle to the axial skeleton. The sacroiliac joint is strengthened and supported by several ligaments that run

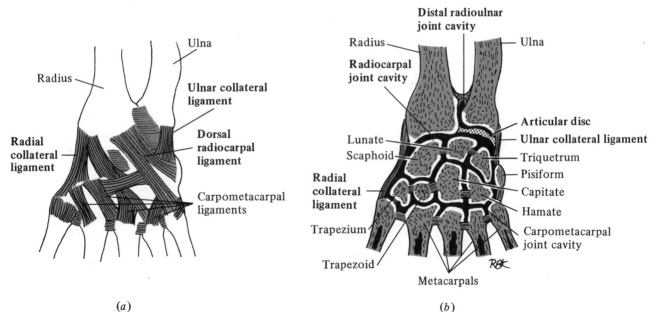

(a)

(b)

Figure 6.14 Wrist Joints. (a) Ligaments of right wrist joint and hand, dorsal view. Articular capsule of wrist joint is strengthened by radial and ulnar collateral ligaments and dorsal radiocarpal ligament. (b) Coronal section through right wrist. Note that ulna does not participate in wrist joint. Distal radioulnar joint cavity is separated from radiocarpal joint cavity by articular disc, which does participate in wrist joint.

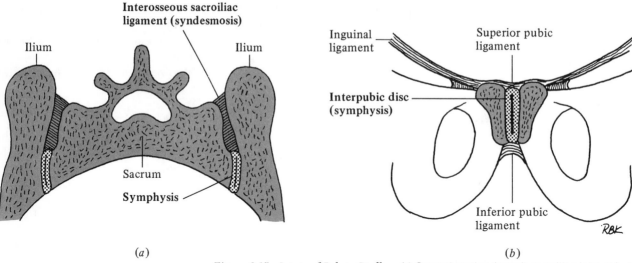

Ilium

Interosseous sacroiliac
ligament (syndesmosis)

Ilium

Inguinal
ligament

Superior pubic
ligament

Interpubic disc
(symphysis)

Sacrum

Symphysis

Inferior pubic
ligament

(a)

(b)

Figure 6.15 *Joints of Pelvic Girdle.* (a) Coronal section through sacroiliac joint at first sacral segment. Note that sacroiliac joint consists of both a syndesmosis (interosseous sacroiliac ligament) and a symphysis. (b) Pubic symphysis is fibrocartilaginous disc that unites pubic portions of hipbones anteriorly.

between the sacrum and the hipbone. The *pubic symphysis* [Fig. 6.15(b)] is a symphysis consisting of the fibrocartilaginous disc that unites the pubes of the two hipbones anteriorly. It is supported from above and below by the pubic ligaments. A hormone, thought to be relaxin, softens the fibrocartilage and loosens the ligaments of the pelvic joints during pregnancy so that the pelvic bones can separate to some extent as the baby passes through the birth canal.

JOINTS OF THE LOWER EXTREMITY
Hip Joint

The hip joint (Fig. 6.16) is a ball-and-socket joint between the acetabulum of the hipbone and the head of the femur. It permits flexion, extension, abduction, adduction, circumduction, and medial and lateral rotation of the thigh. The hip joint combines great stability with a wide range of movements, though these movements are more restricted than those of the shoulder joint. The acetabulum forms a deep socket for the head of the femur. It is further deepened by the *acetabular labrum,* a fibrocartilaginous ring applied to the rim of the acetabulum, which embraces the head of the femur. The thick strong articular capsule is reinforced by *iliofemoral, pubofemoral,* and *ischiofemoral ligaments,* which help check femoral movements.

Since the hip joint is more stable than the shoulder joint, it is much more difficult to dislocate. Congenital hip dislocation in infants, discussed in Chapter 5, results from underdevelopment of the acetabulum and femur. Traumatic hip dislocation, more common in adults, is most commonly the result of an automobile accident in which the flexed knee hits the dashboard.

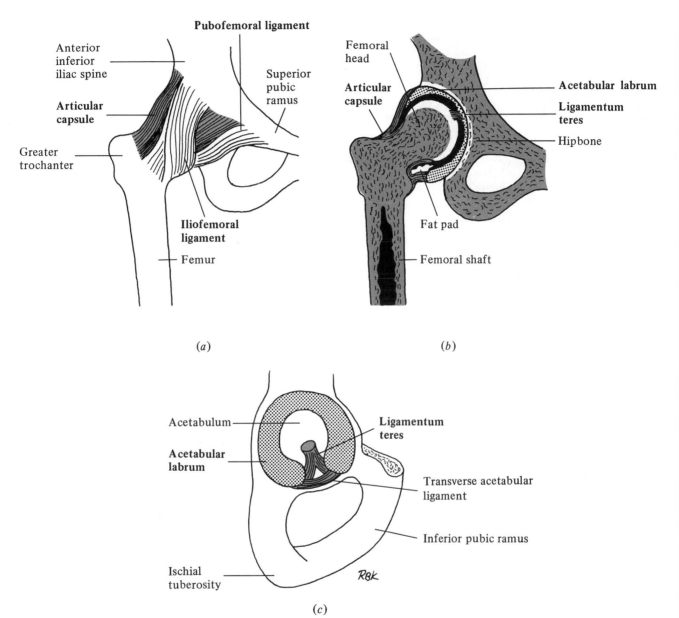

(a)

(b)

(c)

Figure 6.16 *Right Hip Joint.* (a) Anterior view. Articular capsule of hip joint is strengthened anteriorly by iliofemoral and pubofemoral ligaments. (b) Coronal section. Note that acetabulum forms deep socket for femoral head; it is further deepened by acetabular labrum. Ligamentum teres attaches femoral head to acetabulum. (c) Interior view. Hip joint is opened to expose acetabular labrum and ligamentum teres. Note that acetabulum is incomplete inferiorly; acetabular notch is partially covered by transverse acetabular ligament.

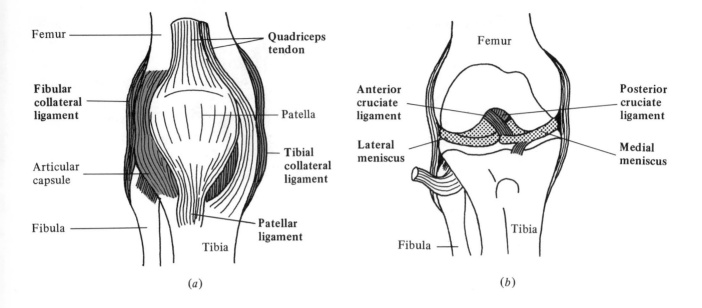

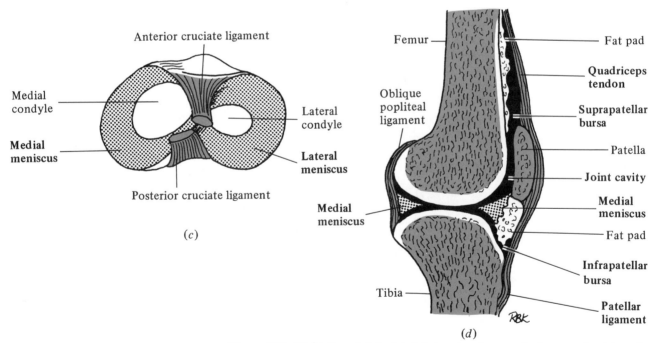

Figure 6.17 Right Knee Joint. (a) Anterior view. Loose articular capsule of knee is strengthened by tibial and fibular collateral ligaments plus quadriceps tendon and patellar ligament. (b) Dissected anteriorly. Knee joint is opened and partially flexed to expose menisci and cruciate ligaments. (c) Right tibia, superior view. Knee joint is opened to expose menisci in position on top of tibial condyles. These fibrocartilaginous discs promote better fit between femoral and tibial condyles. (d) Sagittal section. Note that prepatellar and infrapatellar bursae communicate with joint cavity of knee.

Knee Joint   The knee joint (Fig. 6.17) is a modified hinge joint that unites the thigh to the leg. It consists of the medial and lateral tibiofemoral articulations plus the femoropatellar articulations. The tibiofemoral articulations occur between the tibial and femoral condyles. The rather flat tibial condyles are deepened to fit the femoral condyles by crescent-shaped discs of fibrocartilage called the *medial* and *lateral menisci*. Movements permitted by the tibiofemoral articulations include flexion, extension, hyperextension, and rotation (in certain positions). The femoropatellar articulation occurs between the femoral condyles and the patella. The patella glides on the femoral condyles during flexion and extension of the knee. The tibiofemoral and femoropatellar articulations share a common joint capsule and synovial membrane. The thick strong articular capsule that encloses the knee joint is reinforced by the *tibial* and *fibular collateral ligaments*, which help prevent sideways dislocation of the femur. The *anterior* and *posterior cruciate ligaments* are short strong ligaments that cross each other to form the letter X. They firmly unite the tibia to the femur and help prevent backward and forward dislocation of the femur. A large number of bursae protect structures crossing the knee joint. Several occur between the skin and the patella or its tendon. The large *suprapatellar bursa*, located beneath the quadriceps tendon, communicates with the joint cavity.

Due to its exposed location and weight-bearing function, the knee is subject to frequent structural and functional derangements. Recurrent lateral dislocation of the patella, common in adolescent girls, occurs as the result of trauma to the knee or congenital defects of the joint and its ligaments. Dislocation of the femur is much less common and is generally the result of trauma. Torn menisci are very common in football and basketball players; they are caused by rotary movements at the knee joint. When a torn meniscus is displaced into the joint cavity, it can cause "locking" of the knee joint, a condition the layperson calls "trick knee." Rupture of the tibial collateral ligament occurs when force is applied to the lateral aspect of the knee, as in a football block.

Ankle Joints   The talocrural (ankle) joint (Fig. 6.18), which attaches the foot to the leg, is a hinge joint formed by the tibia and its malleolus, the malleolus of the fibula, and the talus. The weak articular capsule of the ankle joint is reinforced by medial and lateral ligaments. Movements possible at this joint include dorsiflexion and plantarflexion of the foot. Since the ankle joint is somewhat unstable, it is frequently subject to sprains and dislocations.

The talocalcaneal (subtalar) joint is a highly modified gliding joint that occurs between the posterior facets of the talus and calcaneus. This joint is mentioned because eversion and inversion of the foot occur at the subtalar joint rather than the ankle joint, as you might expect.

Figure 6.18 *Ankle Joints.* (a) Ligaments of right ankle and foot, medial view. Articular capsule of ankle is strengthened by deltoid and medial talocalcaneal ligaments. (b) Oblique section through right foot. Note talocrural (ankle), intertarsal, and tarsometatarsal joint cavities. (c) Coronal section through left foot, posterior view. Note talocrural (ankle) and subtalar joint cavities.

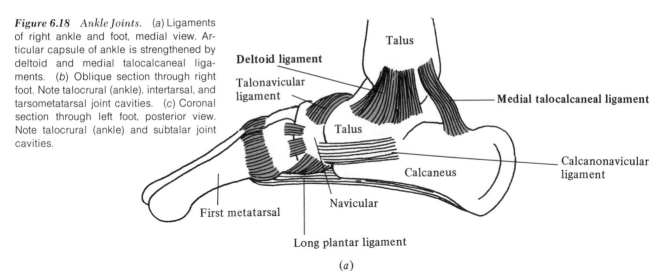

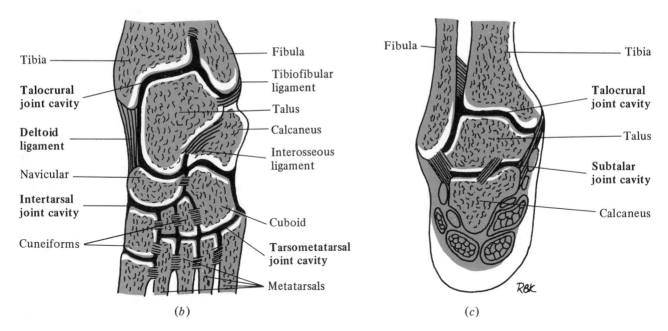

Diseases and Disorders of the Articular System

MECHANICAL INJURIES   In order of increasing severity, mechanical injuries of the articular system include strains, sprains, subluxations, and dislocations.

Strains and Sprains.   The strain is a mild joint injury in which muscles are overstretched but no swelling occurs. A sprain is a more serious injury in which a joint is wrenched or twisted, with partial rupture of its ligaments. Nearby blood vessels, muscles, tendons, or nerves may be damaged. Severe sprains are accompanied by pain and swelling, which prevents use of the joint, plus hemorrhage from ruptured blood vessels.

Subluxations.   A subluxation is a partial dislocation of a joint in which the articular surfaces of the bones are still in contact but are out of proper alignment with one another.

Dislocations.   In a dislocation the articular surface of one of the bones in a joint completely loses contact with the other bone(s). Dislocation of a joint is usually caused by trauma, though congenital bone malformations or laxity of ligaments may predispose certain joints to dislocate when little or no trauma is involved. Dislocations are characterized by pain, swelling, loss of motion of the affected joint, and sometimes shock. Common dislocations include those of a finger, the thumb, shoulder, or patella. Less common are dislocations of the mandible, elbow, hip, or knee.

INFLAMMATORY DISEASE   Among the inflammatory diseases of the articular system are certain types of arthritis and bursitis.

Arthritis.   Arthritis is a disease characterized by inflammation of synovial tissue. Several types of arthritis are listed next.

1. *Pyogenic Arthritis.*   Pyogenic arthritis is an acute infectious joint disease caused by pus-forming microorganisms. The synovial tissue develops an inflammatory response to the microorganisms, characterized by swelling and hypersecretion of synovial fluid.
2. *Rheumatoid Arthritis.*   Rheumatoid arthritis is a chronic systemic disease affecting connective tissue of which arthritis is the dominant clinical manifestation. It is especially common in young women and is thought to be an autoimmune disease (a disease in which the victim produces antibodies against his or her own tissues). Rheumatoid arthritis is characterized by pain, swelling, and stiffness of the affected joints. An invasive synovial tissue membrane destroys the articular cartilage and subchondral bone, which leads to ankylosis (fusion) of the bones of the affected joints.
3. *Gouty Arthritis.*   Gouty arthritis, which occurs mainly in men, results from an inherited defect of uric acid metabolism. In this disease uric acid crystals are frequently deposited in joints, where they cause inflammation and pain. The metatarsophalangeal joint of the big toe is the most frequent site of gouty arthritis attacks.

Bursitis.   Bursitis refers to inflammation of the bursae (the fluid-filled sacs that protect muscles and tendons crossing joints). It can be caused by trauma, prolonged or unusual exercise, infection, or chilling of the joint. Bursitis is very painful and severely limits joint mobility. Treatment commonly consists of injecting hydrocortisone, an antiinflammatory drug, directly into the inflamed bursa. Several examples of occupational bursitis are as follows:

1. *"Housemaid's Knee."* Prepatellar bursitis.
2. *"Student's Elbow."* Olecranon bursitis.
3. *"Postman's Heel."* Calcaneus bursitis.

DEGENERATIVE DISEASE Degenerative disease, called *osteoarthritis,* is the most common of all joint diseases. Osteoarthritis is a disease of middle age, the result of joint wear-and-tear. The disease primarily affects articular cartilages, especially those of weight-bearing bones. Affected articular cartilages soften and fray, exposing the subchondral bone to friction and pressure. Since hyaline cartilage does not regenerate, the subchondral bone hypertrophies in an attempt to protect itself. This causes the formation of bone spurs and surface irregularities that deform the joints and severely limit their mobility. Joints of the fingers, lumbar vertebrae, knee, hip, and ankle are most commonly affected by arthritis. The cheapest and most effective drug for treating osteoarthritis is aspirin, which alleviates joint pain and has an antiinflammatory effect.

Suggested Readings

° GARTLAND, JOHN J. *Fundamentals of Orthopaedics.* Philadelphia: Saunders, 1965.

KAPINDJI, I. A. *The Physiology of the Joints.* Vol. 1: *Upper Limb.* 2d ed. Translated by L. H. Honore. Edinburgh and London: Livingstone, 1970.

_____. *The Physiology of the Joints.* Vol. 2: *Lower Limb.* 2d ed. Translated by L. H. Honore. Edinburgh and London: Livingstone, 1970.

_____. *The Physiology of the Joints.* Vol. 3: *The Trunk and Vertebral Column.* 2d ed. Translated by L. H. Honore. Edinburgh and London: Churchill-Livingstone, 1974.

KOLODNY, A. LEWIS, and ANDREW R. KLIPPER. "Bone and joint diseases in the elderly," *Hosp. Pract., 11,* no. 11 (November 1976), pp. 91–101.

SOKOLOFF, LEON, and JOHN H. BLAND. *The Musculoskeletal System.* Baltimore: Williams & Wilkins, 1975.

° Advanced level.

7
The Muscular System

From the section on muscle tissue in Chapter 3, you may recall that there are three types of muscle fibers: smooth, cardiac, and skeletal. When anatomists talk about the muscular system, they are referring to the voluntary muscles, those muscles of the body that are composed of *skeletal muscle fibers* and are under control of the *somatic* or *voluntary nervous system*. Smooth or visceral muscle fibers, which are found in the walls of many organs, may be classified as muscle tissue but are not considered part of the muscular system proper.

There are over six hundred fifty voluntary muscles in the human body, an astounding figure. Realistically, no anatomy student needs to know the function of each separate muscle unless his or her career is intimately associated with the musculoskeletal system. However, it is important for all beginning anatomy students to have a good working knowledge of the approximately seventy-five pairs of muscles involved in general posture and body movements. In addition, they should have some knowledge of the muscles controlling eye movements, facial expression, speech, and the acts of chewing and swallowing.

The anatomy, development, function, and naming of muscles will be presented in Part A of this chapter. This will be followed by an overview of the major muscle groups, with emphasis on the clinically more important muscles, in Part B.

Part A. Muscles as Organs

Gross Anatomy

Skeletal muscles are organs that consist of a fleshy part or belly, composed primarily of skeletal muscle fibers, and tendons of attachment, composed of dense regular connective tissue (Fig. 7.1).

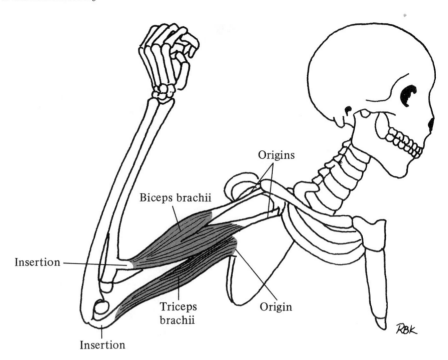

Figure 7.1 Muscle Attachments. Origins and insertions of biceps and triceps brachii indicated. Both muscles originate from scapula and humerus (parts that move least) and insert on radius and ulna (parts that move most).

ATTACHMENT OF MUSCLES

Muscles are attached to movable parts of the body. Most voluntary muscles attach to bones; however, there are some muscles that attach to other muscles, connective tissue septa, or even the skin, as do the muscles of facial expression.

Muscle attachment usually occurs by means of thick cordlike tendons or thin sheetlike tendons called *aponeuroses.* Some muscles do not attach by way of tendons. Instead, short strands of fibrous connective tissue attach individual muscle fibers directly to bone over an area as wide as the cross section of the muscle belly; this is called a *fleshy attachment.*

In anatomical terminology the attachments of muscles are called origins and insertions. The *origin* is the attachment to the body part that moves the least during muscle contraction, while the *insertion* is the attachment to the body part that moves the most. The origin is thus usually proximal, or closer to the trunk than the insertion. Where movement is possible in only one direction, the origins and insertions remain constant. However, where movement is possible in two directions, the origins and insertions are reversible, depending on which attachment site is *fixed* or immovable during a given movement. For example, the trapezius muscle of the back, which has attachment sites on the skull, vertebral column, and scapulae, can extend the head when the scapulae are fixed, elevate the scapulae when the head is fixed, and depress the scapulae when the vertebral column is fixed.

Students should not commit origins and insertions to rote memory; it is time-consuming and impractical. If you know the general location and action(s) of a given muscle, you will have a good working knowledge of that muscle.

INTRINSIC AND EXTRINSIC MUSCLES

Intrinsic muscles are short muscles whose origins and insertions are located within the structure on which they act. For example, the intrinsic muscles of the hand arise and insert on the bones of the hand.

Extrinsic muscles are long muscles whose origins are at some distance from the structure on which they act. For example, the extrinsic muscles of the hand arise from the arm or forearm bones.

MUSCLE PATTERNS

The arrangement of fiber bundles (fascicles) in skeletal muscles are visible to the naked eye. The arrangement of these fascicles with respect to their tendon(s) fall into four basic patterns (Fig. 7.2):

1. *Parallel.* In the parallel pattern, muscle fascicles lie parallel to the long axis of the muscle. *Flat muscles,* such as the epicranius muscle of the scalp, are wide flat sheets of muscle fascicles. *Strap muscles,* such as the sartorius muscle of the thigh, are long narrow bands of muscle fascicles. *Fusiform muscles,* such as the biceps muscle of the arm, are spindle-shaped (thick in the middle and tapered at either end); their fascicles are arranged in overlapping bundles rather than extending the entire distance from origin to insertion.
2. *Pennate.* In the pennate or featherlike pattern, muscle fascicles lie diagonal to the long axis of the muscle. This arrangement of muscle fascicles provides for greater strength of muscle contraction than the parallel arrangement. *Unipennate muscles,* such as the tibialis posterior muscle of the leg, have their fascicles inserted obliquely into one side of a central tendon and resemble half a feather. *Bipennate muscles,* such as the rectus femoris muscle of the thigh, have their fascicles inserted obliquely into both sides of a central tendon and resemble a whole feather. *Multipennate muscles,* such as the deltoid muscle of the shoulder, have their fascicles inserted obliquely into several tendons, which gives the muscle a herringbone appearance.
3. *Radiate.* In the radiate pattern, muscle fascicles from a broad origin converge on a slender central tendon of insertion. Radiate muscles, which are triangular in shape, include the temporalis muscle of the jaw and the pectoralis major muscle of the chest.
4. *Circular.* In the circular pattern, muscle fascicles form a circle around body orifices. Such muscles are called *sphincters.* Sphincters include the orbicularis oculi muscle, which encircles the eye, and the orbicularis oris muscle, which encircles the mouth.

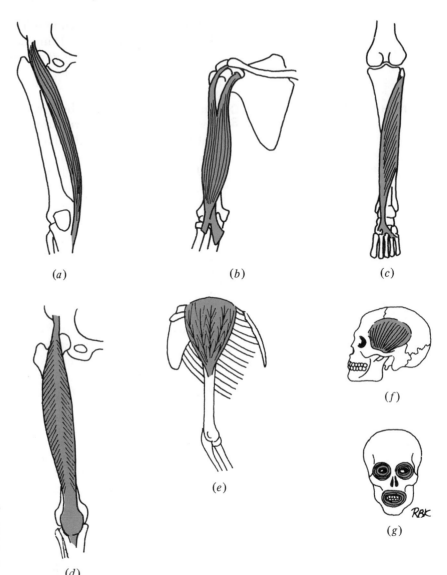

Figure 7.2 Muscle Patterns. (a) Strap (sartorius). (b) Fusiform (biceps brachii). (c) Unipennate (tibialis posterior). (d) Bipennate (rectus femoris). (e) Multipennate (deltoid). (f) Radiate (temporalis). (g) Circular (orbicularis oculi and oris).

ACCESSORY STRUCTURES

Bursae and tendon sheaths are types of synovial sacs that protect soft structures of the body from the effects of friction.

Bursae, previously mentioned in connection with the articular system, are closed sacs formed from a single layer of synovial membrane and are filled with a lubricating fluid. They develop after birth from spaces in the connective tissue at points of friction in the body (between muscle and muscle, muscle and bone, or bone and skin). In addition to friction bursae, there are bursae that facilitate the gliding of muscles or tendons over bony or ligamentous processes. Bursae are

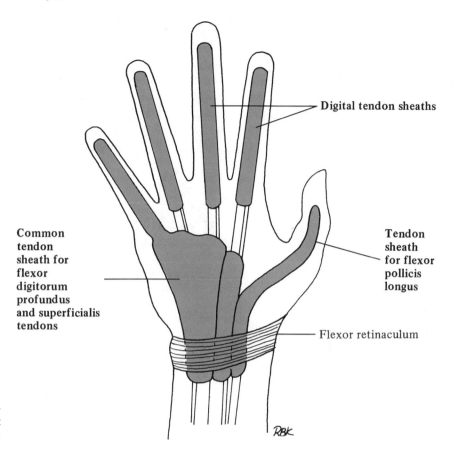

Digital tendon sheaths

Common tendon sheath for flexor digitorum profundus and superficialis tendons

Tendon sheath for flexor pollicis longus

Flexor retinaculum

Figure 7.3 *Palmar Tendon Sheaths.* Synovial tendon sheaths surround and protect digital flexor tendons in palm and fingers.

found for the most part in the extremities and their girdles. For example, many bursae occur around the knee joint. The suprapatellar bursa, which is not a friction bursa, lies between the femur and the tendon of the quadriceps femoris muscle. The prepatellar bursa lies between the patella and the skin. Other bursae occur deep to the gastrocnemius and semimembranosus muscles.

Tendon sheaths are tubular fluid-filled sacs, formed from a double layer of synovial membrane, that wrap around tendons like sleeves. The tendon sheath cushions the tendon and prevents it from fraying where it crosses over bone. Good examples of tendon sheaths are the sheaths of the flexor tendons of the wrist and fingers (Fig. 7.3).

Microscopic Anatomy

THE SKELETAL MUSCLE FIBER

The skeletal muscle fiber (Chapter 3) is a large, multinucleated, cylindrical cell ranging from 1 to 40 millimeters in length and from 10 to 100 millimeters in width. It is surrounded by a cell membrane

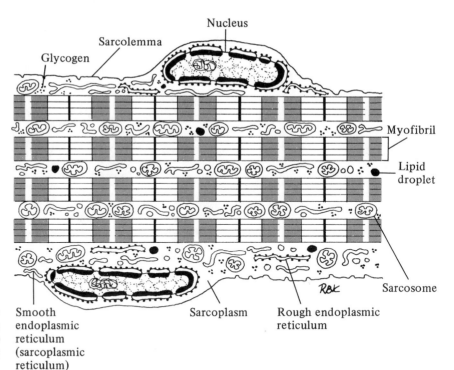

Figure 7.4 *Skeletal Muscle Fiber.* Longitudinal section of short segment showing arrangement of myofibrils in relation to other subcellular organelles. Drawing based on electron microscopic view.

called the *sarcolemma*. Within the muscle fiber cytoplasm, which is called *sarcoplasm*, are various organelles and inclusions (Fig. 7.4). Many flattened oval nuclei lie at the periphery of the fiber, just under the sarcolemma. These nuclei come from the embryonic myoblasts that previously fused to form the skeletal muscle fiber. Approximately 80 percent of the sarcoplasm is occupied by fine protein fibers called *myofibrils*. Individual myofibrils extend the entire length of the cell. Muscle mitochondria, called *sarcosomes*, lie between bundles of myofibrils. The sarcosomes produce energy for muscle contraction in the form of ATP (adenosine triphosphate), a high-energy phosphate molecule that is essential in the contraction process. *Sarcoplasmic reticulum* is a specialized type of smooth endoplasmic reticulum that forms an elaborate sleevelike structure around each myofibril (Fig. 7.5). The sarcoplasmic reticulum releases free calcium ions to the myofibrils following excitation of the sarcolemma, which is important in initiating the contraction process (excitation-contraction coupling). The principal sarcoplasmic inclusion is glycogen, a reserve energy source that is broken down by muscle mitochondria to supply ATP for muscle contraction.

High-magnification electron photomicrographs have revealed that the myofibril has a very complex structure. Each myofibril is made

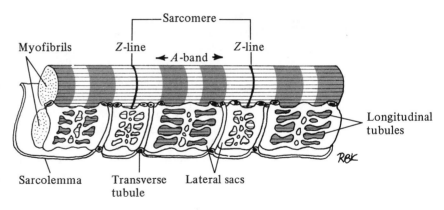

Figure 7.5 *Relationship of Sarcoplasmic Reticulum to Myofibril.* Sarcoplasmic reticulum forms sleevelike wrapping around each myofibril. Transverse tubules, continuous with sarcolemma, contact lateral sacs of sarcoplasmic reticulum at A-band–I-band junctions. They transmit electrical impulses that cause calcium ion release from lateral sacs of sarcoplasmic reticulum.

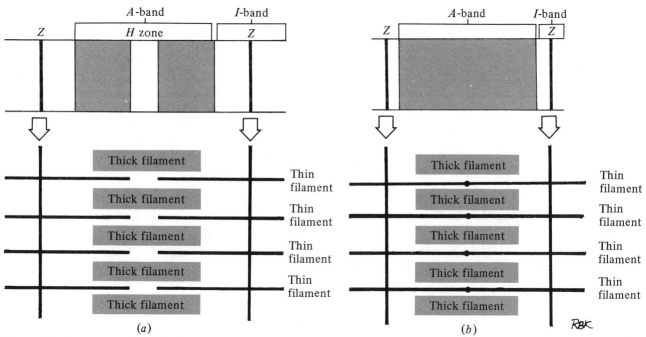

Figure 7.6 *Sarcomere.* (a) Relaxed sarcomere. (b) Contracted sarcomere. Inward movement of thin filaments during contraction brings Z-lines closer together and eliminates H-zone.

up of many short filaments, called *myofilaments*, that lie parallel to one another in overlapping arrays. There are two kinds of myofilaments: (1) thick filaments containing the protein myosin and (2) thin filaments containing the proteins *actin, tropomyosin,* and *troponin.* Thick and thin filaments within a myofibril overlap, giving the myofibril a cross-striated pattern. The myofibrils themselves lie in register with one another so that the striations appear to run across the entire muscle fiber.

THE SARCOMERE AND MUSCLE CONTRACTION

The functional unit of muscle contraction is called the *sarcomere.* There may be as many as forty-five hundred sarcomeres along the length of a single myofibril. Individual sarcomeres are separated from one another by a proteinaceous structure that extends transversely across the myofibril; this structure is referred to as the Z-line. The sarcomere, which extends from one Z-line to the next, is approximately 2.0 to 2.5 microns in length when relaxed.

The arrangement of thick and thin filaments within the sarcomere is shown in Fig. 7.6. The thick filaments lie in the middle of the sarcomere, parallel to the longitudinal axis of the myofibril, and do not reach the Z-lines. The thin filaments extend inward from the Z-lines, which are their attachment sites, but do not reach the center. According to the sliding-filament model of muscle contraction, the thin filaments move inward past the thick filaments, which remain stationary, to reach the center of the sarcomere. Movement of the thin filaments takes place in a series of small steps caused by the making and breaking of cross-bridges between the actin and myosin molecules. As the thin filaments move closer together, the distance between the Z-lines is reduced. The distance between Z-lines in the contracted sarcomere is about 1.65 microns. All the sarcomeres contract at the same time to shorten the myofibril, and all the myofibrils contract at the same time to shorten the muscle fiber. For a more extensive account of the complex sequence of events in muscle contraction, the student is referred to standard physiology and biochemistry textbooks.

THE HIGHER ORGANIZATION OF SKELETAL MUSCLE

The higher organization of skeletal muscle (Fig. 7.7) consists of the packaging of skeletal muscle fibers into progressively larger bundles. Individual muscle fibers are surrounded by a thin layer of reticular connective tissue called *endomysium.* The fibers are then organized into small bundles or fascicles, which are surrounded by a layer of areolar connective tissue called *perimysium.* An entire muscle, which consists of a bundle of muscle fascicles, is surrounded by a tough sheath of fibrous connective tissue called *epimysium.* The fibers of the various connective tissue sheaths of the muscle merge with its tendons, which may thus be considered continuations of the muscle's connective tissue.

Blood Supply

Skeletal muscles are highly vascular organs that are supplied by the branches of nearby medium-sized arteries and drained by their corresponding veins (see Chapter 8). Each artery divides into many smaller arterioles that pierce the epimysium and, in turn, divide into a vast number of small capillaries. These capillaries are dis-

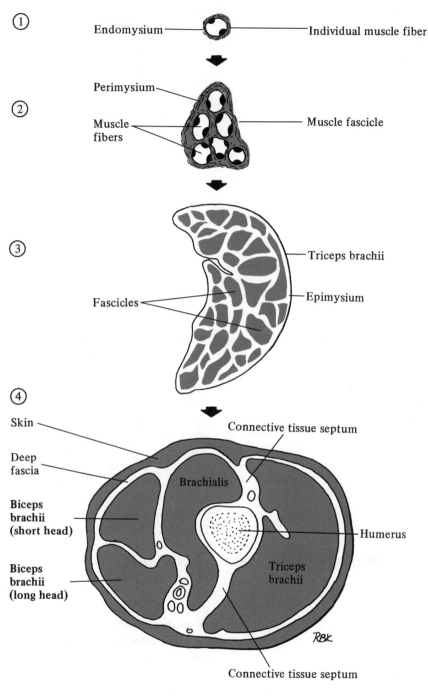

Figure 7.7 *Higher Organization of Skeletal Muscle*. (1) Individual muscle fiber. (2) Muscle fascicle. (3) Entire muscle (triceps brachii). (4) Arm muscles (biceps brachii, brachialis, triceps brachii).

Cross section of arm

tributed throughout the muscle by way of the perimysium and endomysium and ultimately form extensive networks that surround individual muscle fibers. The capillary walls, which are extremely thin, provide for rapid transfer of oxygen and nutrients from the blood to the muscle fiber and the by-products of muscle metabolism from the muscle fiber to the blood.

It is not necessary for beginning anatomy students to memorize the blood supply to individual muscles. General information is given in Chapter 8; specific information can be looked up in anatomy reference books if a more detailed account is necessary.

Innervation

Skeletal muscles are innervated by the peripheral nervous system, which consists of the spinal and cranial nerves. Very early in human embryonic development, each peripheral nerve makes contact with its corresponding muscle mass and maintains that relationship for life. Since muscle and nerve work together, the peripheral nerves are generally discussed along with the muscles they innervate. For this reason it is useful for the anatomy student to gain a general idea of muscle innervation before studying either the muscular or nervous systems in depth.

Peripheral nerves carry information to and from the central nervous system (CNS), which consists of the brain and spinal cord. Information to and from smooth and cardiac muscle is carried by involuntary, or *autonomic,* nerve fibers, while information to and from skeletal muscle is carried by voluntary, or *somatic,* nerve fibers. Nerve fibers carrying information to the CNS are called *afferent* or *sensory* fibers; those carrying information from the CNS are called *efferent* or *motor* fibers. The nerve that supplies a given muscle is called a motor nerve. Motor nerves are mixed; they contain three-fifths efferent (motor) fibers and two-fifths afferent (sensory) fibers.

SENSORY INNERVATION Skeletal muscles communicate with the CNS by way of sensory nerve fibers. Sensory information concerning musculoskeletal movements and posture is detected by a class of specialized nerve endings called *proprioceptors* and is transmitted primarily to the spinal cord and to certain parts of the brain, such as the cerebellum and cerebral cortex. This proprioceptive information permits the CNS to coordinate muscle movements and to help us maintain our equilibrium. Proprioceptors are located within muscle and joint tissues. Those within muscle itself include Golgi tendon organs and muscle spindles. The *Golgi tendon organ* consists of an encapsulated mass of nerve endings located at the junction of a muscle fiber and its tendinous attachment. Golgi tendon organs are stimulated by increases in tendon tension and cause inhibition of their own muscle whenever contraction increases tendon tension to the danger point. The *muscle spindle* consists of several *intrafusal* (within the spindle) muscle fibers and nerve endings enclosed in a connective tissue capsule. Muscle spindles are located between and parallel to the extrafusal fibers of the muscle. Sensory information concerning muscle tone and degree of muscle contraction is detected by specialized afferent nerve endings in the muscle spindle called *annulospiral endings* (Fig. 7.8). The annulospiral ending has as many branches as there are intrafusal fibers, and each branch coils around the midsection of an intrafusal fiber.

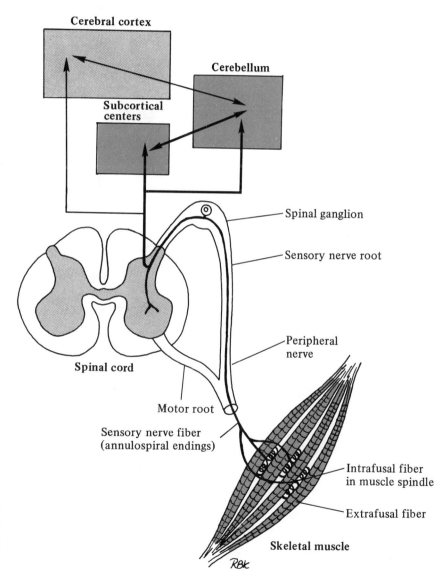

Figure 7.8 Sensory Innervation of Skeletal Muscle. Afferent limb of stretch reflex illustrates how one type of sensory information reaches spinal cord and higher CNS centers. Sensory receptor for stretch detection is annulospiral ending, which wraps around intrafusal fiber in muscle spindle.

Annulospiral endings are stimulated by changes in intrafusal fiber length and are responsible for the stretch reflex, which will be described shortly.

MOTOR INNERVATION   The CNS communicates with skeletal muscles by way of motor nerve fibers. The messages for voluntary muscle contraction originate in motor neurons in the cerebral cortex or other brain centers and are transmitted to motor neurons in the brain stem or spinal cord (see Chapter 15). The axons of these neurons, the motor nerve fibers, leave the CNS and are transported to their respective skeletal muscles by branches of the peripheral nerves (Fig. 7.9). One or more motor nerves

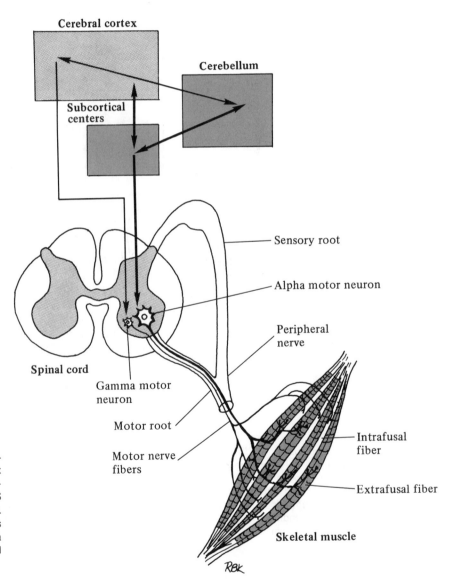

Figure 7.9 Motor Innervation of Skeletal Muscle. Efferent limb of stretch reflex illustrates how messages for muscle contraction from spinal cord and higher CNS centers reach typical skeletal muscle. Alpha motor neuron (large) innervates extrafusal muscle fibers, while gamma motor neuron (small) innervates intrafusal fibers.

pierce the epimysium and enter the substance of each skeletal muscle. These nerves contain a large number of motor nerve fibers, each of which is thought to serve a single muscle fascicle. At the level of the fascicle, the motor nerve fiber divides into a number of terminal branches, each of which terminates on and innervates a single muscle fiber. The motor neuron together with all the skeletal muscle fibers its axon (fiber) innervates constitutes a *motor unit*. The motor unit is the unit of function of the neuromuscular system. Under normal circumstances all the muscle fibers in the motor unit contract at the same time. The number of muscle fibers in a motor unit may vary from less than a hundred to more than a thousand, depending on the precision of the movements made by the particular muscle; the greater the precision of movement, the lower the ratio of muscle fibers to motor neurons.

The axon terminals of the motor neuron do not penetrate the muscle fibers they innervate. Instead, each axon terminal sits in a fluted depression created by infoldings of the sarcolemma. Beneath the sarcolemmal folds, the sarcoplasm contains a rich accumulation of mitochondria and muscle nuclei. This specialized region of the muscle fiber is called the *motor end plate*. There is a minute space or gap about 0.05 micron wide between the surfaces of the axon terminal and the muscle fiber. These structures — axon terminal, gap, and motor end plate — constitute the *neuromuscular* or *myoneural junction* (Fig. 7.10). Within the axon terminal are many vesicles containing *acetylcholine,* a neurotransmitter substance. Excitatory nerve impulses cause the release of acetylcholine from the axon terminal. It subsequently diffuses across the submicroscopic gap to the motor end plate, where it sets off a chain of events that leads to contraction of the muscle fiber.

REFLEXES Much of the information transmitted by sensory receptors does not reach higher brain centers; instead it is processed at the local segmental level in the spinal cord. The mechanism by which sensory information is processed locally and without volition is called a *reflex,* and the anatomical basis for reflex movements is the *reflex arc*. The reflex arc consists of a sensory neuron that enters the spinal cord and synapses with a motor neuron either directly or indirectly through one or more interneurons (connector neurons).

There are two major classes of spinal reflexes: exteroceptive and proprioceptive. *Exteroceptive reflexes* deal with sensory information coming from outside the body (heat, cold, touch, pressure, and pain). A good example is the *flexor reflex*. In the flexor (withdrawal) reflex, a body part is rapidly withdrawn from a source of pain. Since interneurons are located between the sensory and motor limbs of the flexor reflex arc, it is classified as a *polysynaptic reflex*. *Proprioceptive reflexes* deal with sensory information coming from the muscles and joints. A good example is the *stretch reflex* (Fig. 7.11). In a stretch

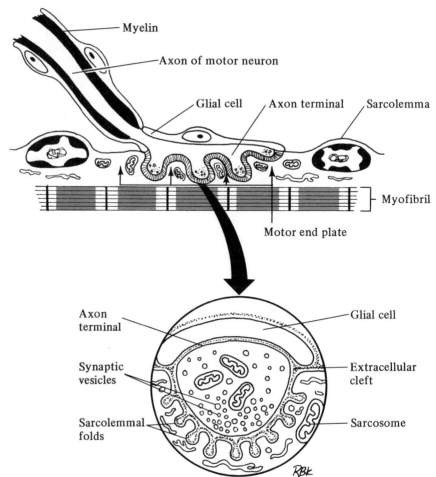

Figure 7.10 *Neuromuscular Junction.* Axon terminal (motor neuron), gap, and motor end plate (skeletal muscle fiber) constitute neuromuscular junction. For muscle contraction to occur, neurotransmitter substance (acetylcholine) must be released from synaptic vesicles in axon terminal, diffuse across submicroscopic gap between nerve and muscle, and attach to receptor sites on motor end plate. Drawing at *top* based on light microscopic view; drawing at *bottom* based on electron microscopic view.

reflex such as the *knee jerk* (patellar tendon) or *elbow jerk* (biceps tendon) reflex, the stretched muscle automatically contracts to a degree roughly proportional to the amount of stretch applied to it, which restores the muscle to its original length. Since no interneurons are located between the sensory and motor limbs of the stretch reflex arc, it is classified as a *monosynaptic reflex.*

Development of Skeletal Muscles

Like bone formation, the formation of skeletal muscles begins before birth (Fig. 7.12). The primordia (rudiments) of the musculoskeletal system consist of segmental blocks of condensed mesenchyme called *somites.* A portion of each somite differentiates into a muscle mass,

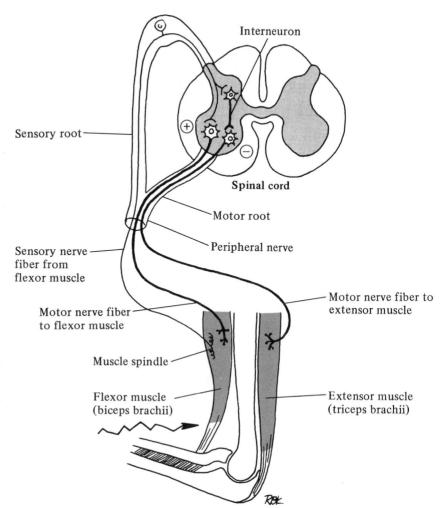

Figure 7.11 *Reflex Arc.* Elbow jerk, a stretch reflex, is initiated by blow on biceps tendon, which puts biceps muscle under stretch. Sensory neuron carries information from stretch receptor in biceps to spinal cord. There it makes excitatory synapse (+) with flexor motor neuron, which causes biceps muscle to contract, and inhibitory synapse (−) with extensor motor neuron, which causes triceps muscle to relax. Simultaneous excitation of one muscle and inhibition of its antagonist is called reciprocal innervation.

or *myotome,* which is innervated by a corresponding cranial or spinal nerve. Most of the skeletal muscles are derived from the myotomes of head somites, branchial (gill) arches, cervical somites, and trunk somites. The rest of the skeletal muscles develop from *lateral plate mesoderm,* which lies lateral to the somites.

Muscles of the head have a complex origin from myotomes of head somites and branchial arches and are innervated by cranial nerves. The extrinsic muscles of the eye and the muscles of the tongue are derived from the myotomes of head somites. Branchial arch myotomes give rise to the muscles of mastication (chewing), facial expression, swallowing, and speech.

Muscles of the neck and trunk are derived from the myotomes of cervical and trunk somites. During embryonic development the myotomic part of each of these somites divides into a small posterior (postaxial) mass called an *epimere* and a larger anterior (preaxial) mass called a *hypomere*. The epimeres give rise to the *epaxial* or extensor muscles of the neck and trunk. In the adult these are known as the intrinsic muscles of the back. The hypomeres give rise to the *hypaxial* or flexor muscles of the neck and trunk plus the intercostals, diaphragm, and muscles acting on the hyoid bone. Muscles of the neck and trunk are segmentally innervated by spinal nerves; extensor muscles are innervated by posterior primary rami and flexor muscles by anterior primary rami.

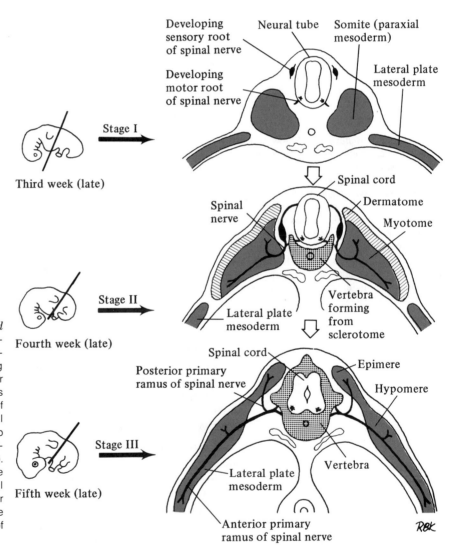

Figure 7.12 Stages in Development and Innervation of Skeletal Muscles. Schematic sections through trunk region of successively older human embryoes, showing development of muscle masses and their innervation. Stage I: formation of somites from paraxial mesoderm; differentiation of motor and sensory nerve roots of spinal nerve. Stage II: differentiation of somite into dermatome (dermis of skin), myotome (skeletal muscle), and sclerotome (vertebra). Stage III: division of myotome into epimere (epaxial muscles) and hypomere (hypaxial muscles). Epimere innervated by posterior primary ramus of spinal nerve; hypomere innervated by anterior primary ramus of spinal nerve.

The muscles of the extremities are derived either from limb bud mesenchyme that has differentiated in place or from mesenchyme that has migrated into the limb buds from adjacent hypomeres. Each limb bud in turn divides internally into an anterior or preaxial compartment and a posterior or postaxial compartment. In general, muscles that act as flexors lie in the anterior compartment of the extremity while muscles that act as extensors lie in the posterior compartment. It should be noted that due to the rotation of the lower extremity during fetal development (Chapter 5), the anterior compartment of the lower extremity ends up lying posteriomedially while the posterior com-

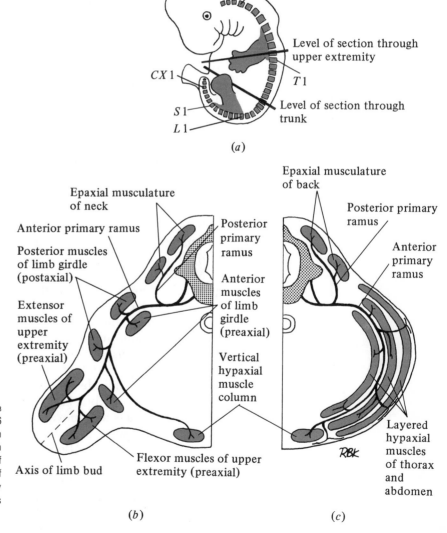

Figure 7.13 Innervation Patterns in Human Embryo. (a) Human embryo, 6 weeks old. (b) Cross section through upper extremity. (c) Cross section through trunk. Note that posterior primary ramus of spinal nerve innervates epaxial muscles of neck and trunk, while anterior primary ramus innervates hypaxial muscles plus muscles of limb girdle and extremity.

partment lies anterolaterally. Muscles of the extremities are innervated by the anterior primary rami of spinal nerves developing opposite the respective limb buds, which weave together to form nerve plexuses. The innervation patterns for major muscle masses in the upper extremity and trunk regions of the embryo are shown in Fig. 7.13.

Muscle Actions

The most important characteristic of muscle tissue is its ability to contract; this is the basis for all muscle actions. The type of movement a contracting muscle produces — flexion, extension, abduction, adduction or rotation — is determined by a number of factors, such as the muscle's attachment sites, the type of joint(s) over which the muscle acts, and the relation of the muscle's line of pull to the joint.

In the past, muscle actions have been deduced by several means: dissection of the muscles and the joints they cross, conjecture and reasoning from knowledge of muscle attachment sites and their relationship to the joints, and palpation and inspection of muscles in normal and paralyzed living subjects. Today, with the advent of sophisticated techniques such as *electromyography* (a technique for recording electrical impulses generated by contracting muscles), we are learning what individual muscles are actually doing instead of deducing what we think they do.

Muscles rarely act independently. Any given body movement is usually the result of coordinated action by a group of muscles. Group action modifies the actions ascribed to individual muscles in the group. For this reason group as well as individual actions are described in the overview of human muscles in Part B of this chapter.

Muscles are also classified on the basis of the roles they play in producing movements. The muscle that is responsible for producing a given movement is called a *mover* or *agonist*. For example, the triceps brachii is the mover in extension of the forearm. Movers are often subdivided into prime movers and assistant movers. The *prime mover* is the muscle that is primarily responsible for a given movement, while the *assistant mover* is the muscle that helps the prime mover. A given muscle can also be the mover in more than one type of movement. Biceps brachii, which acts as a mover in flexion of the forearm, also acts as a mover in supination of the hand. Muscles that act to stabilize some part of the body so that the movers can produce a given movement are called *stabilizers* or *fixators*. A *neutralizer* is a muscle whose action prevents an undesired action of one of the movers. A muscle whose action opposes the movement of the agonist or mover is called the *antagonist* (against the agonist). For example, triceps brachii, the extensor of the forearm, is the antagonist of biceps brachii and other forearm flexors.

Naming of Skeletal Muscles

Although muscle names are long and in Latin, many of them are descriptive; this will help you to remember where they are or what they do. Unfortunately, there is no common method for naming muscles. The early anatomists, who gave muscles their names, used many different criteria and followed no common plan. Some of the criteria used in naming muscles are listed next:

1. *Muscle Shape.* For example, trapezius (trapezoid), rhomboidius (rhomboid), teres (round).
2. *Size Relationships.* For example, gluteus maximus (largest gluteal muscle), gluteus medius (medium-sized gluteal muscle), gluteus minimus (smallest gluteal muscle).
3. *Location.* For example, pectoralis (of the chest), abdominis (of the abdomen), brachii (of the arm), femoris (of the thigh).
4. *Fiber Direction.* For example, rectus (longitudinal or straight), transversus (horizontal), oblique (diagonal).
5. *Number of Heads.* For example, biceps (two heads), triceps (three heads), quadriceps (four heads).
6. *Origins and Insertions.* For example, sternocleidomastoid (originates on the sternum and clavicle and inserts on the mastoid process), coracobrachialis (originates on the coracoid process and inserts on the humerus or arm bone).
7. *Action.* For example, flexor carpi radialis, extensor digitorum longus, adductor magnus, pronator teres, supinator, levator scapulae, depressor anguli oris.

Part B. **Muscles of the Human Body**

The major muscle groups of the human body (Figs. 7.14 and 7.15) are discussed in this part of the chapter. For each muscle group discussed there is a corresponding table that includes pertinent details such as origin, insertion, nerve, and action of all the muscles in the group. Due to limitations of space, only a few individual muscles can be described in any detail. Those muscles described in the body of the text and indicated by an asterisk (*) in the tables are the muscles that beginning anatomy students should know. Practical and relevant details concerning the muscles you should know are presented whenever possible to make the study of muscles more interesting and less of a chore.

For study purposes the muscular system is subdivided into muscles of the axial skeleton and muscles of the appendicular skeleton to parallel the two major divisions of the skeletal and articular systems. Muscle groups are described according to their function whenever

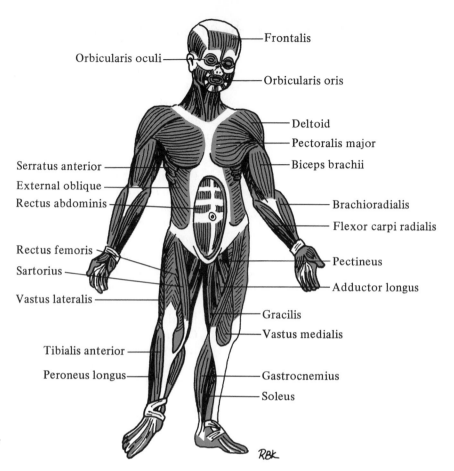

Orbicularis oculi

Frontalis

Orbicularis oris

Deltoid

Pectoralis major

Biceps brachii

Serratus anterior

External oblique

Rectus abdominis

Brachioradialis

Flexor carpi radialis

Rectus femoris

Sartorius

Pectineus

Adductor longus

Vastus lateralis

Gracilis

Vastus medialis

Tibialis anterior

Peroneus longus

Gastrocnemius

Soleus

RBK

Figure 7.14 Muscles of Human Body: Anterior Aspect.

possible; otherwise, they are described according to their anatomical location. Sometimes muscles in the same anatomical location have different functions; when this occurs the different functions will be explained in the text.

If you approach the study of the muscular system with a good background in the skeletal and articular systems, you will have little difficulty in understanding how muscles work. Although you will have visual aids to assist you in studying the muscles, it will be easier for you to understand muscular action by looking at a living body rather than looking at charts, models, or even a good cadaver dissection. You have a living body available for study—your own. Since many of the muscles and their tendons can be seen and felt through the skin, you can observe your own muscles in action when you perform the muscle movements described in the following sections. Stand in front of a mirror as you do this; see what happens when you clench your teeth or flex your forearm.

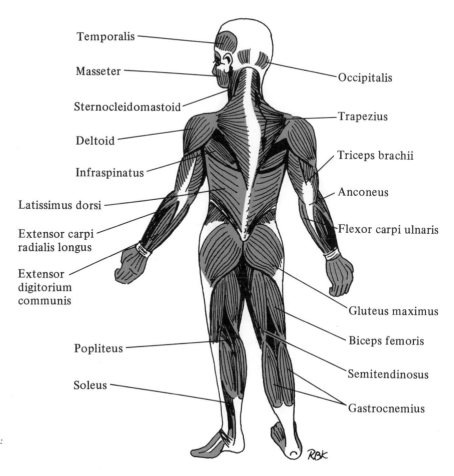

Temporalis

Masseter

Sternocleidomastoid

Deltoid

Infraspinatus

Latissimus dorsi

Extensor carpi
radialis longus

Extensor
digitorium
communis

Popliteus

Soleus

Occipitalis

Trapezius

Triceps brachii

Anconeus

Flexor carpi ulnaris

Gluteus maximus

Biceps femoris

Semitendinosus

Gastrocnemius

Figure 7.15 *Muscles of Human Body:
Posterior Aspect.*

Muscles of the Axial Skeleton

Muscles of the axial skeleton include muscles of the head, neck, and
trunk.

MUSCLES OF THE HEAD   Muscles of the head include the extrinsic muscles of the eye, the
muscles of facial expression, and the muscles of mastication. The
muscles of the tongue, palate, and pharynx will be discussed in
Chapter 12 as part of the digestive tract.

**Extrinsic
Muscles of the Eye**   The extrinsic muscles of the eye (Fig. 7.16) consist of six muscles that
move the eyeball plus a muscle that raises the upper eyelid (Table
7.1). The *superior, inferior, medial,* and *lateral rectus* muscles are
four small strap muscles that originate from the optic foramen and
insert into the sclera, the outermost coat of the eyeball. The *superior*

Superior

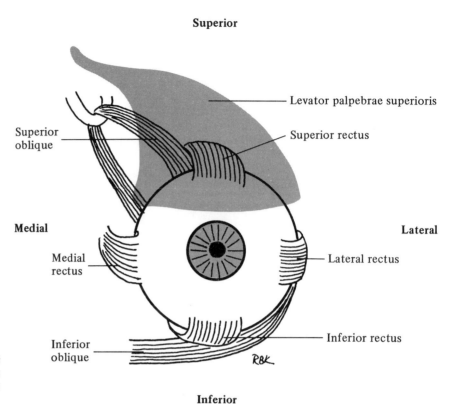

Levator palpebrae superioris

Superior rectus

Superior oblique

Medial

Lateral

Medial rectus

Lateral rectus

Inferior rectus

Inferior oblique

Figure 7.16 Extrinsic Muscles of Left Eye. Levator palpebrae superioris (*shaded*) has same developmental origin as superior rectus.

Inferior

and *inferior oblique* muscles arise from the roof and floor of the orbit, respectively, and also insert into the sclera. Movements of the eyeball include abduction (lateral rectus assisted by the superior and inferior obliques), adduction (medial rectus assisted by the superior and inferior recti), elevation (superior rectus and inferior oblique), depression (inferior rectus and superior oblique), medial rotation (superior oblique and superior rectus), and lateral rotation (inferior oblique and inferior rectus). The *levator palpebrae superioris* (upper eyelid lifter) opens the eye. It arises from the roof of the orbit and inserts into the upper eyelid. The extrinsic muscles of the eye are innervated by three cranial nerves: *oculomotor, trochlear,* and *abducens.* The lateral rectus is innervated by the abducens nerve, the superior oblique by the trochlear nerve, and the rest by the oculomotor nerve. *Ptosis* (drooping of the upper eyelid), caused by paralysis of levator palpebrae superioris, is a clinical sign of injury to the oculomotor nerve.

Muscles of Facial Expression   There are about thirty small muscles located beneath the skin of the scalp and face that act primarily on facial orifices. They arise from fascia or bone, lie in the subcutaneous fascia, and insert into the dermis

Table 7.1   Extrinsic Muscles of Eye

Muscle	Origin	Insertion	Nerve	Action
Superior rectus*	Fibrous cuff around optic foramen	Superior portion of sclera posterior to sclerocorneal junction	Oculomotor	Elevates and adducts eyeball; rotates eyeball medially
Medial rectus*	Fibrous cuff around optic foramen	Medial portion of sclera posterior to sclerocorneal junction	Oculomotor	Adducts eyeball
Inferior rectus*	Fibrous cuff around optic foramen	Inferior portion of sclera posterior to sclerocorneal junction	Oculomotor	Depresses and adducts eyeball; rotates eyeball laterally
Lateral rectus*	Fibrous cuff around optic foramen	Lateral portion of sclera posterior to sclerocorneal junction	Abducens	Abducts eyeball
Superior oblique*	Roof of orbital cavity	Tendon passes through fibrous ring (trochlea) to insert on sclera deep to superior rectus	Trochlear	Depresses eyeball along with inferior rectus and abducts it; rotates eyeball medially
Inferior oblique*	Floor of orbital cavity	On sclera between insertion of superior and lateral recti	Oculomotor	Elevates eyeball along with superior rectus and abducts it; rotates eyeball laterally
Levator palpebrae* superioris	Roof of orbital cavity	Upper eyelid	Oculomotor	Elevates upper eyelid

* Muscles that must be learned.

of the skin. Many of these muscles are indistinct from their neighbors. The muscles of facial expression are an important means of communication between people; they are listed in Table 7.2 and are illustrated in Fig. 7.17 along with the expressions they produce.

The *epicranius* is the principal muscle of the forehead and scalp. It consists of two flat muscles, the *frontalis* (forehead portion) and the *occipitalis* (scalp portion), which are connected by an aponeurosis called the *galea aponeurotica*. Contraction of the frontalis raises the eyebrows and throws the skin of the forehead into horizontal wrinkles, as in the expression of surprise.

The *orbicularis oculi* is the sphincter muscle of the eye and antagonist of levator palpebrae superioris. It closes the eyelids, which protects the eye from excessive light and irritation from foreign bodies. Full contraction of this muscle causes wrinkling of the skin around the eye, as seen in squinting. In old age these wrinkles, or "crow's-feet," become permanent lines radiating out from the corner of the eye.

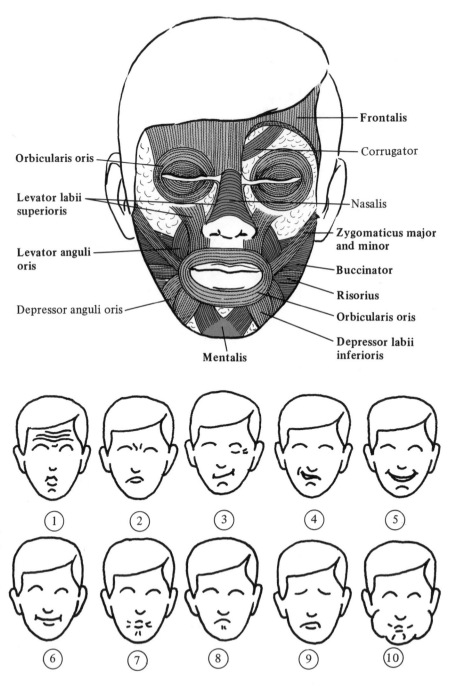

Figure 7.17 Muscles of Facial Expression. Some common facial expressions (cartoon faces) and prime muscles causing them are as follows: (1) surprise, frontalis; (2) frown, corrugator; (3) wink, orbicularis oculi; (4) sneer, levator labii superioris; (5) smile, zygomaticus major and minor; (6) grin, risorius; (7) pucker, orbicularis oris; (8) pout, mentalis; (9) grief, depressor anguli oris; (10) puffing and blowing, buccinator.

Table 7.2 Muscles of Facial Expression

Muscle	Origin	Insertion	Nerve	Action
Epicranius*				
Frontalis	Galea aponeurotica	Skin of forehead	Facial	Wrinkles skin of forehead horizontally; raises eyebrows
Occipitalis	Occipital bone	Galea aponeurotica	Facial	Wrinkles scalp
Procerus	Frontalis muscle	Skin between eyebrows	Facial	Wrinkles skin on bridge of nose; draws eyebrows downward
Corrugator	Frontal bone (superciliary arches)	Skin of eyebrows	Facial	Draws eyebrows together; wrinkles skin of forehead vertically
Orbicularis oculi*	Frontal, maxillary, and zygomatic bones; medial palpebral ligament	Skin encircling rim of orbit	Facial	Closes eyelids
Orbicularis oris*	Muscles inserting into lips	Skin encircling mouth	Facial	Closes mouth and puckers lips
Levator labii superioris	Maxilla and zygomatic bone below orbit	Upper lip	Facial	Elevates upper lip
Depressor labii inferioris	Mandible	Lower lip	Facial	Depresses lower lip
Zygomaticus major and minor*	Zygomatic arch	Angle of mouth	Facial	Draws angle of mouth up and back
Levator anguli oris	Maxilla	Angle of mouth	Facial	Elevates angle of mouth; curls upper lip
Risorius*	Masseter fascia	Angle of mouth	Facial	Retracts angle of mouth
Depressor anguli oris	Mandible	Angle of mouth	Facial	Depresses angle of mouth
Mentalis	Mandible	Skin of chin	Facial	Raises and protrudes lower lip
Buccinator*	Alveolar processes of maxilla and mandible, pterygomandibular raphe	Orbicularis oris	Facial	Compresses cheek against teeth; accessory muscle of mastication
Platysma*	Superficial fascia of chest	Mandible, skin of cheek, angle of mouth, and orbicularis oris	Facial	Depresses lower jaw, tenses skin of neck

* Muscles that must be learned.

The *orbicularis oris* is the sphincter muscle of the mouth. It closes the mouth and compresses the lips over the teeth. The orbicularis oris is also the muscle used to pucker the lips for whistling or kissing.

Many small muscles associated with movements of the mouth and lips converge on the mouth and insert into the orbicularis oris. Depending on their location these muscles raise the upper lip, depress the lower lip, draw the angle of the mouth upwards, as in smiling or laughing (*zygomaticus major*), retract the angle of the mouth, as in grinning (*risorius*), draw the angle of the mouth downward, as in grief (*depressor anguli oris*), or produce the pout (*mentalis*).

The *buccinator* is a deep facial muscle that forms the muscular wall of the cheek. It arises from the maxilla and mandible and inserts into the orbicularis oris at the angle of the mouth. The buccinator (trumpeter) muscles are responsible for compressing the cheeks to force air out of the mouth when one plays muscial instruments such as the clarinet or trumpet. They also assist the chewing process by keeping the cheeks in close contact with the teeth; this prevents food from being pocketed between the teeth and cheeks.

A superficial neck muscle that affects facial expression is the *platysma*. The platysma is a thin broad sheet of muscle extending from the fascia of the superficial chest muscles to the mandible; when contracted, it forms cords in the skin of the neck. It also retracts and depresses the angle of the mouth to produce the expressions of melancholy or horror. Loss of platysma tone is responsible for the sagging chin of middle age.

The muscles of facial expression are innervated by the *facial nerve*. Both central injury (stroke) and peripheral injury (Bell's palsy) to the facial nerve can produce paralysis of these muscles. People with unilateral (one-sided) facial paralysis have the following problems: (a) their mouth droops on the paralyzed side and they tend to drool (orbicularis oris and zygomaticus muscles affected); (b) they are unable to close the eye or elevate the brow on the paralyzed side (orbicularis oculi and frontalis affected); and (c) they have to use their fingers to remove food from between the teeth and cheek on the paralyzed side (buccinator affected). This condition is very embarrassing and upsetting to the affected person.

Muscles of Mastication   The muscles of mastication, which are illustrated in Fig. 7.18 and are described in Table 7.3, arise from the side or base of the skull and insert on the mandible. These muscles, which act across the temporomandibular joint, enable you to open or close your mouth or to clench your teeth as well as bite and chew.

The *masseter*, or chewing muscle, is a large thick quadrilateral muscle that runs diagonally between the zygomatic arch and the angle of the jaw. You can palpate this mouth-closing muscle above the angle of the jaw when you clench your teeth.

The *temporalis*, or biting muscle, is a large fan-shaped muscle that

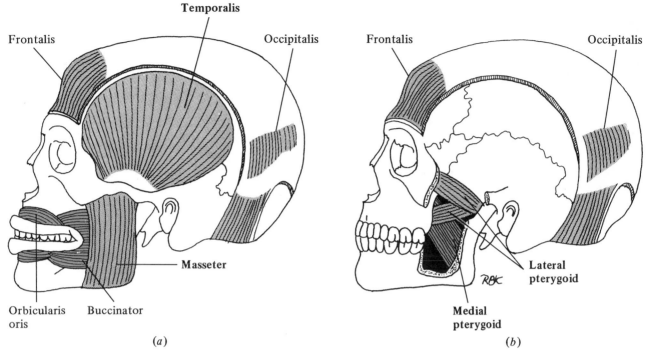

Temporalis

Frontalis   Occipitalis

Masseter

Orbicularis   Buccinator
oris

(a)

Frontalis   Occipitalis

RBK

Lateral
pterygoid

Medial
pterygoid

(b)

Figure 7.18   Muscles of Mastication. (a) Superficial. (b) Deep. Temporalis and masseter muscles plus portions of zygomatic arch and ramus of mandible are cut away to expose pterygoid muscles.

has a fleshy origin from the temporal fossa (a shallow depression on the side of the skull). It runs downward, deep to the zygomatic arch, to insert on the coronoid process of the mandible. You can palpate this powerful biting and mouth-closing muscle by pressing your fingers against your temples as you clench your teeth.

The pterygoid muscles lie deep to the other muscles of mastication; they run from the undersurface of the sphenoid bone to the internal surface of the mandible. The *medial pterygoid,* which inserts on the angle of the jaw, elevates the mandible and assists the masseter and temporalis in closing the mouth, while the *lateral pterygoid,* which inserts on the mandibular condyle, protracts the mandible and assists in opening the mouth. Studies of the muscles of mastication indicate that the masseter and medial pterygoid act together with the temporalis muscle to close the mouth in biting and chewing. Side-to-side grinding movements, such as used in chewing, are produced by alternate side activity of the lateral pterygoids.

The muscles of mastication are innervated by the *trigeminal nerve.* Injury to the motor division of the trigeminal nerve causes the jaw to deviate to the side of the paralyzed muscles when the mouth is opened.

MUSCLES OF THE NECK

Muscles of the neck fall into two major categories: those that move the hyoid bone in the act of swallowing and those that move the head and neck.

Table 7.3 Muscles of Mastication

Muscle	Origin	Insertion	Nerve	Action
Masseter*	Zygomatic arch	Ramus of mandible	Trigeminal	Elevates and protracts mandible
Temporalis*	Temporal fossa	Coranoid process of mandible	Trigeminal	Elevates and retracts mandible
Medial pterygoid	Medial side of lateral pterygoid plate	Angle of jaw (medial surface)	Trigeminal	Elevates and protracts mandible
Lateral pterygoid	Greater wing of sphenoid and lateral side of lateral pterygoid plate	Neck of condyloid process of mandible	Trigeminal	Depresses and protracts mandible; moves mandible from side to side

* Muscles that must be learned.

Muscles Moving the Hyoid Bone The muscles that act on the hyoid bone are located either beneath the mandible or on the anterior aspect of the neck and lie deep to platysma (Fig. 7.19). These muscles can be divided into two groups; the suprahyoid muscles and the infrahyoid muscles (Table 7.4).

Suprahyoid Muscles The suprahyoid muscles lie above the hyoid bone and move the hyoid bone and tongue. They consist of the digastric, mylohyoid, and geniohyoid muscles. The *digastric muscle*, as its name implies, has two bellies: an anterior belly, which is attached to the mandible, and a posterior belly, which is attached to the mastoid process of the temporal bone. The two bellies are connected by an intermediate tendon that passes through a fibrous sling attached to the hyoid bone. As a group the suprahyoid muscles are concerned with the act of swallowing. By first raising the hyoid bone and tongue, then retracting the hyoid bone, these muscles assist the muscles of the tongue and soft palate in passing food into the pharynx and preventing its return into the oral cavity. The digastric also depresses the mandible, which action opens the mouth.

Infrahyoid Muscles The infrahyoid muscles lie below the hyoid bone and move the hyoid bone and larynx. This group consists of the *omohyoid, sternohyoid, sternothyroid,* and *thyrohyoid* muscles. These muscles, the extrinsic muscles of the larynx, are primarily strap muscles. Like the digastric, the *omohyoid* has two bellies; a superior belly, which is attached to the hyoid bone, and an inferior belly, which is attached to the scapula. The two bellies are connected by an intermediate tendon that passes through a fibrous sling attached to the clavicle. Group actions of the infrahyoid muscles consist of depressing the hyoid bone and larynx

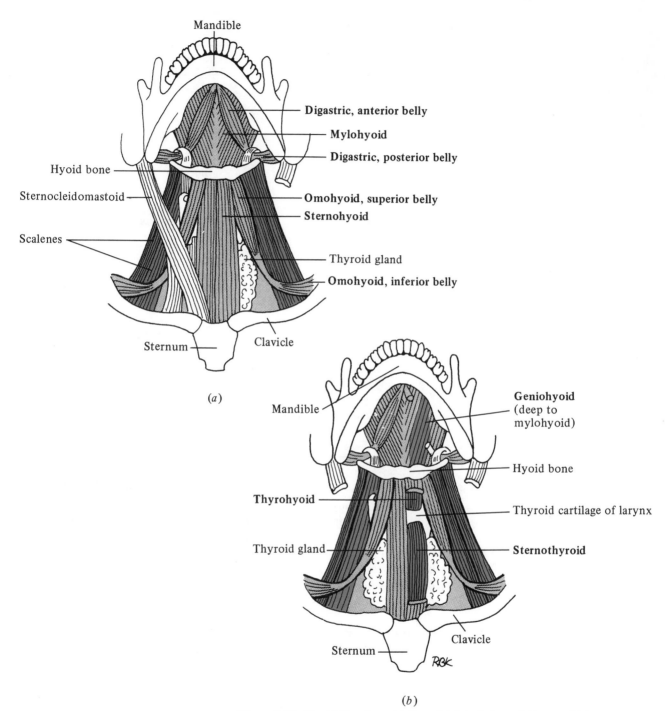

Figure 7.19 *Hyoid Muscles.* (a) Superficial. Left sternocleidomastoid is cut away so that infrahyoids can be seen more clearly. (b) Deep. Left digastric and sternohyoid muscles are cut away to expose deeper-lying muscles.

Table 7.4   Muscles Moving Hyoid Bone and Larynx

Muscle	Origin	Insertion	Nerve	Action
SUPRAHYOID MUSCLES				
Digastric* (two bellies)	Mastoid process of temporal bone	Mandible near mental symphysis	Trigeminal, anterior belly; facial, posterior belly	Elevates hyoid and base of tongue; depresses mandible
Mylohyoid	Mylohyoid line on mandible	Body of hyoid	Trigeminal	Elevates hyoid and floor of oral cavity
Geniohyoid	Genial tubercle of mandible	Body of hyoid	C1–C2 via hypoglossal	Elevates hyoid and draws base of tongue forward
INFRAHYOID MUSCLES				
Omohyoid* (two bellies)	Suprascapular notch	Lower border of body of hyoid	C1–C3 via ansa cervicalis	Depresses hyoid and larynx
Sternohyoid	Manubrium and medial end of clavicle	Lower border of body of hyoid	C1–C3 via ansa cervicalis	Depresses hyoid and larynx
Sternothyroid	Posterior surface of manubrium	Thyroid cartilage (oblique line)	C1–C3 via ansa cervicalis	Depresses thyroid cartilage and larynx
Thyrohyoid	Thyroid cartilage (oblique line)	Greater horn of hyoid	C1–C2 via hypoglossal	Depresses hyoid; elevates thyroid cartilage

* Muscles that must be learned.

after they have been elevated by the suprahyoids in the act of swallowing. They also assist the suprahyoid muscles in opening the mouth and flexing the head and neck.

Muscles Moving the Head and Neck   The principal movements of the head and neck consist of flexion (bending forward), lateral flexion (bending to the side), extension, and rotation. The muscles that perform these movements can be divided into prevertebral, postvertebral, and lateral vertebral groups.

Prevertebral Muscles   The prevertebral, or hypaxial, muscles of the neck are located on the anterior side of the cervical vertebrae, deep to the larynx and trachea (Fig. 7.20). They consist of the *rectus capitis anterior, rectus capitis lateralis, longus capitis,* and *longus colli* (Table 7.5). Group actions of the prevertebral neck muscles consist of flexion, lateral flexion, and rotation of the head and neck. Longus capitis and rectus capitis anterior are the chief antagonists of the postvertebral muscles of the neck, which act as extensors.

Table 7.5 Muscles Moving Head and Neck

Muscle	Origin	Insertion	Nerve	Action
PREVERTEBRAL MUSCLES				
Rectus capitis anterior	Lateral mass of atlas	Occipital bone	C1–C2	Flexes and rotates head
Rectus capitis lateralis	Transverse process of atlas	Occipital bone	C1–C2	Flexes and bends head laterally
Longus capitis	Transverse processes of second to sixth cervical vertebrae	Occipital bone	C1–C4	Flexes and rotates head
Longus colli	Transverse processes and bodies of third cervical to third thoracic vertebrae	Bodies of first four cervical vertebrae and transverse processes of fifth and sixth cervical vertebrae	C2–C6	Flexes and rotates head
POSTVERTEBRAL MUSCLES				
Obliquus capitis superior	Transverse process of atlas	Occipital bone	C1	Extends head
Obliquus capitis inferior	Spine of axis	Transverse process of atlas	C1	Extends and rotates head
Rectus capitis posterior major	Spine of axis	Occipital bone	C1	Extends and rotates head
Rectus capitis posterior minor	Posterior tubercle of atlas	Occipital bone	C1	Extends head
Splenius capitis	Nuchal ligament and spines of upper six thoracic vertebrae	Mastoid process and occipital bone	C2–C5	Extends head and rotates head and neck
Splenius cervicis	Nuchal ligament and spines of third through sixth cervical vertebrae	Transverse processes of first three cervical vertebrae	C2–C5	Extends head and bends it laterally
Semispinalis capitis and cervicis	(See Table 7.6)	—	—	—
LATERAL VERTEBRAL MUSCLES				
Sternocleidomastoid*	Manubrium and medial end of clavicle	Mastoid process of temporal bone	Spinal accessory, C2–C3	Unilateral, bends neck to the side and rotates head to opposite side; bilateral, flexes head
Anterior scalene*	Transverse processes of third through sixth cervical vertebrae	Scalene tubercle on first rib	C5–C7	Elevates first rib; bends neck laterally

Table 7.5 *(Continued)*

Muscle	Origin	Insertion	Nerve	Action
Middle scalene*	Transverse processes of second through seventh cervical vertebrae	Upper surface of first and second ribs	C5–C8	Elevates first and second ribs; bends neck laterally
Posterior scalene*	Transverse processes of fifth through seventh cervical vertebrae	Outer surface of second rib	C7–C8	Elevates second rib; bends neck laterally

* Muscles that must be learned.

Postvertebral Muscles The postvertebral, or epaxial, muscles of the neck are located on the posterior side of the cervical vertebrae. Some of these muscles are intrinsic to the neck while others are upward continuations of the deep muscles of the back.

The *suboccipital* muscles (Fig. 7.21) comprise the deepest group of postvertebral neck muscles. They include rectus capitus posterior major and minor. Group actions of the suboccipital muscles, which cross the atlantooccipital or atlantoaxial joints, consist of lateral

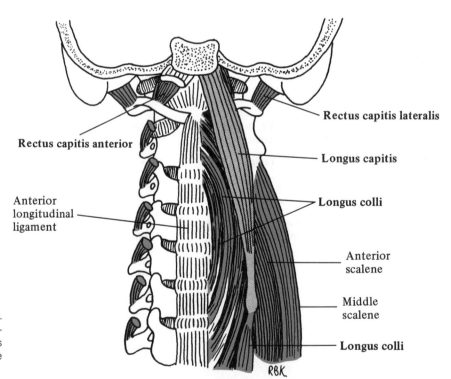

Rectus capitis lateralis

Rectus capitis anterior

Longus capitis

Anterior longitudinal ligament

Longus colli

Anterior scalene

Middle scalene

Longus colli

RBK

Figure 7.20 *Prevertebral Neck Muscles.* Right longus capitis and longus colli muscles are cut away so that rectus capitis anterior and rectus capitis lateralis can be seen more clearly.

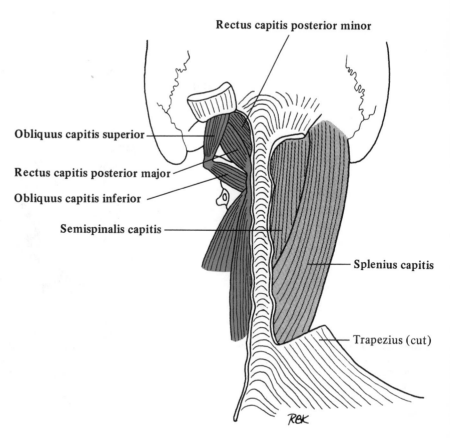

Rectus capitis posterior minor

Obliquus capitis superior

Rectus capitis posterior major

Obliquus capitis inferior

Semispinalis capitis

Splenius capitis

Trapezius (cut)

RBK

Figure 7.21 Postvertebral Neck Muscles. Trapezius is cut away on right side to expose splenius capitis and semispinalis capitis. Splenius capitis and semispinalis capitis are cut away on left side to expose suboccipital muscles.

flexion and rotation of the head to the same side when they act unilaterally and extension of the head when they act bilaterally.

The remaining postvertebral neck muscles lie deep to the trapezius (a back muscle) and superficial to the suboccipital muscles. They include *splenius capitus* and *cervicis* plus *semispinalis capitis* and *cervicis*. These muscles arise from upper thoracic or lower cervical vertebrae and insert into upper cervical vertebrae or the occipital bone. Group actions of these important postural muscles include lateral flexion and rotation of the head and neck to the same side when they act unilaterally and extension of the head and neck when they act bilaterally.

Lateral Vertebral Muscles

The lateral vertebral muscles (Fig. 7.22) are hypaxial muscles located lateral to the cervical vertebrae. They include the sternocleidomastoid and scalene muscles (Table 7.5).

The *sternocleidomastoid* is a superficial muscle that runs diagonally up the neck from the sternum and clavicle to the mastoid process of the temporal bone. You can palpate this muscle for its entire length when you press your chin against the palm of your hand. The sterno-

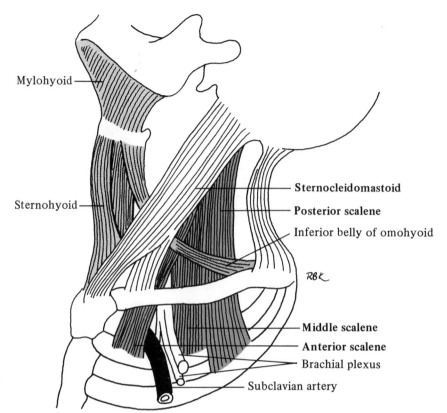

Mylohyoid

Sternohyoid

Sternocleidomastoid

Posterior scalene

Inferior belly of omohyoid

RBK

Middle scalene

Anterior scalene

Brachial plexus

Subclavian artery

Figure 7.22 Lateral Vertebral Neck Muscles. Three scalenes lie deep to sternocleidomastoid. Note that subclavian artery and nerves of brachial plexus exit from neck netween anterior and middle scalenes.

cleidomastoid causes flexion of the head when it acts bilaterally and lateral flexion and rotation of the head to the opposite side when it acts unilaterally. When the head is fixed, it may also act as an assistant respiratory muscle. *Torticollis* or *wryneck* is a condition caused by a spasm of or injury to the sternocleidomastoid muscle. In this condition the head is fixed in lateral flexion and rotation.

The scalene muscles consist of *scalenus anterior, medius,* and *posterior.* Together they form a triangular mass that extends from the transverse processes of the cervical vertebrae to the first and second ribs. When the head and neck are fixed, the scalenes act from above to elevate the first two ribs and increase the vertical diameter of the thoracic cage. When the ribs are fixed, the scalenes act from below to flex the head and neck. The neurovascular trunk to the upper extremity, which consists of the nerves of the brachial plexus and the subclavian artery, passes between the middle and anterior scalenes on its way to the axilla. The *anterior scalene syndrome,* which is characterized by numbness and poor circulation in the upper extremity, is due to compression of the neurovascular trunk against the first rib by a hypertrophic anterior scalene muscle.

MUSCLES OF THE TRUNK Muscles of the trunk include the muscles of the back, muscles of the thorax, and muscles of the abdomen.

Muscles of the Back The muscles of the back can be subdivided into three groups: superficial, intermediate, and deep.

Superficial Back Muscles The superficial group of back muscles includes the trapezius and latissimus dorsi. Since these muscles act primarily on the shoulder girdle and arm, they will be discussed with the muscles of the appendicular skeleton.

Intermediate Back Muscles The intermediate group of back muscles includes the serratus posterior superior and serratus posterior inferior. Since these muscles act on the ribs, they will be discussed with the muscles of the thorax.

Deep Back Muscles The deep group of back muscles (Fig. 7.23) are of epaxial origin, are postvertebral in location, act on the vertebral column, and function as a single unit. They are further subdivided into longitudinal and transversospinal groups on the basis of fiber direction.

The *longitudinal* group of deep back muscles is known collectively as the *erector spinae complex* or *sacrospinalis*. It is a broad thick muscle mass that extends from the sacrum to the skull and whose fibers run longitudinally. At vertebral level *L*5 it splits into lateral, intermediate, and medial columns, which are subdivided further according to their vertebral level of insertion (Table 7.6). The lateral column, called the *iliocostalis*, ascends to the neck. The intermediate column, called the *longissimus*, is the longest of the three columns and ascends to the mastoid process of the temporal bone. The medial and smallest of the three columns, called the *spinalis*, ascends to the occipital bone. The sacrospinalis is important in maintaining our erect posture. When it acts bilaterally, it extends the vertebral column and is the chief antagonist of the abdominal flexors; when it acts unilaterally, it causes lateral flexion of the vertebral column. The sacrospinalis is innervated by the posterior primary rami of segmental spinal nerves.

The *transversospinal* group of deep back muscles consists of the *semispinalis*, *multifidi*, and *rotatores*. These muscles, whose fibers run superomedially as they ascend the vertebral column, lie deep to the sacrospinalis and occupy the gutter between the transverse processes and spines of the vertebrae (Table 7.6). Instead of lying side by side, muscles of the transversospinal group are stacked one on top of the other like layers in a sandwich. The key to understanding these muscles is that the deeper they lie, the fewer vertebral segments they span. The transversospinal muscles extend the vertebral column when they act bilaterally, and they laterally flex and rotate the vertebral column to the opposite side when they act unilaterally.

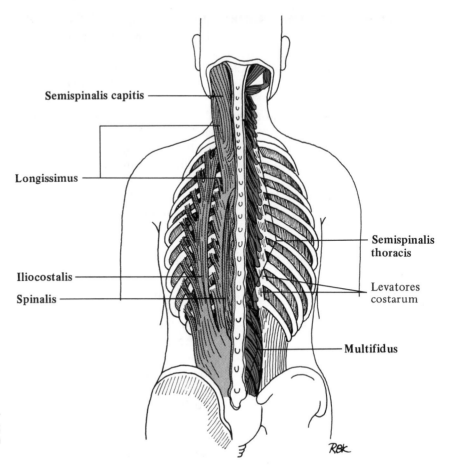

Semispinalis capitis

Longissimus

Iliocostalis

Spinalis

Semispinalis thoracis

Levatores costarum

Multifidus

Figure 7.23 Deep Back Muscles. Longitudinal muscles are shown on left side; transversospinal muscles are shown on right side.

They are also innervated by the posterior primary rami of segmental spinal nerves.

Deep to the transversospinal group are the interspinales and intertransversarii. These paired small muscles run between adjacent vertebrae and assist the longitudinal and transversospinal groups in extension and lateral flexion of the vertebral column.

Muscles of the Thorax   The muscles that form the walls of the thorax or act on the ribs serve the process of respiration by altering the diameter of the thoracic cavity. These muscles include the intermediate back muscles and the muscles of the thorax proper.

Intermediate Back Muscles   The intermediate back muscles (Table 7.7) include two thin quadrilateral sheets of muscle, which are separated from the underlying sacrospinalis by the nuchal and lumbodorsal fasciae. *Serratus pos-*

Table 7.6 Deep Muscles of Back

Muscle	Origin	Insertion	Nerve	Action
LONGITUDINAL GROUP				
Iliocostalis (lumborum, thoracis, cervicis)*	Iliac crest and sacrospinal aponeurosis and third to twelfth ribs	Angles of the ribs and transverse processes of fourth to seventh cervical vertebrae	Posterior primary rami of segmental spinal nerves	Unilateral, bends vertebral column to the side; bilateral, extends vertebral column
Longissimus (thoracis, cervicis, capitis)*	Sacrospinal aponeurosis, transverse processes of upper two lumbar, and all thoracic and cervical vertebrae	Angles of upper seven ribs, transverse processes of all thoracic and cervical vertebrae, and mastoid process of temporal bone	Posterior primary rami of segmental spinal nerves	Unilateral, bends vertebral column to the side; bilateral, extends vertebral column
Spinalis (thoracis, cervicis)*	Spines of upper two lumbar and lower two thoracic vertebrae; spines of upper two thoracic and lower two cervical vertebrae	Spines of second through ninth thoracic vertebrae; spines of second through fourth cervical vertebrae	Posterior primary rami of segmental spinal nerves	Unilateral, bends vertebral column to the side; bilateral, extends vertebral column
TRANSVERSOSPINAL GROUP				
Semispinalis (thoracis, cervicis, capitis)* (span four to six segments)	Transverse processes of all thoracic vertebrae and seventh cervical vertebrae	Spines of upper six thoracic vertebrae and all cervical vertebrae (except atlas) plus occipital bone	Posterior primary rami of segmental spinal nerves	Unilateral, bends head to the side; bilateral, extends vertebral column
Multifidi: (span about three segments)	Sacrum and transverse processes of lumbar, thoracic, and lower cervical vertebrae	Spines of lumbar, thoracic, and lower cervical vertebrae	Posterior primary rami of segmental spinal nerves	Extend and rotates vertebral column
Rotatores (span one segment)	Transverse processes from sacrum to second cervical vertebrae	Lamina above vertebra of origin	Posterior primary rami of segmental spinal nerves	Extend and rotate vertebral column

* Muscles that must be learned.

terior superior, which extends diagonally downward from the upper part of the vertebral column to the second to fifth ribs, helps elevate the ribs and increases the diameters of the thoracic cavity; *serratus posterior inferior,* which extends diagonally upward from the lower part of the vertebral column to the ninth to twelfth ribs, helps depress the ribs and decrease the diameters of the thoracic cavity.

Muscles of the Thorax Proper   The muscles of the thorax proper (Table 7.8) include the levatores costarum, small muscles thought to help elevate the ribs, the intercostal muscles, and the thoracic diaphragm. These muscles all have respiratory functions.

Table 7.7 Intermediate Muscles of Back

Muscle	Origin	Insertion	Nerve	Action
Serratus posterior superior*	Ligamentum nuchae, supraspinous ligament and spines of cervical and upper thoracic vertebrae	Upper borders of second to fifth ribs lateral to angles	Posterior primary rami of T1–T4	Elevates ribs, increases diameters of thoracic cage
Serratus posterior inferior*	Supraspinous ligament and spines of lower two thoracic and upper three lumbar vertebrae	Lower borders of ninth to twelfth ribs lateral to angles	Posterior primary rami of T9–T12	Depresses ribs, decreases diameters of thoracic cage

* Muscles that must be learned.

Table 7.8 Muscles of Thorax

Muscle	Origin	Insertion	Nerve	Action
External intercostals* (11 pairs)	Lower border of rib	Upper border of rib below origin	Intercostals	Elevate ribs and increase diameters of thoracic cavity
Internal intercostals* (11 pairs)	Upper border of rib	Lower border of rib above origin	Intercostals	Depress ribs and decrease diameters of thoracic cavity
Innermost intercostals	Variable in extent; sometimes considered as deep portion of internal intercostals, being separated by intercostal vessels and nerves; origin and insertion same as internal intercostals		Same as above	Same as above
Transversus thoracis	Inner surface of lower third of sternum and xiphoid process	Inner surface of costal cartilages of second to sixth ribs	Intercostals	Draws costal cartilages downward and decreases diameters of thoracic cavity
Subcostals	Inner surface of lower ribs near their angle	Inner surface of second or third rib below rib of origin	Intercostals	Draws adjacent ribs together
Thoracic diaphragm*	Slips from xiphoid process, inner surfaces of ninth through twelfth ribs, and bodies of upper lumbar vertebrae	Central tendon	Phrenic (C3–C5)	Increases vertical diameter of thoracic cavity

* Muscles that must be learned.

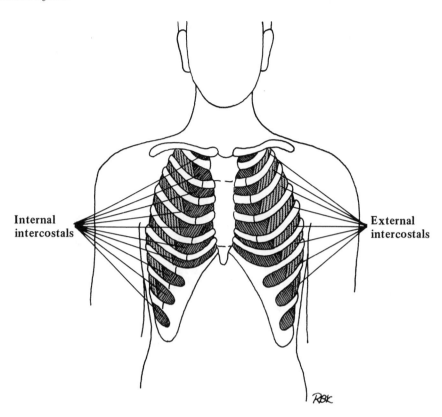

Internal intercostals

External intercostals

Figure 7.24 Intercostal Muscles. Superficial chest muscles are removed to expose external and internal intercostals.

Intercostal Muscles

The *intercostal muscles,* which form the muscular portion of the thoracic wall, lie deep to the pectoral muscles anteriorly and the back and shoulder girdle muscles posteriorly. They are segmentally innervated by the intercostal and subcostal nerves, which are the anterior primary rami of the thoracic spinal nerves. The intercostal muscles are split into superficial, middle, and deep layers (Fig. 7.24).

The superficial layer consists of eleven pairs of *external intercostals,* which fill in the spaces between the ribs from the vertebral column to the costochondral junction; their fasciae extend to the sternum as the anterior intercostal membranes. These muscles, whose fibers run down and inward, arise from the lower border of each rib except the last and insert on the upper border of the rib below. The external intercostals enlarge the vertical and transverse diameters of the thoracic cage by elevating the ribs.

The middle layer of intercostal muscles consists of eleven pairs of *internal intercostals,* which fill in the spaces between the ribs from the sternum to the angle of the rib; their fasciae continue to the vertebral column as the posterior intercostal membranes. These muscles, whose fibers run up and outward, arise from the upper border of each rib and insert on the lower border of the rib above. The internal in-

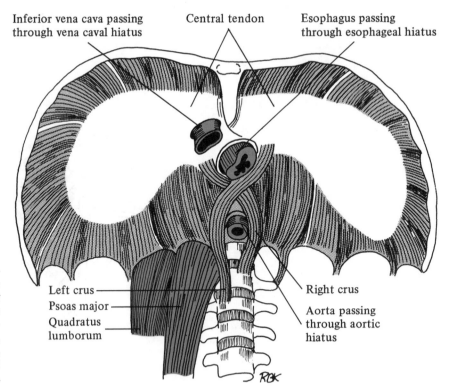

Inferior vena cava passing through vena caval hiatus

Central tendon

Esophagus passing through esophageal hiatus

Left crus
Psoas major
Quadratus lumborum

Right crus

Aorta passing through aortic hiatus

Figure 7.25 Thoracic Diaphragm: Inferior Aspect. Radially arranged diaphragmatic fibers arise from sternum, lower six ribs, and lumbar crura and insert into central tendon. There are three large openings in thoracic diaphragm: aortic hiatus, which transmits aorta, esophageal hiatus, which transmits esophagus, and vena caval hiatus, which transmits inferior vena cava.

tercostals decrease the vertical and transverse diameters of the thoracic cage by depressing the ribs.

The deep layer of intercostal muscles is discontinuous and is described as three separate muscles: transversus thoracis (anterior), innermost intercostal (lateral), and subcostalis (posterior). The lower fibers of the *transversus thoracis* run more or less parallel to those of the transversus abdominis, while the upper fibers radiate upward and outward from the sternum to the ribs.

Thoracic Diaphragm   The thoracic diaphragm (Fig. 7.25) is a dome-shaped musculotendinous organ that separates the thoracic cavity from the abdominal cavity below and is the chief muscle of inspiration (drawing of air into the lungs). The muscular portion of the diaphragm arises from the sternum, lower ribs, and upper lumbar vertebrae; its fibers insert on a central area of fibrous connective tissue called the *central tendon.* Contraction of the muscular portion of the diaphragm depresses the central tendon, which increases the vertical diameter of the thoracic cage. This, plus the increase in both vertical and transverse diameters caused by contraction of the external intercostals, results in increased volume and decreased pressure in the thorax that leads to the intake of air into the lungs. Expiration (the release of air from the lungs) is produced by elastic recoil of the thoracic walls following the relaxation of these muscles. Contraction of the diaphragm together with the muscles of the anterolateral abdominal wall facilitates expulsion of pelvic contents by raising the intraabdominal pressure. The diaphragm is innervated by the phrenic nerve, which is derived from the cervical plexus (Fig. 7.26). This is due to the fact that the diaphragm is a displaced cervical hypaxial muscle. Unilateral phrenic nerve injury causes paralysis of the corresponding side of the diaphragm. Phrenic nerve section may be used in the treatment of lung tuberculosis when there is a need to rest the lung on one side. Bilateral phrenic nerve paralysis can be very serious since the resulting respiratory disturbances encourage the development of pneumonia.

Abdominal Muscles   Muscles of the abdomen include the muscles of the anterolateral and posterior abdominal walls plus the pelvic diaphragm (Table 7.9).

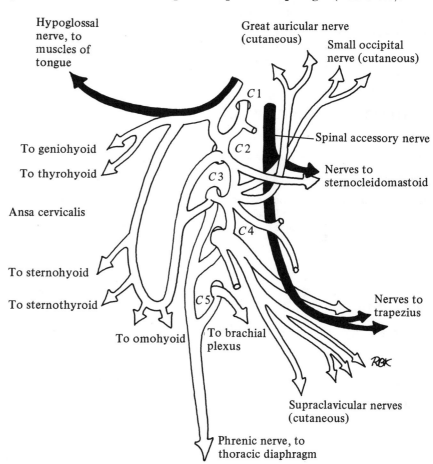

Figure 7.26 Cervical Plexus. Derived from anterior primary rami of spinal nerves C1–C5. Muscular branches of cervical plexus innervate neck muscles of hypaxial origin, namely, infrahyoids and thoracic diaphragm (a displaced cervical muscle).

Table 7.9   Muscles of Abdomen

Muscle	Origin	Insertion	Nerve	Action
ANTEROLATERAL ABDOMINAL WALL				
External abdominal oblique*	External surface of lower eight ribs	Anterior half of iliac crest and linea alba	Lower five inter-costals, subcostal, and iliohypogas-tric	Rotates and flexes vertebral column; com-presses abdom-inal viscera
Internal abdominal oblique*	Inguinal ligament, iliac crest, and lumbodorsal fascia	Lower four ribs, linea alba, and pubis (by way of con-joint tendon)	Lower five inter-costal, iliohypo-gastric, and ilio-inguinal	Rotates and flexes vertebral column; com-presses abdom-inal viscera
Transversus ab-dominis	Inguinal ligament, iliac crest, lumbodorsal fascia, and costal cartilages of lower four ribs	Linea alba and pubis (by way of conjoint tendon)	Same as above	Compresses ab-dominal viscera; depresses ribs
Rectus abdominis*	Xiphoid process and costal cartilages of fifth to seventh ribs	Pubic crest and symphysis	Lower five inter-costals and sub-costal	Unilateral, bends vertebral column to side; bilateral, flexes vertebral column
POSTERIOR ABDOMINAL WALL				
Iliacus	(See Table 7.17)	—	—	—
Psoas major	(See Table 7.17)	—	—	—
Psoas minor	(See Table 7.17)	—	—	—
Quadratus lumborum*	Lumbar vertebrae, lum-bodorsal fascia, and iliac crest	Twelfth rib and transverse processes of upper lumbar vertebrae	Subcostal ($T12$) and $L1$	Unilateral, bends vertebral column to the side or draws hip up to rib cage; bilat-eral, flexes verte-bral column
PELVIC DIAPHRAGM				
Levator ani*	Pubis, ischial spine, and obturator fascia	Coccyx and anococcygeal raphe	$S4$–$S8$	Supports abdom-inal and pelvic viscera
Coccygeus	Ischial spine	Sacrum and coccyx	Same as above	Same as above

* Muscles that must be learned.

Muscles of the Anterolateral Abdominal Wall

Muscles of the anterolateral abdominal wall consist of three layers of flat muscular sheets, corresponding to the muscles of the thoracic wall, which are supported by a pair of strap muscles (Figs. 7.27 and 7.28).

The superficial layer of anterolateral abdominal muscles consists of the paired *external obliques,* whose fibers run down and inward (hands-in-pockets direction) like those of the external intercostals. Each external oblique terminates in a tough aponeurosis that inserts into the linea alba (white line), a strip of fascia that extends from the xiphoid process to the pubic symphysis. The free lower border of the external oblique rolls up upon itself to form the *inguinal ligament,* which extends from the anterior superior iliac spine to the pubic tubercle.

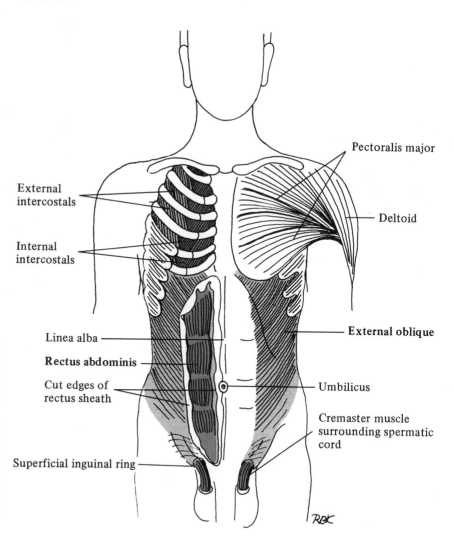

Figure 7.27 Anterolateral Abdominal Wall Muscles: Superficial. Integument is removed on left side of trunk to expose pectoralis major and external oblique. On right side of trunk, pectoral muscles are removed to expose intercostal muscles, and rectus sheath is opened to expose rectus abdominis. Note that fibers of external oblique run in same direction as those of external intercostals. Also note that spermatic cord and its musculature exit from inguinal canal through superficial inguinal ring, a narrow opening in aponeurosis of external oblique.

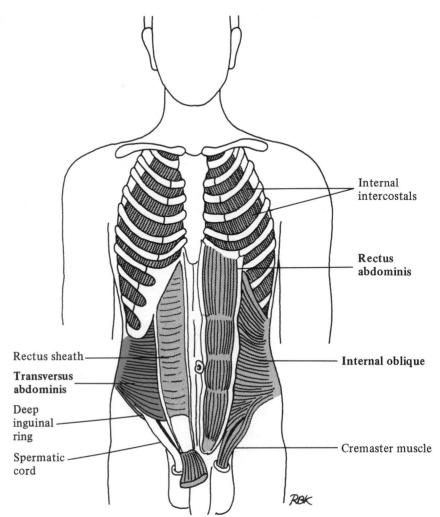

Figure 7.28 Anterolateral Abdominal Wall Muscles: Deep. On left side of the trunk, external intercostals and external oblique are removed to expose internal intercostals and internal oblique. Note that fibers of these muscles run in same direction. Cremaster muscle, a continuation of lower fibers of internal oblique, follows spermatic cord down into scrotal sac. On right side of trunk, internal oblique is removed to expose transversus abdominis, and rectus abdominis is cut away to expose posterior wall of rectus sheath. Also, cremaster muscle is removed to expose spermatic cord. Note that spermatic cord emerges from deep inguinal ring, an opening in fascia deep to transversus abdominis.

The middle layer consists of the paired *internal obliques,* whose fibers run up and outward like those of the internal intercostals. Each internal oblique also terminates in an aponeurosis that inserts into the linea alba.

The deep layer consists of the paired *transversus abdominis* muscles, whose fibers run more or less transversely and are continuous with the lower fibers of the transversus thoracis. They also insert into the linea alba by way of their aponeuroses.

The strap muscles consist of the vertically oriented *rectus abdominis* muscles, which extend from the lowermost costal cartilages to the pubic symphysis on either side of the linea alba. These muscles act as the chief flexors of the vertebral column when they act bilater-

ally and help flex the vertebral column laterally when they act unilaterally. Each rectus muscle is enclosed in a sheath formed from the
aponeurosis of the internal oblique with contributions from the
transversus aponeurosis. The function of the rectus sheath is to hold
the rectus muscle in place and prevent it from springing forward
when the vertebral column is flexed.

The *inguinal canals* are important structural features of the anterolateral abdominal wall. They consist of a pair of oblique tunnels
through the abdominal wall that transmit the spermatic cords in the
male and the round ligaments of the uterus in the female. Each inguinal canal is formed from portions of the transversus and internal
oblique muscles and the aponeurosis of the external oblique. It has
an internal opening, the deep inguinal ring, and an external opening,
the superficial inguinal ring. The deep inguinal ring represents a
weak spot in the abdominal wall, which is a common site of inguinal
hernia (see "Diseases and Disorders of the Muscular System," p. 253).

The anterolateral abdominal muscles act like a "living girdle"; they
support and compress the abdominal viscera. When you wear a girdle
or other external abdominal support, you encourage these muscles
to be lazy and lose tone. The anterolateral abdominal muscles also

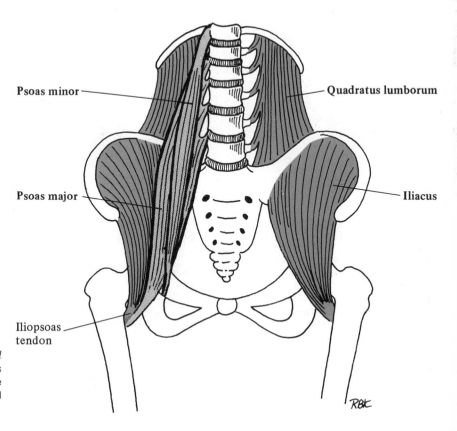

*Figure 7.29 Posterior Abdominal Wall
Muscles.* On left side of trunk, psoas
major and minor are removed to expose
deeper-lying quadratus lumborum and
iliacus muscles.

Psoas minor

Quadratus lumborum

Psoas major

Iliacus

Iliopsoas
tendon

assist in the act of forced expiration and, by contracting simultaneously with the diaphragm, participate in the acts of micturition (urination), defecation, and parturition (childbirth) through elevation of intraabdominal pressure. Other actions of these muscles include flexion, lateral flexion, and rotation of the vertebral column. They are innervated segmentally by the anterior primary rami of thoracic and lumbar spinal nerves.

Muscles of the Posterior Abdominal Wall

The muscles of the posterior abdominal wall (Fig. 7.29) are of hypaxial origin and are prevertebral in location. These muscles, which arise mainly from the lumbar vertebrae and ilia, include psoas major and minor, iliacus, and quadratus lumborum. *Psoas major* and *iliacus*, which flex the thigh as well as the vertebral column, will be discussed with the anterior hip muscles. *Quadratus lumborum* is a quadrilateral muscle that fills in the gap between the last rib and the iliac crest. It causes lateral flexion of the vertebral column when the hip is fixed and draws the hip up to the rib cage when the vertebral column is fixed,

Figure 7.30 is a cross section through the trunk; it will help you to picture the relationships of the muscles of the back and abdomen to the vertebral column and to each other.

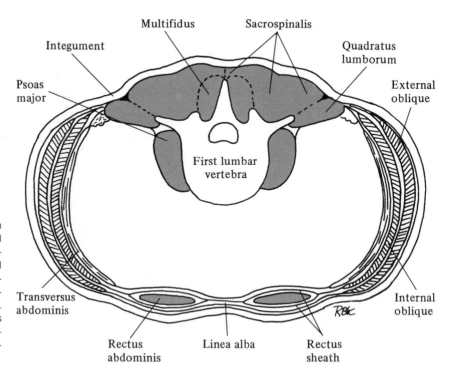

Figure 7.30 Trunk. Cross section through trunk at vertebral level *L*1. Epaxial (postvertebral) trunk muscles include sacrospinalis and multifidus, while hypaxial (prevertebral) trunk muscles include quadratus lumborum, psoas major, layered abdominal muscles, and rectus abdominis. Note that at this level rectus sheath is formed from aponeurosis of internal oblique, which splits to enclose rectus muscle.

Pelvic Diaphragm

The pelvic diaphragm (Fig. 7.31) is a fibromuscular sling, consisting of the levator ani and coccygeus muscles, that forms the floor of the pelvic cavity. *Levator ani* is a broad thin muscular sheet that arises from the pubis and ischial spine. Its posterior fibers insert into the coccyx, while its anterior fibers insert with those from the opposite side into a midline seam, the *anococcygeal raphe*. Gaps in the midline of levator ani transmit the vagina in the female and the urethra and anal canal in both sexes. The *coccygeus* is a small triangular muscle sheet, located posterior to levator ani, that extends from the ischial spine to the sacrum and coccyx. In addition to supporting abdominal and pelvic viscera, the pelvic diaphragm resists the elevated intraabdominal pressure created by simultaneous contraction of the diaphragm and anterolateral abdominal muscles; thus it also plays an active role in micturition, defecation, and parturition.

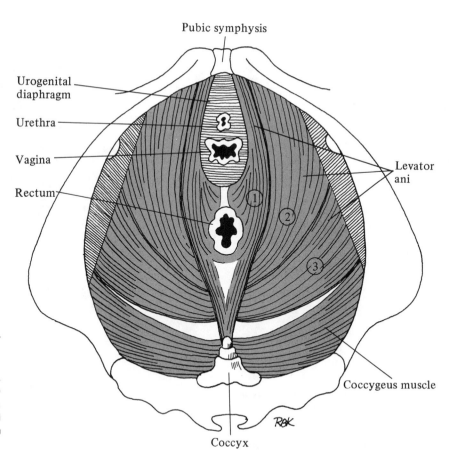

Figure 7.31 *Pelvic Diaphragm (Female): Superior Aspect.* This is a broad sling, composed of coccygeus and levator ani muscles, that forms floor of pelvic cavity. Levator ani consists of three parts: (1) puborectalis, (2) pubococcygeus, and (3) iliococcygeus. In female, urethra and vagina pass through genital hiatus, which is covered by urogenital diaphragm.

Pubic symphysis

Urogenital diaphragm

Urethra

Vagina

Rectum

Levator ani

Coccygeus muscle

Coccyx

Muscles of the Appendicular Skeleton

MUSCLES OF THE SHOULDER GIRDLE AND UPPER EXTREMITY

Muscles of the shoulder girdle and upper extremity are very important to humans; they are used in skilled activities such as writing, manipulating tools, and playing musical instruments. These muscles can be subdivided into the following groups; superficial back muscles, superficial chest muscles, muscles of the shoulder, muscles of the arm, muscles of the forearm, and muscles of the hand.

Superficial Back Muscles

The superficial muscles of the back, which are illustrated in Figs. 7.32 and 7.33 and are described in Table 7.10, act on the scapula or the arm. Most of these muscles attach the scapula to the vertebral column.

The *trapezius* is a large triangular muscle covering the neck and upper back. Together the right and left trapezius muscles form a

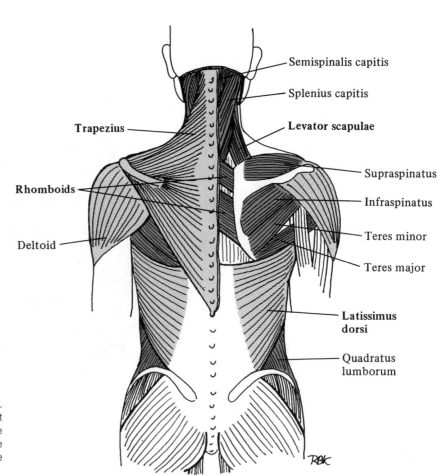

Figure 7.32 *Superficial Back Muscles.* Trapezius and latissimus dorsi, the first layer of superficial back muscles, are shown on left side of back. On right side of back, trapezius is cut away to expose rhomboids and levator scapulae.

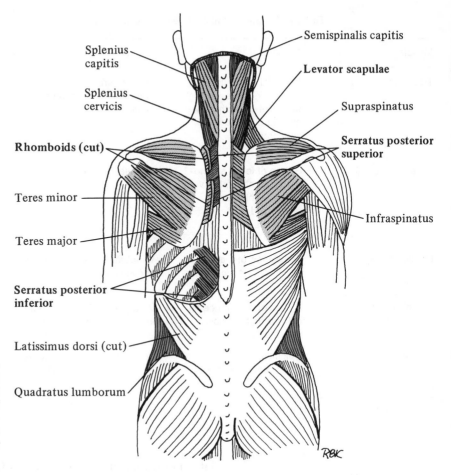

Figure 7.33 Superficial and Interme-diate Back Muscles. Levator scapuli and rhomboids, the second layer of superficial back muscles, are shown on right side of back. On left side of back, superficial muscles are removed to expose intermediate back muscles, serratus posterior superior and serratus posterior inferior.

trapezoid or diamond-shaped muscle mass. Each trapezius arises from the occipital bone, ligamentum nuchae, and vertebral column; it inserts on the clavicle, acromion, and scapular spine. The trapezius can produce different actions on the scapula depending on which of its parts are called into action. Its upper part, whose fibers run down and outward, elevates the scapula; its middle part, whose fibers run transversely, retracts (adducts) the scapula; and its lower part, whose fibers run up and outward, depresses the scapula. When its upper and lower parts work together with serratus anterior, the scapula is ro-tated upward, raising the arm above the head. Paralysis of the trape-zius interferes with shrugging the shoulder (elevation of the scapula) and abducting the arm more than 90° (upward rotation of the scapula).

The *latissimus dorsi* (widest muscle of the back) is a large tri-angular muscle that covers the loin and lower half of the back. It has a broad origin from the lower thoracic and lumbar vertebrae, sacrum, and iliac crest; its fibers run up and outward to converge on a narrow

Table 7.10 Superficial Muscles of Back

Muscle	Origin	Insertion	Nerve	Action
Trapezius*	Occipital bone, ligamentum nuchae, and spines of seventh cervical and all thoracic vertebrae	Lateral third of clavicle, acromion, and scapular spine	Spinal accessory plus *C3* and *C4*	Retracts and rotates scapula upward; upper fibers elevate scapula, lower fibers depress scapula
Latissimus dorsi*	Spines of lower six thoracic and all lumbar vertebrae, lumbodorsal fascia, and posterior iliac crest	Bicipital groove of humerus	Thoracodorsal	Trunk fixed: extends, adducts, and rotates arm medially; arm fixed: elevates trunk
Levator scapulae*	Transverse processes of upper four cervical vertebrae	Vertebral border of scapula above scapular spine	Dorsal scapular	Elevates scapula
Rhomboid minor	Spines of seventh cervical and first thoracic vertebrae	Vertebral border of scapula at root of scapular spine	Dorsal scapular	Elevates, retracts, and rotates scapula downward
Rhomboid major	Spines of second through fifth thoracic vertebrae	Vertebral border of scapular below root of scapular spine	Dorsal scapular	Elevates, retracts, and rotates scapula downward

* Muscles that must be learned.

flat tendon that inserts into the bicipital groove of the humerus. The latissimus dorsi adducts, extends, and medially rotates the arm. When the arms are fixed, the latissimus dorsi muscles draw the trunk upward to the arms; this is what happens when you chin yourself or climb a rope. Paralysis of the latissimus causes forward displacement of the shoulder, due to the unopposed pull of the pectoral muscles, and weakens all downward movements of the arm.

Deep to the trapezius lie three small strap muscles that constitute the second layer of superficial back muscles: *levator scapulae,* *rhomboideus minor,* and *rhomboideus major.* These muscles originate on the vertebral column and run down and outward to insert on the vertebral border of the scapula. Their line of pull from the vertebral column to the scapula make it possible for them to elevate, retract, and rotate the scapula downward. When levator scapulae and the rhomboids act together with the middle part of the trapezius, the shoulders are braced backward (retraction and elevation of the scapulae).

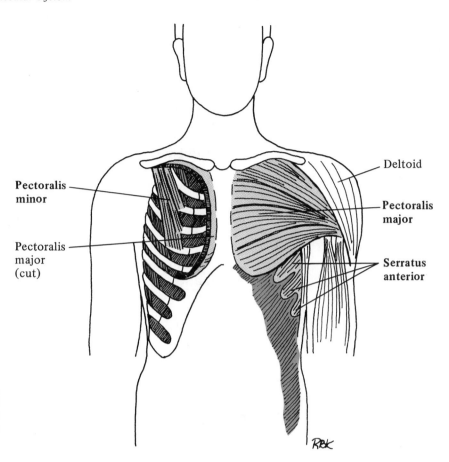

Pectoralis
minor

Pectoralis
major
(cut)

Deltoid

Pectoralis
major

Serratus
anterior

Figure 7.34 Superficial Chest Muscles.
Pectoralis major and serratus anterior are
shown on left side of chest. On right side
of chest, these muscles are removed to
expose pectoralis minor.

Superficial Chest Muscles

The superficial muscles of the chest, or *pectoral muscles,* which are illustrated in Fig. 7.34 and described in Table 7.11, act on the arm or shoulder girdle.

The *pectoralis major* is a large fan-shaped muscle located on the upper anterior surface of the chest. It has a broad origin from the clavicle (clavicular head) and from the sternum and costal cartilages (sternocostal head); its fibers converge on a narrow flat tendon that inserts on the humerus. The clavicular part, acting alone, flexes the arm; the sternocostal part, acting alone, extends and adducts the arm. When both parts of pectoralis major act together, they flex the arm and draw the arm across the chest (medial rotation). Pectoralis major, which lies beneath the breast and contains lymph nodes and vessels draining the breast, is often removed in cases of cancer of the breast by an extensive surgical procedure called a *radical mastectomy.*

The *pectoralis minor* is a thin triangular muscle that lies deep to pectoralis major. It arises from the third to fifth ribs and inserts on the coracoid process of the scapula. Pectoralis minor depresses the scapula when the ribs are fixed and elevates the ribs when the scapula is fixed, thus acting as an accessory respiratory muscle. It also assists serratus anterior in holding the scapula against the chest wall. This muscle is removed during radical mastectomy operations for the same reasons that pectoralis major is removed. The anterior part of the deltoid and coracobrachialis can be made to substitute for the absent pectoral muscles when the radical mastectomy patient regains use of the arm on the operated side.

The *serratus anterior* (sawtooth muscle) is a large muscular sheet whose fibers arise like the teeth of a saw from the upper eight ribs and wrap around the side of the chest to insert into the vertebral border of the scapula. It is a powerful fixator and protractor (abductor) of the scapula and holds the scapula against the chest in forward pushing movements. Since it draws the scapula away from the vertebral column, it is the chief antagonist of the rhomboids. The lower part of the serratus also assists the lower part of the trapezius in rotating the scapula upward so that the arm can be raised above the head. When this important chest muscle is paralyzed, the medial border of the scapula juts out from the thorax; this condition is known as *"winging" of the scapula.*

Table 7.11 Superficial Muscles of Chest

Muscle	Origin	Insertion	Nerve	Action
Pectoralis major*	Clavicular head, medial half of clavicle; sterno-costal head, sternum, and costal cartilages of upper six ribs	Bicipital groove of humerus	Lateral and medial pectoral	Flexes, adducts, and rotates arm medially
Pectoralis minor	Second to fifth ribs	Coracoid process of scapula	Medial pectoral	Draws scapula down and forward; elevates ribs when scapula is fixed
Subclavius	Costochondral junction of first rib	Lower surface of clavicle	Nerve to subclavius	Draws clavicle down and forward
Serratus anterior*	First through eighth ribs	Vertebral border of scapula	Long thoracic	Protracts and rotates scapula upward; holds scapula against the chest

* Muscles that must be learned.

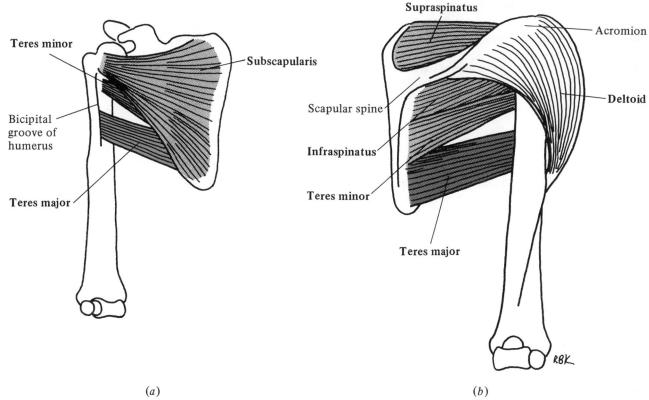

Teres minor

Subscapularis

Bicipital
groove of
humerus

Teres major

(a)

Supraspinatus

Acromion

Scapular spine

Deltoid

Infraspinatus

Teres minor

Teres major

RBK

(b)

Figure 7.35 *Shoulder Muscles.* (*a*) Anterior aspect of scapula and humerus. (*b*) Posterior aspect of scapula and humerus.

Muscles of the Shoulder   The shoulder muscles (Fig. 7.35 and Table 7.12) arise from the scapula and act across the shoulder joint.

The deltoid is a fleshy muscle, triangular in shape, that covers the shoulder joint and creates the rounded contour of the shoulder and upper arm. It arises from the clavicle (anterior part), acromion (middle part), and scapular spine (posterior part) and inserts on the deltoid tuberosity of the humerus. The anterior deltoid flexes the arm and rotates it medially; the posterior deltoid extends the arm and rotates it laterally. The medial deltoid acts as the 30°–90° abductor of the arm; that is, it raises the arm into the horizontal position. Raising the arm above the head (90°–180° abduction) requires upward rotation of the scapula. Because the deltoid is so thick, it is a common site for intramuscular injections. These must be given into the fleshy part of the muscle (about two or three fingerbreadths below the acromion) to avoid the nonyielding tendinous septa above and below.

Table 7.12 Muscles of Shoulder

Muscle	Origin	Insertion	Nerve	Action
Deltoid*	Lateral third of clavicle, acromion, and scapular spine	Deltoid tuberosity of humerus	Axillary	Anterior fibers, flex and rotate arm medially; middle fibers, abduct arm; posterior fibers, extend and rotate arm laterally
Supraspinatus*	Supraspinatus fossa of scapula	Greater tubercle of humerus	Suprascapular	Initiates abduction of arm; rotates arm laterally
Infraspinatus*	Infraspinatus fossa of scapula	Greater tubercle of humerus	Suprascapular	Extends and rotates arm laterally
Teres minor*	Axillary border of scapula	Greater tubercle of humerus	Circumflex scapular	Adducts and rotates arm laterally
Teres major*	Axillary border and inferior angle of scapula	Lesser tubercle of humerus	Lower subscapular	Adducts and rotates arm medially
Subscapularis*	Subscapular fossa	Lesser tubercle of humerus	Upper and lower subscapular	Extends, adducts, and rotates arm medially

* Muscles that must be learned.

The *supraspinatus, infraspinatus,* and *teres minor* originate from the posterior surface and axillary border of the scapula, pass in back of the shoulder joint, and insert on the greater tubercle of the humerus. Together they are known as the SIT muscles (because they "sit" on the greater tubercle of the humerus). The supraspinatus initiates abduction of the arm (0°–30° abduction). When the supraspinatus is paralyzed, the arm must be moved passively into the 30° position before the deltoid can act. The infraspinatus and teres minor act together as extensors and chief lateral rotators of the arm.

Teres major is a small round muscle that originates from the axillary border of the scapula and passes in front of the humerus to insert into the bicipital groove. It acts as an adductor and medial rotator of the arm.

Subscapularis is a large triangular muscle that has a fleshy origin from the costal surface of the scapula. It passes in front of the shoulder joint to insert on the lesser tubercle of the humerus. Subscapularis extends, adducts, and acts as the chief medial rotator of the arm, which makes it the antagonist of infraspinatus and teres minor. Together with the SIT muscles it forms the rotator cuff, a musculotendinous structure that fuses with and strengthens the capsule of the shoulder joint (Chapter 6).

Muscles of the Arm   During prenatal development the extremities divide into anterior (preaxial) and posterior (postaxial) compartments, and these compartments are retained in the adult. In the arm the anterior and posterior compartments are separated by medial and lateral intermuscular septa, which extend inward from the deep fascia to the humerus. The muscles of the arm (Table 7.13) are located within these two compartments; flexor muscles are in the anterior compartment and extensor muscles in the posterior compartment.

Anterior Compartment   The anterior compartment of the arm contains the three flexor muscles: coracobrachialis, biceps brachii, and brachialis [Fig. 7.36(a)]. These muscles act across the elbow or shoulder joint.

    Coracobrachialis is a small muscle located deep to the deltoid and pectoralis major. It stabilizes the shoulder joint and acts as an assistant flexor of the arm.

    Biceps brachii is the large two-headed muscle of the arm whose

Table 7.13 Muscles of Arm

Muscle	Origin	Insertion	Nerve	Action
ANTERIOR COMPARTMENT				
Coracobrachialis	Coracoid process of scapula	Middle of humeral shaft	Musculocutaneous	Flexes and adducts arm
Biceps brachii*	Long head, supraglenoid tubercle; short head, coracoid process	Radial tuberosity and antebrachial fascia	Musculocutaneous	Flexes arm and forearm; supinates hand
Brachialis*	Distal two-thirds of anterior surface of humerus	Coronoid process of ulna	Musculocutaneous	Flexes forearm
POSTERIOR COMPARTMENT				
Anconeus	Lateral epicondyle of humerus	Olecranon process and posterior surface of ulna	Radial	Extends forearm
Triceps brachii*	Long head, infraglenoid tubercle; lateral head, proximal half of humerus; medial head, distal half of humerus	Olecranon process of ulna	Radial	Extends and adducts arm; extends forearm

* Muscles that must be learned.

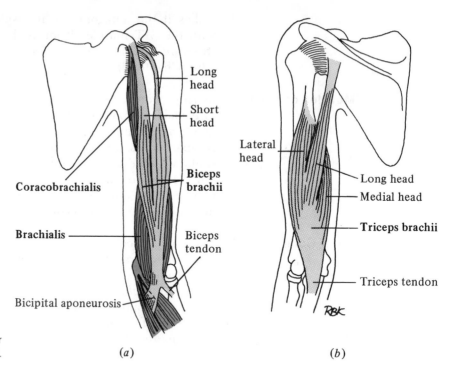

Figure 7.36 *Arm Muscles.* (a) Anterior compartment. (b) Posterior compartment.

(a)

(b)

belly creates a conspicuous bulge when you flex your forearm. It arises from the scapula and inserts on the radial tuberosity by way of a long tendon that can be palpated at the elbow. Since the biceps crosses the front of the shoulder joint as well as the elbow joint, it flexes the arm in addition to the forearm. It also assists in supinating the hand.

Brachialis is a thick fleshy muscle that arises from the humerus and inserts on the ulna. It lies deep to biceps brachii and covers the anterior surface of the elbow joint. Forearm flexion is the sole function of the brachialis, which is equally effective in the supinated, mid, or pronated position.

Posterior Compartment The posterior compartment of the arm [Fig. 7.36(*b*)] contains a single muscle, *triceps brachii,* which acts across both the shoulder and elbow joints. The triceps has three heads of origin, two from the humerus and one from the scapula, and inserts on the olecranon process of the ulna. At the shoulder it extends and adducts the arm; at the elbow it acts as chief extensor of the forearm and antagonist of biceps brachii and brachialis. The *anconeus,* a small triangular muscle that crosses the back of the elbow joint, is considered to be a continuation of the triceps.

Table 7.14   Muscles of Anterior Compartment of Forearm

Muscle	Origin	Insertion	Nerve	Action
SUPERFICIAL GROUP				
Pronator teres*	Medial epicondyle of humerus and coronoid process of ulna	Middle of radial shaft	Median	Pronates hand
Flexor carpi radialis*	Medial epicondyle of humerus	Second and third metacarpals	Median	Flexes and abducts hand
Flexor carpi ulnaris*	Medial epicondyle, olecranon process, and posterior border of ulna	Pisiform, hamate, and fifth metacarpal	Ulnar	Flexes and adducts hand
Palmaris longus (may be absent)	Medial epicondyle of humerus	Flexor retinaculum and palmar aponeurosis	Median	Flexes hand
INTERMEDIATE GROUP				
Flexor digitorum superficialis*	Medial epicondyle, olecranon process, and anterior border of radius	Middle phalanges of fingers	Median	Flexes forearm, hand, and fingers at MP and PIP joints
DEEP GROUP				
Flexor digitorum profundus*	Proximal three-fourths of ulna and interosseous membrane	Distal phalanges of fingers	Median and ulnar	Flexes hand and fingers at MP and IP joints
Flexor pollicis longus*	Shaft of radius, interosseous membrane, and coronoid process of ulna	Distal phalanx of thumb	Median	Flexes thumb at MP and IP joints
Pronator quadratus*	Distal fourth of ulna	Distal fourth of radius	Median	Pronates hand

* Muscles that must be learned.

Antecubital Fossa   The antecubital fossa is the diamond-shaped depression on the anterior aspect of the elbow. Structures of clinical importance that can be palpated in the antecubital fossa are the brachial artery and biceps tendon.

Muscles of the Forearm   The forearm contains muscles that act across the wrist, metacarpophalangeal (MP), and interphalangeal (IP) joints. Like the arm, the forearm is divided into anterior and posterior compartments. The radius, ulna, interosseous membrane, and medial and lateral intermuscular septa separate the flexor muscles in the anterior compartment from the extensor muscles in the posterior compartment.

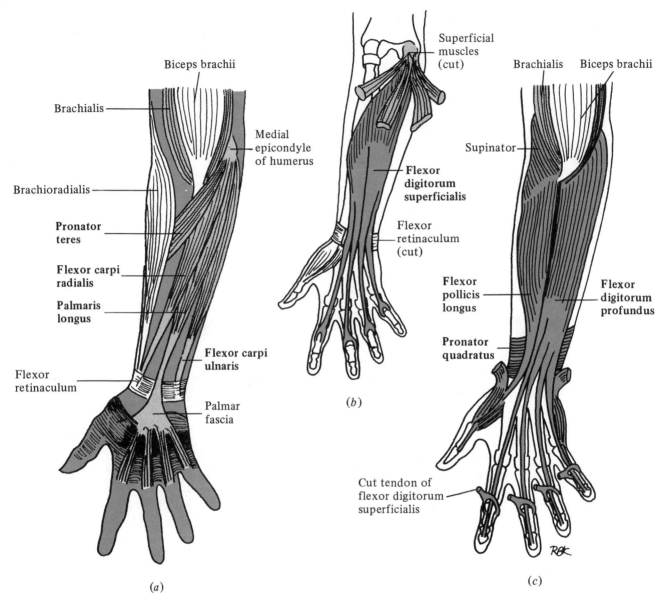

Figure 7.37 *Muscles of Anterior Compartment of Forearm.* (a) Superficial group.
Pronator teres is partially hidden by brachioradialis, a forearm flexor that belongs to posterior compartment. (b) Intermediate group. Consists of flexor digitorum superficialis,
which inserts by four tendons on middle phalanges of fingers. (c) Deep group. Includes
flexor digitorum profundus (which inserts on distal phalanges of fingers), flexor pollicis
longus, and pronator quadratus. Note that tendons of flexor digitorum profundus pass
through slits in superficialis tendons to reach distal phalanges.

Anterior Compartment The anterior compartment of the forearm contains muscles that flex the hand and fingers and pronate the hand. They arise from the medial (flexor) epicondyle of the humerus or the shafts of the radius and ulna and insert by way of long tendons on the carpals of the wrist, metacarpals of the hand, or phalanges of the fingers. Their bellies form a bulge on the medial or ulnar side of the forearm, and their tendons can be palpated on the anterior surface of the wrist. The flexor-pronator muscles of the forearm are subdivided into superficial, intermediate, and deep groups; they are illustrated in Fig. 7.37 and described in Table 7.14. Pronators of the hand include *pronator teres,* which acts across the proximal radioulnar joint, and *pronator quadratus,* which acts across the distal radioulnar joint. The chief flexors of the hand consist of *flexor carpi ulnaris,* which also adducts the hand (ulnar deviation), and *flexor carpi radialis,* which also abducts the hand (radial deviation). Extrinsic flexors of the digits include *flexor digitorum superficialis,* which acts primarily across the proximal IP joints of the fingers, *flexor digitorum profundus,* which acts primarily across the distal IP joints of the fingers, and *flexor pollicis longus,* which acts across the MP and IP joints of the thumb.

Posterior Compartment The posterior compartment of the forearm contains muscles that extend the hand and fingers and supinate the hand. They arise from the lateral (extensor) epicondyle of the humerus or the shafts of the radius and ulna and insert by way of long tendons on the carpals, metacarpals, or phalanges. Their bellies form a bulge on the lateral or radial side of the forearm and their tendons can be palpated on the posterior surface of the wrist. The extensor-supinator muscles of the forearm are subdivided into superficial and deep groups, which are illustrated in Fig. 7.38 and described in Table 7.15. *Brachioradialis,* which is anatomically a part of the posterior compartment, is functionally a flexor of the forearm. The hand is supinated by the *supinator* muscle, which acts together with biceps brachii. The chief extensors of the hand consist of *extensor carpi ulnaris,* which also adducts the hand, and *extensor carpi radialis longus* and *brevis,* which also abduct the hand. Extrinsic extensors of the digits include *extensor digitorum communis,* which acts primarily across the MP and IP joints of the fingers, and *extensor pollicis longus,* which acts across the MP and IP joints of the thumb.

Special Features of the Wrist An important structure on the anterior surface of the wrist is the carpal tunnel. The *carpal tunnel* is a deep groove (the concavity of the carpal bones) roofed over by a band of fibrous connective tissue called the *flexor retinaculum.* Tendons of the extrinsic flexors of the fingers plus the median nerve pass through this osseofibrous tunnel on their way to the hand. Occasionally, fibrosis of the flexor retinaculum reduces the volume of the carpal tunnel, which compresses the median

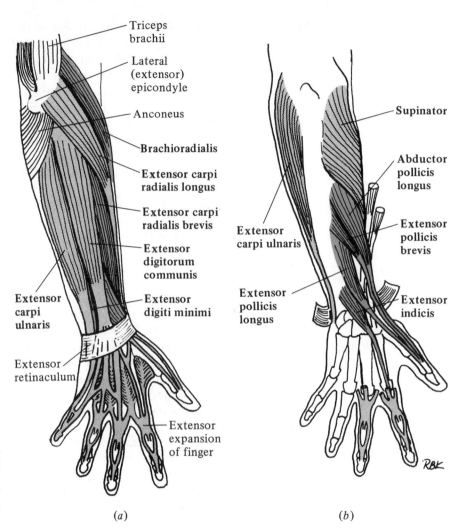

Figure 7.38 *Muscles of Posterior Compartment of Forearm.* (a) Superficial group. Note that extensor digitorum communis inserts on phalanges of fingers by way of extensor expansions. (b) Deep group.

(a)

(b)

nerve (carpal tunnel syndrome). The tendons of the extrinsic extensors of the fingers pass through a series of osseofibrous grooves on the posterior and radial aspects of the wrist; they are held in place by the *extensor retinaculum.*

Muscles of the Hand   The intrinsic muscles of the hand are small muscles that are responsible for the precise finger movements that humans alone are capable of making. They are divided into *thenar, hypothenar,* and *intermediate* groups, which are illustrated in Fig. 7.39 and described in Table 7.16. Muscles of the thenar group form a bulge on the thumb side of the palm called the *thenar eminence.* The most important thenar muscle is the *opponens pollicis,* which opposes the thumb

Table 7.15 Muscles of Posterior Compartment of Forearm

Muscle	Origin	Insertion	Nerve	Action
SUPERFICIAL GROUP				
Brachioradialis*	Lateral supracondylar ridge of humerus	Styloid process of radius	Radial	Flexes forearm
Extensor carpi radialis longus*	Lateral supracondylar ridge of humerus	Second metacarpal	Radial	Extends and abducts hand
Extensor carpi radialis brevis	Lateral epicondyle of humerus	Third metacarpal	Radial	Extends and abducts hand
Extensor digitorum communis*	Lateral epicondyle of humerus	Extensor expansions of fingers	Radial	Extends fingers at MP and IP joints
Extensor digiti minimi	Lateral epicondyle of humerus	Extensor expansion on little finger	Radial	Extends little finger
Extensor carpi ulnaris*	Lateral epicondyle of humerus and posterior border of ulna	Fifth metacarpal	Radial	Extends and adducts hand
DEEP GROUP				
Supinator*	Lateral epicondyle of humerus and proximal end of ulna	Upper third of radius	Radial	Supinates hand
Abductor pollicis longus	Posterior surface of ulna and middle third of radius	First metacarpal	Radial	Abducts hand and thumb
Extensor pollicis brevis	Middle third of radius	Proximal phalanx of thumb	Radial	Abducts hand and extends thumb
Extensor pollicis longus*	Middle third of ulna and interosseous membrane	Distal phalanx of thumb	Radial	Abducts hand and extends thumb
Extensor indicis	Posterior surface of ulna	Extensor expansion on index finger	Radial	Extends index finger

* Muscles that must be learned.

to the palm of the hand. Muscles of the hypothenar group form a bulge on the little finger side of the hand called the *hypothenar eminence*. Muscles of the intermediate group are located in the middle of the palm. They include the *lumbricales*, which flex the fingers at the MP joints and extend them at the IP joints, and the *interossei*, which abduct and adduct the fingers at the MP joints.

Special Features of the Hand The hand is a highly complex structure with several features of clinical importance. The superficial fascia over the dorsum of the hand is very loose, while that over the palm is tight. The *palmar aponeuro-*

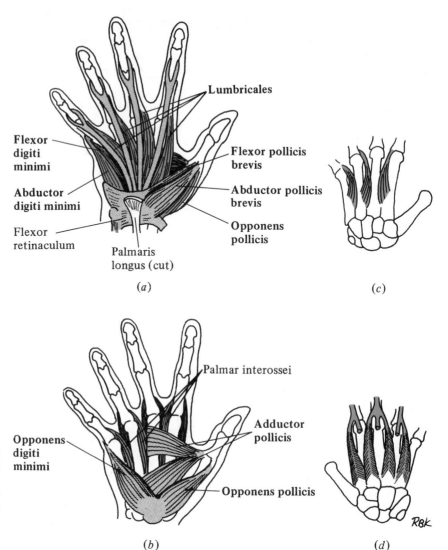

Lumbricales

Flexor digiti minimi

Abductor digiti minimi

Flexor retinaculum

Flexor pollicis brevis

Abductor pollicis brevis

Opponens pollicis

Palmaris longus (cut)

(*a*)

(*c*)

Palmar interossei

Opponens digiti minimi

Adductor pollicis

Opponens pollicis

(*b*)

(*d*)

RBK

Figure 7.39 Intrinsic Muscles of Hand. (*a*) Superficial palmar muscles. (*b*) Deep palmar muscles. (*c*) Palmar interossei. These adduct index, ring, and little fingers, extend at IP joints, and flex at MP joints. (*d*) Dorsal interossei. These abduct index, ring, and little fingers, extend at IP joints, and flex at MP joints.

sis, which fans out across the palm and attaches to the fingers, is part of the deep fascia of the hand. Fibrosis of the palmar aponeurosis causes flexion contracture of the fingers *(Dupuytren's contracture).* Common sheaths surround the tendons of the wrist, while individual sheaths surround the tendons of the fingers. The thenar and mid-palmar spaces are potential spaces deep in the connective tissue beneath the tendon sheaths. In infections of the hand, these spaces become distended with fluid, which is prevented from escaping by the impenetrable palmar fascia.

Table 7.16 Intrinsic Muscles of Hand

Muscle	Origin	Insertion	Nerve	Action
THENAR MUSCLES				
Abductor pollicis brevis	Flexor retinaculum, scaphoid, and trapezium	Proximal phalanx of thumb	Median	Abducts thumb
Flexor pollicis brevis	Flexor retinaculum and trapezium	Proximal phalanx of thumb	Median	Flexes thumb
Opponens pollicis*	Flexor retinaculum and trapezium	First metacarpal	Median	Opposes thumb
INTERMEDIATE MUSCLES				
Adductor pollicis	Oblique head, capitate; transverse head, second and third metacarpals	Proximal phalanx of thumb	Ulnar	Adducts thumb
Lumbricales (4)*	Tendons of flexor digitorum profundus	Extensor expansions of fingers	Median and ulnar	Flex fingers at MP joints; extend fingers at IP joints
Palmar interossei (3)*	Medial side of second metacarpal; lateral sides of fourth and fifth metacarpals	Proximal phalanges of index, ring, and little fingers and extensor digitorum communis	Ulnar	Adduct fingers towards middle finger at MP joints
Dorsal interossei (4)*	Adjacent sides of metacarpals	Proximal phalanges of index and middle fingers (lateral sides) plus proximal phalanges of middle and ring fingers (medial sides) and extensor digitorum communis	Ulnar	Abduct fingers away from middle finger at MP joints
HYPOTHENAR MUSCLES				
Abductor digiti minimi	Pisiform, tendon of flexor carpi ulnaris	Proximal phalanx of little finger	Ulnar	Abducts little finger
Flexor digiti minimi brevis	Flexor retinaculum and hook of hamate	Proximal phalanx of little finger	Ulnar	Flexes little finger
Opponens digiti minimi	Flexor retinaculum and hook of hamate	Fifth metacarpal	Ulnar	Opposes little finger

* Muscles that must be learned.

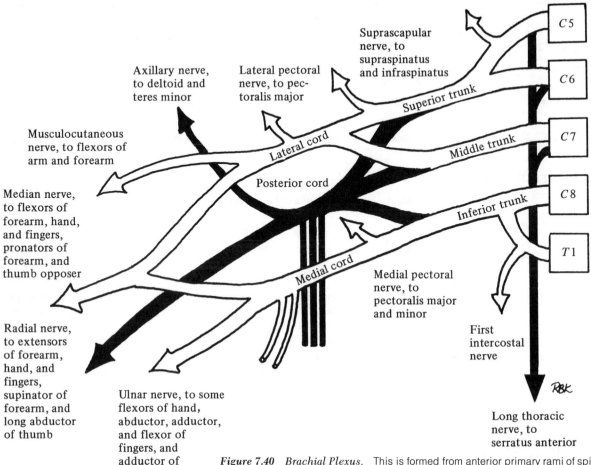

Dorsal scapular nerve, to rhomboids

Suprascapular nerve, to supraspinatus and infraspinatus

Axillary nerve, to deltoid and teres minor

Lateral pectoral nerve, to pectoralis major

Superior trunk

Middle trunk

Inferior trunk

C 5

C 6

C 7

C 8

T 1

Musculocutaneous nerve, to flexors of arm and forearm

Lateral cord

Posterior cord

Median nerve, to flexors of forearm, hand, and fingers, pronators of forearm, and thumb opposer

Medial cord

Medial pectoral nerve, to pectoralis major and minor

First intercostal nerve

Radial nerve, to extensors of forearm, hand, and fingers, supinator of forearm, and long abductor of thumb

Ulnar nerve, to some flexors of hand, abductor, adductor, and flexor of fingers, and adductor of thumb

Long thoracic nerve, to serratus anterior

Figure 7.40 *Brachial Plexus.* This is formed from anterior primary rami of spinal nerves *C*5–*T*1. Five rami unite to form three trunks—superior, middle, and inferior—which in turn divide into anterior divisions *(white)* and posterior divisions *(black)*. Divisions regroup to form posterior, lateral, and medial cords, which give rise to terminal nerves of plexus: radial, axillary, musculocutaneous, median, and ulnar. Nerves derived from posterior divisions innervate postaxial muscles while nerves derived from anterior divisions innervate preaxial muscles.

INNERVATION OF UPPER EXTREMITY MUSCLES

The muscles of the upper extremity are innervated by the nerves of the *brachial plexus* (Fig. 7.40).

The *musculocutaneous nerve* is the nerve of the anterior compartment of the arm. It pierces the coracobrachialis muscle and innervates all three flexor muscles, then pierces the deep fascia of the forearm to innervate a segment of skin on the lateral side of the forearm. Injury to this nerve interferes with forearm flexion.

The *radial nerve* is the largest nerve of the brachial plexus; it innervates the extensor muscles in the posterior compartment of the arm plus the extensor-supinator muscles in the posterior compartment of the forearm. Injury to the radial nerve, such as caused by a fracture of the humeral shaft, prevents a person from extending the forearm, hand, fingers, and thumb. In addition, the hand is pronated, with wrist and fingers flexed *(wristdrop)*.

The *median nerve* is the nerve of the anterior compartment of the forearm; it innervates the flexor-pronator muscles of the forearm and thenar muscles of the hand. When this nerve is injured, a person loses ability to pronate the hand, while the hand shows *apehand* deformity (atrophy of the thenar muscles with the thumb held in the plane of the palm). Inability to oppose the thumb results from paralysis of the opponens pollicis.

The *ulnar nerve* innervates two forearm muscles (flexor carpi ulnaris and part of flexor digitorum profundus) plus the bulk of the intrinsic muscles of the hand. Ulnar nerve injury produces *clawhand,* a condition characterized by atrophy of the hypothenar and interosseous muscles. The ring and little fingers are hyperextended at the MP joints and flexed at the IP joints, which helps give the hand a clawlike appearance.

MUSCLES OF THE PELVIC GIRDLE AND LOWER EXTREMITY

Muscles of the pelvic girdle and lower extremity are used mainly for locomotion and to support the body in the upright position. Movements of the pelvic girdle are greatly restricted compared to movements of the shoulder girdle. These muscles can be subdivided into the following groups: muscles of the hip, muscles of the thigh, muscles of the leg, and muscles of the foot.

Muscles of the Hip
Anterior Hip Muscles

The anterior hip muscles act as flexors of the thigh (Table 7.17) and are located mainly in the pelvis. They include the iliacus, psoas major, and psoas minor muscles, which are muscles of the posterior abdominal wall. The *iliacus* is a triangular muscle that arises from the iliac fossa of the hip bone, while the psoas major is a fusiform muscle that arises from the transverse processes and lumbar vertebrae. They share a common tendon of insertion, the *iliopsoas tendon,* which passes in front of the hip joint to insert on the lesser trochanter of the femur. These muscles, sometimes called the *iliopsoas,* are powerful flexors of the thigh. When the thigh is fixed, they act as flexors of the vertebral column. The iliopsoas is an important postural muscle because it helps hold the trunk upright and keeps it from falling backward when one is standing erect.

Posterior Hip Muscles

The posterior hip muscles act as extensors and rotators of the thigh (Fig. 7.41 and Table 7.17). They include the three gluteal muscles, tensor fasciae latae, and the six lateral rotators.

Table 7.17   Muscles of Hip

Muscle	Origin	Insertion	Nerve	Action
ANTERIOR HIP MUSCLES				
Iliacus*	Iliac fossa	Lesser trochanter of femur via iliopsoas tendon	Femoral	Flexes thigh and rotates it laterally; flexes vertebral column when thigh is fixed
Psoas major*	Transverse processes and bodies of lumbar vertebrae	Lesser trochanter of femur via iliopsoas tendon	Femoral	Flexes thigh and rotates it laterally; flexes vertebral column when thigh is fixed
Psoas minor	Bodies and intervertebral discs of twelfth thoracic and first lumbar vertebrae	Pectineal line and iliopectineal eminence	L1	Flexes vertebral column
POSTERIOR HIP MUSCLES				
Gluteus maximus*	Upper ilium (external), sacrum, and coccyx	Gluteal tuberosity of femur and iliotibial band	Inferior gluteal	Extends and rotates thigh laterally
Gluteus medius*	Middle ilium (external)	Greater trochanter of femur	Superior gluteal	Abducts thigh; anterior part rotates thigh medially; posterior part rotates thigh laterally
Gluteus minimus*	Lower ilium (external)	Greater trochanter of femur	Superior gluteal	Abducts thigh and rotates it medially
Tensor fasciae latae*	Iliac crest and anterior superior iliac spine	Iliotibial band	Superior gluteal	Tenses fascia lata; assists in flexing and rotating leg medially
Piriformis	Pelvic surface of sacrum and sacrotuberous ligament	Greater trochanter of femur	Nerve to piriformis	Rotates thigh laterally
Superior gemellus	Ischial spine	Greater trochanter of femur	Nerve to obturator internus	Abducts and rotates thigh laterally
Obturator internus	Pelvic surface of pubis and ischium; obturator membrane	Greater trochanter of femur	Nerve to obturator internus	Abducts and rotates thigh laterally

Table 7.17   (Continued)

Muscle	Origin	Insertion	Nerve	Action
Inferior gemellus	Ischial tuberosity	Greater trochanter of femur	Nerve to quadratus femoris	Abducts and rotates thigh laterally
Quadratus femoris	Ischial tuberosity	Greater trochanter of femur	Nerve to quadratus femoris	Rotates thigh laterally
Obturator externus	Outer surface of pubis and ischium; obturator membrane	Greater trochanter of femur	Obturator	Rotates thigh laterally

* Muscles that must be learned.

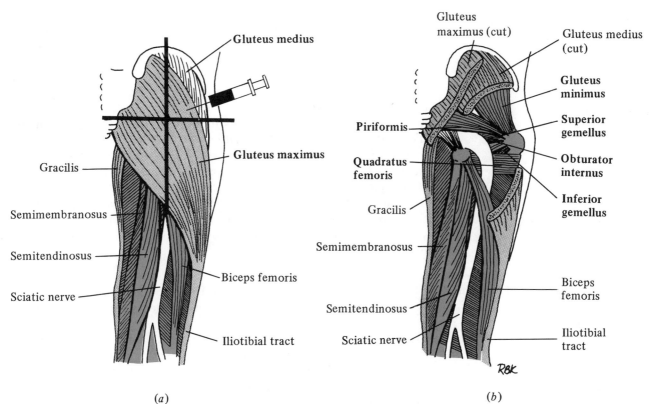

(a)   *(b)*

Figure 7.41   *Posterior Hip Muscles.*   (a) Superficial muscles. These also include tensor fasciae latae (not shown). Proper gluteal injection site is indicated (upper outer quadrant). (b) Deep muscles. Gluteus maximus and medius are cut away to expose gluteus minimus and six lateral rotators of hip. Note that sciatic nerve emerges below piriformis and runs deep to gluteus maximus.

The three gluteal muscles are large powerful muscles that form the muscle mass known as the buttock. *Gluteus maximus* is a thick fleshy quadrilateral muscle that forms the bulk of the buttock. It arises from the ilium, sacrum, and coccyx, and its coarse fibers run down and outward to insert on the femur and into the iliotibial band, part of the deep fascia of the thigh. Gluteus maximus extends and is the chief lateral rotator of the thigh. Through the iliotibial band it also stabilizes the knee joint when the leg is extended. In humans this muscle keeps the body upright in the act of walking. You use your gluteus maximus muscles when you get up from a sitting position, walk, run, or climb stairs. They also cushion your ischial tuberosities when you sit down. Since gluteus maximus is so fleshy, it is frequently used as a site for intramuscular injections. However, only the upper outer quadrant should be used, to avoid injecting into the gluteal vessels or the sciatic nerve, which runs deep to gluteus maximus. *Gluteus medius* is a fan-shaped muscle lying deep to gluteus maximus. It runs down and outward from the ilium to the greater trochanter of the femur. Deep to gluteus medius lies *gluteus minimus,* a smaller fan-shaped muscle that also arises from the ilium and inserts on the greater trochanter. These two muscles act together as abductors and rotators of the thigh.

Tensor fasciae latae is a flat muscle that tenses the deep fascia on the lateral aspect of the thigh (fascia lata) and also flexes and rotates the thigh medially. It inserts into the iliotibial band, a ligament-like portion of the fascia lata that crosses the knee joint and attaches to the lateral tibial condyle. Contraction of the tensor fasciae latae tightens the iliotibial band, which stabilizes the knee joint.

Deep to the gluteal muscles are six small muscles that arise from the posterior portion of the pelvis and insert on the greater trochanter of the femur: piriformis, quadratus internus and externus, gemellus superior and inferior, and quadratus femoris. These muscles act as a unit to rotate the thigh laterally.

Muscles of the Thigh

The adult thigh is divided into anterior, medial, and posterior compartments by intermuscular septa that extend inward from the deep fascia of the thigh to the femur. Due to the rotation of the lower extremity during prenatal development, the anterior compartment of the adult thigh corresponds to the posterior (postaxial) compartment of the embryo, while the medial and posterior compartments of the adult correspond to the anterior (preaxial) compartment of the embryo.

Anterior Compartment

The anterior compartment of the thigh is delimited by lateral and medial intermuscular septa. Muscles of the anterior compartment include the quadriceps femoris and sartorius muscles (Fig. 7.42), which act across the hip or knee joints (Table 7.18).

Quadriceps femoris is a large fleshy muscle mass that covers the front and sides of the femur. It has four heads of origin: rectus femoris, vastus lateralis, vastus intermedius, and vastus medialis. *Rectus femoris* arises from the ilium, while the three vasti arise from the shaft of the femur. They have a common tendinous insertion into the base of the patella, the patellar tendon. The apex of the patella, in turn, is attached to the tibial tuberosity by the patellar ligament. The quadriceps is the chief extensor of the leg. It is a very useful muscle in sports, especially for kicking a ball. Rectus femoris is the only part of the quadriceps group that crosses the front of the hip joint as well as the knee joint; therefore, it acts as a flexor of the thigh as well as an extensor of the leg.

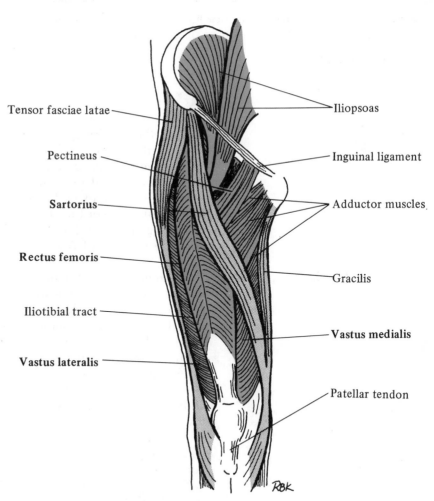

Figure 7.42 Anterior Compartment of Thigh. Muscles located here include sartorius and quadriceps femoris. Four heads of quadriceps femoris are rectus femoris, vastus lateralis, vastus medialis, and vastus intermedius (not seen because it lies deep to rectus femoris).

Table 7.18   Muscles of Anterior Compartment of Thigh

Muscle	Origin	Insertion	Nerve	Action
Quadriceps femoris*				
Rectus femoris	Anterior inferior iliac spine and upper part of rim of acetabulum	Base of patella and tibial tuberosity via patellar ligament	Femoral	Flexes thigh and extends leg
Vastus lateralis	Intertrochanteric line and linea aspera of femur	Base of patella and tibial tuberosity via patellar ligament	Femoral	Extends leg
Vastus intermedius	Upper half of femoral shaft	Base of patella and tibial tuberosity via patellar ligament	Femoral	Extends leg
Vastus medialis	Intertrochanteric line and linea aspera of femur	Base of patella and tibial tuberosity via patellar ligament	Femoral	Extends leg
Sartorius*	Anterior superior iliac spine	Medial side of proximal end of tibia	Femoral	Flexes thigh and leg

* Muscles that must be learned.

The *sartorius,* or "tailor's muscle," is a long strap muscle, superficial to the quadriceps, that extends obliquely downward from the ilium to the medial aspect of the upper tibia. Because the sartorius crosses the front of the hip joint, it flexes the thigh; because it passes behind the medial femoral condyle before inserting on the tibia, it flexes rather than extends the leg. The sartorius, whose name comes from the Latin word for tailor, is the muscle that enables you to sit cross-legged on the floor as Middle and Far Eastern tailors used to do and as students of Yoga still do.

Medial Compartment   The medial compartment of the thigh (Fig. 7.43) is delimited by medial and posterior intermuscular septa. Muscles of the medial compartment, which include pectineus, the three adductors, and the gracilis, act primarily across the hip joint to adduct, flex, and rotate the thigh laterally (Table 7.19).

The three triangular adductor muscles arise from the pubis and insert onto the femur; *adductor longus* is superficial, *adductor brevis* is intermediate, and *adductor magnus* is deep. The upper part of adductor magnus flexes and rotates the thigh medially, while the lower part extends and rotates the thigh laterally. The adductors, especially adductor magnus, draw the thighs together. If you go horseback riding, these muscles enable you to grip the horse's sides with your knees.

The *gracilis,* or "graceful," muscle is a slender superficial muscle that extends down the medial side of the thigh from the pubis to the

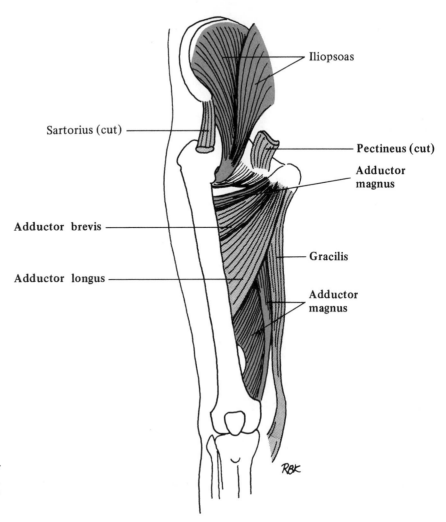

Iliopsoas

Sartorius (cut)

Pectineus (cut)

Adductor
magnus

Adductor brevis

Gracilis

Adductor longus

Adductor
magnus

Figure 7.43 Medial Compartment of Thigh. All muscles of anterior compartment are removed to expose muscles of medial compartment.

upper tibia. It acts across the hip joint to adduct the thigh and across the knee joint to flex and rotate the leg medially.

Posterior Compartment

The posterior compartment of the thigh is delimited by posterior and lateral intermuscular septa. The three muscles of the posterior compartment are collectively known as the *hamstrings*, "ham" referring to the fleshy part of the thigh and "strings" referring to the long tendons of these muscles (Fig. 7.44). The hamstrings, described in Table 7.20, all arise from the ischial tuberosity. The tendon of *biceps femoris*, the lateral hamstring, inserts on the lateral side of the tibia, while the tendons of *semitendinosus* and *semimembranosus*, the medial hamstrings, insert on the medial side of the tibia. The three

Table 7.19   Muscles of Medial Compartment of Thigh

Muscle	Origin	Insertion	Nerve	Action
Pectineus	Pectineal line of pubis	Femur between lesser trochanter and linea aspera	Obturator and femoral	Adducts, flexes, and rotates thigh laterally
Adductor brevis*	Body of pubis below origin of adductor longus	Upper third of linea aspera	Obturator	Adducts, flexes, and rotates thigh laterally
Adductor longus*	Body of pubis	Middle third of linea aspera	Obturator	Adducts, flexes, and rotates thigh laterally
Adductor magnus*	Adductor head, ischiopubic ramus; extensor head, ischial tuberosity	Linea aspera and adductor tubercle of femur	Obturator and sciatic	Adducts and flexes thigh; upper part flexes and lower part extends thigh
Gracilis*	Inferior pubic ramus and pubic symphysis	Upper medial part of tibial shaft	Obturator	Adducts thigh; flexes and rotates leg medially

* Muscles that must be learned.

hamstrings assist gluteus maximus in extending the thigh and are the chief flexors of the leg, the antagonists of quadriceps femoris.

Popliteal Fossa   The popliteal fossa is the diamond-shaped depression on the posterior aspect of the knee. Structures of clinical importance in the popliteal fossa are the popliteal blood vessels plus the distal end of the sciatic nerve and its branches.

Muscles of the Leg   The leg contains muscles that act across the ankle, metatarsophalangeal (MP), and interphalangeal (IP) joints. The adult leg is divided into anterior, lateral, and posterior compartments by the tibia, fibula, interosseous membrane, and intermuscular septa that extend inward from the deep fascia of the leg. Also, due to prenatal rotation of the lower extremity, the anterior compartment of the adult leg corresponds to the posterior (postaxial) compartment of the embryo, while the lateral and posterior compartments of the adult leg correspond to the anterior (preaxial) compartment of the embryo.

Anterior Compartment   The anterior compartment of the leg is delimited by the tibia and anterior intermuscular septum. The three muscles of the anterior compartment of the leg, illustrated in Fig. 7.45 and described in Table 7.21, act as dorsiflexors and invertors of the foot and extensors of the toes.

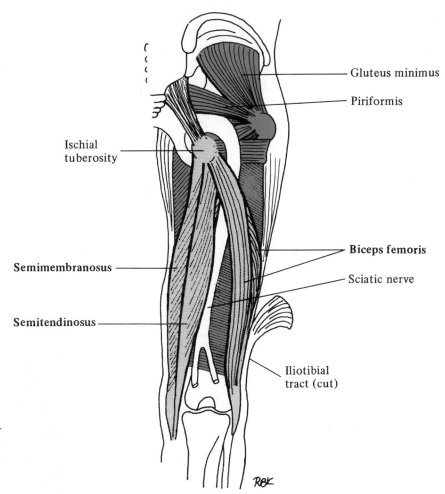

Gluteus minimus

Piriformis

Ischial
tuberosity

Biceps femoris

Sciatic nerve

Iliotibial
tract (cut)

Semimembranosus

Semitendinosus

Figure 7.44 Posterior Compartment of Thigh. Muscles here include three hamstrings: biceps femoris (lateral), semitendinosus (medial), and semimembranosus (medial).

Table 7.20 Muscles of Posterior Compartment of Thigh

Muscle	Origin	Insertion	Nerve	Action
Biceps femoris*	Short head, linea aspera; long head, ischial tuberosity	Head of fibula, lateral tibial condyle	Sciatic	Extends thigh; flexes and rotates leg laterally
Semitendinosus*	Ischial tuberosity	Upper part of tibia (medial side)	Sciatic	Extends thigh; flexes and rotates leg medially
Semimembranosus*	Ischial tuberosity	Medial tibial condyle	Sciatic	Extends thigh; flexes and rotates leg medially

* Muscles that must be learned.

Tibialis anterior, a long slender muscle whose fleshy belly can be palpated lateral to the shin, arises from the lateral surface of the tibia. Its tendon crosses over to the medial side of the ankle, turns under the medial border of the foot, and inserts on the plantar surface. Because tibialis anterior crosses the front of the ankle joint, it is a powerful dorsiflexor of the foot; because its tendon turns under the medial border of the foot, it also acts as an invertor of the foot.

Two muscles arising from the tibial or fibular shafts serve as the extrinsic extensors of the digits. *Extensor digitorum longus* inserts on the phalanges of the lateral toes, while *extensor hallucis longus* inserts on the distal phalanx of the big toe. Because these two muscles cross the front of the ankle joint, they assist in dorsiflexing the foot in addition to extending the toes at the MP and IP joints.

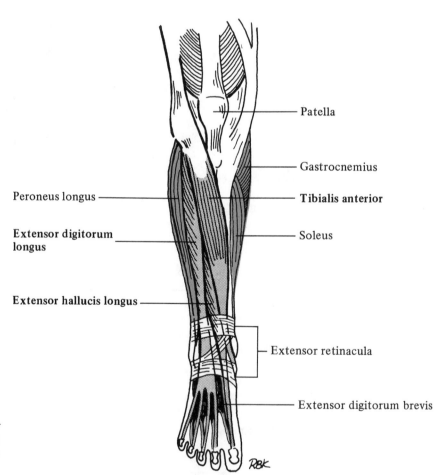

Peroneus longus

Extensor digitorum longus

Extensor hallucis longus

Patella

Gastrocnemius

Tibialis anterior

Soleus

Extensor retinacula

Extensor digitorum brevis

Figure 7.45 Anterior Compartment of Leg. Note how extensor retinacula bind down tendons of these muscles as they cross front of ankle joint.

Lateral Compartment

The lateral, or peroneal, compartment of the leg is delimited by the anterior and posterior intermuscular septa. The two muscles contained in this compartment (Fig. 7.46 and Table 7.21) act as evertors and assistant plantarflexors of the foot.

The two peroneal muscles arise from the fibula. Their tendons pass behind the lateral malleolus of the ankle and turn under the foot to gain the plantar surface; that of *peroneus longus* inserts on the medial side of the foot, while that of *peroneus brevis* inserts on the lateral side of the foot. The peroneal muscles are the chief evertors of the foot and the antagonists of tibialis anterior and posterior; they also help plantarflex the foot. The tendon of peroneus longus along with that of tibialis anterior forms a protective sling for the longitudinal arch of the foot.

Posterior Compartment

The posterior compartment of the leg is delimited by the tibia, fibula, interosseous membrane, and posterior intermuscular septum. The muscles contained in the posterior compartment of the leg, described

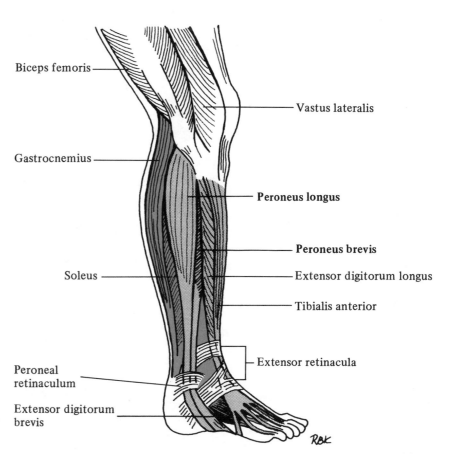

Figure 7.46 Lateral Compartment of Leg.

Table 7.21 Muscles of Anterior and Lateral Compartments of Leg

Muscle	Origin	Insertion	Nerve	Action
ANTERIOR COMPARTMENT				
Tibialis anterior*	Upper two-thirds of tibial shaft and interosseous membrane	Medial cuneiform and first metatarsal	Deep peroneal	Inverts and dorsiflexes foot
Extensor digitorum longus*	Tibia, upper three-fourths of fibular shaft, and interosseous membrane	Lateral toes by way of extensor expansions	Deep peroneal	Extends toes at MP and IP joints; dorsiflexes foot
Extensor hallucis longus*	Middle of fibular shaft and interosseous membrane	Distal phalanx of big toe	Deep peroneal	Extends big toe at MP and IP joints; dorsiflexes foot
Peroneus tertius	Lower fourth of fibular shaft and interosseous membrane	Fifth metatarsal	Deep peroneal	Dorsiflexes and everts foot
LATERAL COMPARTMENT				
Peroneus longus*	Lateral tibial condyle; head and upper two-thirds of fibula	Medial cuneiform and first metatarsal	Superficial peroneal	Everts and plantarflexes foot
Peroneus brevis*	Lower two-thirds of fibular shaft	Fifth metatarsal	Superficial peroneal	Everts and plantarflexes foot

* Muscles that must be learned.

in Table 7.22, act primarily as plantarflexors of the foot and flexors of the toes. They are further subdivided into superficial and deep groups.

The muscles of the superficial group form a prominent mass called the calf of the leg [Fig. 7.47(a)]. The *gastrocnemius* arises by two heads from the femoral condyles, while the *soleus,* which lies beneath the gastrocnemius, arises from the shafts of the tibia and fibula. These two muscles insert on the calcaneus by way of a common tendon known as the *calcaneus* or *Achilles tendon.* The calf muscles are the chief flexors of the foot and are very important in walking, running, and jumping. These are the muscles that enable ballet dancers to dance on the tips of their toes. The gastrocnemius, which crosses the back of the knee joint as well as the ankle joint, also flexes the leg.

The muscles of the deep group [Fig. 7.47(b)] are separated from those of the superficial group by a transverse intermuscular septum. *Flexor digitorum longus* arises from the tibial shaft, passes behind the medial malleolus of the ankle, and inserts on the distal phalanges of the four lateral toes. *Flexor hallucis longus* arises from the fibular

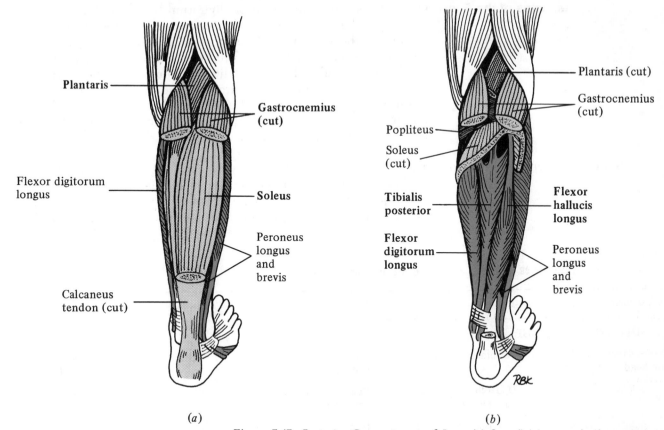

Plantaris

Gastrocnemius (cut)

Flexor digitorum longus

Soleus

Peroneus longus and brevis

Calcaneus tendon (cut)

(a)

Plantaris (cut)

Gastrocnemius (cut)

Popliteus

Soleus (cut)

Tibialis posterior

Flexor hallucis longus

Flexor digitorum longus

Peroneus longus and brevis

(b)

Figure 7.47 *Posterior Compartment of Leg.* (a) Superficial group (calf muscles). Gastrocnemius is cut away to expose deeper-lying plantaris and soleus. (b) Deep group. Calf muscles are cut away to expose deep muscles of posterior compartment: tibialis posterior, flexor hallucis longus, and flexor digitorum longus.

shaft, passes behind the medial malleolus of the ankle, and inserts on the distal phalanx of the big toe. Because these two muscles run behind the ankle joint, they assist in plantarflexing the foot as well as in flexing the toes at the MP and IP joints. *Tibialis posterior* is the deepest muscle in the posterior group. Its tendon also passes behind the medial malleolus of the ankle, then turns under the foot and inserts broadly on the plantar surface. Together with tibialis inferior it is an invertor of the foot, and, because it crosses the back of the ankle joint, acts as an assistant plantarflexor.

Special Features of the Ankle The structures of clinical interest at the ankle are the long tendons that pass over the ankle joint. Three sets of retinacula are present to hold these tendons in place: the superior and inferior extensor retinacula and the peroneal retinaculum.

Muscles of the Foot One intrinsic muscle, extensor digitorum brevis, is located on the dorsum of the foot. The rest of the intrinsic muscles of the foot are located on the sole or plantar surface (Fig. 7.48 and Table 7.23). In general, the plantar muscles correspond to those of the palm of the hand. Individually, they are of little importance; collectively, they are important in supporting the arches of the foot, in maintaining good posture, and in locomotion. Remember that the foot has become highly specialized for weight bearing and locomotion and has lost many movements possible to the hand.

Special Features of the Foot The dorsum of the foot is relatively simple. It is covered with a thin flexible layer of deep fascia that covers and provides a sheath for the extensor tendons. The sole of the foot is very complicated, with the intrinsic muscles being arranged in four layers. The deep fascia of

Table 7.22 Muscles of Posterior Compartment of Leg

Muscle	Origin	Insertion	Nerve	Action
SUPERFICIAL GROUP				
Gastrocnemius* (2 heads)	Femoral condyles	Calcaneus by way of calcaneus tendon	Tibial	Flexes leg and plantarflexes foot
Soleus*	Upper third of fibula, soleal line of tibia	Calcaneus by way of calcaneus tendon	Tibial	Flexes leg and plantarflexes foot
Plantaris (often absent)	Lower part of linea aspera and popliteal ligament	Medial side of calcaneus tendon	Tibial	Flexes leg and plantarflexes foot
DEEP GROUP				
Popliteus	Lateral femoral condyle	Proximal fourth of tibia	Tibial	Rotates thigh laterally when leg is extended
Flexor digitorum longus*	Tibia below soleal line	Distal phalanges of four lateral toes	Tibial	Flexes four lateral toes at MP and IP joints; assists in plantarflexing foot
Flexor hallucis longus*	Lower two-thirds of fibula and intermuscular septum	Distal phalanx of big toe	Tibial	Flexes big toe at MP and IP joints; assists in plantarflexing foot
Tibialis posterior*	Interosseous membrane and adjoining shafts of tibia and fibula	Navicular, cuboid, cuneiforms, and second through fourth metatarsals	Tibial	Inverts and plantarflexes foot

* Muscles that must be learned.

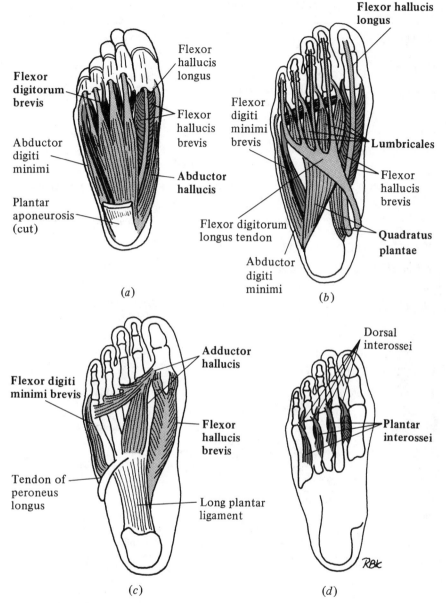

Flexor hallucis longus

Flexor hallucis longus

Flexor digitorum brevis

Flexor hallucis brevis

Flexor digiti minimi brevis

Lumbricales

Abductor digiti minimi

Flexor hallucis brevis

Plantar aponeurosis (cut)

Abductor hallucis

Flexor digitorum longus tendon

Quadratus plantae

Abductor digiti minimi

(a)

(b)

Adductor hallucis

Dorsal interossei

Flexor digiti minimi brevis

Flexor hallucis brevis

Plantar interossei

Tendon of peroneus longus

Long plantar ligament

(c)

(d)

Figure 7.48 Intrinsic Muscles of Foot (Plantar Surface). (a) First layer. This includes flexor digitorum brevis, which inserts on middle phalanges of four lateral toes. (b) Second layer. This includes quadratus plantae and four lumbricales, which arise from tendons of flexor digitorum longus. Quadratus plantae and lumbricales are assistant flexors of MP and IP joints of four lateral toes. (c) Third layer. (d) Fourth layer. This includes plantar interossei, which adduct toes at MP joints and flex them at proximal IP joints. Deep to plantar interossei are dorsal interossei, which abduct toes at MP joints and flex them at proximal IP joints.

the foot forms the plantar aponeurosis, which is attached to the calcaneus posteriorly and the metatarsals anteriorly. It binds down the intrinsic muscles and supports the longitudinal arch of the foot.

INNERVATION OF LOWER EXTREMITY MUSCLES

The muscles of the lower extremity are innervated by the nerves of the lumbar and sacral plexuses (Figs. 7.49 and 7.50).

Table 7.23 Intrinsic Muscles of Foot

Muscle	Origin	Insertion	Nerve	Action
DORSAL				
Extensor digitorum brevis	Dorsal surface of calcaneus	Distal phalanges of four lateral toes	Deep peroneal	Extends four lateral toes
PLANTAR–FIRST LAYER				
Abductor hallucis	Calcaneus	Proximal phalanx of big toe	Medial plantar	Abducts and flexes big toe
Flexor digitorum brevis	Calcaneus, plantar aponeurosis	Proximal phalanges of four lateral toes	Lateral plantar	Flexes four lateral toes
Abductor digiti minimi	Calcaneus	Proximal phalanx of little toe	Lateral plantar	Abducts little toe
PLANTAR–SECOND LAYER				
Quadratus plantae	Calcaneus and plantar fascia	Tendons of flexor digitorum longus	Lateral plantar	Assists in flexing four lateral toes
Lumbricales (4)	Tendons of flexor digitorum longus	Extensor expansions on four lateral toes	Medial and lateral plantar	Flex four lateral toes at MP joints and extends them at IP joints
PLANTAR–THIRD LAYER				
Flexor hallucis brevis	Cuboid and lateral cuneiform	Proximal phalanx of big toe	Medial plantar	Flexes big toe
Adductor hallucis	Oblique head, long plantar ligament; transverse head, capsules of the MP joints	Proximal phalanx of big toe	Lateral plantar	Adducts and flexes big toe
PLANTAR–FOURTH LAYER				
Plantar interossei (3)	Medial side of third, fourth, and fifth metatarsals	Medial side of proximal phalanges of third, fourth, and fifth toes	Lateral plantar	Adducts toes at MP joints toward second toe
Dorsal interossei (4)	Adjacent sides of metatarsals	Proximal phalanx of second toe (both sides) plus proximal phalanges of third and fourth toes (lateral sides)	Lateral plantar	Abducts toes at MP joints away from second toe

Lumbar Plexus

The *femoral nerve* is the largest nerve of the lumbar plexus. Above the inguinal ligament it innervates the iliopsoas muscle; below the inguinal ligament it innervates muscles of the anterior compartment of the thigh. The femoral nerve can be injured by fractures of the pelvis or femur and by use of obstetrical forceps during childbirth. Iliopsoas paralysis interferes with thigh flexion. Paralysis of quadriceps femoris forces a person to walk with the affected leg flexed. The *obturator nerve,* the other major nerve of the lumbar plexus, innervates muscles of the medial compartment of the thigh. When this nerve is injured, the adductor muscles are paralyzed; the person with adductor muscle paralysis has difficulty in crossing the legs.

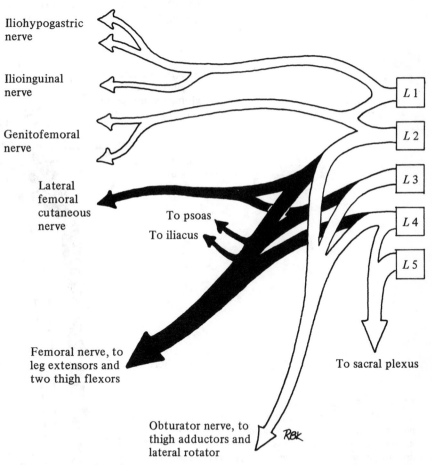

Iliohypogastric nerve

Ilioinguinal nerve

Genitofemoral nerve

Lateral femoral cutaneous nerve

To psoas

To iliacus

L 1
L 2
L 3
L 4
L 5

To sacral plexus

Femoral nerve, to leg extensors and two thigh flexors

Obturator nerve, to thigh adductors and lateral rotator

RBK

Figure 7.49 Lumbar Plexus. This is formed from anterior primary rami of spinal nerves *L*1–*L*4. Rami divide and regroup to form terminal branches that contain either anterior fibers *(white)* or posterior fibers *(black)*. Nerves containing anterior fibers innervate muscles that were originally preaxial in location, while those containing posterior fibers innervate muscles that were originally postaxial in location. Major terminal nerves of lumbar plexus are femoral and obturator nerves. Distribution of lumbar plexus is above knee, with exception of lateral femoral cutaneous nerve.

Sacral Plexus

The *sciatic nerve* is the largest nerve of the sciatic plexus and of the entire body. As it descends from the greater sciatic notch to the knee, it innervates the muscles of the posterior compartment of the thigh. In the popliteal fossa it gives rise to the *tibial* and *common peroneal nerves*. The *deep peroneal nerve*, one branch of the common peroneal nerve, innervates muscles of the anterior compartment of the leg; the *superficial peroneal nerve*, the other branch of the common peroneal nerve, innervates muscles of the lateral compartment of the leg. The tibial nerve innervates muscles of the posterior compartment of the leg. The medial and lateral plantar nerves, terminal branches of the tibial nerve, innervate the muscles of the sole of the foot. The sciatic nerve can be injured by a herniated lumbar disc, by use of obstetrical forceps, or by accidental injection with neurotoxic substances such as alcohol, mercury, or penicillin. Injury to the sciatic nerve produces

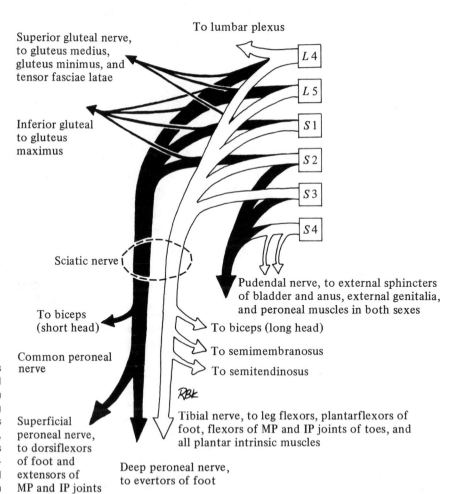

Figure 7.50 Sacral Plexus. This is formed from anterior primary rami of spinal nerves *L4–S4* and is thus continuous with lumbar plexus. Terminal nerves containing anterior fibers *(white)* innervate muscles that were originally preaxial in location, while those containing posterior fibers *(black)* innervate muscles that were originally postaxial in location. Major terminal nerves of sacral plexus are tibial, common peroneal, and pudendal.

hamstring paralysis plus paralysis of all the muscles of the leg and foot. Injury to the common peroneal nerve, as by a fracture of the fibular head, produces *footdrop,* which is due to paralysis of the dorsiflexors and evertors of the foot. Injury to the tibial nerve causes paralysis of the plantarflexors plus atrophy of the intrinsic muscles of the foot *(clawfoot).*

Diseases and Disorders of the Muscular System

Diseases and disorders of the muscular system can be divided into the following categories: muscle agenesis, hernia, atrophy and degeneration, inflammatory disease, and neuromuscular disease.

MUSCLE AGENESIS Occasionally, all or part of a given muscle fails to develop embryologically. This frequently happens to the pectoralis major. Incomplete development of the thoracic diaphragm is the most common cause of diaphragmatic hernia. Muscles of the lower portion of the anterolateral abdominal wall may fail to develop; in extreme cases the bladder may be exposed through a defect in the anterolateral abdominal wall. Small muscles such as the palmaris longus (forearm) or the plantaris (leg) may fail to develop, but they are scarcely missed.

The term "hernia," used in connection with the muscular system, refers to protrusion of abdominal viscera through a defect or weak spot in the abdominal musculature. The most common type of hernia is the *inguinal hernia.* In this condition a portion of the small intestine either pushes straight through the weakest spot in the anterolateral abdominal wall (direct inguinal hernia) or down through the inguinal canal (indirect inguinal hernia). Hernias can also occur through the diaphragm or around the umbilicus. Factors leading to herniation are heavy coughing, improper lifting of heavy objects, or straining during defecation.

ATROPHY AND DEGENERATION *Atrophy.* Muscle atrophy refers to the reduction in size of previously normal skeletal muscles through loss of innervation, disuse, infection, or malnutrition.

Muscular Dystrophy. Muscular dystrophy is a degenerative disease of skeletal muscle. Sex-linked or Duchenne-type muscular dystrophy is a recessive inherited disease that occurs almost entirely in males. In this disease the skeletal muscle fibers first appear large and swollen; they later atrophy and are replaced by fibrous connective tissue. Duchenne-type muscular dystrophy has its onset during childhood or puberty and initially affects the proximal muscles of the extremities. It is slowly progressive, with the affected individuals becoming weaker and weaker. Pneumonia, the result of respiratory muscle involvement, is the most frequent cause of death in muscular dystrophy.

INFLAMMATORY DISEASE The term *myositis* refers to inflammation of skeletal muscles. Myositis is generally caused by trauma or infection by any of the common types of microorganisms.

NEUROMUSCULAR DISEASE *Tetanus.* Tetanus, commonly known as lockjaw, is a generally fatal neuromuscular disease caused by a toxin produced by the tetanus bacillus. Tetanus toxin is carried from the site of the primary lesion to the CNS, where it blocks the release of neurotransmitters from inhibitory synapses on motor neurons in the brain stem and spinal cord. The increased excitability of these motor neurons results in the convulsions and muscle spasms that characterize tetanus. Muscles of the abdomen, back, neck, and face are most likely to go into spasm. Risus sardonicus, the sardonic smile of the tetanus victim, is caused by spasms of the risorius muscles. Asphyxiation during a convulsive seizure is a common cause of death for tetanus victims. Tetanus can be prevented by an initial immunization with tetanus toxoid followed by booster immunizations at four-year intervals. Mild to moderate cases of tetanus have been successfully treated with a combination of antitoxin, antibiotics, tranquilizers, and respiratory support systems.

Myasthenia Gravis. Myasthenia gravis is a baffling neuromuscular disease characterized by weakness and fatigue of certain muscles (most frequently those controlling eye movements, facial expression, chewing, swallowing, and breathing) upon sustained effort. The condition is thus worse in the evening than in the morning.

In severe cases, death usually follows involvement of the respiratory or swallowing muscles. Myasthenic muscles show a decreased ability to respond to repetitive nerve stimuli, apparently due to failure of adequate amounts of acetylcholine in reaching receptor sites on the motor end plates. Current therapy consists of the administration of drugs that either prevent rapid breakdown or stimulate production of acetylcholine. Myasthenia gravis appears to have an immunological basis since removal of the thymus gland (an important part of the immune system) is followed by marked improvement in many myasthenic patients.

Suggested Readings

° BASMAJIAN, J. V. *Muscles Alive.* 3d ed. Baltimore: Williams & Wilkins, 1974.

° GOODGOLD, JOSEPH. *Anatomical Correlates of Clinical Electromyography.* Baltimore: Williams & Wilkins, 1974.

HOLLINGSHEAD, W. H. *Functional Anatomy of the Limbs and Back*. 3d ed. Philadelphia: Saunders, 1969.

HOYLE, GRAHAM. "How is muscle turned on and off?" *Sci. Am., 222,* no. 4 (April 1970), pp. 84–93.

HUXLEY, H. E. "The mechanism of muscular contraction," *Sci. Am., 213,* no. 6 (December 1965), pp. 18–27.

KATZ, BERNARD. *Nerve, Muscle, and Synapse*. New York: McGraw-Hill, 1966.

KONIGSBERG, IRWIN R. "The embryological origin of muscle," *Sci. Am., 211,* no. 2 (August 1964), pp. 61–66.

MERTON, P. A. "How we control the contraction of our muscles," *Sci. Am., 226,* no. 5 (May 1972), pp. 30–37.

MURRAY, JOHN M., and ANNEMARIE WEBER. "The cooperative action of muscle proteins," *Sci. Am., 230,* no. 2 (February 1974), pp. 58–71.

PORTER, KEITH R. "The sarcoplasmic reticulum," *Sci. Am., 212,* no. 3 (March 1965), pp. 72–80.

RAUSCH, PHILIP J., and RODGER K. BURKE. *Kinesiology and Applied Anatomy*. 4th ed. Philadelphia: Lea & Febiger, 1971.

ROSSE, C., and D. K. CLAWSON. *Introduction to the Musculoskeletal System*. New York: Harper & Row, 1970.

SOKOLOFF, LEON, and JOHN H. BLAND. *The Musculoskeletal System*. Baltimore: Williams & Wilkins, 1975.

THOMPSON, CLEM W. *Manual of Structural Kinesiology*. St. Louis: Mosby, 1969.

WELLS, KATHARINE F., and KATHRYN LUTTGENS. *Kinesiology: Scientific Basis of Human Motion*. Philadelphia: Saunders, 1976.

° Advanced level.

8

The Cardiovascular System

The cardiovascular system consists of a muscular pump, *the heart*, which pumps a fluid, *the blood,* throughout the body in a system of closed tubes, the *blood vessels*. This chapter is divided into three parts: Part A deals with the heart and its diseases, Part B deals with the blood and its diseases, and Part C deals with the vascular system and its diseases.

Part A. The Heart

The heart is a hollow muscular pump located in the *mediastinum,* the space in the thoracic cavity between the two pleural sacs. The heart lies between the lungs, behind the sternum and costal cartilages, in front of the esophagus and thoracic aorta, and on top of the thoracic diaphragm. A third of the heart lies to the right of the midline of the sternum and two-thirds lies to the left (Fig. 8.1).

Blood Flow Through the Adult Heart

The adult heart consists of four chambers: the right and left *atria* (singular, "atrium") and the right and left *ventricles* (Fig. 8.2). The right atrium receives deoxygenated blood from the systemic (body) veins and passes it to the right ventricle, which pumps it into the pulmonary (lung) arteries. The pulmonary arteries carry deoxygenated blood to the lungs for oxygenation. The left atrium receives oxygenated blood from the pulmonary veins and passes it to the left

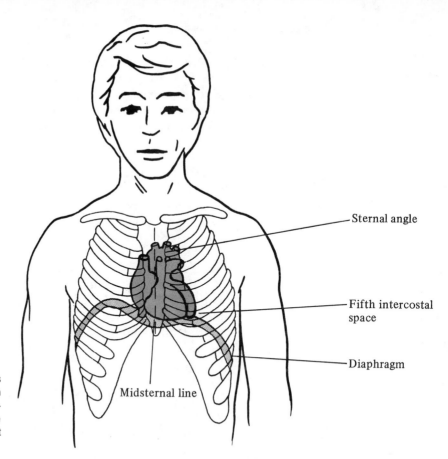

Sternal angle

Fifth intercostal space

Diaphragm

Midsternal line

Figure 8.1 *Location of Heart.* Heart is shown in thoracic cavity, behind sternum and on top of diaphragm. Note that two-thirds of heart lies to left of midsternal line and that its apex is at level of the left fifth intercostal space.

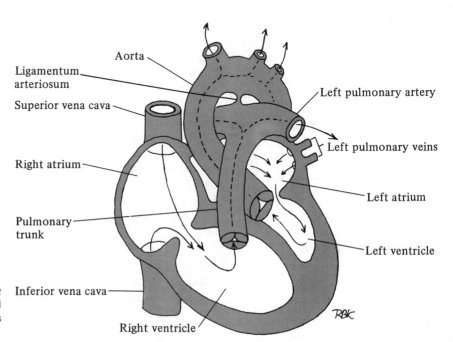

Aorta

Ligamentum arteriosum

Superior vena cava

Right atrium

Pulmonary trunk

Inferior vena cava

Right ventricle

Left pulmonary artery

Left pulmonary veins

Left atrium

Left ventricle

Figure 8.2 *Blood Flow Through Adult Heart.* *Arrows* indicate course of blood flow through chambers and great vessels of adult heart.

ventricle, which pumps it into the systemic arteries. The systemic arteries carry oxygenated blood to the tissues of the body, where it gives up oxygen and picks up carbon dioxide. Normally there is no communication between the right and left sides of the heart and no mixing of oxygenated blood with deoxygenated blood.

Gross Anatomy of the Heart

EXTERNAL ANATOMY The heart is a pyramidal structure, about the size of a person's clenched fist. It has a base, an apex, two surfaces, two borders, and several grooves (Fig. 8.3).

Base and Apex The *base* of the heart is directed to the right, upward, and backward; it is formed by the left atrium and a portion of the right. The *apex* is directed to the left, forward, and downward; it is formed mainly by the left ventricle. The apical heartbeat can be felt and heard at the left fifth intercostal space, about $3\frac{1}{2}$ inches from the midsternal line.

Surfaces The convex *anterior (sternocostal) surface* of the heart is formed by the anterior aspects of the right atrium and ventricle. The flattened *posterior (diaphragmatic) surface* is formed by the posterior aspects of the right and left ventricles plus a portion of the right atrium.

Borders The *right border* of the heart is long and angular; it is formed by the right atrium and ventricle. The *left border,* which is short and rounded, is formed mainly by the left ventricle.

Grooves The *atrioventricular (coronary) sulcus* marks the division between the atria and ventricles. The *anterior* and *posterior interventricular sulci* mark the division between the right and left ventricles.

INTERNAL ANATOMY The interior of the heart is partitioned into two upper chambers (atria) and two lower chambers (ventricles). The *interatrial septum* separates the right and left atria, the *interventricular septum* separates the right and left ventricles, and the *atrioventricular valves* partially separate the atria from their corresponding ventricles. The chambers of the heart are best studied in the order in which they are traversed by the blood (Fig. 8.4).

Right Atrium and Auricle The right atrium is a thin-walled chamber with a hollow flaplike appendage called the right auricle. The inner surface of the right atrium is smooth, while that of the auricle is thrown into parallel ridges by bundles of muscle fibers called *musculi pectinati.* The *fossa ovalis,* an oval depression on the interatrial septum, marks the

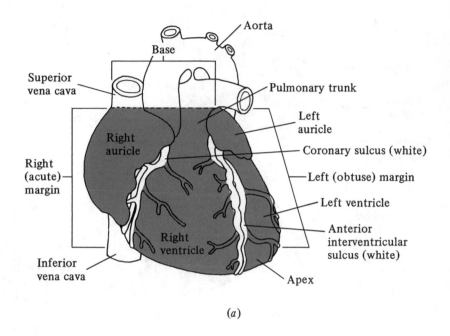

(a)

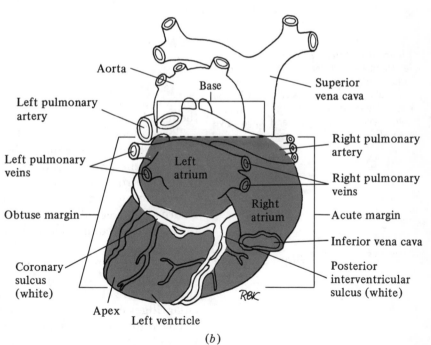

(b)

Figure 8.3 *External Anatomy of Heart.* (a) Sternocostal (anterior) surface. (b) Diaphragmatic (posterior) surface.

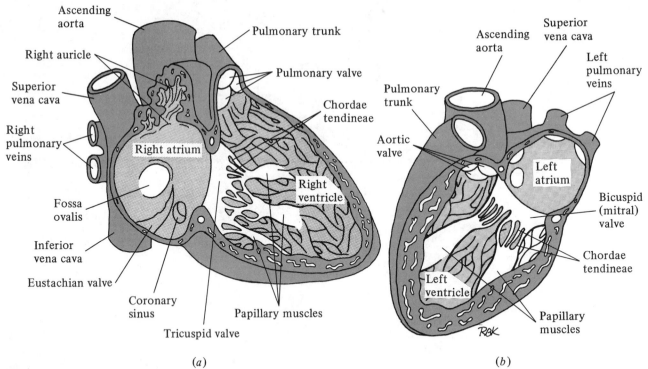

Ascending aorta

Right auricle

Superior vena cava

Right pulmonary veins

Fossa ovalis

Inferior vena cava

Eustachian valve

Coronary sinus

Tricuspid valve

Pulmonary trunk

Pulmonary valve

Chordae tendineae

Right atrium

Right ventricle

Papillary muscles

(a)

Ascending aorta

Superior vena cava

Left pulmonary veins

Pulmonary trunk

Aortic valve

Left atrium

Bicuspid (mitral) valve

Chordae tendineae

Left ventricle

Papillary muscles

(b)

Figure 8.4 *Internal Anatomy of Heart.* (a) Right heart. Heart is cut so that right atrium, tricuspid valve, right ventricle, and pulmonary valve are visible. (b) Left heart. Heart is cut so that left atrium, mitral valve, left ventricle, and aortic valve are visible.

former opening of the foramen ovale, an important structure in the fetal heart.

Three blood vessels enter the right atrium. The *superior vena cava*, which returns deoxygenated blood from the upper half of the body, enters on the upper part of the posterior wall. The *inferior vena cava*, which returns deoxygenated blood from the lower half of the body. enters the lower part of the posterior wall. The *coronary sinus*, which returns deoxygenated blood from the heart muscle, enters between the atrioventricular valve and the opening of the inferior vena cava.

Right Ventricle   Since it functions as a pumping rather than receiving chamber, the right ventricle has a thicker wall then the right atrium. The inner surface of the right ventricle is thrown into a series of irregular muscular ridges called *trabeculae carneae* ("fleshy bundles").

The right atrioventricular valve regulates blood flow into the right ventricle. Because it has three triangular cusps, it is called the *tricuspid valve*. The apex of each valve cusp is anchored by fine tendinous cords *(chordae tendineae)* to a *papillary muscle*. The three papillary muscles are projections from the wall of the right ventricle. This arrangement prevents reversal of the valve cusps and backflow of blood into the atrium when the ventricle contracts.

The *pulmonary trunk* is the outflow vessel of the right ventricle. After it emerges from the heart, it divides into right and left pulmonary arteries, which carry blood to the lungs for oxygenation and removal of carbon dioxide. Blood flow through the pulmonary trunk is regulated by the *pulmonary valve*. This type of heart valve is known as a *semilunar* valve because its three cuplike cusps are shaped like half-moons. During ventricular contraction the force of blood ejected from the right ventricle opens the valve by pushing the cusps against the vessel wall. During relaxation of the heart, backflow of blood in the pulmonary trunk fills the cusps and closes the valve, preventing the blood from reentering the chamber it has just left.

Left Atrium and Auricle

The left atrium is a thin-walled chamber that has a flaplike appendage called the left auricle. Aside from the auricle, which contains musculi pectinati, the inner surface of the left atrium is smooth. Four *pulmonary veins*, which return oxygenated blood from the lungs, enter the posterior wall of the left atrium.

Left Ventricle

The wall of the left ventricle is approximately $\frac{3}{5}$ inch thick, three times as thick as the wall of the right ventricle. This is due to the fact that the left ventricle pumps blood into a much more extensive circuit than the right ventricle does. In cross section (Fig. 8.5) the cavity of the left ventricle is circular in outline while that of the right is cresent-shaped; this is due to the bulging of the thick-walled left ventricle into the thinner-walled right ventricle. The inner surface of the left ventricle contains many trabeculae carneae and gives rise to two large papillary muscles.

The left atrioventricular valve regulates blood flow into the left ventricle. Because it has two triangular cusps, it is called the *bicuspid valve*. However, it is more commonly known as the *mitral valve* due to its resemblance to a bishop's miter (ceremonial hat). The

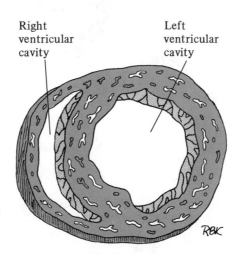

Right
ventricular
cavity

Left
ventricular
cavity

Figure 8.5 *Cross Section Through Ventricles of Heart.* Note that cavity of right ventricle is crescent-shaped in outline while that of left ventricle is circular.

apex of each valve cusp is attached to one of the papillary muscles by very strong chordae tendineae.

The *aorta* is the outflow vessel of the left ventricle. It is the major artery of the systemic circuit, the circuit that delivers oxygenated blood to all parts of the body except the lungs. Blood flow through the aorta is regulated by the *aortic valve*, a semilunar valve like the pulmonary valve. However, the cusps of the aortic valve, which are called *aortic sinuses*, are much larger than those of the pulmonary valve.

Microscopic Anatomy of the Heart

STRUCTURE OF THE HEART WALL   The heart wall is composed of three coats (Fig. 8.6): endocardium, myocardium, and epicardium.

Endocardium   The endocardium is the inner coat of the heart wall. It consists of a sheet of endothelium overlying a base of connective tissue. The

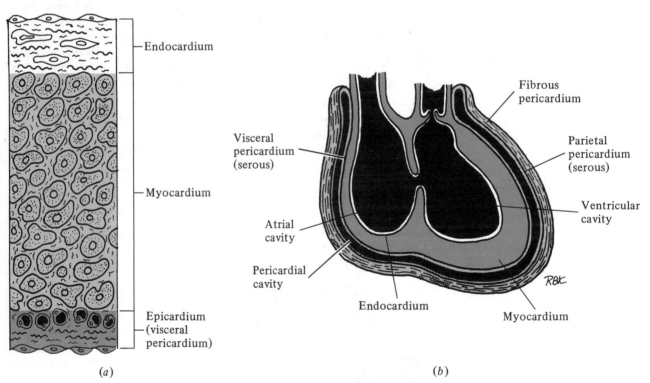

(a)

(b)

Figure 8.6 *Structure of Heart Wall.* (a) Schematic section through heart wall indicates relative thicknesses of endocardium *(white)*, myocardium *(light gray)*, and epicardium *(dark gray)*. (b) Schematic diagram of heart with coats of heart wall and pericardial sac indicated.

endocardium of the heart is continuous with the tunica intima of the blood vessels. Inflammation of the endocardium is called *endocarditis*.

Myocardium

The myocardium is the middle and thickest coat of the heart wall. It is thinner in the atria, where the work requirement is minimal, and thicker in the ventricles, where the work requirement is maximal. The myocardium consists of cardiac muscle fibers, which were described in Chapter 3, plus connective tissue for support. The cardiac muscle fibers have their origin and insertion in the skeleton of the heart, which consists of rings of fibrous connective tissue surrounding the four cardiac valves. These fibers are organized into complex spiral sheets for greater strength; when they contract they literally squeeze the blood out of the ventricles. Inflammation of the myocardium is called *myocarditis*.

Epicardium

The epicardium, also called the *visceral pericardium*, is a serous membrane that forms the outer coat of the heart wall. It contains fat plus the coronary vessels.

Pericardium

The visceral pericardium is continuous with a serous membrane called the *parietal pericardium*. Together the two pericardial membranes form the *serous pericardium*. The serous pericardium is enclosed in the *fibrous pericardium*, a flask-shaped connective tissue sac that surrounds and protects the heart. The *pericardial space* is a potential space between the two serous membranes. It normally contains about 15 milliliters of *pericardial fluid*, a serous fluid that allows the heart to move without friction in the pericardial sac. *Pericarditis* is an inflammation of the pericardial membranes accompanied by excessive secretion of pericardial fluid. The end result of pericarditis is the formation of fibrous adhesions between the serous membranes, which constrict the heart.

BLOOD SUPPLY OF THE HEART WALL
Arterial Supply

The myocardium receives oxygenated blood from the two coronary arteries [Fig. 8.7(*a*)], which arise from the right and left aortic sinuses.

The *right coronary artery* enters the coronary sulcus and travels to the posterior surface of the heart. There it gives rise to the *posterior descending branch*, which runs down the posterior interventricular sulcus to the apex of the heart. The right coronary artery supplies the right atrium and ventricle plus the roots of the aorta and pulmonary trunk.

The *left coronary artery* divides into two branches. The *anterior descending branch* runs down the anterior interventricular sulcus to the apex of the heart where it anastomoses with the right coronary artery. This branch is the main supplier of the interventricular septum. Interruption of the septal blood supply may have the serious consequence of damaging the atrioventricular bundle and causing heart

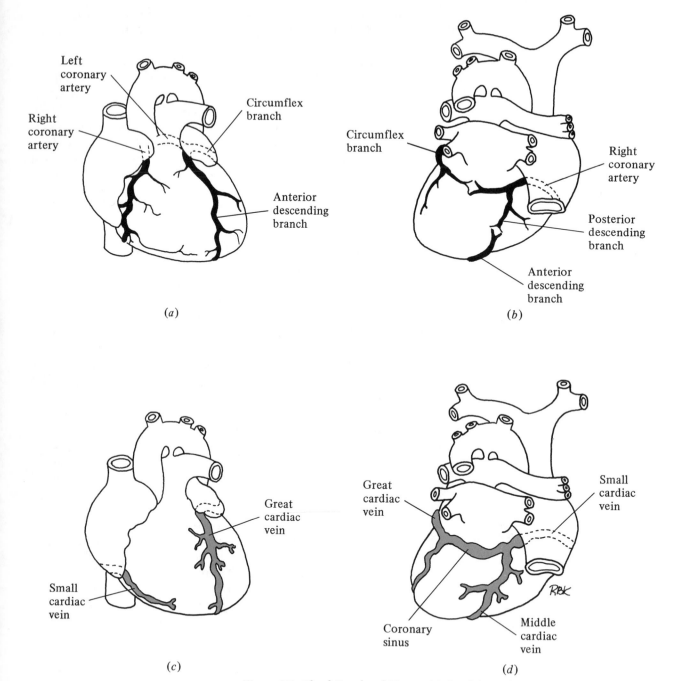

Figure 8.7 Blood Supply of Heart. (a) Arterial supply, anterior aspect. (b) Arterial supply, posterior aspect. (c) Venous drainage, anterior aspect. (d) Venous drainage, posterior aspect.

block. The *circumflex branch* enters the coronary sulcus and travels to the posterior surface of the heart, where it anastomoses with the right coronary artery. This branch supplies the left atrium and ventricle.

Venous Drainage

The *coronary sinus*, a vessel that lies in the coronary sulcus on the posterior surface of the heart, receives deoxygenated blood from the major veins draining the myocardium. These vessels, illustrated in Fig. 8.7(c), include the *great, middle,* and *small coronary veins.*

**INNERVATION
OF THE HEART WALL
Intrinsic Innervation**

The heart has an intrinsic (internal) impulse-conducting system that initiates and coordinates contractions of the cardiac muscle fibers. This conduction system, called the *Purkinje system,* is a nervelike network composed of highly modified cardiac muscle fibers. It consists of the following structures (Fig. 8.8):

1. *Sinoatrial (SA) Node.* A small group of modified cardiac muscle fibers located in the right atrium near the opening of the superior vena cava. Since this node initiates and regulates the rate of cardiac muscle fiber contraction, it is known as the *pacemaker* of the heart. Impulses from the SA node spread out over both atria, causing atrial contraction.
2. *Atrioventricular (AV) Node.* A small group of modified cardiac muscle fibers located in the interatrial septum near the opening of the coronary sinus. This node is activated by impulses that orig-

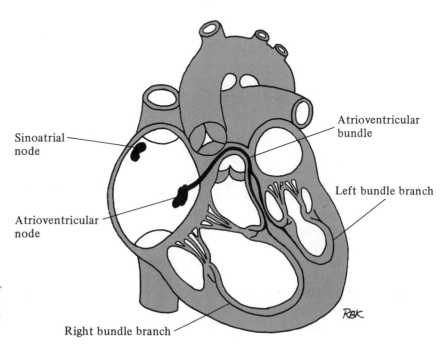

Figure 8.8 Intrinsic Innervation of Heart. Locations of sinoatrial node, atrioventricular node, and atrioventricular bundle plus its branches are indicated.

Sinoatrial node

Atrioventricular node

Atrioventricular bundle

Left bundle branch

Right bundle branch

inate in the SA node; the impulses that the AV node subsequently generates are transmitted to the atrioventricular bundle.

3. *Atrioventricular (AV) Bundle.* A bundle of modified cardiac muscle fibers *(Purkinje fibers)* that begins at the AV node and enters the interventricular septum, where it divides into *right* and *left bundle branches.* Each bundle branch innervates the septum, ventricular wall, and papillary muscles on its side. Impulses transmitted from the AV bundle cause contraction of the ventricles.

Extrinsic Innervation

Extrinsic (external) control of the heart's impulse-conducting system by the autonomic nervous system alters the heart rate and cardiac output to fit changing physiological needs of the body. *Sympathetic* stimulation of the SA and AV nodes accelerates the heart rate and increases cardiac output, while *parasympathetic* stimulation slows down the heart rate and reduces cardiac output.

THE CARDIAC CYCLE

Contractions of the myocardium force blood to move through the heart and into the pulmonary and systemic circuits. They occur at an average rate of 70 per minute. Between each contraction the myocardium rests for a short period. Each cardiac cycle consists of the following phases:

1. *Diastole.* The period in which the heart rests. The atria fill with blood, which flows into the ventricles through the wide open atrioventricular valves.
2. *Systole.* The period in which both atria contract to force extra blood into the ventricles, followed by contraction of both ventricles to force blood into the aorta and pulmonary trunk.

Development of the Heart

It is essential for paramedical students to know how the heart develops in order to understand the makeup and function of the adult heart and to understand how congenital heart malformations arise.

FORMATION OF THE HEART TUBE

At the end of the third week of embryonic development, the heart is represented by a pair of endothelial tubes that have formed from the mesenchyme in the pericardial region of the embryo. The two tubes come together and fuse to form a single median *endocardial heart tube.* The tube is soon surrounded by a thick coat of mesenchymal cells that gives rise to the myocardium and epicardium.

GROWTH AND PARTITIONING OF THE HEART TUBE

The heart tube begins to elongate and, at the same time, develops five dilatations; from caudal to cephalic end they are known as the (a) *sinus venosus,* (b) *atrium,* (c) *ventricle,* (d) *bulbus cordis,* and (e) *truncus arteriosus.* Blood from the venous system is received by the sinus venosus, passes through the heart tube, and is discharged into the arterial system by the truncus arteriosus.

As the heart tube elongates, it must bend back on itself to accommodate its length to that of the pericardial cavity (Fig. 8.9). First the heart tube forms a U-shaped loop, then an S-shaped loop. As the heart tube bends, the atrium comes to lie dorsal to the ventricle, bulbus cordis, and truncus arteriosus.

In the middle of the fourth week, *endocardial cushions* (thickenings of subendocardial tissue) grow out from the heart wall in the re-

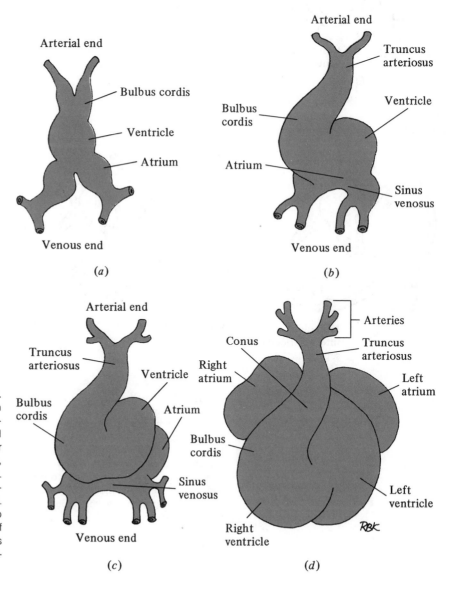

Figure 8.9 Development of Heart. (a) Fusion of paired heart tubes. Fusion begins at cephalic (arterial) end. (b) U-shaped heart tube. Due to differential growth rates between bulboventricular portion of heart tube and pericardial cavity, heart tube bends to form U-shaped loop. (c) S-shaped heart loop. As bending continues, heart tube forms S-shaped loop. (d) Maturing heart. At end of cardiac loop formation, note that truncoconal portion of heart tube comes to lie in front of ventricles and that atria come to lie cephalic to ventricles.

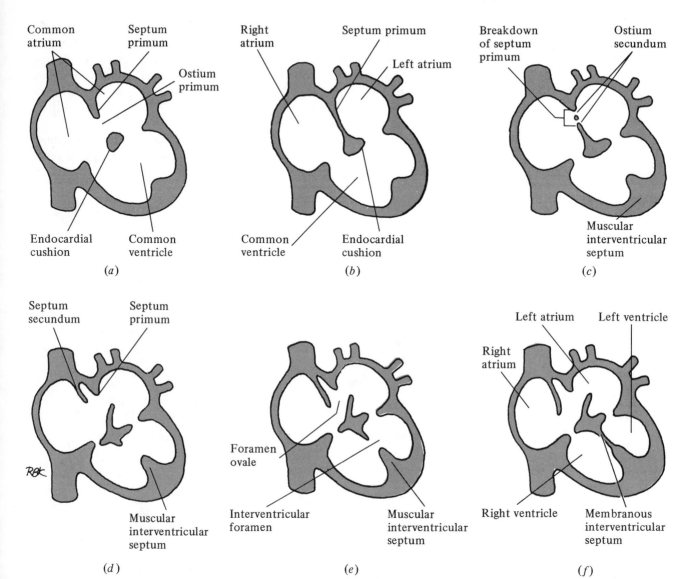

Figure 8.10 *Partitioning of Heart.* Schematic representation of septum formation in development of heart. (*a*) Formation of endocardial cushions and downgrowth of septum primum from atrial roof. (*b*) Septum primum fuses with endocardial cushions to form interatrial septum. (*c*) Portion of septum primum breaks down to create ostium secundum. (*d*) Septum secundum grows downward to right of septum primum. (*e*) Free margin of septum secundum overlaps ostium secundum. Oblique cleft thus formed between two atrial cavities is known as foramen ovale. Meanwhile, muscular portion of interventricular septum has been completed. (*f*) Formation of membranous portion of interventricular septum by outgrowth from aorticopulmonary septum and endocardial cushions completes partitioning of ventricles.

gion of the atrioventricular canal. The fused endocardial cushions partition the atrioventricular canal into right and left portions and give rise to the atrioventricular valves. A membranous downgrowth from the roof of the atrium fuses with the endocardial cushions to form the interatrial septum. The *foramen ovale,* an opening that forms in the interatrial septum, permits blood to pass from the right atrium into the left atrium until birth. The interventricular septum begins as a muscular outgrowth from the floor of the ventricle. A crescent-shaped gap between the endocardial cushions and the muscular septum permits communication between the right and left ventricles for a few weeks. Opposing ridges in the walls of the bulbus cordis and truncus arteriosus fuse to form the spiral *aorticopulmonary septum.* The aorticopulmonary septum partitions the common outflow tract into the aorta and pulmonary trunk. A proliferation of connective tissue from the base of this septum forms the membranous portion of the interventricular septum. Completion of the interventricular septum insures that the right ventricle communicates with the pulmonary trunk and the left ventricle communicates with the aorta. Partitioning of the developing heart is illustrated in Fig. 8.10.

BLOOD FLOW THROUGH THE FETAL HEART

Blood flow through the fetal heart is quite different from blood flow through the adult heart (Fig. 8.11). Since the fetus grows in a fluid environment in the mother's body, its lungs do not function. Fetal

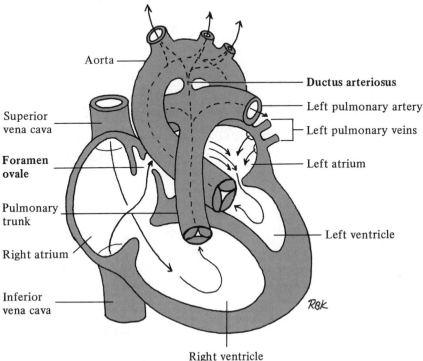

Figure 8.11 Blood Flow Through Fetal Heart. Arrows indicate course of blood flow through chambers and great vessels of fetal heart. Shunts to bypass nonfunctional pulmonary circuit include foramen ovale and ductus arteriosus.

Aorta

Ductus arteriosus

Left pulmonary artery

Left pulmonary veins

Superior vena cava

Left atrium

Foramen ovale

Pulmonary trunk

Left ventricle

Right atrium

Inferior vena cava

Right ventricle

blood is oxygenated by the mother's body instead of the fetus's lungs. There are two structures, present only in the fetal heart, that permit blood to bypass the nonfunctioning lungs:

1. *Foramen Ovale.* An oval hole in the interatrial septum that permits blood to flow from the right atrium into the left. It is guarded by a crescent-shaped tissue flap that acts as a one-way flutter valve.
2. *Ductus Arteriosus.* A short wide blood vessel that connects the pulmonary trunk with the aorta.

Oxygen-rich blood reaches the fetus through the umbilical vein, is transported to the right atrium by the inferior vena cava, and is directed toward the foramen ovale by that vessel's valve. A pressure difference between the right and left sides of the fetal heart forces this blood to flow from the right atrium into the left without passing through the pulmonary circuit. Oxygen-poor blood, entering the right atrium from the superior vena cava, passes into the right ventricle and enters the pulmonary trunk. The greater part of the blood in the pulmonary trunk is shunted into the aorta through the ductus arteriosus.

When the lungs begin to function at birth, blood is no longer diverted from the pulmonary circuit. The ductus arteriosus constricts, and blood that previously passed through this shunt now passes through the lungs. The increased volume of blood in the pulmonary circuit causes pressure in the left atrium to become higher than that in the right. This causes the foramen ovale to close, which shuts off communication between the two atria. These changes in the heart after birth result in complete separation of the pulmonary and systemic circuits.

Diseases and Disorders of the Heart

CONGENITAL MALFORMATIONS If congenital defects are present in the heart, the two blood circuits may not be completely separated after birth and many serious problems result. The most common types of congenital heart malformations are septal defects, patent ductus arteriosus, and tetralogy of Fallot (Fig. 8.12).

Septal Defects. If the interatrial septum is incomplete, the condition is referred to as an *atrial septal defect (ASD)*. In an ASD there is a backflow of blood from the left atrium to the right atrium after birth. If the defect is large enough, the right ventricle is forced to work harder and hypertrophy. If not corrected, a large ASD can lead to right heart failure and death. Sometimes the foramen ovale does not close after birth as it normally should. This condition, which is known as *patent* (open) *foramen ovale*, has the same consequences as an ASD. If the interventricular septum is incomplete, the condition is referred to as

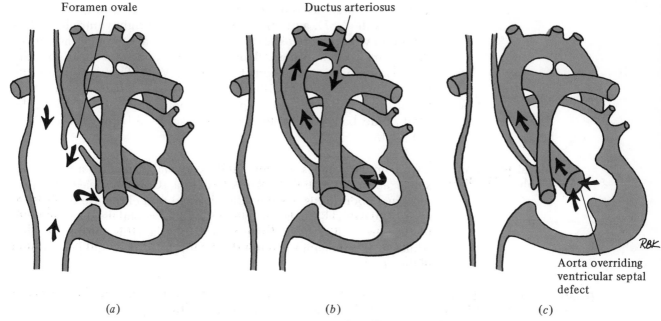

Figure 8.12 *Congenital Malformations of Heart.* (a) Patent foramen ovale. Due to pressure differences between two chambers after birth, blood flows through patent foramen ovale from left atrium to right atrium. (*b*) Patent ductus arteriosus. Due to pressure differences in systemic and pulmonary circuits after birth, blood flows through patent ductus from aorta into pulmonary trunk. (*c*) Tetralogy of Fallot. Blood from both ventricles enters aorta, which overrides a defect in membranous portion of interventricular septum. Very little blood enters stenotic pulmonary trunk.

an *interventricular septal defect (VSD)*. In a VSD the membranous portion of the interventricular septum usually fails to form. Due to postnatal pressure changes, blood is shunted from the left ventricle into the right. This condition, if uncorrected, can also lead to right ventricular hypertrophy, heart failure, and death.

Patent Ductus Arteriosus. If the ductus arteriosus does not close after birth, the resulting condition is referred to as a *patent ductus arteriosus (PDA)*. Before birth, blood passed from the pulmonary trunk into the aorta. After birth, blood passes through the open ductus in the opposite direction. There is an extra amount of blood passing through the lungs as well as a loss of blood from the systemic circuit. This increases blood pressure in the lungs (pulmonary hypertension) and causes both ventricles to work harder and hypertrophy. PDA can lead to heart failure and death.

Tetralogy of Fallot. Tetralogy of Fallot is a congenital heart malformation consisting of four problems (hence the name *tetra*logy). These problems include (a) narrowing of the pulmonary trunk, (b) absence of the membranous interventricular septum, (c) a large aorta that overrides the VSD, and (d) hypertrophy of the right ventricle. In this condition very little deoxygenated blood can pass through the narrow pulmonary trunk. Instead, deoxygenated blood from the right ventricle is pumped into the aorta through the VSD. Thus blood entering the systemic circuit is partially deoxygenated. A baby whose blood is not fully oxygenated may appear cyanotic at times; therefore, the term "blue baby" has been used to describe a baby with tetralogy of Fallot.

Causes of Congenital Heart Malformations

Two common causes of congenital heart malformations are chromosomal aberrations and congenital rubella (German measles). Chromosomal aberrations, such as the absence of a chromosome or presence of an extra chromosome, generally cause septal defects. Infection of the developing heart tissues with rubella virus during the first trimester of pregnancy can interfere with the growth and partitioning of the heart.

Treatment of Congenital Heart Malformations

In many instances congenital heart malformations can be corrected surgically. Techniques for visualizing the defects, such as angiocardiography and cardiac catherization, plus radically new surgical techniques have saved the lives of many persons with congenital heart malformations who normally would never have survived infancy or childhood.

RHEUMATIC HEART DISEASE

Rheumatic fever is a disease that develops as an allergic response to infection by certain strains of streptococcal bacteria. In about 60 percent of rheumatic fever patients, the heart valves, especially the mitral valve, become inflamed. The inflamed valve cusps tend to thicken and stick together, and the valve opening may become excessively narrow *(mitral stenosis)*. Mitral stenosis prevents adequate blood flow into the left ventricle, so the blood is held back in the left atrium and even back in the pulmonary circuit. This forces the right ventricle to work harder and hypertrophy. Eventually the right ventricle fails and death results. Mitral *commissurotomy* is an operation designed to relieve this condition; the narrowed valve is opened by a specially designed surgical knife worn on the surgeon's finger.

CORONARY ARTERY DISEASE

A common cause of sudden death or serious disability is obstruction of one of the coronary arteries or its branches. The portion of the heart wall suddenly deprived of oxygenated blood subsequently becomes necrotic. The area of dead cardiac muscle is referred to as a *myocardial infarct*. If a major coronary artery is obstructed, the individual suffers a massive myocardial infarct and may die; if a small branch is obstructed, the individual suffers a smaller myocardial infarct and usually recovers. The victim of a myocardial infarct must be put on complete bed rest during the time the dead cardiac muscle tissue is being replaced by a connective tissue scar. During the healing process there is always the danger that the heart wall may rupture at the infarct site; this is usually fatal.

Part B. **The Blood**

Blood is a type of special connective tissue consisting of formed elements (blood cells and platelets), a fluid medium (blood plasma), and fibers under certain conditions.

Functions of the Blood

As it circulates through the cardiovascular system, blood performs the following vitally important group of functions:

1. Transportation of oxygen to the tissues and removal of carbon dioxide.
2. Transporation of nutrients to the tissues and removal of metabolic waste products.
3. Transportation of hormones and other cellular secretions from site of production to site(s) of utilization.
4. Transportation of phagocytic cells, immunologically active cells, and circulating antibodies to distant parts of the body to attack disease-producing microorganisms, cancer cells, or foreign tissue grafts.

Formed Elements

The formed elements of the blood consist of erythrocytes (red blood cells), leukocytes (white blood cells), and platelets, which are illustrated in Fig. 8.13. The formed elements make up 38 to 52 percent of the total blood volume.

Undifferentiated mesodermal cells called *hemocytoblasts* give rise to the precursors of all the different types of blood cells. The earliest blood cells are formed in the yolk sac, a temporary embryonic structure, during the fourth week of development. During the second month of development, the liver takes over the manufacture of blood cells; later this function is transferred to the spleen. From the fifth month of development on through the rest of an individual's life, blood cells are manufactured by the red bone marrow. However, in certain blood diseases and following massive hemorrhage, blood cell formation can revert back to the liver and spleen.

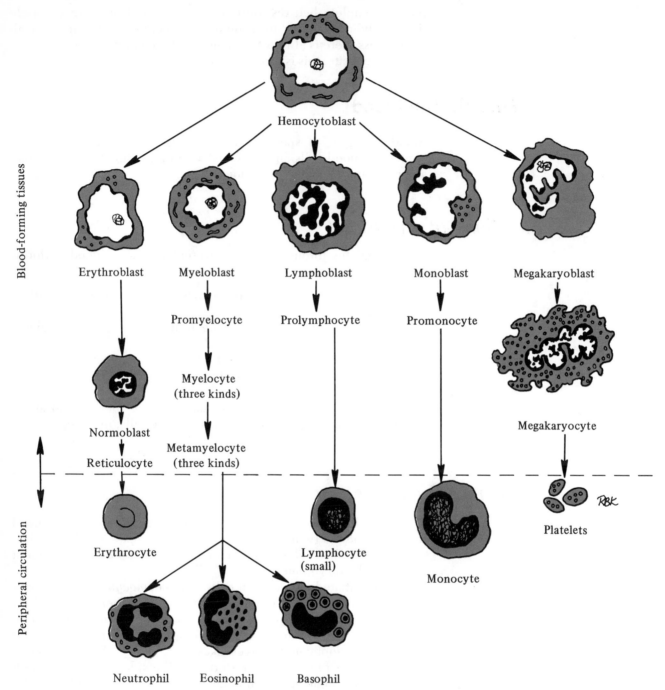

Blood-forming tissues

Peripheral circulation

Hemocytoblast

Erythroblast Myeloblast Lymphoblast Monoblast Megakaryoblast

Promyelocyte Prolymphocyte Promonocyte

Myelocyte
(three kinds)

Megakaryocyte

Normoblast

Reticulocyte Metamyelocyte
(three kinds)

Erythrocyte Lymphocyte
(small)

Platelets

Monocyte

Neutrophil Eosinophil Basophil

Figure 8.13 *Morphology of Formed Blood Elements.* As shown here, all formed elements are derived from primitive stem cells called hemocytoplasts. Note that all precursor (blast) forms are large nucleated cells. Forms indicated above *dashed line* are normally found in blood-forming tissues, while those indicated below *dashed line* are found in peripheral circulation.

ERYTHROCYTES   Mature erythrocytes (red blood cells) are nonnucleated biconcave discs. These discs, approximately 7 to 8 microns in diameter, are very flexible; they are able to squeeze through capillaries of only half their diameter and then regain their original shape. Red blood cells are specialized to transport oxygen and carbon dioxide. To perform this vital function, they produce a substance called hemoglobin, which consists of the protein *globin* plus the iron-containing pigment *heme*. Hemoglobin combines readily with oxygen molecules to form *oxyhemoglobin*. It is oxyhemoglobin that gives arterial blood its bright red color. In tissue capillaries oxyhemoglobin gives up its oxygen molecules to the tissues and is referred to as *deoxyhemoglobin*. It is deoxyhemoglobin that gives a bluish color to blood found in veins. Determination of the hemoglobin content of peripheral blood is now a routine part of every physical examination. In adult males the blood normally contains 16 grams of hemoglobin per 100 milliliters, while in adult females it averages 14 grams of hemoglobin per 100 milliliters.

Red blood cells are formed in red bone marrow from nucleated precursors called *erythroblasts* under the stimulus of *erythropoietin*, a hormone produced by the kidneys. Vitamin B_{12}, the *maturation factor*, causes erythroblasts to mature and begin manufacturing hemoglobin. As erythroblasts accumulate hemoglobin in their cytoplasm, they begin to lose their subcellular organelles and are referred to as *normoblasts*. Normoblasts extrude their nuclei just before being released into peripheral circulation. The presence of large numbers of nucleated red blood cells in peripheral blood is abnormal; the outpouring of immature cells represents an attempt to meet an urgent need for red blood cells.

After a period of 120 days in circulation, red blood cells become very fragile. Worn out red blood cells are removed from circulation and destroyed by the liver and spleen. Hemoglobin is released from fragmented red blood cells; the iron is removed and incorporated into new red blood cells, while the remainder of the hemoglobin molecule is converted into *bilirubin*, a pigment that is excreted from the liver with the bile.

There are approximately 5 million red blood cells per cubic millimeter of blood in the adult male and 4.5 million per cubic millimeter of blood in the adult female. Under normal conditions the rate of red blood cell production equals the rate of destruction. A deficiency of red blood cells results from too rapid a rate of destruction or too slow a rate of production.

LEUKOCYTES   Leukocytes (white blood cells) are nucleated nonpigmented cells with amoeboid properties. White blood cells can easily pass through unruptured blood vessel walls by amoeboid movement (*diapedesis*), enter the tissue spaces, and then travel through the tissues. These cells are attracted to areas of infection or inflammation by chemical substances released during tissue destruction. An important function of white blood cells is *phagocytosis*. Phagocytic white blood cells protect the body by engulfing disease-producing microorganisms and digesting them with lysosomal digestive enzymes.

The average number of white blood cells in a healthy person ranges from 5,000 to 10,000 per cubic millimeter of whole blood. An increase in the number of white blood cells above 10,000 is called *leukocytosis*. Leukocytosis is an indication that an active infectious or inflammatory process, such as appendicitis, is present in the body. A decrease in the number of white blood cells below 5,000 is called *leukopenia*. Leukopenia occurs in persons with certain diseases, such as tuberculosis, in persons taking steroid hormones, such as cortisone, and in persons who have been exposed to high doses of ionizing radiation.

White blood cells are subdivided into two categories, granular and nongranular leukocytes, based on the presence or absence of cytoplasmic granules.

Granular Leukocytes

Granular leukocytes, or granulocytes, are white blood cells that have lobulated nuclei and contain cytoplasmic granules that are visible under the light microscope. Granulocytes arise in red bone marrow from precursor cells called *myeloblasts.* As myeloblasts mature they begin manufacturing cytoplasmic granules. Immature granulocytes, called *bands,* have horseshoe-shaped nuclei. Nuclear lobulation normally takes place before granulocytes are released into peripheral circulation. When bands appear in peripheral blood, it is an indication that there is an increased demand for white blood cells, which causes early release of the immature forms.

Granulocytes are classified as neutrophils, eosinophils, or basophils on the basis of the staining reactions of their granules.

Neutrophils

Neutrophils comprise 50 to 80 percent of all white blood cells in peripheral circulation. They are about 10 to 12 microns in diameter and contain nuclei that have from three to five lobes. The cytoplasm of the mature neutrophil is filled with fine granules (lysosomes) that stain lavender with Wright's blood stain. The major function of neutrophils is phagocytosis and digestion of pathogenic bacteria during *acute* infections. At the first appearance of foreign organisms, neutrophils rapidly leave the capillaries for the tissues, where they attempt to destroy the invaders. Neutrophils seldom survive more than a few days after leaving the bloodstream. If the infection is massive or the bacteria are very virulent, many neutrophils die; an accumulation of dead neutrophils and bacteria is known as *pus*.

Eosinophils

Eosinophils comprise only 1 to 3 percent of all white blood cells. They are about 10 to 15 microns in diameter, have bilobed nuclei, and contain large granules (lysosomes) that stain red with Wright's stain. Eosinophils phagocytize antigen-antibody complexes; the number of eosinophils in the peripheral blood increases in allergic diseases, such as asthma or hay fever, and in parasitic infections, such as trichinosis.

Basophils   Basophils are even rarer and make up only about 0.5 percent of all white blood cells. They are about 8 to 10 microns in diameter, have bilobed or kidney-shaped nuclei, and contain large granules that stain dark blue with Wright's blood stain. These granules contain histamine and heparin, which are released into the bloodstream under conditions of stress or allergy. Basophils are nonphagocytic; they are similar to the mast cells in loose connective tissue.

Nongranular Leukocytes   Nongranular leukocytes, in contrast to the granule-containing white cells, have spherical nuclei and do not contain cytoplasmic granules. There are two types of nongranular leukocytes: lymphocytes and monocytes.

Lymphocytes   Lymphocytes make up about 25 percent of all white blood cells. Small lymphocytes, which constitute a majority of all lymphocytes, have a diameter of 8 microns; large lymphocytes have a diameter of 12 microns. Both small and large lymphocytes have a spherical nucleus surrounded by a thin rim of cytoplasm. Lymphocytes arise in lymphoid tissues and organs from precursor cells called *lymphoblasts.* Mature lymphocytes are released into the lymphatic system and ultimately enter the cardiovascular system, which they leave very rapidly to enter the tissues. Once in the tissues, lymphocytes live only a few days; they may survive for years in the lymphatic system. Lymphocytes play an important role in the body's immunological defense system, which is discussed in greater detail in Chapter 9. They take part in cell-mediated immune responses, manufacture humoral antibodies, and are the principal cell type found at the site of a chronic infection.

Monocytes   Monocytes make up about 5 percent of all white cells. With an average diameter of 15 microns, they are the largest white cells in the peripheral blood. Monocytes have a large horseshoe-shaped nucleus surrounded by abundant cytoplasm. They arise in the red bone marrow from precursor cells called *monoblasts.* Mature monocytes leave the capillaries soon after being released into peripheral circulation. Once in the tissues, monocytes differentiate into *macrophages,* scavenger cells that wander about phagocytizing particulate matter, parasites, and worn-out neutrophils. Monocytes live as macrophages for several months in the tissues and are commonly found at sites of chronic infection or inflammation.

Differential Count   A differential count is a laboratory test done to estimate the percentage of each type of white cell present per 100 to 200 white cells counted in a stained peripheral blood smear. The percentages of the different types of white cells present change in certain diseases; for example, a person with asthma will have a higher-than-normal percentage of eosinophils.

PLATELETS   Platelets are nonnucleated, granular oval discs about 2 to 5 microns in diameter, which makes them the smallest of the formed elements. Platelets are derived from giant cells called *megakaryocytes*. While still in the red bone marrow, megakaryocyte cytoplasm fragments into platelets, which are released into peripheral circulation; they survive for about a week in peripheral circulation before they are removed by the lungs and spleen. Platelets initiate the blood-clotting process at the site of vascular injury. As soon as blood vessels are injured, platelets plug up the holes to reduce blood loss. They release *serotonin*, a vasoconstrictor that constricts blood vessel walls to further reduce blood loss. Platelets also release substances that help plasma-clotting factors convert prothrombin to *thrombin*, an enzyme that in turn converts *fibrinogen*, a plasma protein, into *fibrin*. Fibrin fibers form the framework of blood clots, which ultimately stop bleeding.

There are approximately 150,000 to 300,000 platelets per cubic millimeter of blood. A marked decrease in the number of circulating platelets is called *thrombocytopenia*. Persons with thrombocytopenia have a tendency to bleed from capillaries all over their bodies; their skin is covered with many small purple blotches, called *petechiae*, which represent hemorrhages of skin capillaries. When the platelet count drops below 70,000 per cubic millimeter of blood, spontaneous bleeding becomes a serious problem.

Blood Plasma

Blood plasma is the yellowish, slightly alkaline fluid portion of the blood in which the formed elements are suspended. It is part of the extracellular fluid of the body and is similar to tissue fluid, except for its higher protein content. Approximately 91 percent of blood plasma is water. The remaining 9 percent consists of the following substances:

1. *Plasma Proteins.*   Plasma proteins maintain the colloid osmotic pressure of blood plasma. They include albumin, alpha and beta globulins, which act as carrier proteins; gamma globulin (in the the form of circulating antibodies); and clotting factors, such as prothrombin and fibrinogen.
2. *Nutrients.*   Nutrients include glucose (blood sugar), amino acids, and lipids.

3. *Waste Products.* Waste products include urea, uric acid, lactic acid, creatinine, and ammonium salts.
4. *Elements.* Major plasma elements include copper, iron, manganese, and zinc.
5. *Electrolytes.* Plasma electrolytes include cations, such as sodium (Na^+), potassium (K^+), calcium (Ca^{2+}), and magnesium (Mg^{2+}), and anions, such as chloride (Cl^-), bicarbonate (HCO_3^-), phosphate (HPO_4^{2-} or $H_2PO_4^-$), and sulfate (SO_4^{2-}).

Serum is a clear amber-colored fluid. It represents blood plasma minus fibrinogen.

Diseases and Disorders of the Blood

ANEMIA Anemia is a condition in which red blood cells are decreased in number, deficient in hemoglobin, or both. The clinically important anemias are described next.

Iron-Deficiency Anemia. Iron-deficiency anemia is due to lack of iron. Important causes of iron deficiency are hemorrhage and lack of dietary iron. Women have a higher iron requirement than men during their reproductive years because they lose iron through menstruation, pregnancy, and lactation; many women of childbearing age are mildly anemic and do not know it.

Pernicious Anemia. Pernicious anemia is a disease of the gastric mucosa that occurs primarily in middle-aged people. In pernicious anemia the gastric mucosa no longer produces *intrinsic factor*, so vitamin B_{12} is not absorbed. Since vitamin B_{12} is essential for red blood cell maturation, persons with this disease must receive vitamin B_{12} injections for the rest of their lives.

Sickle-Cell Anemia. Sickle-cell anemia is a type of hereditary anemia that occurs mainly in East African and American blacks. The red blood cells of persons with this disease contain an abnormal hemoglobin, which precipitates into long crystals whenever oxygen concentration is low. The crystals distort the red blood cells (*sickle cells*) and damage their membranes so that rupture and release of hemoglobin into the bloodstream (*hemolysis*) occurs. It is usually a fatal disease for those who are homozygous for the sickle-cell gene.

LEUKEMIA Leukemia is a type of blood cancer characterized by abnormal proliferation of white blood cells. Leukemias are classified according to the type of white cell involved and the clinical course of the disease. Acute leukemias are common in children; these conditions have a sudden onset characterized by fever, sore throat, and mucosal bleeding. The white cell count becomes elevated, with many immature forms present in peripheral circulation, while red cell and platelet counts are depressed. Despite drug-induced remissions, the course of the disease is generally short. *Chronic myeloid leukemia* is a disease of granulocytes that occurs in young adults and middle-aged people. This disease, which has a slow onset and long course, is characterized mainly by an abnormal increase in the number of circulating neutrophils. *Chronic lymphatic leukemia,* which often occurs in men of late middle age, is characterized mainly by an abnormal increase in the number of circulating small lymphocytes. This disease also has a slow onset and long course.

To date, leukemia is largely a fatal disease, though life expectancy can be prolonged by chemotherapy, blood transfusions, and, in some instances, bone marrow transplants.

HEMORRHAGIC DISORDERS

Hemorrhagic disorders are characterized by a tendency to abnormal bleeding; this may result from defects in or lack of one of the clotting factors, thrombocytopenia, or some abnormality of the capillaries that makes them leaky. Classic *hemophilia* is a hemorrhagic disorder due to a hereditary (X-chromosome-linked) deficiency of a clotting factor called *antihemophilic factor* (AHF). Due to the mode of transmission of the hemophilia gene (from carrier female to affected male), its victims are almost always males. Hemophiliacs suffer from persistent bleeding following minor cuts and bruises. Bleeding into joint cavities, such as the knee joint, is common and causes crippling.

Part C. **The Vascular System**

The vascular system consists of two separate and closed circuits—pulmonary and systemic—that will be described in detail after general features of the blood vessels are presented.

The Blood Vessels

The blood vessels are a series of flexible tubes that transport blood to and from the heart. They consist of arteries, veins, and capillaries, and their functions will be easy to understand if you keep these simple principles in mind:

1. Arteries are like *trees;* they have *branches.* An artery is any blood vessel that carries blood *away* from the pumping chambers of the heart.
2. Veins are like *rivers;* they have *tributaries.* A vein is any blood vessel that carries blood *toward* the receiving chambers of the heart.
3. Capillaries are tiny vessels that connect the smallest arteries (arterioles) to the smallest veins (venules).

Microscopic Anatomy of Blood Vessels

All blood vessels have an endothelial lining. However, vessels larger than capillaries have accessory wrappings of smooth muscle and connective tissue that are organized into the three coats or tunics described below:

1. *Tunica Intima.* The tunica intima, or inner coat, of a blood vessel consists of endothelium surrounded by a variable amount of fibroelastic connective tissue. A sheet of elastic fibers, the *internal elastic membrane,* separates the tunica intima from the tunica media in certain types of blood vessels.
2. *Tunica Media.* The tunica media, or middle coat, of a blood vessel consists of sheets of smooth muscle fibers plus elastic fibers (often arranged in the form of elastic membranes).
3. *Tunica Adventitia.* The tunica adventitia, or outer coat, of a blood vessel consists of fibrous connective tissue. A sheet of elastic fibers, the *external elastic membrane,* may separate the tunica adventitia from the tunica media. The tunica adventitia of larger blood vessels contains small blood vessels (*vasa vasorum),* nerves, and lymphatics.

Classification and Function of Blood Vessels

ARTERIES Arteries carry blood at high pressure. They have thick walls, which prevent them from collapsing and help them withstand sudden pressure variations. There are four classes of arteries: large (elastic) arteries, medium-sized (muscular) arteries, small arteries, and arterioles. They differ in size, wall structure, and function.

Elastic Arteries The largest arteries of the body are known as *elastic* arteries because of their high content of elastic fibers. They arise directly from the heart (aorta, pulmonary trunk) or from the arch of the aorta (brachiocephalic trunk, subclavian arteries, and common carotid arteries). Elastic arteries have a wide diameter and a relatively thin wall for their size [Fig. 8.14(a)]. The thin tunica intima is separated from the media by an internal elastic membrane that is split into two or more layers. The tunica media is a moderately thick coat of alternating elastic membranes and sheets of spirally arranged smooth muscle fibers. The tunica adventitia is a relatively thin coat of fibrous connective tissue that lacks an external elastic membrane.

The large amount of elastic fibers in the walls of elastic arteries permits them to withstand variations in the force of blood pumped under pressure from the heart. They passively expand and recoil with the ejection of blood from the ventricles and convert the pulsatile flow into a continuous flow. Functionally, elastic arteries are classified

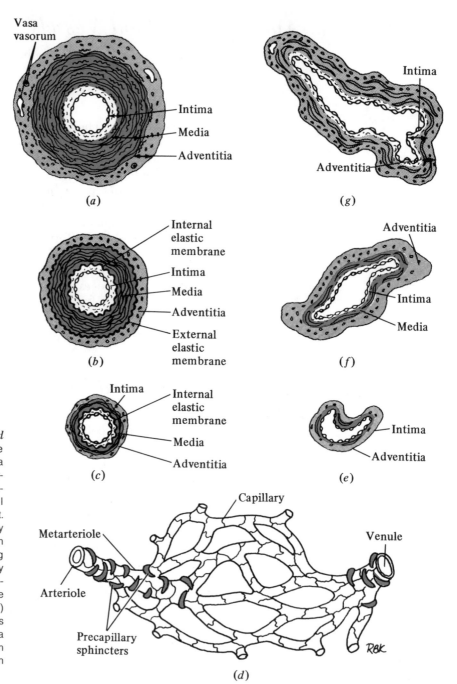

Figure 8.14 *Arteries, Capillaries, and Veins.* (a) Large (elastic) artery. Note large amount of elastic fibers in tunica media. (b) Medium-sized (muscular) artery. Note presence of internal and external elastic membranes. (c) Small artery. Internal elastic membrane present. (d) Capillary bed. Blood enters capillary bed through arteriole and exits through venule. Smooth muscle fibers surrounding openings of true capillaries (precapillary sphincters) regulate flow of blood into capillary bed. (e) Small vein. Note presence of smooth muscle fibers in adventitia. (f) Medium-sized vein. Note that adventitia is thicker than media. (g) Large vein. Tunica media is virtually nonexistent, especially in venae cavae. Smooth muscle is found in adventitia.

as *conducting arteries* because they conduct blood to the muscular arteries.

Muscular Arteries

Medium-sized arteries are known as *muscular* arteries because of their high content of smooth muscle fibers. This class of arteries includes most of the named arteries in the human body. Muscular arteries have thick walls due to the relative thickness of the tunica media [Fig. 8.14(b)]. The tunica media, which is the thickest coat of muscular arteries, consists of many spirally arranged sheets of smooth muscle fibers interspersed with scattered elastic fibers. Prominent internal and external elastic membranes separate the tunica media from the tunica intima and tunica adventitia, respectively.

Muscular arteries are functionally classified as *distributing arteries* because they distribute blood to smaller arterial vessels in organs and body walls. Although muscular arteries contain a large quantity of smooth muscle in their walls, there is apparently little fluctuation in muscle fiber activity and little change in their diameters. The major regulation of blood flow occurs at the level of the arterioles.

Small Arteries and Arterioles

Small arteries and arterioles show a progressive reduction in diameter and loss of structural components as they get closer to the capillary beds. A small artery [Fig. 8.14(c)] becomes an arteriole when it reaches a diameter of 300 microns or less. The walls of arterioles are relatively thick and their lumens are narrow. Internal and external elastic membranes are absent, and the media contains no more than one or two layers of smooth muscle fibers. The terminal portion of an arteriole is sometimes called a *metarteriole*. The metarteriole consists of an endothelial tube surrounded by one or two smooth muscle fibers.

Smooth muscle fibers regulate the flow of blood into capillary beds. *Precapillary sphincters* are smooth muscle fibers that surround the openings of the true capillaries on their upstream side. Blood flow is excluded from those capillary beds whose precapillary sphincters are maximally constricted and shunted into others whose precapillary sphincters are more dilated. Thus the pattern of blood flow distribution depends primarily on the degree of constriction of the precapillary sphincters in each organ and tissue.

CAPILLARIES

Capillaries are endothelial tubes with an average diameter of 7 to 10 microns. The capillary wall is approximately 0.2 to 0.08 microns thick and is supported by a delicate layer of reticular fibers and an occasional pericyte, an undifferentiated smooth muscle cell. The extreme thinness of the capillary wall makes possible the exchange of respiratory gases, nutrients, and waste products between the blood and the tissues. Gases pass through endothelial cell cytoplasm by diffusion while small molecules pass through by pinocytosis. Red blood cells and plasma proteins normally do not leave the lumen of a capillary unless there is structural damage to the endothelium. White blood cells, on the other hand, pass with great ease through the junctions of endothelial cells by amoeboid movement.

The capillaries form extensive networks or *capillary beds* [Fig. 8.14(*d*)] throughout the body. The sizes of capillary beds are determined by the functional activities of the structures they supply. Only a small portion of the total number of capillary beds are open to blood flow at any given time; the remaining capillary beds are collapsed and empty of blood.

VEINS Veins carry blood at low pressure. Their walls are not as thick as those of corresponding arteries and their lumens are larger. Since veins are thin-walled, they readily collapse under slight external pressure. Most veins of medium size have many pairs of one-way valves that aid in moving blood toward the heart. Valves are most numerous in the veins of the lower extremities since blood in these vessels must flow against the force of gravity. There are four classes of venous vessels: venules and small veins, medium-sized veins, large veins, and sinusoids and sinuses.

Venules and Small Veins Venules are small vessels that drain the capillary beds; *postcapillary venules* are the vessels immediately connected to capillaries. Venules consist of a tube of endothelium surrounded by occasional fibroblasts and reticular fibers. The transition from venule to small vein is very gradual and is marked by the appearance of smooth muscle fibers between the endothelium and connective tissue [Fig. 8.14(*e*)].

Medium-sized Veins Medium-sized veins [Fig. 8.14(*f*)] include most of the named veins in the human body; they have thinner walls and larger lumens than their companion arteries. In medium-sized veins larger than 2 millimeters in diameter, the intima is thrown into folds that form paired valve cusps. The free margins of venous valves are directed toward the heart, permitting a one-way flow of blood in that direction. The tunica media consists of a few layers of smooth muscle and elastic fibers. The tunica adventitia, which consists of fibrous connective tissue plus smooth muscle fibers, is thicker than the media. Distinct internal and external elastic membranes are generally absent.

Large Veins Large veins [Fig. 8.14(*g*)] are venous vessels having a diameter greater than 10 millimeters. This class of veins includes the venae cavae plus the pulmonary, portal, brachiocephalic, azygos, renal, splenic, and superior mesenteric veins. Valves are absent in large veins. The tunica media is a thin ill-defined coat containing a small number of smooth muscle fibers. The adventitia forms the thickest coat of large veins; it consists of fibrous connective tissue plus a large number of smooth muscle fibers.

Sinusoids and Sinuses

Sinusoids and sinuses are unusual types of venous channels lacking the tubular shape and some of the structural components of veins. Sinusoids are large-caliber, thin-walled, irregularly-shaped venous channels found in visceral organs such as the liver, spleen, lymph nodes, bone marrow, and certain endocrine glands. They lack an adventitia and are lined with reticuloendothelium; *fixed* macrophages, such as the Kupffer cells of the liver, substitute for some of the endothelial cells. Sinuses are very large irregular venous channels lacking an adventitia. Included in this category are the dural sinuses, which are located between the two layers of dura mater covering the brain.

The Pulmonary Circuit

The pulmonary circuit (Fig. 8.15) consists of the right ventricle, the pulmonary trunk, the right and left pulmonary arteries, the pulmonary capillary bed, the four pulmonary veins, and the left atrium. This circuit is concerned with transporting blood to the lungs, where carbon dioxide is exchanged for oxygen, and returning it to the heart.

PULMONARY ARTERIES

Pulmonary arteries carry deoxygenated blood from the heart to the lungs. The *pulmonary trunk* is the chief arterial vessel of the pulmonary circuit. It is a short elastic artery, about 2 inches long, that arises from the base of the right ventricle. Beneath the arch of the

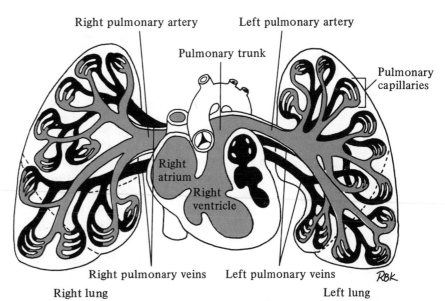

Figure 8.15 Vessels of Pulmonary Circuit. Vessels and chambers containing oxygenated blood are indicated in *black;* those containing deoxygenated blood are indicated in *gray*. Pumping chamber for pulmonary circuit is right ventricle.

aorta it divides into two branches of approximately equal size. The *right pulmonary artery* is the longer of the two branches. It runs horizontally behind the ascending aorta to enter the hilum of the right lung. The shorter *left pulmonary artery* passes horizontally in front of the descending aorta to the hilum of the left lung. The left pulmonary artery is connected to the aortic arch by a fibrous cord, the *ligamentum arteriosum,* which is the remnant of the ductus arteriosus.

PULMONARY CAPILLARIES Blood from the pulmonary arteries and their branches enters the pulmonary capillaries, which are in close proximity to the air sacs of the lungs. Carbon dioxide in the pulmonary capillaries is exchanged for oxygen in the air sacs.

PULMONARY VEINS There are four pulmonary veins that return oxygenated blood to the lungs; two emerge from the hilum of the right lung and two emerge from the hilum of the left lung. The pulmonary veins open separately into the posterior wall of the left atrium.

The Systemic Circuit

The systemic circuit consists of the left ventricle, the aorta and its branches, the systemic capillaries, the venae cavae and their tributaries, and the right atrium. This circuit is concerned with transporting blood to the tissues, where oxygen and nutrients are exchanged for carbon dioxide and waste products, and returning it to the heart.

SYSTEMIC ARTERIES Systemic arteries (Fig. 8.16) carry oxygenated blood from the heart to all parts of the body except the air sacs of the lungs. The *aorta* is the chief artery of the systemic circuit and the largest and longest artery in the body. It is a large elastic artery that extends from the left ventricle to vertebral level *L4*. For descriptive purposes the aorta is divided into three regions: the *ascending aorta*, the *aortic arch*, and the *descending aorta*. The descending aorta is further subdivided into thoracic and abdominal segments.

Ascending Aorta and Its Branches The ascending aorta arises from the base of the left ventricle, where it gives off the right and left *coronary arteries*. It passes behind and to the right of the pulmonary trunk and extends upward to the level of the sternal angle. The coronary arteries, which supply the myocardium of the heart, were discussed in Part A of this chapter.

Aortic Arch and Its Branches The aortic arch (Fig. 8.17) is the continuation of the ascending aorta; it loops over the right pulmonary artery and passes backwards and to the left of the trachea and vertebral column. At the level of the

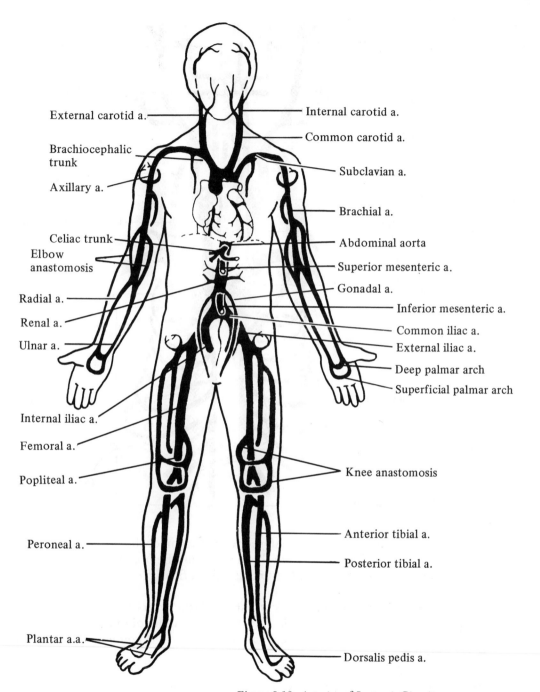

External carotid a.

Internal carotid a.

Common carotid a.

Brachiocephalic trunk

Subclavian a.

Axillary a.

Brachial a.

Celiac trunk

Abdominal aorta

Elbow anastomosis

Superior mesenteric a.

Gonadal a.

Radial a.

Inferior mesenteric a.

Renal a.

Common iliac a.

Ulnar a.

External iliac a.

Deep palmar arch

Superficial palmar arch

Internal iliac a.

Femoral a.

Popliteal a.

Knee anastomosis

Peroneal a.

Anterior tibial a.

Posterior tibial a.

Plantar a.a.

Dorsalis pedis a.

Figure 8.16 *Arteries of Systemic Circuit.*

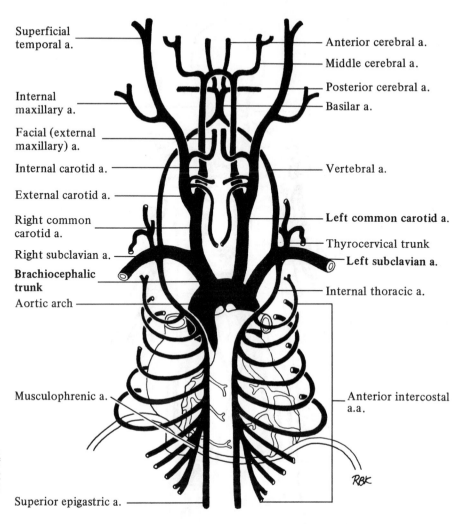

Superficial temporal a.

Internal maxillary a.

Facial (external maxillary) a.

Internal carotid a.

External carotid a.

Right common carotid a.

Right subclavian a.

Brachiocephalic trunk

Aortic arch

Musculophrenic a.

Superior epigastric a.

Anterior cerebral a.

Middle cerebral a.

Posterior cerebral a.

Basilar a.

Vertebral a.

Left common carotid a.

Thyrocervical trunk

Left subclavian a.

Internal thoracic a.

Anterior intercostal a.a.

Figure 8.17 Arteries of Aortic Arch. Schematic diagram shows arteries of aortic arch (brachiocephalic trunk, left common carotid, and left subclavian) and their major branches.

sternal angle, it begins to descend and becomes the descending aorta. Three branches arise from the aortic arch: the *brachiocephalic trunk,* the *left common carotid artery,* and the *left subclavian artery.*

Brachiocephalic Trunk   The brachiocephalic trunk is a short elastic artery that arises from the first part of the aortic arch. It ascends to the level of the right sternoclavicular joint, where it gives rise to the *right common carotid* and *right subclavian arteries.*

Common Carotid Arteries   The common carotid arteries are elastic arteries that supply the head and neck. The right common carotid arises from the brachiocephalic trunk, while the left common carotid arises directly from the arch of the aorta. Once these arteries reach the neck, they run a similar course. Each common carotid ascends alongside the trachea to vertebral level *C3*, where it gives rise to the *internal* and *external carotid arteries*. The *carotid body*, a small glandlike structure, is closely associated with each common carotid artery at the point of division. The carotid body is a chemoreceptor whose nerve endings notify the respiratory centers in the brain about increases in the carbon dioxide concentration and hydrogen ion concentration in the blood.

Internal Carotid Arteries. The two internal carotid arteries form one major blood supply to the brain. The internal carotid artery is dilated at its origin. This dilation contains a pressure receptor called the *carotid sinus*, whose nerve endings notify the cardiovascular centers in the brain about changes in the blood pressure. Each internal carotid artery ascends the neck to the base of the skull. After entering the skull through the carotid canal, the internal carotid artery gives off branches that supply the pituitary gland and eye and contribute to forming the *circle of Willis* (Fig. 8.18). The circle of Willis is an arterial ring at the base of the brain that gives rise to the cerebral arteries.

External Carotid Arteries. The external carotid arteries and their branches supply the superficial structures of the head and neck plus the deep structures of the head below the floor of the skull. As each external carotid artery ascends the neck and side of the head, it gives off many branches. The most important branches of the external carotid artery are listed next and illustrated in Fig. 8.19:

1. *Superior Thyroid Artery.* Supplies the thyroid gland and larynx.
2. *Lingual Artery.* Supplies the tongue.
3. *Facial Artery.* Supplies superficial facial structures such as the lips and nose.
4. *Occipital Artery.* Supplies the posterior part of the scalp.
5. *Maxillary Artery.* Supplies deep facial structures including the teeth; it is one of the terminal branches of the external carotid artery.
6. *Superficial Temporal Artery.* Supplies the temple and anterior portion of the scalp; it is the other terminal branch of the external carotid artery.

Subclavian Arteries   The two subclavian arteries have different origins; the right subclavian arises from the brachiocephalic trunk, while the left arises directly from the aortic arch. Each subclavian artery enters the root of the neck and crosses over to the outer border of the first rib, where it becomes the *axillary artery*. Before changing its name, the subclavian artery gives off the following important branches to the head and neck:

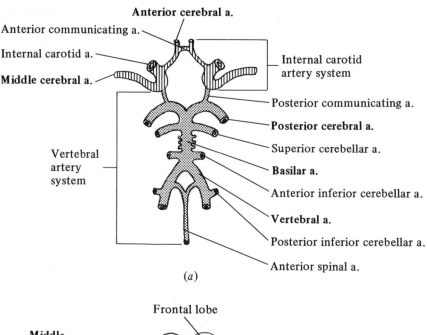

Figure 8.18 *Arterial Blood Supply of Brain.* (*a*) Circle of Willis. Vertebral and internal carotid artery contributions are indicated. (*b*) Basal surface of brain showing circle of Willis and its major branches. Portion of cerebellum and temporal lobe are removed on right side of brain to expose middle and posterior cerebral arteries.

1. *Vertebral Artery.* An artery that ascends through the transverse foramina of the cervical vertebrae to enter the skull, where it unites with its opposite member to form the *basilar artery.* The basilar artery joins the circle of Willis to help supply the brain.
2. *Thyrocervical Trunk.* A short arterial trunk whose branches supply the thyroid gland, structures of the shoulder, and muscles of the scapula.
3. *Internal Thoracic Artery.* An artery that passes down the anterior wall of the thorax parallel to the sternum. It supplies the pectoral

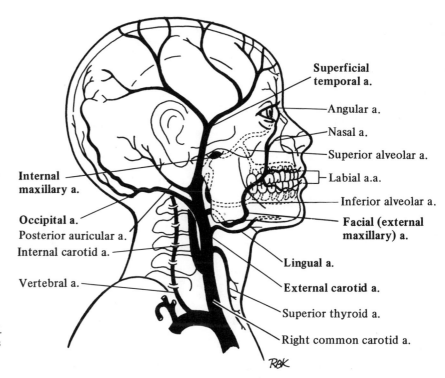

Superficial temporal a.

Angular a.

Nasal a.

Superior alveolar a.

Labial a.a.

Inferior alveolar a.

Facial (external maxillary) a.

Lingual a.

External carotid a.

Superior thyroid a.

Right common carotid a.

Internal maxillary a.

Occipital a.
Posterior auricular a.
Internal carotid a.

Vertebral a.

Figure 8.19 Arterial Supply of Head. Carotid arteries and their major branches are shown on right side of head and neck.

muscles and mammary gland and gives off anterior intercostal branches. At the level of the sixth intercostal space, it gives off the *superior epigastric artery,* which supplies the upper part of the anterolateral abdominal wall.

Axillary Artery. The axillary artery (Fig. 8.20) is the great arterial trunk of the upper extremity. It passes through the axillary space, giving off branches to the upper part of the thorax and shoulder. At the lower border of the teres major muscle, it becomes the *brachial artery.*

The brachial artery passes down the anterior surface of the arm, giving off branches to the humerus and arm muscles and contributing to the arterial anastomosis around the elbow. Just below the elbow joint it divides into two terminal branches that supply the forearm, wrist, and hand. The *ulnar artery,* the larger of the two branches, passes down the medial side of the forearm to the wrist, giving off branches to the elbow anastomosis (arterial communications around the elbow joint) and muscles of the anterior compartment of the forearm. At the wrist it passes lateral to the pisiform bone and enters the palm, where it gives rise to the superficial palmar arch. The *radial artery,* the other branch of the brachial artery, passes down the lateral side of the forearm to the wrist, giving off branches to the elbow anastomosis and muscles of the posterior compartment of the fore-

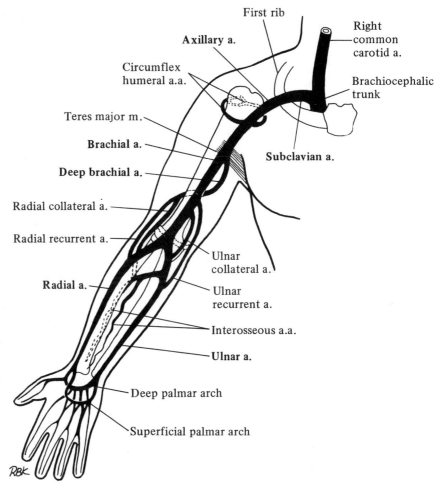

First rib

Axillary a.

Circumflex
humeral a.a.

Right
common
carotid a.

Brachiocephalic
trunk

Teres major m.

Brachial a.

Deep brachial a.

Subclavian a.

Radial collateral a.

Radial recurrent a.

Ulnar
collateral a.

Radial a.

Ulnar
recurrent a.

Interosseous a.a.

Ulnar a.

Deep palmar arch

Superficial palmar arch

Figure 8.20 Arterial Supply of Upper Extremity. Axillary artery and its branches are shown on anterior aspect of right upper limb. Vessels located on posterior side of extremity are indicated by *dashed lines.* Note that radial and ulnar collateral and recurrent arteries form anastomosis around elbow. Elbow anastomosis provides alternative routes for blood to reach forearm and hand should brachial artery be obstructed at elbow.

arm. At the wrist it curves around the base of the thumb and enters the palm, where it gives rise to the deep palmar arch.

The *superficial palmar arch* is formed by the terminal portion of the ulnar artery plus an anastomotic branch from the radial artery, while the *deep palmar arch* is formed by the terminal portion of the radial artery plus an anastomotic branch from the ulnar artery. The two palmar arches (Fig. 8.21) give off metacarpal and digital branches, which supply structures of the hand and fingers.

Thoracic Aorta   The thoracic aorta (Fig. 8.22) is the portion of the descending aorta that extends from the inferior border of the fourth thoracic vertebra to the diaphragm; after passing through the diaphragm, it becomes the abdominal aorta. At its superior end the thoracic aorta lies to the left of the vertebral column but swings medially so that its inferior

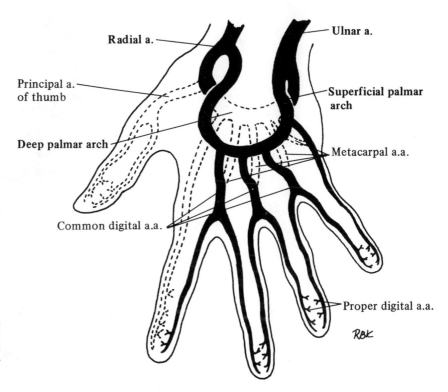

Figure 8.21 Arterial Supply of Hand. Palmar aspect of right hand. Arterial supply is derived from superficial and deep palmar arches. Branches of deep palmar arch are indicated by *dashed lines.*

end lies on top of the vertebral column. Within the thoracic cavity the aorta gives off branches to the thoracic walls and viscera.

Parietal Branches

Branches to the thoracic walls include nine pairs of *posterior intercostal arteries* and a pair of *subcostal arteries*, which supply the intercostal muscles and the muscles and skin of the back, plus a pair of *superior phrenic arteries*, which supply the superior surface of the diaphragm.

Visceral Branches

Branches to the thoracic viscera include *pericardial arteries*, which supply the posterior part of the pericardial sac; *bronchial arteries* (about three), which supply the bronchi and substance of the lungs; *esophageal arteries* (about four or five), which supply the esophagus; and *mediastinal arteries*, which supply structures in the posterior mediastinum.

Abdominal Aorta

The abdominal aorta (Fig. 8.23) begins at the aortic hiatus of the diaphragm and descends in front of the vertebral column to the level of the fourth lumbar vertebra, where it divides into two terminal branches. As it passes through the abdominal cavity, the aorta gives off branches to the abdominal walls and viscera.

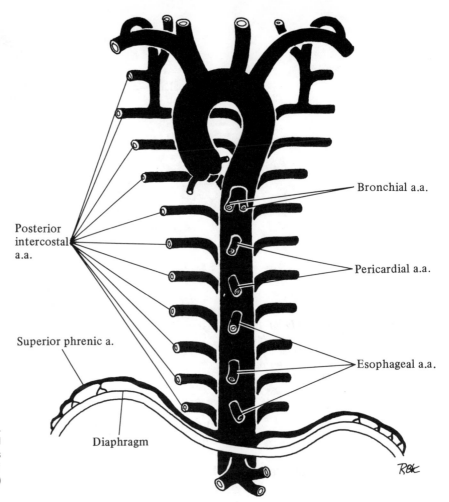

Bronchial a.a.

Pericardial a.a.

Posterior intercostal a.a.

Esophageal a.a.

Superior phrenic a.

Diaphragm

RBK

Figure 8.22   Branches of Thoracic Aorta. Schematic diagram shows visceral and parietal branches. Visceral branches *(unpaired)* arise from anterior surface of aorta, while parietal branches *(paired)* arise from posterolateral surface.

Parietal Branches

The aorta gives off the following branches to the abdominal walls:

1. *Inferior Phrenic Arteries.*   A pair of arteries that supply the inferior surface of the diaphragm.
2. *Lumbar Arteries.*   Four pairs of arteries that supply the posterior abdominal wall.

Visceral Branches

The aorta gives rise to the following paired and unpaired branches to the abdominal viscera; these branches are described in the order in which they arise from the aorta.

1. *Celiac Trunk.*   A short thick vessel that arises from the front of the aorta just below the diaphragm (Fig. 8.24). Almost immediately it divides into three branches—the left gastric, splenic, and common hepatic arteries. The *left gastric artery* is a small branch

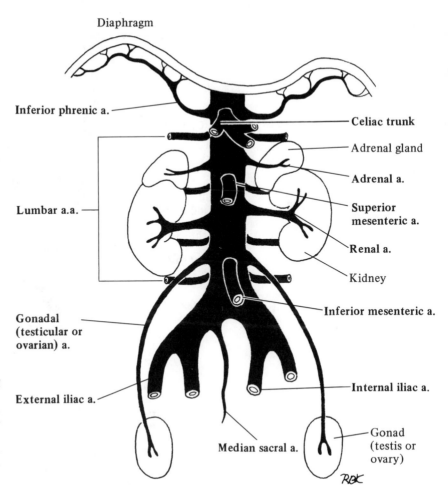

Diaphragm

Inferior phrenic a.

Celiac trunk

Adrenal gland

Adrenal a.

Superior
mesenteric a.

Lumbar a.a.

Renal a.

Kidney

Inferior mesenteric a.

Gonadal
(testicular or
ovarian) a.

External iliac a.

Internal iliac a.

Median sacral a.

Gonad
(testis or
ovary)

Figure 8.23 Branches of Abdominal Aorta. Schematic diagram shows visceral and parietal branches. Anterior visceral branches *(unpaired)* arise from anterior surface of aorta, while lateral visceral branches *(paired)* arise from lateral surfaces and parietal branches arise from posterolateral surface.

that runs from left to right along the lesser curvature of the stomach; it supplies the abdominal portion of the esophagus and part of the stomach. The *splenic artery* is a large branch that runs horizontally to the left behind the stomach to reach the spleen; it supplies the spleen and gives off branches to the stomach and pancreas. The *common hepatic artery* is a medium-sized branch that runs to the right to enter the substance of the liver; it supplies the liver and gives off branches to the stomach, pancreas, duodenum, and gall bladder.

2. *Superior Mesenteric Artery.* A large vessel that arises from the front of the aorta $\frac{1}{2}$ inch below the celiac trunk and runs anteriorly and downward to the lower right quadrant of the abdomen, where it terminates by anastomosing with one of its own branches (Fig. 8.25). The superior mesenteric artery supplies all the intestinal tract except for the upper part of duodenum and the distal one-half of the colon.

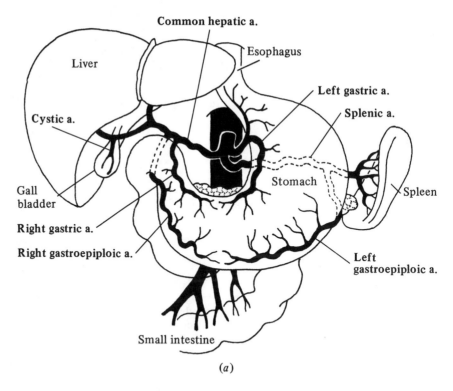

Common hepatic a.

Liver

Esophagus

Cystic a.

Left gastric a.

Splenic a.

Gall
bladder

Stomach

Spleen

Right gastric a.

Right gastroepiploic a.

Left
gastroepiploic a.

Small intestine

(a)

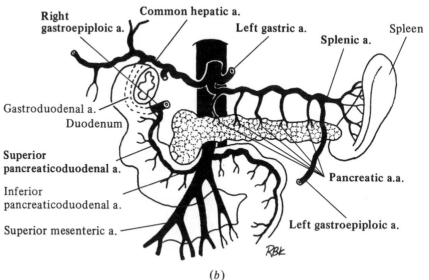

Right
gastroepiploic a.

Common hepatic a.

Left gastric a.

Spleen

Splenic a.

Gastroduodenal a.

Duodenum

Superior
pancreaticoduodenal a.

Pancreatic a.a.

Inferior
pancreaticoduodenal a.

Superior mesenteric a.

Left gastroepiploic a.

RBK

(b)

Figure 8.24 Branches of Celiac Trunk.
(a) Branches shown in relation to liver and
stomach. (b) Liver and stomach are removed to show branches in relation to
pancreas, spleen, and proximal part of
small intestine.

3. *Adrenal Arteries (Middle).* A pair of slender vessels that arise laterally from the aorta at the level of the superior mesenteric artery; they partially supply the adrenal glands.

4. *Renal Arteries.* A pair of large vessels that arise laterally from the aorta below the level of the superior mesenteric artery; they supply the kidneys.

5. *Gonadal Arteries.* A pair of long slender vessels that arise laterally from the aorta below the level of the renal arteries; in the male they supply the testes (testicular arteries), while in the female they supply the ovaries (ovarian arteries).

6. *Inferior Mesenteric Artery.* A medium-sized vessel that arises from the front of the aorta $1\frac{1}{2}$ inches above its termination; it runs downward to the lower left quadrant of the abdomen, where it terminates as the superior rectal artery (Fig. 8.25). The inferior mesenteric artery supplies the distal one-half of the colon plus part of the rectum.

At the level of the fourth lumbar vertebra, the aorta gives rise to its terminal branches, the *median sacral* and *common iliac arteries.*

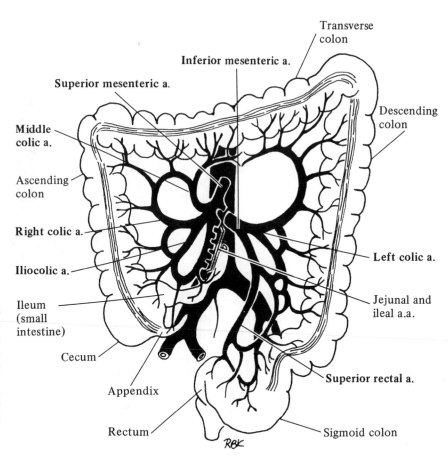

Figure 8.25 Branches of Mesenteric Arteries. Superior and inferior mesenteric arteries are shown in relation to large intestine. Small intestine is removed to expose colic and rectal branches of these arteries. Note that branches of both mesenteric arteries anastomose to form a continuous arterial pathway along border of large intestine; this is so-called "marginal artery."

Median Sacral Artery   The median sacral artery is a small slender vessel that arises from the back of the aorta at its bifurcation. It runs down the sacrum to the coccyx, giving off branches to the posterior pelvic wall and rectum.

Common Iliac Arteries   The right and left common iliac arteries are short vessels formed when the aorta bifurcates (splits). At the level of the sacroiliac joints, the common iliac arteries each give rise to *internal* and *external iliac arteries.*

Internal Iliac Artery. The internal iliac, or hypogastric, artery (Fig. 8.26) enters the pelvis, where it gives off parietal branches that supply the pelvic walls, gluteal branches that supply the gluteal muscles and external genitalia, and visceral branches that supply pelvic viscera such as the rectum, bladder, prostate gland in the male, and uterus and vagina in the female.

External Iliac Artery. The external iliac artery, the great arterial trunk of the lower extremity (Fig. 8.27), is the other branch of the common iliac artery. It extends downward and laterally through the pelvis, passes under the inguinal ligament, and enters the thigh, where it becomes the femoral artery. This artery gives rise to a large

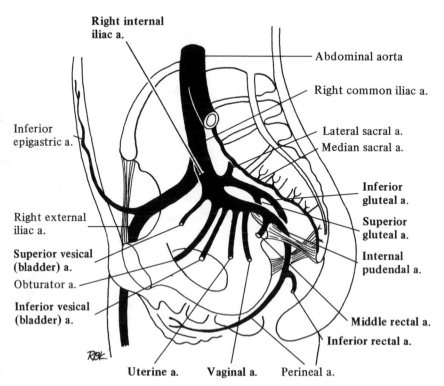

Figure 8.26 Branches of Internal Iliac Artery. Female subject. Right internal iliac (hypogastric) artery and its major branches are shown in sagittal section of female pelvis. Pelvic viscera are removed to aid visualization of vessels.

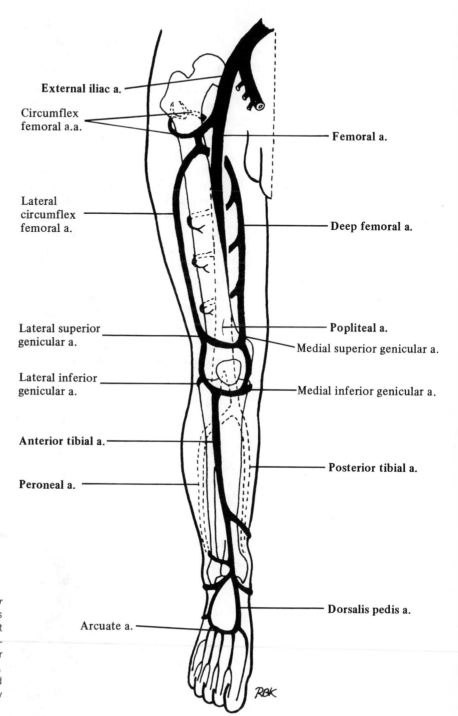

External iliac a.

Circumflex femoral a.a.

Femoral a.

Lateral circumflex femoral a.

Deep femoral a.

Lateral superior genicular a.

Popliteal a.

Medial superior genicular a.

Lateral inferior genicular a.

Medial inferior genicular a.

Anterior tibial a.

Posterior tibial a.

Peroneal a.

Dorsalis pedis a.

Arcuate a.

Figure 8.27 Arterial Supply of Lower Extremity. External iliac artery and its branches are shown on anterior aspect of lower limb. Posterior branches are indicated by *dashed lines.* Note that genicular arteries form an anastomosis around knee, which provides alternative routes for blood to reach leg and foot should popliteal artery become obstructed at knee.

branch, the inferior epigastric artery, that supplies the lower part of the anterolateral abdominal wall.

The *femoral artery* is the artery of the thigh. It extends obliquely downward and posteriorly through a fascial canal in the thigh, passes through a hiatus in the adductor magnus muscle, and enters the popliteal fossa, where it is called the popliteal artery. The femoral artery gives off branches to the femur plus the muscles of the thigh and contributes to the arterial anastomosis around the knee.

The *popliteal artery* passes through the popliteal fossa, contributing branches to the knee anastomosis. Just below the knee joint it gives rise to two terminal branches, the anterior and posterior tibial arteries, which supply the leg, foot, and ankle. The *anterior tibial artery* passes between the tibia and fibula to enter the anterior compartment of the leg. As it descends the front of the leg to the ankle, it gives off branches to the muscles of the anterior compartment. At the ankle it becomes the *dorsalis pedis artery,* which gives off branches to the dorsum of the foot and the toes (Fig. 8.28). The *posterior tibial artery* is the artery of the posterior compartment of the leg. At its proximal end it gives off an important branch, the *peroneal artery,* which supplies the lateral compartment of the leg. The posterior tibial artery descends the back of the leg to the ankle and passes around the medial malleolus to enter the sole of the foot.

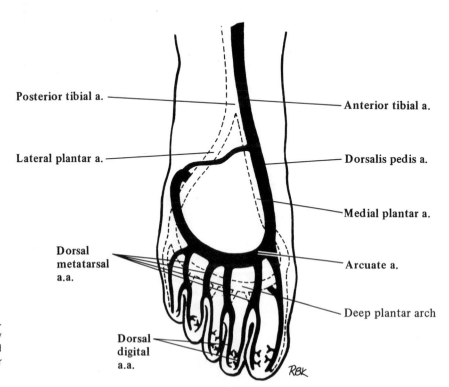

Posterior tibial a.

Lateral plantar a.

Dorsal metatarsal a.a.

Dorsal digital a.a.

Anterior tibial a.

Dorsalis pedis a.

Medial plantar a.

Arcuate a.

Deep plantar arch

Figure 8.28 Arterial Supply of Foot. Dorsal aspect of right foot. Arterial supply of foot is derived from dorsalis pedis and plantar arteries. Plantar arteries and their branches are indicated by *dashed lines.*

There it divides into two terminal branches, the *medial* and *lateral plantar arteries*. The lateral plantar artery anastomoses with a branch from the dorsalis pedis artery to form the *deep plantar arch*. The deep plantar arch gives off metatarsal and digital branches to the sole of the foot and the toes, respectively.

SYSTEMIC CAPILLARIES   Capillaries of the systemic circuit form vast beds in the skeletal muscles and viscera. Not all capillary beds are filled to capacity at the same time. For example, the skeletal muscles usually receive very little of the cardiac output, while vital organs such as the brain, kidneys, and liver continuously receive a large part of the cardiac output in proportion to their size. During exercise, blood is diverted from the capillary beds of the digestive tract and shunted to the capillary beds of the skeletal muscles, hence the concern about engaging in vigorous exercise just after a heavy meal.

SYSTEMIC VEINS   Veins of the systemic circuit (Fig. 8.29) can be divided into three groups: tributaries of the coronary sinus, tributaries of the venae cavae, and tributaries of the portal vein. In addition, there are both superficial and deep systemic veins. *Superficial veins* run in the subcutaneous fascia and return blood from superficial areas of the body. These veins, mostly unnamed, have a variable pattern of distribution and communicate freely with deep veins. *Deep veins* accompany arteries and usually have the same names as the arteries they accompany. The larger arteries are accompanied by a single vein, while the smaller arteries are accompanied by a pair of veins, one on either side of the artery; these vein pairs are called *venae comitantes*.

Tributaries of the Coronary Sinus   The coronary sinus is a wide venous channel that drains the myocardium of the heart and opens into the right atrium. Its principal tributaries include the *great, middle,* and *small cardiac veins;* these vessels were described in Part A.

Tributaries of the Superior Vena Cava   There are, in fact, two great systemic veins that correspond to the aorta: the superior vena cava (considered here) and the inferior vena cava (considered later). There is a gap between these two vessels in the thoracic cavity, which is covered by the azygos system of veins.

The superior vena cava is a large venous trunk, located in the superior mediastinum, that corresponds to the ascending aorta and aortic arch. Tributaries of the superior vena cava include the *right* and *left brachiocephalic veins* and the *azygos vein*.

Brachiocephalic Veins   The right and left brachiocephalic veins are large venous vessels located on either side of the root of the neck. They unite below the level of the first rib and to the right of the sternum to form the superior vena cava. Chief tributaries of the brachiocephalic vein are the *internal jugular* and *subclavian veins*. In addition, it receives the

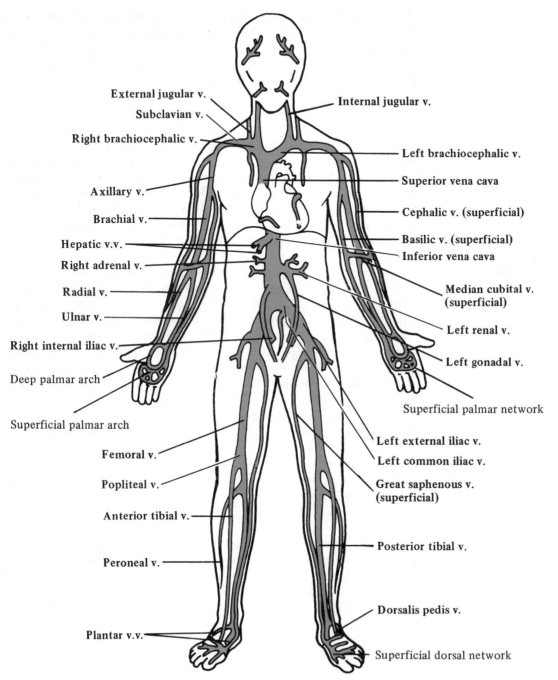

External jugular v.

Subclavian v.

Right brachiocephalic v.

Internal jugular v.

Left brachiocephalic v.

Superior vena cava

Axillary v.

Cephalic v. (superficial)

Brachial v.

Basilic v. (superficial)

Hepatic v.v.

Inferior vena cava

Right adrenal v.

Median cubital v. (superficial)

Radial v.

Left renal v.

Ulnar v.

Right internal iliac v.

Left gonadal v.

Deep palmar arch

Superficial palmar network

Superficial palmar arch

Left external iliac v.

Femoral v.

Left common iliac v.

Popliteal v.

Great saphenous v. (superficial)

Anterior tibial v.

Posterior tibial v.

Peroneal v.

Dorsalis pedis v.

Plantar v.v.

Superficial dorsal network

Figure 8.29 Veins of Systemic Circuit.

vertebral and *internal thoracic veins*, which drain the territories supplied by their corresponding arteries.

Veins of the Head and Neck. The jugular veins (Fig. 8.30) drain the territory supplied by the carotid arteries. The *external jugular veins* are a pair of superficial veins that receive blood from the superficial veins of the scalp and the deep veins of the face. Each external jugular vein empties into the corresponding subclavian vein. The *internal jugular veins* are a pair of deep veins that receive blood from the brain and deep structures of the head and neck. Each internal jugular vein begins at the base of the skull and descends the neck alongside the common carotid artery. At the root of the neck it joins the subclavian vein to form the brachiocephalic vein.

Veins of the Upper Extremity. The deep veins of the upper extremity (Fig. 8.31) have the same names as their companion arteries. The *ulnar veins* receive tributaries from the superficial palmar arch, while

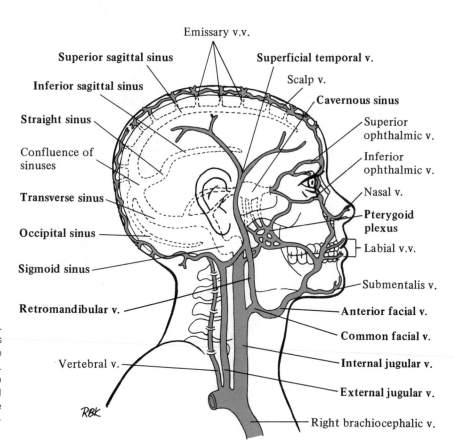

Figure 8.30 *Venous Drainage of Head.* Jugular veins and dural venous sinuses are shown on right side of head. Veins deep to skull are indicated by *dashed lines*. Note that emissary veins connect scalp veins and cavernous sinus connects facial veins to dural venous sinuses. These are important routes for transmission of infections from superficial structures to brain.

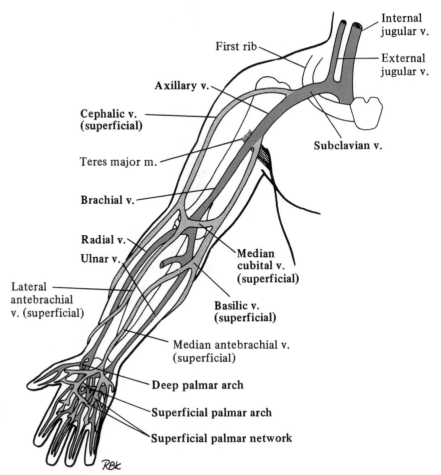

Internal
jugular v.

First rib

External
jugular v.

Axillary v.

Cephalic v.
(superficial)

Subclavian v.

Teres major m.

Brachial v.

Radial v.

Ulnar v.

Median
cubital v.
(superficial)

Lateral
antebrachial
v. (superficial)

Basilic v.
(superficial)

Median antebrachial v.
(superficial)

Deep palmar arch

Superficial palmar arch

Superficial palmar network

RBK

Figure 8.31 Venous Drainage of Upper Extremity. Anterior aspect of right upper limb. Superficial veins are indicated in *light gray*, while deep veins are indicated in *dark gray*.

the *radial veins* receive tributaries from the deep palmar arch. They unite in the antecubital fossa to form the *brachial vein*. At the inferior border of the teres major muscle, the brachial vein becomes the *axillary vein*. At the outer border of the first rib, the axillary vein becomes the *subclavian vein*. Each subclavian vein extends from the lateral border of the first rib to the sternal end of the clavicle, where it unites with the internal jugular vein to form the brachiocephalic vein.

The superficial veins of the upper extremity (Fig. 8.32) include two large veins that drain blood from the superficial venous networks in the upper extremity and convey it to the axillary vein. The *cephalic vein* begins in the dorsal venous network lateral to the base of the thumb. It extends up the lateral side of the forearm and arm to the shoulder, then goes deep to join the axillary vein just below the clavicle. The *basilic vein* begins in the dorsal venous network on the

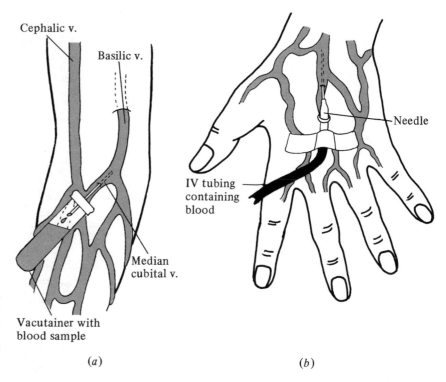

Figure 8.32 Clinical Uses of Superficial Veins of Upper Extremity. (a) Withdrawal of blood sample from median cubital vein. (b) Administration of blood transfusion into superficial vein on dorsum of hand.

back of the hand, curves around the medial border of the forearm, and extends up the medial side of the arm. Halfway up the arm it goes deep to the medial side of the biceps brachii to join the axillary vein at its origin. The *median cubital vein* is a short superficial vein in the antecubital fossa that connects the cephalic and basilic veins. It is commonly used for withdrawing blood samples by venipuncture.

Azygos System

The azygos vein and its tributaries (Fig. 8.33) drain most of the thoracic cavity and its viscera. It serves as a connecting link between the superior and inferior venae cavae and roughly corresponds to the thoracic aorta.

The *azygos vein* originates in the abdominal cavity as the right ascending lumbar vein and ascends along the right side of the vertebral column to the level of *T*4, where it passes over the root of the right lung to enter the superior vena cava. As it passes through the thoracic cavity, the azygos vein receives the right intercostal and subcostal veins plus some of the bronchial, esophageal, and pericardial veins.

The *hemiazygos vein* originates in the abdominal cavity as the left ascending lumbar vein and ascends along the left side of the vertebral column to the level of *T*8, where it crosses over the vertebral column to join the azygos vein. Before it crosses, the hemiazygos vein

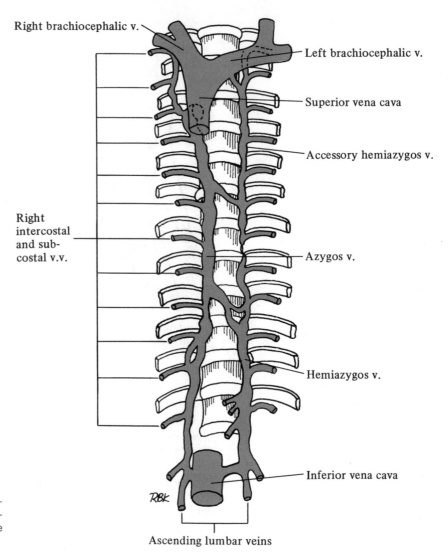

Right brachiocephalic v.

Left brachiocephalic v.

Superior vena cava

Accessory hemiazygos v.

Right
intercostal
and sub-
costal v.v.

Azygos v.

Hemiazygos v.

Inferior vena cava

RBK

Ascending lumbar veins

Figure 8.33 Azygos System. This system drains thoracic cavity. Note connections between azygos veins and venae cavae.

joins the *accessory hemiazygos vein,* which ascends the left side of·
the vertebral column to join the superior vena cava. The hemiazygos
and accessory hemiazygos veins receive the left intercostal and subcostal veins plus the remaining bronchial, esophageal, and pericardial
veins.

It is important to note that when the inferior vena cava is obstructed,
the azygos system forms an alternative pathway for blood from the
lower extremities to reach the heart. This is therefore a lifesaving
backup system.

Tributaries of the Inferior Vena Cava

The inferior vena cava (Fig. 8.34) is the largest vein in the human body. It originates in the abdominal cavity at vertebral level *L*5 and ascends the vertebral column to the right of the aorta. It passes through the vena caval hiatus in the diaphragm to gain the thoracic cavity, where it enters the right atrium of the heart. This vessel, which corresponds to the abdominal aorta, receives tributaries from the abdominal walls and viscera plus two tributaries of origin, the common iliac veins.

Parietal Tributaries

The inferior vena cava receives the following tributaries from the abdominal walls:

1. *Lumbar Veins.* Four pairs of veins that drain the posterior abdominal walls.

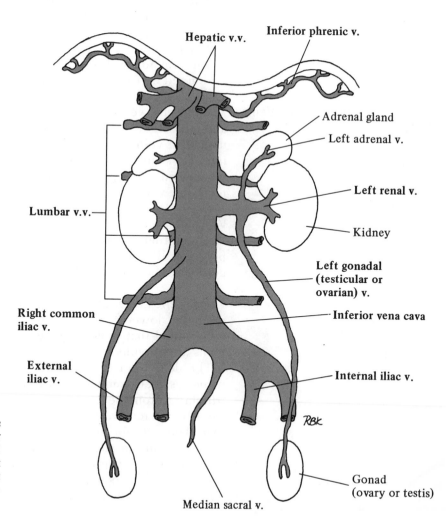

Figure 8.34 *Tributaries of Inferior Vena Cava.* Schematic diagram of inferior vena cava and its anterior visceral, lateral visceral, and parietal tributaries. Note that left adrenal and gonadal veins are direct tributaries of left renal vein rather than inferior vena cava.

2. *Inferior Phrenic Veins.* A pair of veins that drain the inferior surface of the diaphragm.

Visceral Tributaries   The inferior vena cava receives the following tributaries from the abdominal viscera; these vessels are described in the order in which they join the inferior vena cava.

1. *Gonadal Veins.* A pair of long slender veins that drain the testes in the male and the ovaries in the female. The right gonadal vein is a tributary of the inferior vena cava, while the left gonadal vein is a tributary of the left renal vein.
2. *Renal Veins.* A pair of large veins that drain the kidneys. The left renal vein, which crosses over the aorta, is the longer of the two tributaries.
3. *Adrenal Veins.* A pair of slender veins that drain the adrenal glands. The right adrenal vein is a tributary of the inferior vena cava, while the left adrenal vein is a tributary of the left renal vein.
4. *Hepatic Veins.* Several short veins that drain the liver. The spleen and remainder of the digestive organs in the abdominal cavity are drained by tributaries of the portal vein.

Common Iliac Veins   The inferior vena cava receives two tributaries of origin, the right and left common iliac veins, at vertebral level *L*5. In turn, each common iliac vein is formed by the union of the *internal* and *external iliac veins* in front of the sacroiliac joint. The *median sacral vein,* which drains the posterior pelvic wall, is a tributary of the left common iliac vein.

Veins of the Pelvis. The internal iliac veins drain pelvic structures. Each internal iliac vein receives parietal tributaries from the posterior pelvic wall, gluteal tributaries from the gluteal region and external genitalia, and visceral tributaries from the pelvic viscera.

Veins of the Lower Extremity. Both deep and superficial sets of veins drain the lower extremity. Since the veins of the lower extremity carry blood that is flowing against the force of gravity, both sets of veins are provided with numerous valves.

The deep veins of the lower extremity (Fig. 8.35) have the same names as their companion arteries. The *anterior tibial veins* receive tributaries from the dorsal venous arch of the foot, while the *posterior tibial veins* receive tributaries from the plantar venous arch of the foot. The anterior and posterior tibial veins unite in the popliteal fossa to form the *popliteal vein.* The popliteal vein is a short vessel that extends upward through the popliteal fossa; it then passes through the adductor hiatus to become the *femoral vein.* The femoral vein, in turn, extends up through the thigh to the groin. It then passes

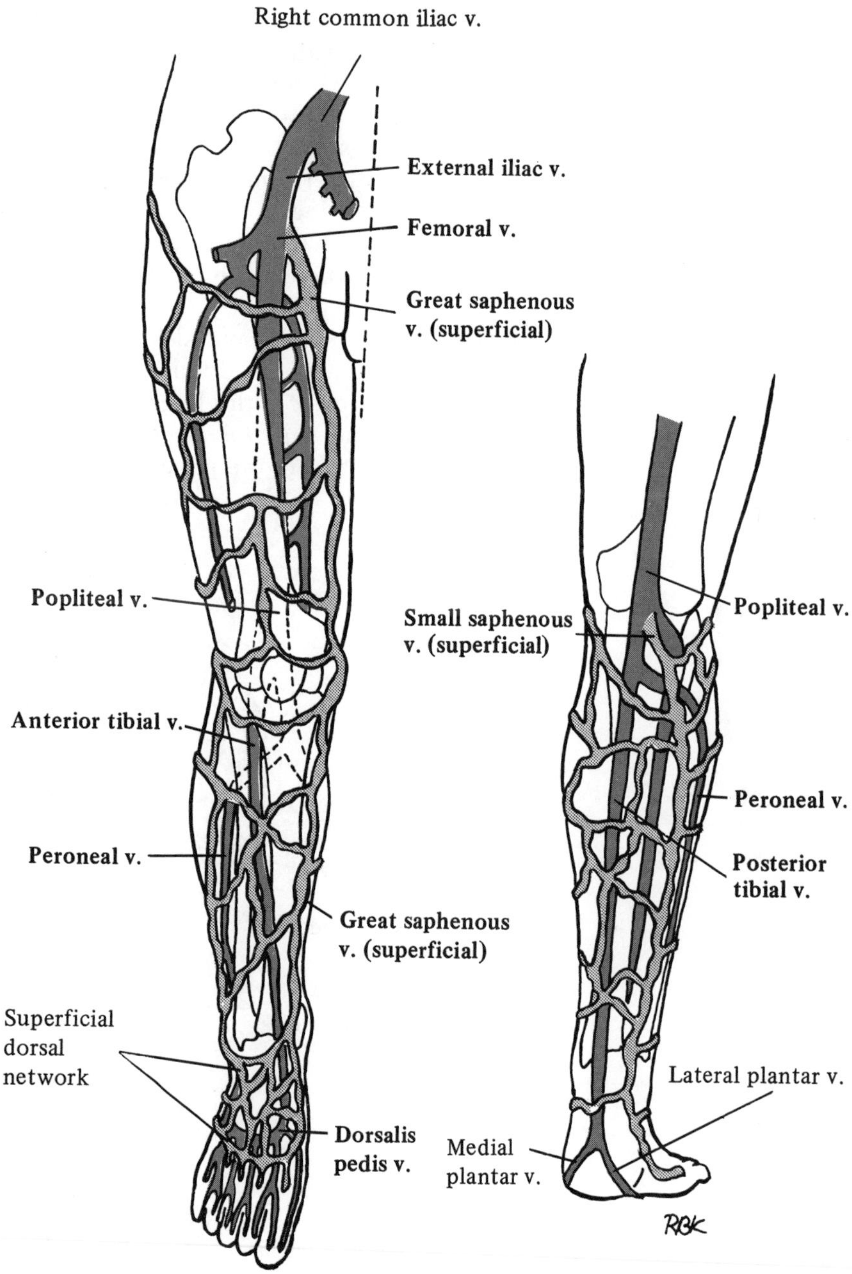

Figure 8.35 Venous Drainage of Lower Extremity. Superficial veins are indicated in *light gray;* deep veins are indicated in *dark gray.* (a) Anterior aspect of right lower limb. (b) Posterior aspect of right leg.

under the inguinal ligament and, as the *external iliac vein,* enters the abdominal cavity to unite with the internal iliac vein.

The superficial veins of the lower extremity consist of the two saphenous veins and their tributaries. The *great saphenous vein,* the longest vein in the human body, arises on the medial side of the dorsum of the foot, passes in front of the medial malleolus of the ankle, and extends up the medial side of the leg and thigh. In the groin it goes deep to unite with the femoral vein. The *small saphenous vein* arises on the lateral side of the dorsum of the foot, passes behind the lateral malleolus of the ankle, and ascends straight up the back of the leg to the popliteal fossa. In the popliteal fossa it goes deep to unite with the popliteal vein. The saphenous veins are of clinical importance due to the fact that they readily develop varicosities.

Hepatic Portal System   The hepatic portal system consists of the portal vein and its tributaries, which drain blood from the digestive organs and spleen and convey it to the liver, as shown in Fig. 8.36.

The *superior mesenteric vein* is a long vessel that receives tributaries from the pancreas, stomach, small intestine, and proximal two-thirds of the colon. It is formed from intestinal tributaries in the

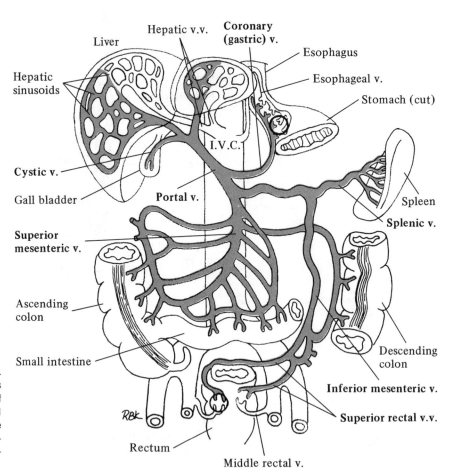

Figure 8.36 Hepatic Portal System. This system drains all unpaired organs in abdominal cavity except liver. Sites of communication between hepatic portal and vena caval systems are *circled*. These sites provide for collateral venous circulation in event of portal vein or liver obstruction.

lower right quadrant of the abdomen and ascends vertically to the level of the pancreas, where it unites with the splenic vein.

The *splenic vein* is a short vessel that passes transversely behind the body of the pancreas. It receives small tributaries from the pancreas and stomach plus the *inferior mesenteric vein,* a major tributary that drains the distal one-third of the colon and the upper part of the rectum. It unites with the superior mesenteric vein behind the neck of the pancreas to form the portal vein.

The *portal vein* is a large short vessel that ascends to the hilum of the liver, where it divides into two branches. Direct tributaries of the portal vein include the coronary vein from the stomach, the cystic vein from the gall bladder, and the paraumbilical veins from the anterolateral abdominal wall. Branches of the portal vein accompany branches of the hepatic arteries into the substance of the liver, where they ultimately empty into *hepatic sinusoids.* From the hepatic sinusoids blood is conveyed by progressively larger vessels into the hepatic veins, which empty into the inferior vena cava.

When either the hepatic sinusoids or the portal vein are obstructed, blood backs up in the portal system; this creates *portal hypertension.* Collateral routes for returning portal blood to the inferior vena cava consist of:

1. Communications between the gastric and esophageal veins, which drain into the azygos system.
2. Communications between the superior rectal vein and the middle and inferior rectal veins, which drain into the internal iliac veins.
3. Communications between the parumbilical veins and superficial epigastric veins, which drain into the great saphenous veins.

Veins are not designed to carry blood under high pressure, and the collateral veins listed above develop dilations, or *varices,* if forced to do so. Esophageal varices that result from portal hypertension are prone to rupture, causing massive blood loss.

Fetal Circulation

Fetal circulation, illustrated in Fig. 8.37, is different from adult circulation due to the fact that the lungs are not functional. The *placenta* (see Chapter 14), the fetal respiratory organ, provides for exchange of oxygen and carbon dioxide between the fetal and maternal bloodstreams. It is connected to the fetal abdomen by the *umbilical cord,* a structure that contains the umbilical blood vessels.

The *umbilical vein,* which transports oxygenated blood from the placenta to the fetus, enters the abdominal cavity through the umbilicus and travels to the undersurface of the liver. It gives off several branches to the substance of the liver plus the *ductus venosus,* which

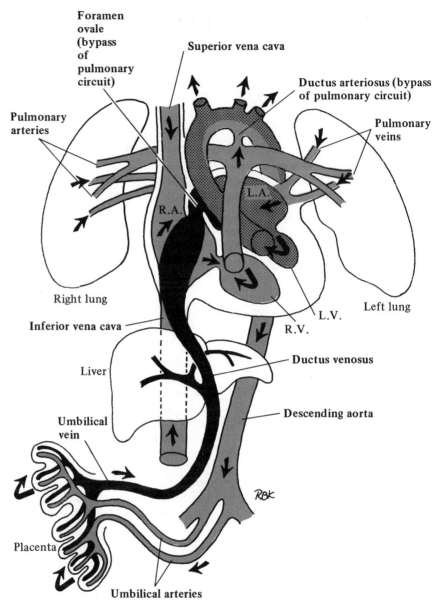

Foramen ovale (bypass of pulmonary circuit)

Superior vena cava

Ductus arteriosus (bypass of pulmonary circuit)

Pulmonary arteries

Pulmonary veins

L.A.

R.A.

Right lung

Left lung

Inferior vena cava

L.V.

R.V.

Ductus venosus

Liver

Descending aorta

Umbilical vein

Placenta

Umbilical arteries

***Figure* 8.37** *Fetal Circulation.* Oxygenated blood is indicated in *black,* partially oxygenated blood in *stipple,* and deoxygenated blood in *gray.*

enters directly into the inferior vena cava. In the inferior vena cava, oxygenated blood from the umbilical vein mixes with blood returning from the lower extremities. Upon entering the right atrium, most of this oxygen-rich blood is shunted into the left atrium through the foramen ovale. Deoxygenated blood from the superior vena cava enters the right atrium and is pumped into the pulmonary circuit by the right ventricle. However, the greater part of this blood is shunted into the systemic circuit through the ductus arteriosus. Blood from the left atrium is pumped into the systemic circuit by the left ventricle. This blood is distributed to the extremities, brain, and other viscera. Most of the deoxygenated blood in the systemic circuit is returned to the placenta by the two *umbilical arteries,* which are almost direct extensions of the internal iliac arteries.

When the lungs begin to function at birth, the fetal shunts close down, and circulation follows the adult pattern. The umbilical vein and ductus venosus are converted to fibrous ligaments, as are the distal parts of the umbilical arteries.

Anastomoses

A communication between two or more blood vessels is called an *anastomosis.* Be means of anastomoses, blood reaches vital body parts through more than one vessel. Examples of arterial anastomoses include the circle of Willis, which is formed by anastomoses between branches of the internal carotid and vertebral arteries, the anastomoses between branches of the right and left coronary arteries in the myocardium, and the anastomoses between the deep and superficial arches in the hands and feet. Examples of venous anastomoses include the communications between tributaries of the portal system and tributaries of the inferior vena cava.

Collateral Circulation

Collateral circulation refers to a series of alternative routes for blood to take when the principal vessel to a vital body part is occluded by disease or surgical ligation. Collateral circulation occurs by way of anastomotic channels around important vessels. For example, when the brachial artery is occluded, blood can reach the radial and ulnar arteries by way of the elbow anastomosis. In some parts of the body, the collateral vessels are sufficiently large so that collateral circulation occurs as soon as the principal vessel is occluded. In other parts of the body the collateral vessels are very small and collateral circulation occurs only if the principal vessel is occluded slowly. There are parts of the body where collateral circulation is limited or entirely lacking; the principal arteries to these parts are known as *end arteries.* When end arteries are occluded, the tissues they supply are deprived of blood and *infarction* results.

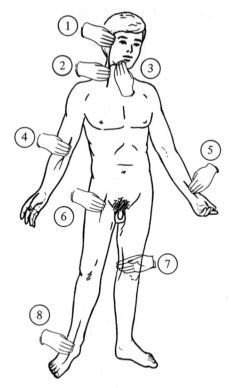

Figure 8.38 *Pulse Points.* (1) Temple, temporal artery; (2) side of neck, common carotid artery; (3) side of jaw, facial artery; (4) front of elbow, brachial artery; (5) lateral side of wrist, radial artery; (6) groin, femoral artery; (7) back of knee, popliteal artery; (8) instep of foot, dorsalis pedis artery.

Pulse Points

The ventricles pump blood into the arteries about 70 to 80 times per minute. The force of ventricular contraction starts a wave of increased pressure that begins at the heart and travels along the arteries; this wave is called the *pulse.* The pulse can be felt in arteries that are relatively close to the surface, especially where the artery can be pressed against bone. Places where the pulse is usually taken are illustrated in Fig. 8.38. When taking the pulse of another person, use your index or middle finger. If you use your thumb, you may find that you are feeling your own pulse, which is quite prominent in the thumb.

Hemorrhage

A massive loss of blood from ruptured or severed blood vessels is known as *hemorrhage.* Hemorrhages may occur in any part of the body, may be internal or external, and may involve vessels of any size. While loss of blood from a severed vein may be slow and oozing in nature, the loss of blood from a severed artery is usually rapid and spurting; immediate action is needed in stopping arterial hemorrhage to prevent massive blood loss. To stop external arterial hemorrhage, certain arteries can be pressed against underlying bone. The major pressure points are similar to or identical with places where the pulse is taken.

Diseases and Disorders of the Blood Vessels

ANEURYSM   An aneurysm is a bulge in a vessel wall due to structural weakness. The structural weakness may be present at birth or may be due to degenerative cardiovascular disease. Eventually the weakened area may yield to pressure and burst like a balloon, causing a hemorrhage. A ruptured aortic aneurysm usually causes death due to the resulting massive blood loss.

ARTERIOSCLEROSIS   Arteriosclerosis, or hardening of the arteries, is a condition in which there is a loss of elasticity or an irregular thickening of the arterial wall so that lumen becomes narrower.

Medial Sclerosis.   Medial sclerosis is a disease of medium-sized arteries with thick muscular walls. In this disease the tunica media degenerates and becomes calcified. Calcification of the media makes the arterial wall as rigid as a lead pipe and less able to accommodate an increase in blood volume.

Atherosclerosis.   Atherosclerosis is a disease that affects the tunica intima in certain arteries, particularly the aorta and the femoral, coronary, and cerebral arteries. In these arteries collections of lipid material ("fatty streaks") accumulate beneath the intimal surface. The streaks, which contain large amounts of cholesterol, are then converted to fibrous plaques, called *atheromas,* that often project into the lumen. In addition to narrowing the lumen, atheromas serve as sites of blood clot formation (*thrombosis*). Most of the deaths from atherosclerosis result from thrombotic occlusion of essential arteries.

PHLEBITIS   Inflammation of a vein is called *phlebitis*. The inflammatory process involves the entire vein wall and is accompanied by considerable pain and swelling. A blood clot may form within the inflamed vein; this condition is called *thrombophlebitis*. Thrombophlebitis in deep veins is quite dangerous because a piece of the clot could break loose and become an *embolus* (a clot that moves). A *pulmonary embolus* (an embolus that passes through the right heart and enters the pulmonary circuit) can cause lung damage or even death. Early ambulation after surgical procedures and use of anticoagulants have greatly reduced the incidence of thrombophlebitis and pulmonary embolism in hospitalized patients.

VARICOSE VEINS   Varicose veins are veins that have become swollen and tortuous mainly because their valves have been destroyed. Venous valve destruction occurs when the veins are overstretched for a long period of time by excessive venous pressure. The veins most frequently involved are the saphenous veins of the lower extremities. Varicose veins are commonly seen in people who spend a lot of time on their feet, such as nurses, teachers, and salespersons. Pregnant women often develop varicose veins due to obstruction of the pelvic veins by the enlarged uterus. Varicose veins in the rectum and anal canal are called *hemorrhoids* (or *piles*). They usually result from straining during bowel movements and cause considerable discomfort.

Suggested Readings

ADOLPH, E. F. "The heart's pacemaker," *Sci. Am., 216,* no. 3 (March 1967), pp. 32–37.

°° BRAUNWALD, E. *The Myocardium: Failure and Infarction.* New York, HP Publishing Co., 1974.

CERAMI, ANTHONY, and CHARLES M. PETERSON. "Cyanate and sickle-cell disease," *Sci. Am., 232,* no. 4 (April 1975), pp. 41–50.

COOPER, LOUIS Z. "German measles," *Sci. Am., 215,* no. 1 (July 1966), pp. 30–37.

DEBAKEY, MICHAEL E., and LEONARD ENGEL. "Blood vessel surgery," *Sci. Am., 204,* no. 4 (April 1961), pp. 88–101.

EFFLER, DONALD B. "Surgery for coronary disease," *Sci. Am., 219,* no. 4 (October 1968), pp. 36–43.

FREI, EMIL, III, and EMIL J. FREIREICH. "Leukemia," *Sci. Am., 210,* no. 5 (May 1964), pp. 88–96.

FREIMER, EARL H., and MACLYN MCCARTY. "Rheumatic fever," *Sci. Am., 213,* no. 6 (December 1965), pp. 67–74.

HAMMOND, E. CUYLER. "The effects of smoking," *Sci. Am., 207,* no. 1 (July 1962), pp. 39–51.

HARARY, ISAAC. "Heart cells in vitro," *Sci. Am., 206,* no. 5 (May 1962), pp. 141–152.

KOLFF, WILLEM J. "An artificial heart inside the body," *Sci. Am., 213,* no. 5 (November 1965), pp. 38–46.

LAKI, KOLOMAN. "The clotting of fibrinogen," *Sci. Am., 206,* no. 3 (March 1962), pp. 60–66.

LILLEHEI, C. WALTON, and LEONARD ENGEL. "Open-heart surgery," *Sci. Am., 202,* no. 2 (February 1960), pp. 76–90.

MCKUSICK, VICTOR A. "The royal hemophilia," *Sci. Am., 213,* no. 2 (August 1965), pp. 88–95.

NETTER, FRANK H. *The Ciba Collection of Medical Illustrations.* Vol. 5: *Heart.* Summit, N.J.: Ciba, 1969.

PHIBBS, BRENDAN. *The Human Heart: A Guide to Heart Disease.* 3d ed. St. Louis: Mosby, 1975 (paperback).

PONDER, E. "The red blood cell," *Sci. Am., 196,* no. 1 (January 1957), pp. 95–102.

** RUSHMER, R. F. *Structure and Function of the Cardiovascular System.* Philadelphia: Saunders, 1972.

SCHER, ALLEN M. "The electrocardiogram," *Sci. Am., 205,* no. 5 (November 1961), pp. 132–141.

SPAIN, DAVID M. "Atherosclerosis," *Sci. Am., 215,* no. 2 (August 1966), pp. 48–56.

STARR, ALBERT, and CECILLE O. SUNDERLAND. "One-stage correction of tetralogy of Fallot," *Hosp. Pract.,* 8, no. 6 (June 1973), pp. 61–68.

WIGGERS, C. J. "The heart," *Sci. Am., 196,* no. 5 (May 1957), pp. 74–87.

WISSLER, ROBERT W. "Development of the atherosclerotic plaque," *Hosp. Pract.,* 8, no. 3 (March 1973), pp. 61–72.

WOOD, J. EDWIN. "The venous system," *Sci. Am., 218,* no. 1 (January 1968) pp. 86–96.

ZUCKER, MARJORIE. "Blood platelets," *Sci. Am., 204,* no. 2 (February 1961), pp. 58–64.

ZWEIFACH, B. J. "The microcirculation of the blood," *Sci. Am., 200,* no. 1 (January 1959), pp. 54–60.

** Highly advanced level.

The Lymphatic System, Lymphoid Organs, and Immunity

The lymphatic system is a one-way drainage system that collects excess tissue fluid and transports it as *lymph* to the cardiovascular system. The lymphatic system begins peripherally as a series of blind-ended capillaries forming networks in the tissue spaces. Centrally converging lymphatic vessels empty into the great veins near the heart. Before lymph enters the cardiovascular system, it passes through a series of filters called *lymph nodes,* where particulate matter is removed and lymphocytes are added. The lymphoid tissues and organs associated with the lymphatic system form the *immune system,* an important part of the body's defense mechanism.

The Lymphatic System

LYMPH Lymph is a clear, faintly straw-colored fluid, similar to blood plasma, that is formed by the filtration of tissue fluid. Lymph contains white blood cells (lymphocytes and a few granular leukocytes) but no red blood cells or platelets. Lymph will clot, but much more slowly than blood. It also contains enzymes and antibodies (protective protein substances that defend the body from invasion by harmful microorganisms).

LYMPHATIC VESSELS
Gross Anatomy Vessels involved in the collection and transportation of lymph include lymphatic capillaries, collecting vessels, and lymphatic ducts.

Lymphatic Capillaries Lymphatic capillaries are microscopic endothelial tubes that form complex networks in the tissue spaces. Unlike the blood capillaries, lymphatic capillaries begin blindly; the blind end lies free in a pool of tissue fluid while the other end communicates with a larger lymphatic vessel. These capillaries resemble blood capillaries in being very thin-walled, but they are more porous; therefore, the larger particles of waste products found in the tissue fluid can be picked up by the lymphatic capillaries but not by blood capillaries. Specialized lymphatic capillaries called *lacteals* are found in the walls of the small intestine. Lacteals (pertaining to milk) are so named because they pick up a white milklike lymph called *chyle*, which is rich in absorbed lipids.

Collecting Vessels

Small lymphatic collecting vessels are thin-walled tubes formed by the union of lymphatic capillaries. They unite in turn to form progressively larger vessels. Collecting vessels are distributed in superficial and deep sets. *Superficial lymphatics* are found in the subcutaneous fascia or on the surfaces of organs. *Deep lymphatics* are larger in size than the superficial lymphatics and accompany the deep veins. Collecting vessels have valves; these valves occur in pairs, with their free edges directed toward the current flow. Lymphatic valves prevent backflow of lymph the same way venous valves prevent backflow of blood. Collecting vessels are indented in the regions where their valves are located and are swollen with lymph between the valves; this gives collecting vessels a beaded appearance.

Lymphatic Ducts

There are two main lymphatic trunks, the thoracic duct and the right lymphatic duct; all the lymph in the body drains into either of these two trunks (Fig. 9.1).

The *thoracic duct* is the larger of the two main lymphatic ducts. It drains all parts of the body except those located on the right side above the diaphragm (Fig. 9.2). The thoracic duct is about 16 inches long and begins in the abdominal cavity at vertebral level *L*2. The first part of this vessel is enlarged to form a temporary storage sac called the *cisterna chyli*. Chyle comes from the intestinal lacteals by way of intestinal and mesenteric collecting vessels. All the lymph and chyle from below the diaphragm empties into the cisterna chyli by way of the right and left lumbar vessels and the intestinal vessels. The thoracic duct enters the thoracic cavity by passing through the aortic hiatus in the diaphragm and ascends along the right side of the vertebral column between the azygos vein and the aorta. At the level of the aortic arch, it crosses over to the left side of the vertebral column, where it receives the left thoracic, subclavian, and jugular vessels. The thoracic duct empties into the cardiovascular system at the junction of the left internal jugular and subclavian veins. At its termination the thoracic duct contains a pair of valves that prevent backflow of venous blood into the lymphatic system. From the thoracic

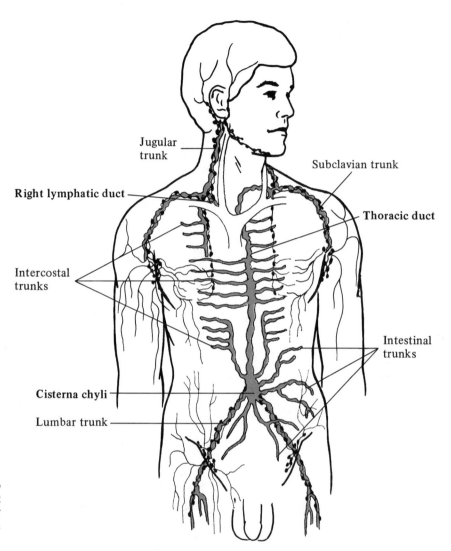

Jugular trunk

Subclavian trunk

Right lymphatic duct

Thoracic duct

Intercostal trunks

Intestinal trunks

Cisterna chyli

Lumbar trunk

Figure 9.1 Lymphatic Vessels. Two main lymphatic trunks are thoracic duct and right lymphatic duct. Note that lumbar and intestinal trunks empty into cisterna chyli, dilated origin of thoracic duct.

duct lymph, lymphocytes, lipids, and waste products enter the bloodstream to be distributed to all parts of the body for use or for excretion.

The *right lymphatic duct* is a short vessel, about ½ inch in length, that is formed by several tributaries on the right side of the thoracic cavity. It receives the right thoracic, subclavian, and jugular vessels. The right lymphatic duct empties into the cardiovascular system at the junction of the right internal jugular and subclavian veins.

Microscopic Anatomy

It is not necessary for you to know the structure of lymphatic vessels in detail. The important features of these vessels can be conveyed in the following general statements:

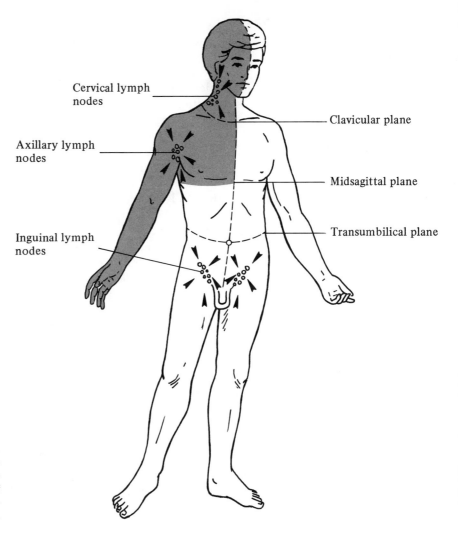

Cervical lymph nodes

Clavicular plane

Axillary lymph nodes

Midsagittal plane

Inguinal lymph nodes

Transumbilical plane

Figure 9.2 Patterns of Lymph Drainage. Shaded portion of human figure indicates areas of body drained by right lymphatic duct; *unshaded* portion indicates areas of body drained by thoracic duct. Arrows indicate drainage to regional lymph nodes.

1. Lymphatic vessels have three coats, just like blood vessels: tunica intima, tunica media, and tunica adventitia.
2. All lymphatic vessels, like blood vessels, are lined with endothelium.
3. The coats of lymphatic vessels are less well defined than those of blood vessels.
4. The collecting vessels and main lymphatic ducts resemble veins of the same size except that their walls may be thinner.
5. Lymphatic vessels contain an appreciable amount of smooth muscle in their walls to aid in propelling the lymph along its way. The main lymphatic ducts contain more smooth muscle in their walls than veins of corresponding size.

LYMPH NODES Lymph nodes are small round to bean-shaped lymphoid organs situated at various points in the course of lymphatic vessels. Their two major functions include the production of blood lymphocytes and tissue plasma cells and the filtration of pathogenic (disease-producing) microorganisms and foreign particles transported in the lymph. The location and structure of the lymph nodes will be described along with the other lymphoid organs.

REGULATION OF LYMPH FLOW Since there is no "lymph heart" to pump lymph through the lymphatic system, other mechanisms must be employed to move lymph. One mechanism is the continuous production of new lymph, which pushes the old lymph ahead of it. Rhythmic peristaltic contractions of smooth muscle in the walls of lymphatic vessels also help move lymph toward the venous system. However, the most important mechanism is the squeezing action of skeletal muscles, which forces lymph forward as they alternately contract and relax.

The average flow rate of lymph in the lymphatic system is about 1 to 2 milliliters per minute; it varies with the tissue fluid pressure and amount of exercise. An increase in amount of tissue fluid results in an increased production of lymph, while increased body activity pushes lymph through the lymph vessels faster.

Lymphoid Tissue and Organs

Accumulations of lymphoid tissue occur in many regions of the body. The degree of organization varies greatly with location and function. The simpler types of lymphoid tissue are *nonencapsulated* and occur in or near mucous membranes. The more complex lymphoid tissues are *encapsulated* and comprise lymphoid organs such as the tonsils, lymph nodes, spleen, and thymus gland.

STRUCTURE OF LYMPHOID TISSUE All lymphoid tissues, whether encapsulated or nonencapsulated, contain the following structural components:

1. *Reticular Fibers.* Fibers that form the stroma (framework) of lymphoid tissue.
2. *Fixed Cells*
   a. *Reticular Cells.* Fixed cells that produce the reticular fibers and give rise to the lymphoid cells.
   b. *Fixed Macrophages.* Fixed cells that act as phagocytes.
3. *Free Cells*
   a. *Free Macrophages.* Free cells that act as phagocytes.
   b. *Lymphocytes.* White blood cells that play a major role in the immune response.
   c. *Plasma Cells.* Lymphocyte-derived cells that produce humoral (circulating) antibodies.

FUNCTIONS OF LYMPHOID TISSUE

In general, the functions of lymphoid tissue are twofold. Lymphoid tissue attempts to prevent the entry of pathogenic microorganisms, particulate matter, or cancer cells into peripheral circulation. Lymphoid tissue also stores large lymphocytes, which originate in the bone marrow and migrate to the lymphoid tissue to complete their development. The derivatives of these precursor cells, mainly small lymphocytes, are released into peripheral circulation. The lymphocytes and the plasma cells derived from them that remain within lymphoid tissue produce antibodies that usually act locally.

NONENCAPSULATED LYMPHOID TISSUE
Diffuse Lymphoid Tissue

Diffuse lymphoid tissue has no special organization; it consists of collections of lymphoid cells and fibers in the mucous membranes particularly those of the respiratory and digestive tracts.

Lymph Nodules

Lymph nodules consist of dense aggregations of lymphoid tissue arranged in spherical masses. They can occur singly or serve as the functional units of more complex lymphoid organs.

A typical lymph nodule (Fig. 9.3) has a loose irregular framework of reticular fibers, which supports the cellular elements. The core of the nodule, called the *germinal center*, contains large reticular cells, large lymphocytes, and macrophages. The following functions are performed in the germinal center: phagocytosis of patho-

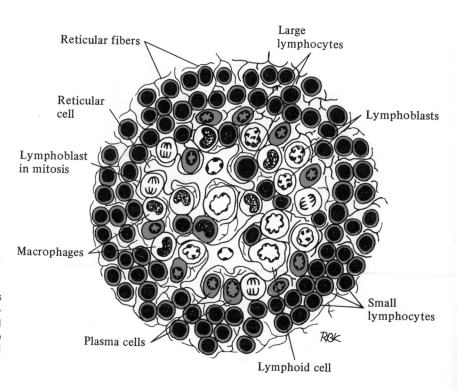

Figure 9.3 Lymph Nodule. This is functional unit of all lymphoid organs except thymus gland. It consists of germinal center *(white)*, which is rich in immature lymphocytes and plasma cells *(gray)*, and a peripheral shell of small lymphocytes.

genic microorganisms, production of small lymphocytes and plasma cells from precursor cells, and the synthesis of antibodies by lymphocytes and plasma cells. The germinal center is surrounded by the *cortex*, a concentric shell consisting almost entirely of small lymphocytes derived by mitotic division from the large lymphocytes in the germinal center.

Aggregate Nodules

Aggregate nodules are closely packed groups of lymph nodules that occur in the intestinal mucosa, mainly in the ileum (distal portion of the small intestine), where they are known as Peyer's patches. Peyer's patches are drained by efferent lymphatics and serve as local filters of tissue fluid. The clinical importance of Peyer's patches lies in the fact that they are the chief sites of intestinal ulceration in diseases such as typhoid fever and intestinal tuberculosis.

ENCAPSULATED LYMPHOID TISSUE (LYMPHOID ORGANS)

Tonsils are masses of lymphoid tissue that show more advanced organization than aggregate nodules.

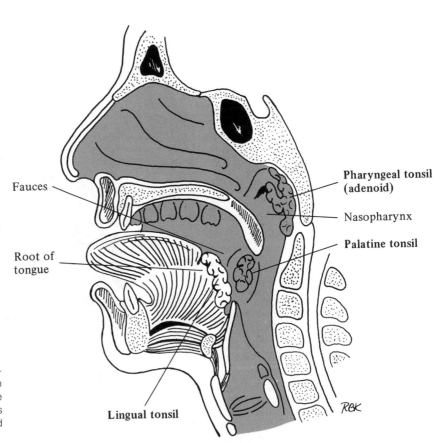

Figure 9.4 Location of Tonsils. Nasopharyngeal tonsil (adenoid) is situated in posterior wall of nasopharynx, palatine tonsil is situated in lateral wall of fauces (between oral cavity and oropharynx), and lingual tonsil is situated in root of tongue.

Tonsils
Gross Anatomy

Three groups of tonsils form a protective ring of lymphoid tissue (Waldyer's ring) around the entrance to the respiratory and digestive tracts (Fig. 9.4):

1. *Palatine Tonsil.* Masses of lymphoid tissue located in the lateral walls of the archway between the mouth and throat. These are the "tonsils" that are removed during a tonsillectomy
2. *Pharyngeal Tonsil.* A mass of lymphoid tissue located in the posterior wall of the pharynx (throat). The clinical name for the pharyngeal tonsil is "adenoids," especially when it is enlarged.
3. *Lingual Tonsil.* A mass of lymphoid tissue located in the root of the tongue near the entrance to the throat.

Microscopic Anatomy

The tonsillar lymphoid tissue, or parenchyma (Fig. 9.5), is partially enclosed in a fibrous connective tissue capsule. Inward projections from the capsule, called *septa,* subdivide the tonsillar parenchyma into lobules. The surface of each tonsil is covered with stratified squamous epithelium. Epithelial pockets called *crypts* penetrate into the interior of the palatine tonsils. The tonsillar crypts collect bacteria and occasionally become inflamed through bacterial activity. Lymphocytes produced in the underlying lymphoid tissue pass through the epithelium of the crypts and are observed in the saliva, where they are called *salivary corpuscles.*

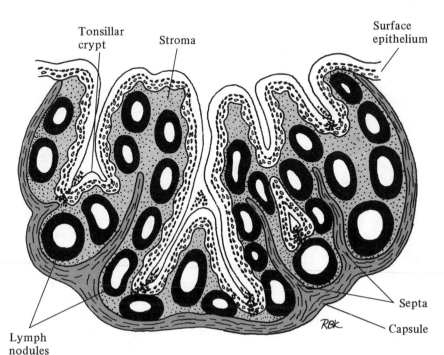

Tonsillar crypt · Stroma · Surface epithelium · Septa · Capsule · Lymph nodules

Figure 9.5 Microscopic Structure of Palatine Tonsil. Histological section. Only palatine tonsil has crypts. Within lumens of tonsillar crypts are desquamated epithelial cells, bacteria, and live and dead lymphocytes. Tonsillar crypts are frequently sites of infection.

Function   Tonsils filter local tissue fluid around the pharynx and produce lymphocytes and circulating antibodies to defend this vulnerable region of the body against invasion by pathogenic microorganisms. Any or all of the tonsils may become so loaded with pathogenic bacteria that the bacteria get the upper hand and cause tonsillar inflammation and enlargement. Slight tonsillar enlargement is not serious, but a massive enlargement may interfere with breathing and swallowing. All lymphoid organs are larger in children than adults, so a doctor must consider the patient's age in determining whether or not the tonsils are significantly enlarged. Since recent studies have indicated that tonsils and adenoids play an important role in protecting the respiratory and digestive tracts from infection, doctors today are less likely to perform routine tonsillectomies and adenoidectomies than in the past. However, frequent attacks of tonsillitis, with or without respiratory tract obstruction, are an indication that a tonsillectomy is necessary.

Lymph Nodes   Lymph nodes are completely encapsulated masses of lymphoid tissue found in the course of lymphoid vessels. They possess a more advanced type of organization than the tonsils.

Gross Anatomy   Lymph nodes are scattered throughout the body, both singly and in groups. Groups of lymph nodes are arranged in deep and superficial sets that filter lymph from deep and superficial lymphatics. Since lymph nodes that filter lymph from infected regions of the body often become swollen and tender to touch, knowledge of the regions drained by particulate groups of lymph nodes is useful in determining sites of acute infection. The clinically important lymph nodes in major body regions and the areas that they drain are described next and illustrated in Fig. 9.6.

Lymph Nodes of the Head and Neck.   The main lymph nodes draining the head and neck are the *deep cervical nodes*. They form a chain that extends along the internal jugular vein from the base of the skull to the root of the neck. In addition, there is a horizontal series of superficial nodes that surrounds the junction of the head and neck. This group includes the *superficial cervical nodes*, which are associated with the external jugular vein.

Lymph Nodes of the Upper Extremity.   The most important group of lymph nodes in the upper extremity lie in the axilla. The superficial and deep *axillary lymph nodes* filter lymph drained from the upper extremity and the breast. Since malignant cells from cancerous breasts often spread to this group of lymph nodes, the axillary nodes are usually removed along with the diseased breast at the time of surgery.

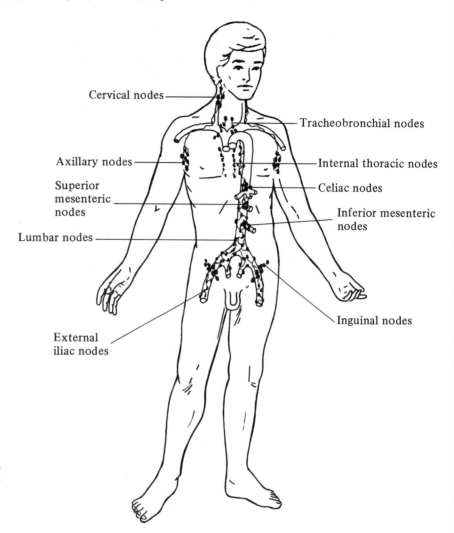

Figure 9.6 Location of Deep Lymph Nodes. Major groups are indicated. Deep lymph nodes are found in axilla, in groin, along great vessels of neck, in thorax, and along abdominal aorta and its branches.

Lymph Nodes of the Thorax. The *thoracic lymph nodes* lie deep within the thoracic cavity. The *intercostal* and *internal thoracic nodes* filter lymph drained from the thoracic wall, while the *mediastinal* and *tracheobronchial nodes* filter lymph drained from the thoracic viscera. The tracheobronchial nodes of coal miners and cigarette smokers become so full of carbon particles that they are black in appearance.

Lymph Nodes of the Abdomen. Certain deep abdominal lymph nodes form chains that course along the aorta and its major branches. These *lumbar nodes*, which follow the abdominal aorta, filter lymph drained from the deep parietal and visceral nodes, the kidneys and adrenal glands, and the posterior abdominal wall. The *celiac nodes,* which follow the celiac trunk, filter lymph drained from the stomach and spleen, part of the liver, the pancreas, and part of the duodenum. The *superior mesenteric nodes,* which follow the course of the superior mesenteric artery, filter lymph drained from the rest of the small intestine and the proximal half of the colon. The *inferior mesenteric nodes,* which follow the course of the inferior mesenteric artery, filter lymph drained from the distal half of the colon plus the rectum.

Lymph Nodes of the Pelvis.   The pelvic lymph nodes, which are continuous with the paraortic nodes, form chains that course along the pelvic arteries. The *external iliac nodes,* which follow the external iliac artery, and the *internal iliac nodes,* which follow the internal iliac artery, filter lymph drained from the pelvic organs, the perineum, the gluteal region, and the posterior pelvic wall.

Lymph Nodes of the Lower Extremity.   The most important group of lymph nodes in the lower extremity are the superficial and deep *inguinal lymph nodes,* which are located in the groin (inguinal region). These nodes filter lymph drained from the lower extremity, external genitalia, perineum, and gluteal region. This lymph is then passed on to the external iliac nodes. When the inguinal nodes are enlarged, they are often referred to as *bubos.* From the term "bubos" we get the name for the bubonic plague, which killed so many people in the Middle Ages.

Microscopic Anatomy   The lymph node (Fig. 9.7) is an ovoid body, convex in shape except for an indented region called the *hilus.* The lymph node is enclosed in a fibrous connective tissue capsule; partitions called trabeculae project from the capsule into the lymphoid tissue (parenchyma). The parenchyma is supported by a fine-meshed framework of reticular fibers that makes up the stroma. It is subdivided into two regions: an outer region called the *cortex* and an inner region called the *medulla.* The cortex lies just beneath the capsule and consists of lymph nodules surrounded by diffuse lymphoid tissue. The medulla lies in the interior of the lymph node and consists of cords of diffuse lymphoid tissue continuous with the diffuse lymphoid tissue of the cortex.

Function   *Afferent lymphatic vessels* bring lymph to the lymph node. They enter the node at multiple points on its convex surface and communicate with internal lymph channels called *lymph sinuses.* The lymph sinuses surround and penetrate the lymphoid parenchyma, allowing lymph to percolate through the lymphoid tissue. As lymph percolates through the lymphoid tissue, bacteria and particulate matter are filtered out of the lymph and lymphocytes are added.

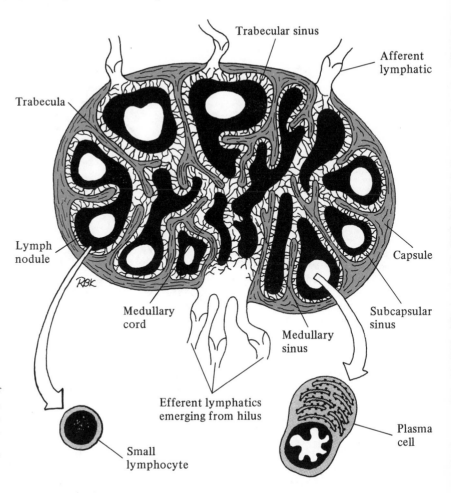

Figure **9.7** *Microscopic Structure of Lymph Node.* Histological section. Cortical portion contains lymph nodules, while medullary portion contains cords of lymphoid tissue. Lymph node filters lymph and produces antibodies (by means of plasma cells and small lymphocytes).

Efferent lymphatic vessels carry lymph away from the lymph node. They emerge from the hilus of the node. At the hilus, blood vessels and nerves also enter and leave the node.

The Spleen

The spleen is an accessory lymphoid organ designed to filter blood.

Gross Anatomy

The spleen (Fig. 9.8) is located in the upper left quadrant of the abdominal cavity, to the left of and slightly behind the stomach, and is normally protected by the lower part of the rib cage.

The spleen is about 5 to 6 inches long and 2 to 3 inches wide. Its diaphragmatic surface is convex to fit the undersurface of the diaphragm, and its visceral surface is concave, marked by impressions of the stomach, pancreas, left kidney, and left flexure of the colon. There is an indentation, or *hilus*, on the visceral surface where splenic blood and lymph vessels and nerves enter and leave

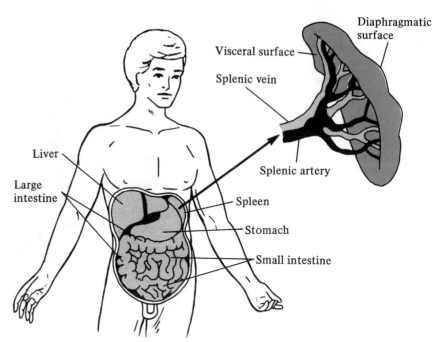

Figure 9.8 *Location of Spleen.* Shown in position in upper left quadrant of abdominal cavity. Enlargement *(right)* shows its diaphragmatic and visceral surfaces. Note that splenic vessels enter organ through depression (hilus) on its visceral surface.

the substance of the organ. Because of its high degree of vascularity, the spleen is purplish red in color.

Microscopic Anatomy

The splenic parenchyma, or *splenic pulp,* is enclosed in a tough, relatively thick, fibrous connective tissue capsule containing both elastic and smooth muscle fibers. Connective tissue trabeculae project inward from the capsule to subdivide the splenic pulp into many pyramidal compartments (Fig. 9.9).

The splenic pulp, which is supported by a framework of reticular fibers, is of two distinct types: *white pulp* and *red pulp.* White pulp consists of scattered areas of lymph nodules and diffuse lymphoid tissue that surround certain small arteries like a sheath. The rest of the splenic pulp is called red pulp because of its high content of red blood cells. It consists of irregular cords of lymphoid and connective tissue cells that are separated by wide blood-filled venous sinusoids. The pulp cords are infiltrated by circulating peripheral blood cells, especially red blood cells.

Blood Supply

Blood is supplied to the spleen by the splenic artery, a branch of the celiac trunk, which enters the spleen at the hilus. It gives rise to a number of branches, the *trabecular arteries,* which travel in the connective tissue trabeculae. These arteries give rise to the *central arteries,* which are the arteries of the white pulp. The central arteries in turn give rise to the *pulp arteries* and *arterioles* of the red pulp.

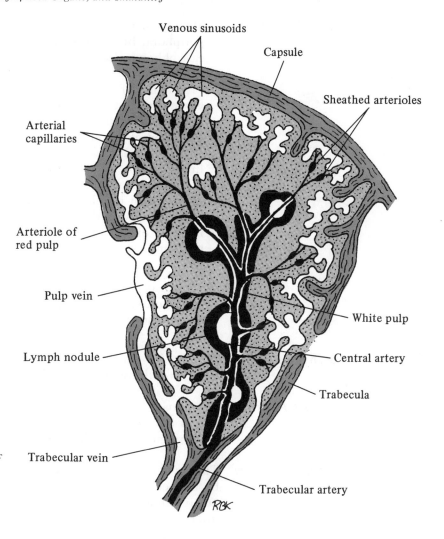

Figure 9.9 Microscopic Structure of Splenic Lobule. Histological section. Note distribution of white pulp *(black)* and red pulp *(light gray)* with relation to splenic vessels within lobule.

These arterioles are either connected to *venous sinusoids* by capillaries (a closed system of circulation) or empty directly into the cords of red pulp, from which sites blood slowly percolates into the splenic sinusoids (an open system of circulation). Blood from the splenic sinusoids enter *pulp veins*, which are tributaries of the *trabecular veins*. The trabecular veins unite to form the splenic vein, a large vessel that leaves the spleen at the hilus to enter into formation of the portal vein.

Function   The functions of the spleen are listed next; most of them are based on the spleen's ability to draw cells from the peripheral bloodstream and provide a chamber for their storage, proliferation, transformation, or destruction.

1. Storage of red blood cells, which can then be released into the peripheral blood stream by contraction of the splenic capsule in case of hemorrhage or other emergency.
2. Production of blood cells before birth.
3. Production of antibodies by lymphocytes and plasma cells in the white pulp following contact with antigen and their subsequent transformation into antibody-producing cells.
4. Filtration of blood by phagocytosis; this is performed by the free and fixed macrophages of the red pulp and by the macrophages lining the splenic sinusoids.
5. Destruction of worn-out red blood cells, with release of hemoglobin. Iron is released from the hemoglobin to be used by the bone marrow in the manufacture of new red blood cells. The remainder of the hemoglobin is returned to the liver to be used in the manufacture of bile.

Removal of the spleen in adulthood in cases of accidental rupture or to control certain diseases does not seriously impair health, as lymphoid and hematopoietic tissue elsewhere in the body can take over the spleen's functions.

The Thymus Gland

The thymus gland is a very important lymphoid organ that is much larger in children than in adults.

Gross Anatomy

This broad, flat, bilobed lymphoid organ (Fig. 9.10) lies partly in the root of the neck and partly in the superior mediastinum (upper part of the thorax) behind the sternum.

Microscopic Anatomy

The thymus gland consists of two lateral lobes united in the midline by connective tissue. Each lobe is enclosed in a thin fibrous connective tissue capsule. Connective tissue trabeculae extend inward from the capsule, subdividing the thymic parenchyma into lobules (Fig. 9.11). A framework of star-shaped reticulo-epithelial cells supports the thymic parenchyma; there are no reticular fibers in the thymus. The parenchyma of each thymic lobule consists of two distinct regions: *cortex* and *medulla*. The cortex is the dark-staining outer portion of the thymic lobule; it contains numerous densely packed small lymphocytes that have infiltrated the reticulo-epithelial cell framework. The medulla is the light-staining inner portion of the thymic lobule in which reticulo-epithelial cells predominate, though some lymphocytes, macrophages, and plasma cells are present in this region. Nests of keratinized epithelial cells, called *thymic corpuscles,* are a diagnostic feature of the medulla. Their significance, if any, is still unknown. There are neither lymph nodules nor lymph sinuses present in either cortex or medulla.

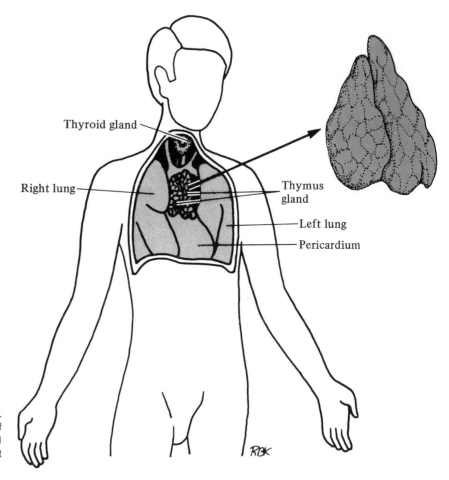

Figure 9.10 *Location of Thymus Gland.* Shown in position in thoracic cavity of adolescent boy before marked postpubertal involution has occurred. Enlargement *(right)* shows its gross appearance.

Function The thymus gland plays a key role in the development of the immune system. Starting in prenatal life, the thymus is colonized by lymphocyte stem cells that have originated elsewhere (in the liver of the fetus and in the bone marrow of the adult) and migrated to the thymus, where they multiply. Some of these cells become mature T-cells (cells that are responsible for the cell-mediated immune response). Mature T-cells may leave the thymus through medullary blood vessels and populate certain areas of other lymphoid tissues (the cortical areas of lymph nodes and periarterial areas in the white pulp of the spleen and Peyer's patches in the small intestine). In these areas the T-cells can be stimulated by antigen (any substance not normally present in the body) and become "committed" to respond to a particular antigen if exposed to it again. Areas of lymphoid tissue not

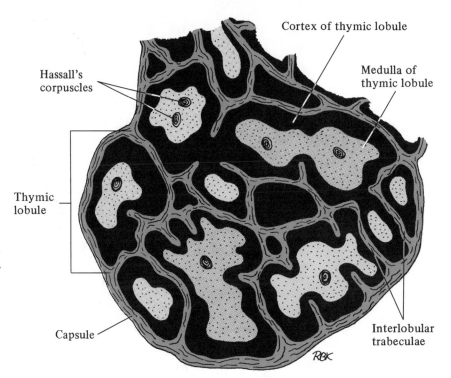

Figure 9.11 Microscopic Structure of Thymic Lobe. Histological section. Each lobule contains cortical tissue *(black)* and medullary tissue *(light gray)*. Hassell's corpuscles are keratinized epithelial structures found only in thymic medullary tissue. Lymph nodules are not found in normal thymic tissue.

occupied by T-cells contain B-cells (cells that are responsible for the humoral antibody response). B-cells are lymphocytes whose stem cells originate in the bone marrow and migrate to an organ as yet unknown in humans (in chickens it is the bursa of Fabricius, a part of the gut) where they complete their maturation. Mature B-cells that come into contact with antigen give rise to plasma cells, which synthesize humoral antibodies.

The thymus gland is highly active in infancy, when the immune system is still being established, and is relatively inactive in adulthood. At puberty the thymus gland starts to involute (grow smaller) in response to the rising level of sex hormones, and its lymphoid tissue is largely replaced by adipose connective tissue in the adult.

Immunity and Antibody Production

The human body must be able to defend itself from invasion by pathogenic microorganisms and cancerous changes in its own cells if it is to survive. The immune system, which consists mainly of lymphoid tissues and organs, has evolved to help the body defend itself through the production of *antibodies.* Antibodies are proteins that react against foreign substances called *antigens* (macromolecules

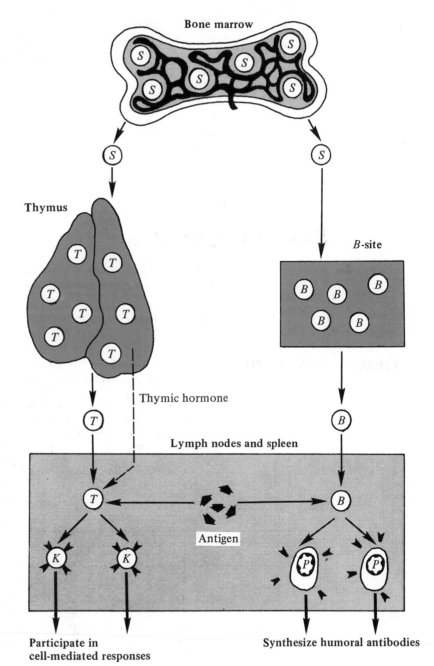

Figure 9.12 *Origin of Antibodies.* Cell-bound antibodies are produced by T-lymphocytes *(T),* which are derived from bone marrow stem cells *(S)* that have passed through thymus gland. When T-lymphocytes are stimulated (mainly by viral, fungal, or cellular antigens), they are committed to cell-mediated immune responses. Committed T-lymphocytes are sometimes called "killer" lymphocytes *(K).* Humoral (circulating) antibodies are produced by plasma cells *(P),* which are derived from B-lymphocytes *(B).* B-lymphocytes are in turn derived from bone marrow stem cells *(S)* that have passed through B-sites, whose location(s) in humans are as yet unknown. When B-lymphocytes are stimulated (mainly by bacterial antigens), they transform into plasma cells and are committed to make humoral antibodies. Humoral antibodies (immunoglobulin or gamma globulin) are not bound to cell surface but are released into blood or lymph.

that may or may not be located on the surface of microorganisms). Each antibody is specific for the particular antigen that stimulates its production.

The immune system produces two types of responses: humoral and cellular (Fig. 9.12). Humoral antibodies, or immunoglobulins, are secreted by sensitized B-cells and plasma cells and are released into peripheral circulation to attack their corresponding antigens. These antibodies defend the body against most bacteria. Cell-mediated reactions are produced by T-cells and the *lymphokines* they secrete. Sensitized T-cells defend the body against viruses, fungi, parasites, and a few bacteria; they are also responsible for the destruction of cancer cells and foreign tissue grafts. These immunocompetent cells react with their stimulating antigens and inactivate them in various ways, one of which is by cellular destruction ("killer" T-cells).

Tissue Transplantation

For a long time doctors have dreamed of replacing damaged tissues and organs with tissues and organs from healthy donors. During the last two decades, many types of tissue transplants have been attempted, and medical researchers have taken an active interest in the methods and mechanisms of tissue transplantation.

TYPES OF TRANSPLANTS The following types of tissue and organ transplants have been made in experimental animals and in humans:

1. *Autograft.* The donor and recipient are the same person; for example, a skin graft from one part of the body to another or the reunion of a severed limb.
2. *Isograft.* The donor and recipient are genetically identical; for example, an organ graft between identical twins or highly inbred strains of animals.
3. *Homograft.* The donor and recipient are of the same species but have different genetic backgrounds.
   a. *Homostructural Graft.* A homograft in which the graft does not have to be alive in order to function but merely serves as a framework for the regrowth of the recipient's own tissues (e.g., homografts of arteries or bone).
   b. *Homovital Graft.* A homograft in which the graft has to be alive in order to function (e.g., homografts of skin, endocrine glands, kidneys, liver, heart, or lungs).
4. *Heterograft.* The donor and recipient are of different species; for example, an organ graft from a chimpanzee to a human being.

TRANSPLANTATION BARRIER With rare exceptions, tissues or organs grafted between genetically dissimilar individuals of the same species or between individuals of two different species are doomed to die. This transplantation barrier exists because each individual is genetically unique down to the proteins on the membranes of his or her cells. When confronted with a homograft or heterograft, the immune system recognizes these proteins as foreign and proceeds to destroy the alien cells of the graft.

HOMOGRAFT REJECTION   When tissues or organs are grafted between two persons who are not identical twins, the graft is generally rejected, as is illustrated in Fig. 9.13. When a graft is homotransplanted into a genetically dissimilar recipient, some of the proteins become dissociated from the membranes of the graft's cells. These proteins, called *transplant antigens*, enter the recipient's lymphatics and are transported to lymph nodes in the region of the graft. Within the regional lymph nodes, the transplant antigens stimulate resident T-cells and B-cells to produce antigens. Committed T-cells ("killer lymphocytes") and humoral antibodies produced by B-cell-derived plasma cells enter peripheral circulation. Reaching the graft by way of the bloodstream, the T-cells

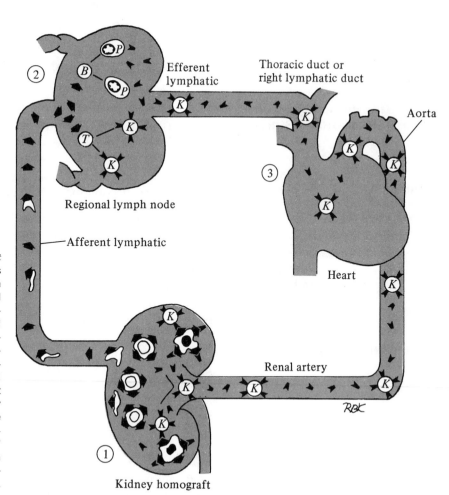

Figure 9.13 Mechanism of Homograft Rejection. Kidney homograft used as example. (1) Antigens escape from graft either in soluble form or bound to cell fragments. (2) Antigen is transported by afferent lymphatics to regional lymph node, where it encounters immunocompetent T-lymphocytes and B-lymphocytes. "Killer" T-cells (K) and humoral antibodies are released from regional lymph node and transported by efferent lymphatics to their junction with circulatory system, pass through heart, and are transported back to graft by systemic arteries. Back in graft, antibodies and "killer" T-cells carry out immunological attack on kidney cells. When sufficient cells are destroyed, grafted kidney ceases to function.

and humoral antibodies attack and destroy its cells. This is called *homograft rejection*. The endpoint of homograft rejection is the loss of function of the grafted tissue or organ.

EXCEPTIONS TO HOMOGRAFT REJECTION

There are several exceptions to the general rule that all homografts are rejected; these exceptions are as follows:

1. Identical human twins and highly inbred strains of animals have such a high degree of genetic similarity that they readily accept homografts.
2. Embryos and fetuses accept homografts because their immune system is immature.
3. Humans and animals deprived of their thymus glands before or just after birth accept homografts.
4. Persons with certain diseases resulting in diminished antibody production will accept homografts.
5. Chimeras (organisms with two or more genetically different tissues) accept homografts. Natural chimerism occurs in nonidentical cattle and human twins whose placentas were fused during fetal development, leading to the exchange of blood-forming tissues.

MEANS OF PROLONGING HOMOGRAFT SURVIVAL

There are several means of prolonging homograft survival; some are impractical in human beings, some are being used today, and some may be useful in the future.

Transplantation to an Immunologically Privileged Site

The anterior chamber of the eye is an immunologically privileged site because it contains no blood vessels or lymphatics. This is the reason why corneal grafts are so successful. Unfortunately, there are few such privileged sites in the body, and they cannot receive large grafts.

Production of Acquired Tolerance

Producing acquired tolerance in human beings, such as by removing the thymus gland at birth, is not feasible since it leaves the body vulnerable to pathogenic microorganisms from without and cancerous changes from within.

Modification of the Recipient

The approach most often used today is modification of the recipient by *immunosuppression* so that he or she cannot make antibodies against the graft. A combination of immunosuppressive drugs such as Imuran®, which suppresses DNA synthesis, and prednisone, which suppresses lymphocyte formation, are used to keep the recipient from producing large numbers of lymphocytes sensitized to the graft. Anti-lymphocyte serum (ALS) and globulin (ALG), which are produced by injecting human lymphocytes into horses, are sometimes given to graft recipients along with other immunosuppressive drugs to further reduce the number of sensitized lymphocytes.

Tissue Typing Human lymphocytes contain cell surface antigens, called HL–A antigens, that are genetically determined. Before a homotransplant is performed, the recipient and potential graft donors are typed for HL–A antigens. The closer the match between donor and recipient in terms of HL–A antigens, the more successful the graft is and the smaller the amount of immunosuppressive drugs needed to protect the graft from rejection. This is why grafts between parent and child or between children within a family "take" much more successfully than grafts between unrelated donors.

Graft Modification In the future it might be possible to modify the graft by such techniques as cultivating the graft in tissue culture for a period of time or by using some form of genetic engineering to transform cell membrane lipoproteins of the graft to those of the recipient.

Diseases and Disorders of the Lymphatic System

LYMPHANGITIS Lymphangitis is the term used to describe inflammation of the lymphatic vessels. The presence of red streaks extending along an extremity from the site of an infected or neglected injury is a clear indication of lymphangitis. Lymphangitis is a danger signal that pathogenic microorganisms within the inflamed lymphatic vessels might soon enter the bloodstream and cause *septicemia* (blood poisoning) if the lymph node filters fail to overcome the infection.

LYMPHEDEMA Lymphedema refers to edema of the extremities resulting from obstruction of their lymphatic drainage. Lymphedema of the upper extremity is often seen in women who have had a radical mastectomy operation for breast cancer. During this operation the axillary lymph nodes on the affected side are removed and the lymphatic vessels are damaged. *Filariasis* is an infectious tropical disease in which lymphedema of the extremities results from blockage of their lymphatic vessels by small worms. The swelling of the legs or the scrotum in men may be so great that the victims become incapacitated.

LYMPHADENITIS Lymphadenitis refers to inflammation of the lymph nodes, which become enlarged and tender to the touch. For example, cervical lymphadenitis occurs frequently along with the common cold and is also seen with measles, scarlet fever, and diphtheria. Chronic lymphadenitis may be due to tuberculosis, since the lymph nodes are frequently involved in tubercular infections.

LYMPHOID TUMORS *Lymphosarcoma.* Lymphosarcoma is a malignant tumor of lymphoid tissue that is usually fatal in a short period of time. Since it has so many sites of involvement, intensive radiation therapy cannot be used (for high doses of radiation over so many areas would be dangerous and impractical). Low doses of radiation combined with chemotherapy are used in an attempt to prolong life.

Hodgkin's Disease.   Hodgkin's disease is a strange chronic disease considered to be a malignant tumor of reticular cells. It is characterized by anemia and by a generalized enlargement of the lymph nodes plus enlargement of the spleen. The cause of this disease is still unknown. It is ultimately fatal, though radiation and chemotherapy may prolong survival for many years.

Suggested Readings

* BACH, MARILYN L., and FRITZ H. BACH. "The genetics of histocompatibility," *Hosp. Pract.*, 5, no. 8 (August 1970), pp. 34–44.

BURNET, SIR MACFARLANE. "The mechanism of immunity," *Sci. Am.*, *204*, no. 1 (January 1961), pp. 58–67.

———. "The thymus gland," *Sci. Am.*, *207*, no. 5 (November 1962), pp. 50–57.

COOPER, MAX D., and ALEXANDER R. LAWTON, III. "The development of the immune system," *Sci. Am.*, *231*, no. 5 (November 1974), pp. 59–72.

GLASSER, RONALD J. *The Body Is the Hero.* New York: Random House, 1976.

GOOD, R. A., and D. W. FISHER, EDS. *Immunobiology.* Stamford, Conn.: Sinauer, 1971.

* GORDON, BENJAMIN LEE, II. *Essentials of Immunology.* 2nd ed. Philadelphia: F. A. Davis, 1974.

HUME, DAVID M. "Organ transplants and immunity," *Hosp. Pract.*, *3*, no. 5 (May 1968), pp. 27–35.

JERNE, NIELS KAJ. "The immune system," *Sci. Am.*, *229*, no. 1 (July 1973), pp. 52–60.

LERNER, RICHARD A., and FRANK J. DIXON. "The human lymphocyte as an experimental animal," *Sci. Am.*, *228*, no. 6 (June 1973), pp. 82–91.

MAYERSON, H. S. "The lymphatic system," *Sci. Am.*, *208*, no. 6 (June 1963), pp. 80–90.

MEDAWAR, PETER B. "Antilymphocyte serum: its properties and potential," *Hosp. Pract.*, *4*, no. 5 (May 1969), pp. 26–33.

———. "Skin transplants," *Sci. Am.*, *196*, no. 4 (April 1957), pp. 62–66.

MOORE, FRANCIS D. *Give and Take: The Development of Tissue Transplantation.* Philadelphia: Saunders, 1964.

NOSSAL, G. J. V. "How cells make antibodies," *Sci. Am.*, *211*, no. 6 (December 1964), pp. 106–115.

PORTER, R. R. "The structure of antibodies," *Sci. Am.*, *217*, no. 4 (October 1967), pp. 81–90.

RAFF, MARTIN C. "Cell surface immunology," *Sci. Am.*, *234*, no. 5 (May 1976), pp. 30–39.

REISFELD, RALPH A., and BARRY D. KAHAN. "Markers of biological individuality," *Sci. Am., 226,* no. 6 (June 1972), pp. 28–37.

SPEIRS, ROBERT S. "How cells attack antigens," *Sci. Am., 210,* no. 2 (February 1964), pp. 58–64.

* WEISS, LEON. *The Cells and Tissues of the Immune System.* Englewood Cliffs, N.J.: Prentice-Hall, 1972.

* Advanced level.

10

The Urinary System

The urinary system consists of organs (specifically the kidneys) that regulate the composition and volume of body fluids and excrete the waste products of cell metabolism in the form of urine. The organs of the urinary system (Fig. 10.1) are as follows:

1. Two kidneys that form urine.
2. Two ureters that conduct urine from the kidneys to the bladder.

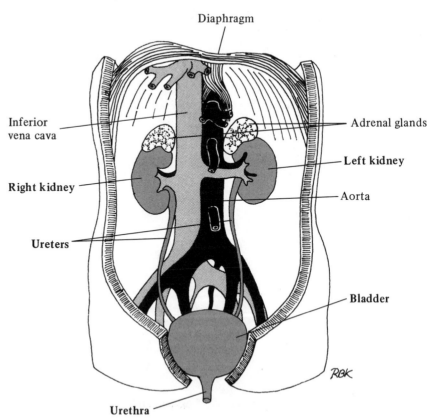

Figure 10.1 *Urinary System.* Abdominopelvic cavity shows organs of urinary system and major blood vessels.

3. The urinary bladder, which receives and stores urine.
4. The urethra, which conducts urine from the bladder to the exterior of the body.

The Kidneys

GROSS ANATOMY
Location

The kidneys are paired bean-shaped organs that are about 4 inches long, 2 inches wide, and 1 inch thick. They lie in the lumbar region of the abdomen against the muscles of the posterior abdominal wall. Since the kidneys lie behind the peritoneum (the serous membrane that lines the abdominal cavity), they are said to be *retroperitoneal*. Both kidneys are not at the same level; the right kidney is slightly lower than the left kidney due to the presence of the liver in the upper right quadrant of the abdomen.

Connective Tissue Coverings

The *perirenal fat*, a thick adipose connective tissue capsule, surrounds the true capsule of the kidney. It acts as a shock absorber to protect the kidney from jolting and jarring. A condensation of areolar connective tissue, called the *renal fascia*, surrounds both the kidney and adrenal gland, holding these organs in place.

External Structure

Each kidney (Fig. 10.2) has anterior and posterior surfaces, medial and lateral borders, and upper and lower poles. An adrenal gland sits on the upper pole of each kidney. The lateral border of the kidney is convex, while the medial border is concave. On the medial border of each kidney is a wedge-shaped depression called the *hilus*. The hilus is a door that leads into an internal chamber called the *renal sinus*. Blood vessels, nerves, and lymphatics that enter and leave the substance of the kidney pass through the hilus.

The surface of the adult human kidney is smooth. Adult human kidneys show no external evidence of lobation, whereas fetal human kidneys do. Some adult animal kidneys show external evidence of lobation and some do not. The next time you are in a supermarket, notice that beef kidneys have lobated surfaces, while lamb and pork kidneys are smooth.

Internal Structure

If a kidney is sliced longitudinally through its margins and one of the cut surfaces is examined (Fig. 10.3), three general regions can be seen: the *renal sinus, medulla*, and *cortex*.

Renal Sinus

The renal sinus is an internal chamber in the substance of the kidney that opens onto the medial surface through the hilus. *Medullary pyramids*, small cone-shaped masses of renal parenchyma, project into the renal sinus. The *renal pelvis* (the dilated upper end of the ureter) plus renal blood vessels, lymphatics, and nerves are also lo-

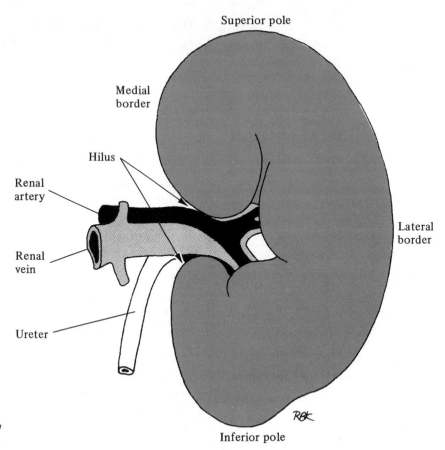

Superior pole

Medial
border

Hilus

Renal
artery

Lateral
border

Renal
vein

Ureter

Inferior pole

Figure 10.2 *Left Kidney: Gross Anatomy
(External).*

cated in the renal sinus. The renal pelvis gives rise to several tubular
extensions called *major calyces*. The major calyces in turn give rise
to 7 to 13 cuplike branches called *minor calyces*. Each minor calyx
is in contact with one or more *renal papillae* (the apices of medullary
pyramids) from which it collects urine. Adipose connective tissue fills
the spaces between structures in the renal sinus.

Medulla   The medulla is the dark brown inner portion of the renal parenchyma.
The medullary tissue is organized into 8 to 18 cone-shaped medullary
pyramids that correspond to the number of lobes found in the fetal
kidney. The bases of the medullary pyramids face outward, touching
the cortex, while their apices are free and project into the renal sinus.
The medullary pyramids have a striated appearance in cut section because they contain many longitudinally oriented ducts that open onto
the tips of the renal papillae. Each medullary pyramid and its overlying cortical tissue constitutes a *renal lobe*.

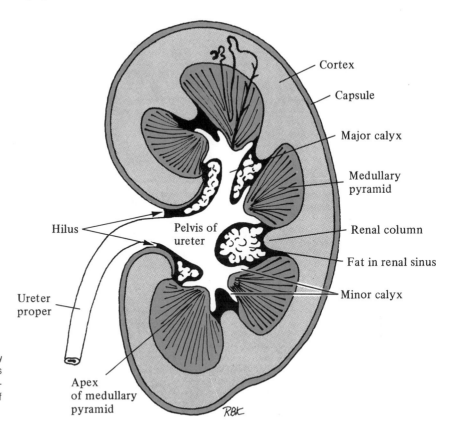

Figure 10.3 *Left Kidney: Gross Anatomy (Internal).* Longitudinal section shows capsule, cortex, medulla (medullary pyramids), renal sinus, and proximal part of ureteral system.

Cortex, Capsule, Major calyx, Medullary pyramid, Renal column, Fat in renal sinus, Minor calyx, Pelvis of ureter, Hilus, Ureter proper, Apex of medullary pyramid

Cortex   The cortex, the reddish brown outer portion of the renal parenchyma, is located between the medulla and the capsule. It arches over the bases of the medullary pyramids and extends down between them as the *renal columns.* The cut surface of the cortex is paler than that of the medulla and has a fine granular appearance due to the presence of many small round bodies called renal corpuscles.

MICROSCOPIC ANATOMY   The kidney is a solid glandular organ. One can think of the kidney as a compound tubular gland whose parenchyma consists almost entirely of *uriniferous tubules* and blood vessels supported by a scanty connective tissue stroma. A thin but tough fibrous connective tissue capsule encloses the renal parenchyma.

Uriniferous Tubules   Each uriniferous tubule (Fig. 10.4) consists of two portions of different embryological origin: the *nephron* and *collecting tubule.* The nephron is roughly comparable to the secretory portion of a gland, while the collecting tubule is roughly comparable to its excretory duct.

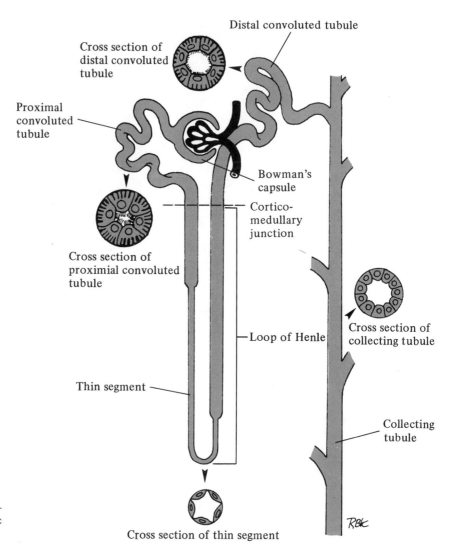

Distal convoluted tubule

Cross section of distal convoluted tubule

Proximal convoluted tubule

Bowman's capsule

Cortico-medullary junction

Cross section of proximial convoluted tubule

Loop of Henle

Cross section of collecting tubule

Thin segment

Collecting tubule

Cross section of thin segment

Figure 10.4 *Uriniferous Tubule.* Cross-sectional insets indicate microscopic structure of various segments of tubule.

The Nephron

The nephron, which actively participates in the formation of urine, is the functional unit of the uriniferous tubule. It has been estimated that there are one to three million nephrons per kidney. There are two types of nephrons, which are named according to their location in the renal lobe (Fig. 10.5). Nephrons that lie almost entirely in the cortex are called *cortical nephrons.* Other nephrons, called *juxtamedullary nephrons,* begin near the corticomedullary junction and extend deep into the medulla.

Structurally, the nephron is a long, tortuous, epithelial tubule consisting of four successive segments: *Bowman's capsule,* the *prox-*

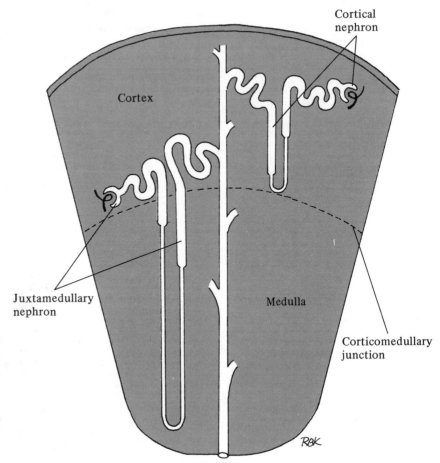

Cortical nephron

Cortex

Juxtamedullary nephron

Medulla

Corticomedullary junction

Figure 10.5 Nephron Types. Schematic diagram of renal lobe shows appearance and relative positions of cortical and juxtamedullary nephrons.

imal convoluted tubule, the *loop of Henle,* and the *distal convoluted tubule.* These segments are described next.

Renal Corpuscle.   The renal corpuscle is a spherical structure consisting of *Bowman's capsule* and a *glomerulus* (Fig. 10.6). Bowman's capsule is the dilated proximal end of the nephron. This thin-walled structure is invaginated into the shape of a cup by the glomerulus, a globular tuft of capillaries that arises from an afferent arteriole and empties into an efferent arteriole. Each renal corpuscle has two poles: a vascular pole and a urinary pole. The vascular pole is the region where the afferent and efferent arterioles enter and leave the renal corpuscle. The urinary pole is opposite the vascular pole and is the region where Bowman's capsule becomes continuous with the proximal convoluted tubule.

Bowman's capsule consists of two layers of simple squamous epithelium separated by a slitlike cavity called the *urinary space.*

The outer or *parietal layer* is continuous with the epithelium of the proximal convoluted tubule. The inner or *visceral layer* is closely applied to the endothelium of the glomerulus. The highly specialized cells of the visceral layer are called *podocytes* (foot cells). Podocytes have long radiating processes that terminate in end feet on the basement membrane of the glomerulus. The spaces between adjacent end feet from the same or neighboring podocytes are called *slit pores;* they permit small molecules and ions to pass into the urinary space.

The glomerular endothelial cells, the basement membrane of the glomerulus, and the podocyte end feet constitute a *filtration* or *blood-urine barrier,* which is shown at the electron microscopic level in Fig.

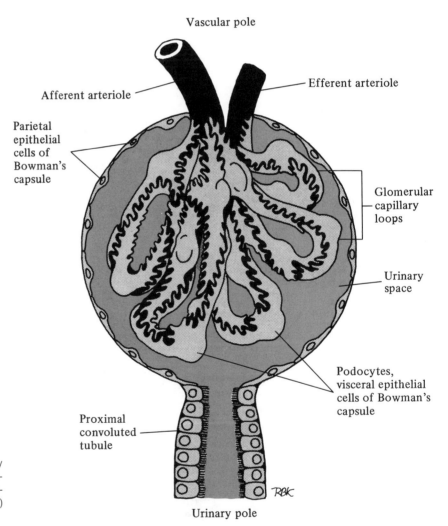

Vascular pole

Efferent arteriole

Afferent arteriole

Parietal epithelial cells of Bowman's capsule

Glomerular capillary loops

Urinary space

Podocytes, visceral epithelial cells of Bowman's capsule

Proximal convoluted tubule

Urinary pole

Figure 10.6 Renal Corpuscle. Highly enlarged and sectioned to show its microscopic structure. Note peculiar appearance of visceral epithelial cells (podocytes) that cover glomerular capillary loops.

10.7. The filtration barrier permits an ultrafiltrate of blood plasma, which is free of red blood cells and large plasma proteins such as albumin, to pass from the glomerular lumen into the urinary space. When the filtration barrier is damaged, red blood cells and albumin appear in the urine.

Proximal Convoluted Tubule. The proximal convoluted tubule is continuous with Bowman's capsule at the urinary pole of the renal corpuscle. It forms a series of tortuous loops that lie adjacent to the renal corpuscle from which it originates then straightens out and dips down toward the medulla.

The proximal convoluted tubule is lined with a single layer of large pyramidal cells (Fig. 10.8). The free surfaces of these cells bear long, closely packed microvilli that are visible under the light microscope as a *brush border*. The basal surfaces of these cells are deeply indented by cytoplasmic membrane infoldings. Both the brush border and the basal infoldings greatly increase the surface area available for diffusion and active transport.

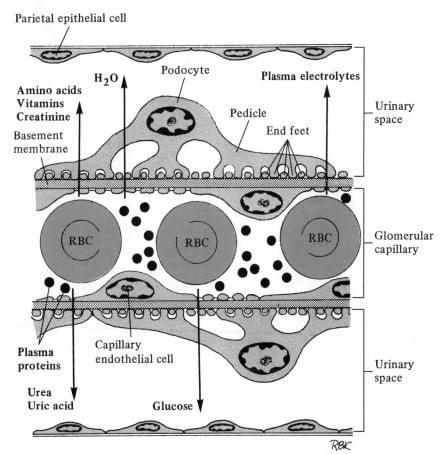

Figure 10.7 Filtration Barrier. Schematic representation of glomerulus in Bowman's capsule. Filtration barrier consists of glomerular capillary endothelium, basement membrane, and visceral epithelium (podocyte end feet) of Bowman's capsule. Note that some substances are able to pass through filtration barrier while others are not. Drawings based on electron microscopic view.

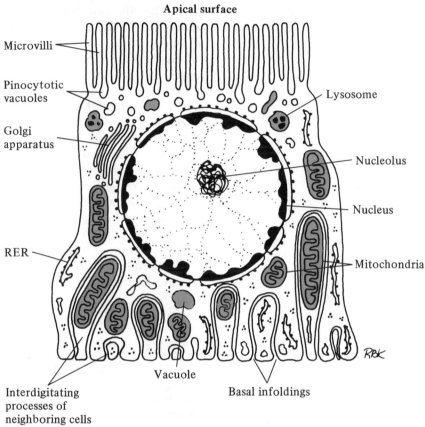

Apical surface

Microvilli

Pinocytotic vacuoles

Golgi apparatus

RER

Lysosome

Nucleolus

Nucleus

Mitochondria

Vacuole

Interdigitating processes of neighboring cells

Basal infoldings

Basal surface

Figure 10.8 Proximal Tubule Absorptive Cell. Note abundant microvilli at apical end of cell and basal infolding. Both these structures increase cell's surface area for transepithelial transport. Drawing based on electron microscopic view.

Loop of Henle.   The loop of Henle is the U-shaped part of the nephron that extends from the straight portion of the proximal tubule through the straight portion of the distal tubule. This structure descends into the medulla, makes a hairpin bend, and reascends into the cortex. The loop of Henle can be subdivided into three segments: *descending limb, thin segment,* and *ascending limb.* The descending limb consists of the descending straight portion of the proximal tubule. It is lined with a simple cuboidal epithelium whose cells are freely permeable to water. The thin segment in juxtamedullary nephrons is shaped like a hairpin and is located in the medulla. It is lined with a thin simple squamous epithelium whose cells are also freely permeable to water. The ascending limb consists of the ascending straight portion of the distal tubule. It is lined with a simple cuboidal epithelium whose cells are impermeable to water.

Distal Convoluted Tubule.   The last segment of the nephron is known as the distal convoluted tubule. The straight distal tubule ascends into the cortex and approaches its own renal corpuscle, where it breaks into tortuous loops, then becomes continuous with the arched collecting tubule.

The distal convoluted tubule is lined with a single layer of cuboidal cells. The free surfaces of these cells bear short scattered microvilli that do not form a brush border. The basal surfaces, however, do have many infoldings.

Juxtaglomerular Apparatus. As the distal convoluted tubule returns to the vicinity of its own renal corpuscle, it pushes in between the afferent and efferent arterioles at the vascular pole. Where the distal convoluted tubule comes in contact with the afferent arteriole, a complex structure called the *juxtaglomerular apparatus* is formed (Fig. 10.9). At the point of contact, the smooth muscle fibers in the

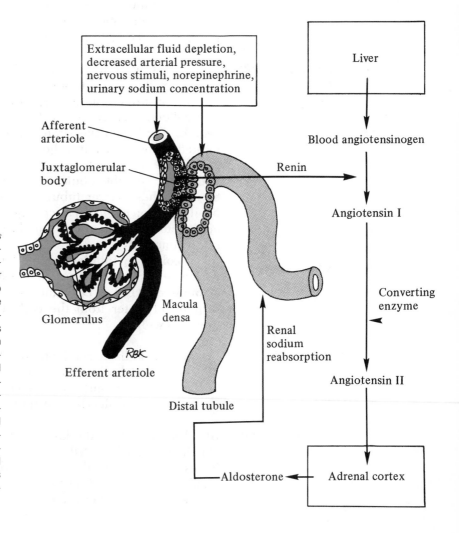

Figure 10.9 Juxtaglomerular Apparatus and Renin-Angiotensin System. Schematic diagram shows structures of juxtaglomerular apparatus (juxtaglomerular body and macula densa) and relationship to renin-angiotensin system. When the renal blood flow or urinary sodium concentration is low, the juxtaglomerular cells secrete renin. Renin is enzyme that acts on plasma protein, angiotensinogen, to produce decapeptide (ten amino acids) called angiotensin I. Plasma enzyme subsequently converts angiotensin I into octapeptide (eight amino acids) called angiotensin II. Angiotensin II stimulates adrenal cortex to secrete hormone called aldosterone. Aldosterone in turn promotes increased reabsorption of sodium by distal tubule. Resulting sodium retention causes elevation of blood pressure, which increases renal blood flow.

tunica media of the afferent arteriole resemble cuboidal epithelial cells. These modified cells, called *juxtaglomerular cells,* produce secretory granules that contain the enzyme *renin.* Renin is released into the bloodstream whenever blood flow through the kidney is reduced or sodium reabsorption from the nephron is low. The effect of increased circulating renin, acting through the renin-angiotensin system, is to stimulate the production of aldosterone. Aldosterone is an adrenocortical hormone that promotes increased reabsorption of sodium from the distal tubule. The resulting sodium retention causes elevation of the blood pressure and increases the rate of renal blood flow. The cells of the distal convoluted tubule are also modified at the point of contact with the afferent arteriole. These enlarged cells form a thickening on the wall of the tubule called the *macula densa.* The macula densa acts as a sodium sensor that detects changes in the sodium concentration of the tubular fluid. Whenever the sodium concentration is low, the cells of the macula densa signal the juxtaglomerular cells in the afferent arteriole to release renin.

The Collecting Tubule

The collecting tubule is the portion of the uriniferous tubule that is the site of the final dilution or concentration of urine. It also serves to transport urine from the nephron to a minor calyx in the renal sinus. The collecting tubule consists of two segments: the *arched* or *connecting tubule* and the *straight tubule.* The arched tubule, which is located in the cortex, is a short curved tubule that connects the distal part of the nephron to a straight tubule. The straight tubule unites with other straight tubules in the medulla to form larger collecting tubules. These larger tubules unite with one another in the renal papilla to form *papillary ducts.* Approximately 10 to 25 papillary ducts open onto the apex of each medullary pyramid and discharge urine into a minor calyx.

The collecting tubules are lined with a simple cuboidal epithelium whose cells are freely permeable to water in the presence of large amounts of ADH. ADH (antidiuretic hormone) promotes the reabsorption of water from the collecting tubule (and from the distal tubule as well).

BLOOD SUPPLY

Because the kidneys rid the blood of the accumulated waste products of cellular metabolism, they have a very rich blood supply. In order for you to understand how the kidney functions, it is important for you to know how blood circulates through this vital organ.

Arterial Supply

Blood is brought to the kidneys by the right and left *renal arteries,* which are branches of the abdominal aorta. Each renal artery enters its respective kidney through the hilus. After entering the kidney, the renal artery divides into two or three branches that course through the renal sinus. These vessels in turn give rise to the *interlobar*

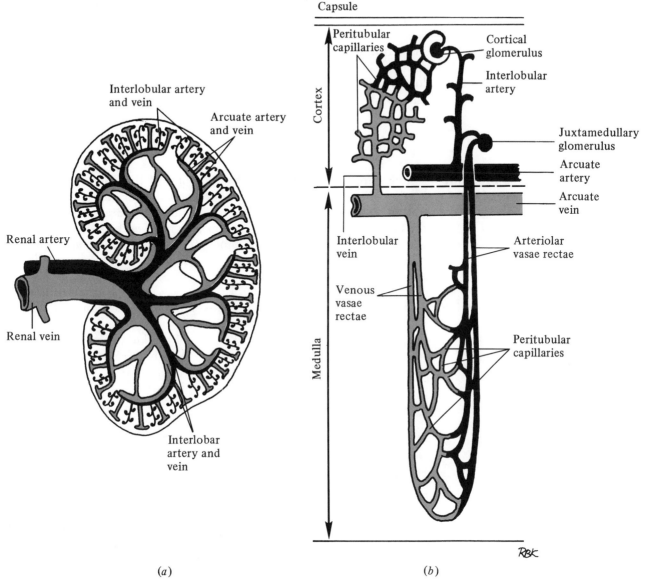

(a)   (b)

Figure 10.10 *Blood Supply of Kidney.*
(a) Macro supply. Longitudinal section through left kidney shows major renal blood vessels. (b) Micro supply. Schematic representation of minor renal blood vessels such as cortical and medullary peritubular networks and medullary vasae rectae.

arteries, which enter the substance of the kidney and ascend the renal columns. At the corticomedullary junction, the interlobar arteries give rise to the *arcuate arteries*, which arch over the bases of the medullary pyramids. The arcuate arteries give off the *interlobular arteries*, vertical branches that course upwards and terminate in a capillary network just beneath the capsule [Fig. 10.10(a)].

Each interlobular artery gives off several lateral branches that become *afferent arterioles*. Each afferent arteriole gives rise to a tuft of capillaries or glomerulus. The glomerular capillaries reunite to form *efferent arterioles*, which carry filtered blood away from the renal corpuscles. Efferent arterioles from cortical glomeruli (glomeruli located high in the cortex) give rise to capillaries that supply cortical tubules. Efferent arterioles from juxtamedullary glomeruli (glomeruli located at the corticomedullary junction) give rise to capillaries that supply medullary tubules and to *vasa recta*, capillarylike vessels that form hairpin loops at various levels in the medulla. The arterial limbs of the vasa recta also contribute to the medullary network [Fig. 10.10(*b*)].

Venous Drainage Venules from the capsule and upper part of the cortex empty into small radiating veins called *stellate veins*. The stellate veins are tributaries of the *interlobular veins*, which in turn are tributaries of the *arcuate veins*. The arcuate veins also receive the venous limbs of the vasa recta, which return blood from the medulla and cortico-medullary junction. The arcuate veins join the *interlobar veins*, which converge to form the *renal vein.*

INNERVATION The kidneys are innervated by the sympathetic division of the autonomic nervous system. Nerve fibers from the celiac plexus follow the renal blood vessels into the substance of the kidney, where they innervate the walls of the blood vessels. Whether or not they innervate the uriniferous tubules is still a matter of controversy.

Figure 10.11 Urine Formation. Schematic diagram of juxtamedullary nephron and related structures. Their role in urine formation is as follows: (1) Blood reaches glomerulus by way of afferent arteriole and leaves by way of efferent arteriole. (2) Ultrafiltrate of blood plasma is formed and passes into Bowman's capsule. (3) Filtrate flows from Bowman's capsule into proximal tubule, where reabsorption of water and solutes occurs. (4) Water and urea are reabsorbed by passive means while sodium, glucose, amino acids, uric acid, and others are reabsorbed by active means. These substances are returned to general circulation by blood flowing through cortex. Filtrate next flows through descending and ascending limbs of Henle's loop. Descending limb is permeable to water and sodium from medullary interstitium (interstitial tissue), while ascending limb is impermeable to water and actively transports sodium into interstitium. Net result is that filtrate leaving ascending limb contains less sodium and more water than that entering descending limb. (5) Continuous recycling of sodium between filtrate in Henle's loop, medullary interstitium, and vasa recta creates osmotic gradient in medullary interstitium that is responsible for final concentration of urine. (6) As filtrate flows through distal tubule, it loses sodium (by active transport) and water (by diffusion). It also gains potassium, hydrogen, and ammonium ions (by tubular secretion). (7) Filtrate that enters collecting tubule is subject to forces created by osmotic gradient in medullary interstitium. When sufficient ADH (antidiuretic hormone) is present, filtrate in collecting tubule loses water (by osmosis) into interstitium, and resulting fluid (urine) is highly concentrated.

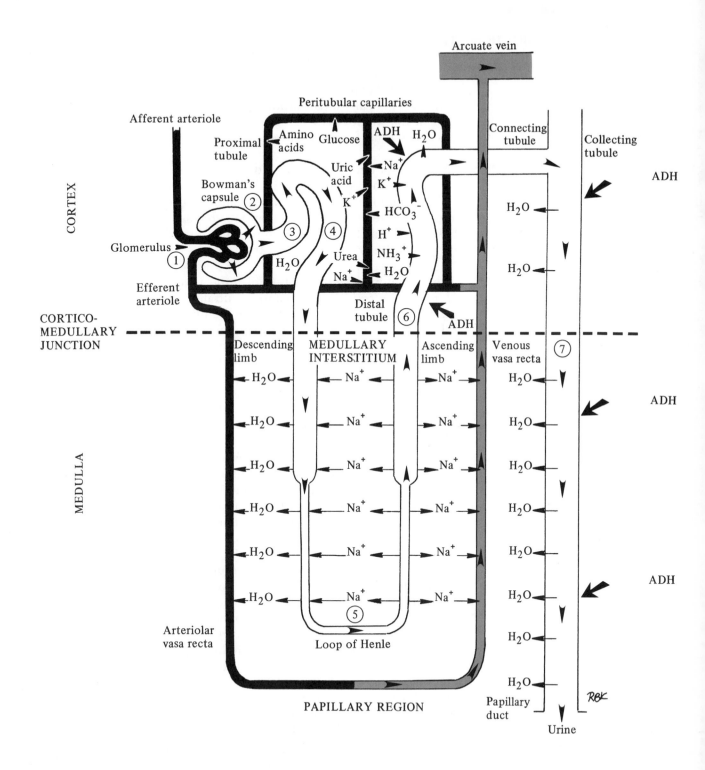

FUNCTION The major functions of the kidneys include formation of urine, the secretion of renin (an enzyme that influences blood pressure) and erythropoietin (a hormone that stimulates red blood cell production), and the biochemical alteration of vitamin D to a state in which it can stimulate gut calcium absorption. A brief description of urine formation in the kidney will be given here. A more comprehensive discussion of renal function can be found in a human physiology textbook.

Urine Formation In the formation of urine, the kidneys retain sufficient water, electrolytes, and other substances needed by the body to maintain homeostasis (a stable internal environment) and dispose of foreign substances and the waste products of cellular metabolism. The kidney performs these functions through a combination of filtration, tubular reabsorption, and tubular secretion (Fig. 10.11).

First Step The first step in the formation of urine involves the production of *glomerular ultrafiltrate*. Glomerular ultrafiltrate is formed from blood plasma that has been forced through the filtration barrier by pressure differences between fluid in the glomerular capillaries (capillary pressure minus colloid osmotic pressure) and fluid in Bowman's capsule. The composition of glomerular ultrafiltrate is similar to blood plasma minus albumin and other large plasma proteins. It contains small molecules such as the smaller plasma proteins, amino acids, glucose, vitamin C, creatinine, uric acid, urea, and sodium, potassium, and chloride ions.

Second Step The second step in the formation of urine involves the conversion of the dilute glomerular ultrafiltrate described above into the concentrated, highly modified fluid known as urine as the ultrafiltrate passes through various portions of the uriniferous tubule. Normally, of the 125 milliliters of fluid formed per minute by the glomeruli, all but 1 milliliter is reabsorbed by the time the final product leaves the uriniferous tubule. Approximately 85 percent of the water and sodium chloride in the proximal convoluted tubule is reabsorbed into the peritubular tissue fluid. Under normal conditions all or almost all of the glucose, vitamin C, amino acids, uric acid, and some urea are reabsorbed from the proximal convoluted tubule and returned to the blood.

As the tubular fluid passes through the loop of Henle in the medullary pyramid, functional differences between the descending and ascending limbs of the loop of Henle lead to an exchange of sodium chloride and water between the tubular fluid, the peritubular tissue fluid, and the blood flowing through the vasa recta (Fig. 10.12). The net result is that the fluid leaving the loop of Henle contains less sodium and more water than the fluid that entered. In addition, an osmotic gradient created by variations in sodium chloride concentration is established in the medullary interstitium; this gradient is directly responsible for the final concentration of tubular fluid.

About 14 percent more water is reabsorbed from the tubular fluid as it passes through the distal convoluted and collecting tubules. Here water reabsorption occurs under the influence of ADH, the antidiuretic hormone produced by the hypothalamus (a part of the brain). ADH makes these portions of the uriniferous tubule more permeable to water, which is drawn into the medullary tissue fluid by the osmotic gradient mentioned above. In addition, a number of substances such as hydrogen, ammonium, and potassium ions are actively secreted into the fluid in the distal convoluted tubule.

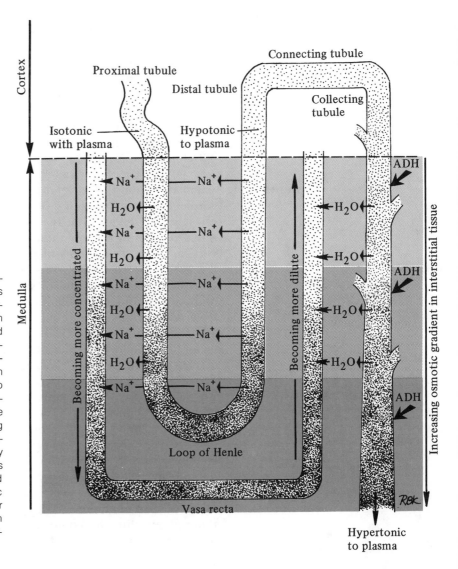

Figure 10.12 Countercurrent Mechanism. The uriniferous tubule produces concentrated urine by complex countercurrent mechanism that resides in both loop of Henle and vasa recta. It is called countercurrent because fluid flows in opposite directions in two limbs of each U-shaped structure. Due to differences in permeability of two limbs of Henle's loop to sodium and water and to active transport of sodium from ascending limb, there is constant flow of sodium from ascending limb to descending limb. This creates osmotic gradient in interstitium (indicated by increasing *degrees of grayness*), which is maintained by countercurrent flow of blood through two limbs of vasa recta. Osmotic gradient is responsible for flow of water from collecting tubule into interstitium under stimulus of ADH (antidiuretic hormone).

After passing through the collecting tubules, the remaining 1 percent of the tubular fluid (now called urine) is delivered to the ureter by the papillary ducts.

The Ureters

GROSS ANATOMY The ureters are paired muscular tubes, about 10 to 12 inches long, that conduct urine from the kidneys to the bladder. Like the kidneys, the ureters are retroperitoneal in location. For descriptive purposes the ureter is subdivided into two regions: the *pelvis* and the *ureter proper* [Fig. 10.13(*a*)].

Pelvis The pelvis is the funnel-shaped portion of the ureter that lies partially within the renal sinus. As previously mentioned, the intrarenal portion of the pelvis is subdivided into *major* and *minor calyces*. There are approximately 7 to 13 minor calyces per kidney. Each minor calyx is a cuplike structure that encircles one or more renal papillae. The minor calyces unite to form several short tubes, called major calyces, that join to form the pelvis. The pelvis emerges from the hilum of the kidney, decreases in diameter, and becomes continuous with the ureter proper.

Ureter Proper Situated behind the parietal peritoneum, the ureter proper descends vertically on the muscles of the posterior abdominal wall until it reaches the pelvic brim. It passes over the pelvic brim and the terminal portion of the common iliac artery, then continues to descend vertically on the muscles of the lateral pelvic wall. At the level of the greater sciatic notch, the ureter swings medially and anteriorly until it reaches the posterior surface of the bladder and then passes obliquely through the bladder's posterolateral wall. In the female the ureter lies in close proximity to the reproductive organs, especially the uterine cervix. Therefore, it is occasionally injured or mistaken for a blood vessel and tied off during surgical procedures such as a hysterectomy (removal of the uterus).

MICROSCOPIC ANATOMY The ureter is a slender muscular tube with a relatively thick wall and a small lumen [Fig. 10.13(*b*)]. Its wall is similar in construction to that of a blood vessel. From inside out, the ureteral wall consists of three coats: a *mucosa*, a *muscularis*, and an *adventitia*. The mucosa contains a layer of transitional epithelium, about four to five cells deep when the ureter is not distended, which is supported by a layer of fibrous connective tissue. The muscularis consists of a coat of smooth muscle fibers. In the pelvis the muscularis is organized into inner longitudinal and outer circular layers; in the ureter proper it is organized into inner longitudinal, middle circular, and outer longi-

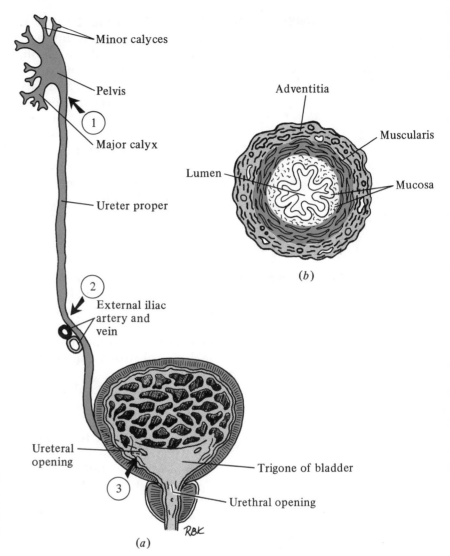

Figure 10.13 Right Ureter. (a) Gross anatomy. Note that ureter is constricted at three places: (1) at ureteropelvic junction, (2) at crossing of iliac vessels, and (3) at junction with bladder. (b) Microscopic anatomy. Cross section through ureter showing three coats of ureteral wall.

tudinal layers. The adventitia consists of a coat of fibrous connective tissue continous with the renal fascia at its proximal end and with the adventitia of the bladder at its distal end.

BLOOD SUPPLY   The ureter receives blood from branches of arteries that lie along its course; these vessels include the renal, gonadal, internal iliac, and inferior vesical arteries. The ureter is drained by tributaries of veins having the same names as the arteries listed above.

INNERVATION   The ureters are innervated by autonomic nerve fibers from the inferior mesenteric, testicular, and pelvic plexuses. Nerve impulses from sympathetic fibers inhibit muscular contractions of the ureters, whereas impulses from parasympathetic fibers stimulate muscular contractions. The ureters are also richly supplied with pain receptors, as anyone who has experienced the passage of a kidney stone will testify. The pain caused by passage of a kidney stone can be excruciating.

FUNCTION   The ureters conduct urine formed by the kidneys to the bladder for temporary storage. Contractions of ureteral musculature produce peristaltic (wormlike) movements that propel urine down the ureter and into the bladder.

The Bladder

GROSS ANATOMY
Location   The bladder is a hollow muscular organ that serves as a reservoir for urine. It lies in the pelvic cavity, between the pubic symphysis and rectum, and is supported by the pelvic diaphragm. The bladder is separated from the rectum by the seminal vesicles and terminal portions of the deferent ducts in the male and by the uterus and vagina in the female (see Chapter 14).

External Structure   The bladder is divided into three regions: a *neck* (constricted region), *body* (main portion), and *fundus* (base). The shape and position of the bladder depend on whether it is empty or distended with urine. When empty, the bladder is shaped like a tetrahendron (a geometrical figure with four triangular sides) and extends up to the level of the pubic symphysis; when full, the bladder is shaped like a sphere and extends up into the abdominal cavity.

Four Surfaces   The empty bladder has four triangular surfaces [Fig. 10.14(*a*)]: one *superior,* two *inferolateral,* and one *posterior.* The superior surface faces the abdominal cavity. The two inferolateral surfaces face the lateral pelvic walls and meet in the midline to form the anterior border of the bladder. The posterior surface faces the rectum and forms the fundus of the bladder; the superior surface and about $\frac{1}{2}$ inch of the fundus are covered by peritoneum.

Four Angles   The empty bladder also has four angles: an *anterior angle* or *apex,* two *posterolateral angles,* and an *inferior angle.* The ureters enter the bladder at its posterolateral angles, while the urethra exits from the bladder at its inferior angle.

Internal Structure   If you examine the interior of an empty bladder, you will notice a smooth area shaped like an upside-down triangle on its posterior wall

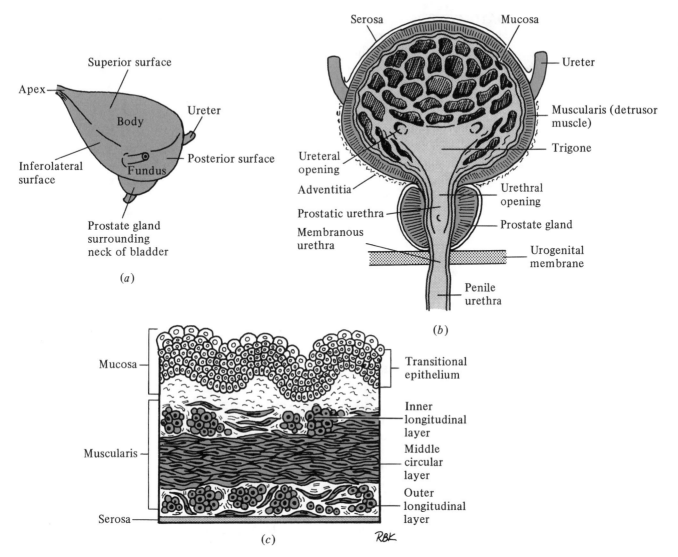

Figure 10.14 Bladder. (a) Gross anatomy, external. (b) Gross anatomy, internal. Male bladder sectioned in frontal plane to expose trigone and its openings plus proximal portion of urethra. Note that mucosa of trigone is smooth, while that of rest of bladder is thrown into rugae (folds). (c) Microscopic anatomy. Segment of cross section of bladder wall with its three coats indicated.

[Fig. 10.14(b)]. This triangular area is known as the *trigone* of the bladder. The ureteral openings mark the posterolateral angles of the trigone, while the urethral opening marks the inferior angle. Except in the area of the trigone, the lining of the empty bladder has a wrinkled appearance. These wrinkles, or *rugae*, are caused by the loose attachment of the lining to the underlying muscular coat; they appear when the bladder wall is contracted and disappear when the bladder wall is stretched. The lining of the trigone is always smooth since it is firmly bound to the underlying muscle.

MICROSCOPIC ANATOMY The wall of the bladder, like the ureter, is composed of a *mucosa, muscularis,* and *adventitia* [Fig. 10.14(*c*)], part of which is replaced by a *serosa* in the bladder wall. The mucous membrane lining the bladder contains a layer of transitional epithelium supported by a relatively thick layer of loose connective tissue. The number of rows of cells and the shapes of cells in the epithelial layer depend on whether the bladder is distended or contracted. The looseness of the underlying connective tissue layer allows rugae to form in all regions of the bladder, except the trigone, whenever the bladder is contracted. The muscularis consists of a thick coat of smooth muscle fibers known as the *detrussor muscle.* These muscle fibers are organized into indistinctly separated inner (longitudinal) middle (circular) and outer (longitudinal) layers. Distinct from the detrussor muscle is the *trigonal muscle,* a coat of smooth muscle fibers that encircles the openings of the ureters and urethra. The trigonal muscle also gives rise to the internal sphincter of the urethra. The bulk of the bladder, which is covered with fibrous connective tissue, has an adventitia. However, the superior surface and a small portion of the posterior surface, which are covered with peritoneum, have a serosa.

BLOOD SUPPLY Blood is supplied to the bladder by the anterior, middle, and inferior vesical (bladder) arteries, which arise from branches of the internal iliac arteries. In the female the bladder also receives arterial blood from small branches of the uterine and vaginal arteries. Venous drainage is provided by the vesical veins, which form an extensive plexus along the inferior border and fundus of the bladder. This plexus in turn drains into the internal iliac veins.

INNERVATION The bladder is innervated by the autonomic nervous system. Sympathetic nerve fibers arise from spinal cord segments *L*5 to *S*2. Their impulses cause relaxation of the detrussor muscle and contraction of the internal sphincter, which leads to retention of urine in the bladder. Parasympathetic nerve fibers innervating the bladder arise from spinal cord segments *S*2 to *S*3. Impulses from these fibers cause contraction of the detrussor muscle plus relaxation of the internal sphincter, which leads to expulsion of urine from the bladder. Stretch receptors in the bladder wall form the afferent (sensory) limb of the micturition reflex. This reflex is adjusted by excitatory and inhibitory centers in the brain.

FUNCTION The bladder acts as a reservoir for urine. Under normal circumstances the bladder can hold from 300 to 350 milliliters of urine without great discomfort. Emptying of the bladder, or *micturition,* is an autonomic reflex that can be voluntarily controlled by higher centers in the brain. As the bladder fills with urine, stretch receptors in the bladder wall send messages to the spinal cord to induce the urge to

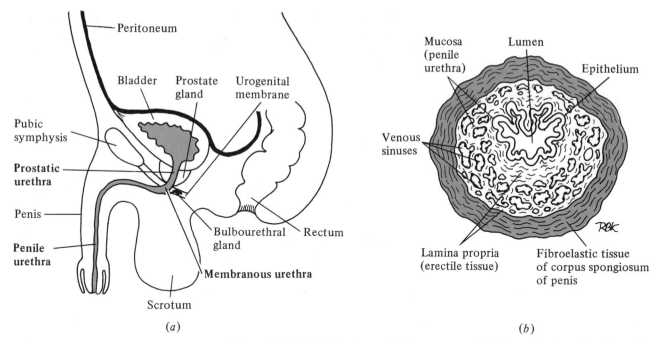

Figure 10.15 *Male Urethra.* (a) Gross anatomy. Sagittal section through male pelvis and external genitalia shows location of bladder and urethra. (b) Microscopic anatomy. Cross section through corpus spongiosum of penis shows location of penile urethra and surrounding tissues.

empty the bladder. Efferent fibers coming from the spinal cord stimulate the detrussor muscle to contract and the internal sphincter to relax. Normally, the external sphincter of the urethra, which is under voluntary control, must receive impulses from the brain to relax before micturition can occur.

The Urethra

The urethra is a flexible tube that extends from the trigone of the bladder to the exterior of the body. Since its structure and function are so different in the two sexes, it will be described separately for each sex.

THE MALE URETHRA
Gross Anatomy

The male urethra is approximately 8 inches long and is divided into *prostatic*, *membranous*, and *cavernous (penile)* portions (Fig. 10.15).

Prostatic Portion

The prostatic portion is approximately 1 inch long and extends from the trigone of the bladder to the upper border of the urogenital diaphragm. For most of its length, it is surrounded by the *prostate gland*, an accessory sex gland that empties its secretions into this portion of the urethra through many small ducts. Sperm enter the prostatic urethra through the paired *ejaculatory ducts*, the terminal portions of the deferent ducts.

Membranous Portion

The membranous portion is approximately $\frac{1}{2}$ inch long and passes through the urogenital diaphragm. Between the two membranes that form the urogenital diaphragm the urethra is encircled by a voluntary muscle that constitutes the external sphincter of the urethra.

Penile Portion

The penile portion is approximately 6 inches long and extends from the lower border of the urogenital diaphragm to the tip of the penis. Throughout its entire length, this portion of the urethra is enclosed in a cavernous body called the *corpus spongiosum* of the penis. The *bulbourethral glands,* a pair of accessory sex glands located in the urogenital diaphragm, empty their secretions into the penile urethra.

Microscopic Anatomy

The male urethra is a membranous tube consisting of a mucous membrane supported primarily by external structures. The mucosa contains an epithelium that varies from transitional near the bladder to stratified columnar to stratified squamous near the external urethral opening. The prostatic portion of the urethra is supported by the fibromuscular tissue of the prostate gland, the membranous portion is supported by the striated muscle fibers of the external urethral sphincter, and the penile portion is supported by the fibroelastic tissue of the corpus spongiosum.

Blood Supply

The male urethra receives blood from branches of the prostatic vessels (prostatic portion) and from the urethral artery (membranous and cavernous portions). Blood from the urethra drains into the pudendal plexus and the deep vein of the penis.

Innervation

The male urethra receives both voluntary and autonomic innervation by way of branches of the pudendal nerves.

Function

The male urethra has a dual function. It serves as a duct for conducting urine from the bladder to the exterior of the body and for transferring semen to the female reproductive tract.

THE FEMALE URETHRA
Gross Anatomy

The female urethra is a relatively straight tube, approximately $1\frac{1}{2}$ inches long, that corresponds to the prostatic and membranous portions of the male urethra (Fig. 10.16). It extends from the trigone of the bladder to the *perineum* (the diamond-shaped region between the thighs). On route to the perineum, the female urethra perforates the urogenital diaphragm, where it is encircled by the external urethral sphincter. The external urethral opening is located in the

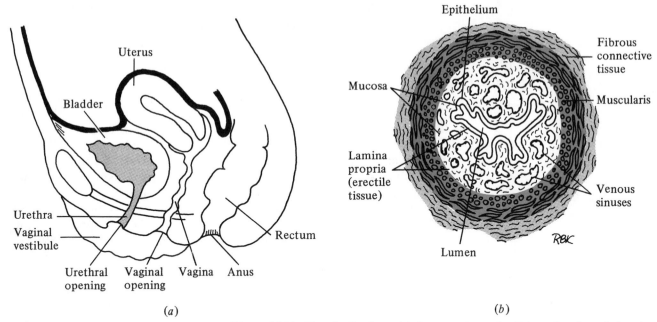

Figure 10.16 *Female Urethra.* (a) Gross anatomy. Sagittal section through female pelvis and external genitalia shows location of bladder and urethra. (b) Microscopic anatomy. Cross section through wall of female urethra shows its three coats.

vaginal vestibule (a female perineal structure) directly in front of the external vaginal opening and about an inch behind the tip of the clitoris (the female counterpart of the penis). This relationship is important to remember if you have to insert a urinary catheter in a female patient.

Microscopic Anatomy
The female urethra is a muscular tube consisting of a *mucosa* and a *muscularis*. The mucosa contains transitional epithelium near the bladder and stratified squamous epithelium the remainder of its length. This epithelium rests on a thick layer of loose connective tissue that resembles erectile tissue in the cavernous bodies of the penis. The muscularis consists of a relatively thick coat of smooth muscle fibers organized into inner (longitudinal) and outer (circular) layers. The female urethra has no distinct adventitia since the fibrous connective tissue that surrounds it is fused with that of the vagina.

Blood Supply
The female urethra receives blood from branches of the inferior vesical and internal pudendal arteries. Blood from the urethra drains into the vesical and vaginal veins.

Innervation
The female urethra receives both voluntary and autonomic innervation by way of branches of the pudendal nerves.

Function   The female urethra has a single function. It serves as a duct to conduct urine from the bladder to the exterior of the body.

The Artificial Kidney

The artificial kidney is a machine used to treat persons with acute or chronic kidney failure. Though several types of artificial kidneys are available, all employ the principle of exposing blood to a dialyzing fluid across a semipermeable membrane in order to remove waste products or toxic substances. The basic components of the artificial kidney include a coil of cellophane tubing, the patient's blood, and a dialyzing solution. Blood is usually removed from the patient through the radial artery, circulated through the cellophane tubing (which is immersed in a dialyzing solution), and returned to the patient through a radial vein. The constituents of the dialyzing solution are carefully chosen so that those substances in excess in the blood can pass rapidly into the dialyzing solution while those substances in normal concentration can remain essentially the same. Dialysis is continued for up to 12 hours until equilibrium is achieved between the patient's blood and the dialyzing solution. Through two or three sessions per week on the artificial kidney, doctors are now able to keep patients without kidneys alive for about five years. The artificial kidney is also used to maintain patients that are waiting to receive kidney transplants or to assist patients with temporary kidney failure.

Kidney Transplantation

In the last 20 years a growing number of patients have received kidney transplants from live or cadaver donors. The donor kidney is placed in the recipient's iliac fossa; its renal vessels are then anastomosed to the recipient's iliac vessels, while its ureter is implanted into the recipient's bladder. Long-term survival of the transplanted kidney is usually quite good following the use of immunosuppressive drugs. Chances of survival are also improved when the recipient and donor share most of the same HL–A antigens (see Chapter 9). For this reason kidney transplants between close relatives, such as brother-sister or parent-child, are more successful than those between unrelated persons. However, since cadaver kidneys are more readily available than live kidneys, most transplant recipients receive cadaver kidneys from unrelated donors. If a transplanted kidney is rejected, it is removed and the patient is maintained on an artificial kidney until another donor kidney is available. Because the quality of life is so much better with a kidney transplant, a number of kidney patients have willingly undergone three or four transplants before receiving a kidney that was not rejected.

Diseases and Disorders of the Urinary System

CONGENITAL MALFORMATIONS

Kidney Agenesis. Sometimes one or both of the kidneys fail to develop (agenesis). When both kidneys fail to develop, the affected individual dies shortly after birth. However, a person with only one kidney can lead a normal life. The single kidney becomes much larger and heavier than a normal kidney as the result of compensatory hypertrophy.

Horseshoe Kidney. The horseshoe kidney is a congenital malformation caused by fusion of the lower poles of the developing kidneys. The fused kidneys form a horseshoe-shaped mass surrounding the root of the inferior mesenteric artery.

Polycystic Kidney. Polycystic kidney is a relatively common congenital malformation in which many nephrons fail to unite with collecting tubules. These isolated nephrons develop in a normal manner and secrete urine, which causes them to form urine-filled *cysts*. As these cysts get larger, they encroach on functional renal parenchyma and eventually cause kidney failure.

DISEASES AND DISORDERS OF THE KIDNEY

Glomerulonephritis. Glomerulonephritis (Bright's disease) is an inflammatory disease of the glomeruli caused in many instances by an allergic reaction to certain strains of streptococcal bacteria. Damage to glomerular basement membranes impairs the filtration barrier, permitting albumin and red blood cells to pass out into the urine. Glomerulonephritis occurs in acute and chronic forms. Acute glomerulonephritis usually follows a streptococcal infection and is reversible in most cases with proper treatment. Chronic glomerulonephritis may develop immediately after an attack of accute glomerulonephritis or may appear many years later. It then progresses slowly but steadily on to *uremia* (kidney failure). During the course of the disease, the kidneys become scarred and shrunken as the result of nephron destruction and eventually are unable to remove urea and other waste products from the blood. Some patients with chronic glomerulonephritis can be helped by treatments with an artificial kidney and live many years with their disease.

Nephrosis. Nephrosis is a noninflammatory kidney disease in which degenerative changes occur in the nephrons, especially in the highly sensitive proximal and distal convoluted tubules. One of the many causes of nephrosis is exposure to toxic chemicals, such as mercury, that poison the tubular cells. Symptoms of nephrosis include *edema* (retention of fluid in the tissues) and marked loss of protein into the urine. The outcome of the disease varies with the cause and the degree of tubular damage.

Pyelonephritis.   Pyelonephritis is an inflammatory kidney disease resulting from bacterial infection of the renal parechyma. This disease is most frequently caused by *Escherichia coli,* an intestinal bacterium that is a common contaminant of the urinary tract. Any interference with the outflow of urine from the bladder or urethra may encourage bacterial growth in the excretory passages and the ascent of these microorganisms into the renal parenchyma.

Kidney Tumors.   Adenocarcinoma is the most common malignant tumor of renal origin. It arises from the epithelium of the uriniferous tubules and is twice as common in men as in women.

DISEASES AND DISORDERS OF THE URETERS

Kidney Stones.   Kidney stones are concretions formed from uric acid or other substances that have precipitated out of the urine instead of remaining in solution. They range in size from tiny grains (kidney gravel) to large masses that fill the renal pelvis and calyces (stag-horn calculi). Kidney stone formation is caused by urinary tract infections, stagnation of urine, or excessive excretion of relatively insoluable substances. When a large kidney stone moves into and down the ureter proper, it causes severe pain and generally obstructs the ureter. If the ureter is obstructed, the stone must either be dissolved chemically or removed surgically to prevent hydronephrosis from occurring.

Hydronephrosis.   Hydronephrosis is a ureteral disease characterized by distention of the renal pelvis and calyces with urine. The most common cause of hydronephrosis is obstruction in the urinary tract. If the obstruction is not relieved, the progressive enlargement of the ureteral structures causes destruction of the renal parenchyma.

Ureteral Stricture.   A ureteral stricture is a narrowing or complete obstruction of a ureter. The stricture may be the result of a structural abnormality, the presence of a kidney stone, or kinking of the ureter when the kidney is displaced (dropped kidney).

Ureteral Tumors.   The most common type of malignant ureteral tumor is the papillary carcinoma. This tumor generally arises from the transitional epithelium of the renal pelvis and is frequently associated with the presence of kidney stones.

DISEASES AND DISORDERS OF THE BLADDER

Cystitis. Cystitis refers to an inflammation of the lining of the bladder. This type of bladder disease may result from a perineal infection ascending by way of the urethra or from a kidney infection descending by way of the ureter. The use of an indwelling urinary catheter or a cystocope (a lighted tube to examine the interior of the bladder) is another cause of cystitis. Women are more prone to develop cystitis than men. This is due to the fact that the anal and urethral openings are very close to one another in women. Improper personal hygiene results in the transfer of intestinal bacteria into the urinary tract and the subsequent development of an ascending infection.

Bladder Tumors. Papillary carcinoma, which arises from the epithelial layer of the mucosa, is the most common malignant tumor of the bladder. There is a high incidence of carcinoma of the bladder in industrial workers exposed to aniline dyes.

DISEASES AND DISORDERS OF THE URETHRA

A *urethral stricture* is a narrowing or obstruction of the urethra. It may be present as a congenital defect or be the result of infection or injury. In the male the pressure produced by an enlarging prostate gland can cause partial or complete obstruction of the urethra.

Suggested Readings

DAVIS, JAMES O. "Control of renin release," *Hosp. Pract.*, 9, no. 4 (April 1974), pp. 55-65.

DIXON, FRANK J. "Glomerulonephritis and immunopathology," *Hosp. Pract.*, 2, no. 11 (November 1967), pp. 35–43.

Kidney and Urinary Tract Infections. Indianapolis: Eli Lilly, 1971.

MERRILL, JOHN P. "The artificial kidney," *Sci. Am.*, 205, no. 1 (July 1961), pp. 56–64.

NETTER, FRANK H. *The Ciba Collection of Medical Illustrations.* Vol. 6: *Kidneys, Ureters, and Urinary Bladder.* Summit, N.J.: Ciba, 1973.

POLLAK, VICTOR E. "Proteinuria I: Mechanisms," *Hosp. Pract.*, 6, no. 6 (June 1971), pp. 49–56.

SMITH, H. W. "The kidney," *Sci. Am.*, 188, no. 1 (January 1953), pp. 40–48.

11

The Respiratory System

The respiratory system is concerned with *external respiration*, the exchange of oxygen and carbon dioxide between alveolar air in the lungs and the blood. The subsequent exchange of oxygen and carbon dioxide between the blood and the cells of the body is known as *internal respiration*. In addition to external respiration, the respiratory system serves accessory functions such as olfaction and voice production.

The organs that directly or indirectly serve the function of external respiration will be considered in the order in which they are traveled by air that is taken into the respiratory tract (Fig. 11.1): the *nose* and *paranasal sinuses, pharynx, larynx, trachea, bronchi,* and *lungs.*

The Nose

The nose is the first portion of the tract that conducts air from the exterior of the body to the lungs. The three major subdivisions of the nose are the external nose, the internal nose (nasal cavity), and the paranasal sinuses.

EXTERNAL NOSE
Gross Anatomy

The external nose (Fig. 11.2) is a pyramidal structure that protrudes from the middle of the face. It consists of a *base* (the inferior surface), an *apex* (the tip of the nose), and two lateral surfaces that meet in the midline to form the *dorsum.* The *nostrils,* or *anterior nares,* are paired openings in the base of the nose that are bounded medially by the *septum* and laterally by the *alae* (wings of the nose).

Microscopic Anatomy

The external nose consists of a framework of bone and cartilage held together by fibrous connective tissue. The nasal cartilages include the *septal cartilage,* which is attached to the bony septum, and the paired *lateral* and *alar cartilages,* which are unattached to bone. This framework is covered with thin skin and lined with a mucous membrane whose epithelium is mostly stratified squamous. Several small

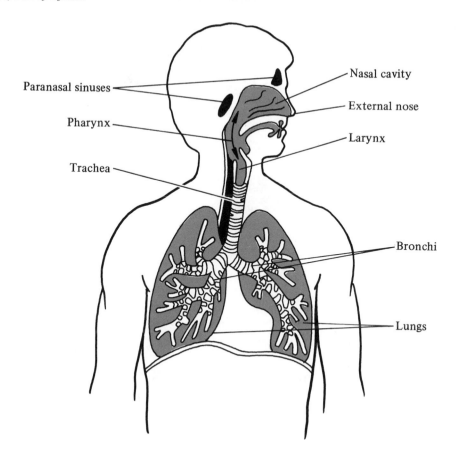

Paranasal sinuses

Pharynx

Trachea

Nasal cavity

External nose

Larynx

Bronchi

Lungs

Figure 11.1   Respiratory Tract.

skeletal muscles act on the external nose either to dilate or contract the nostrils. Each nostril is guarded by *vibrissae,* coarse nasal hairs that filter out large foreign objects.

Blood Supply   The external nose receives blood from branches of the facial, ophthalmic, and maxillary arteries. It is drained by tributaries of the anterior facial and ophthalmic veins.

Innervation   The skin covering the nose is innervated by the ophthalmic and maxillary branches of the trigeminal nerve; the nasal muscles are innervated by branches of the facial nerve.

Function   The function of the external nose is to convey partially filtered air into the nasal cavity.

INTERNAL NOSE
Gross Anatomy   The internal nose, or nasal cavity (Fig. 11.3), lies mainly within the skull. It communicates with the exterior through the nostrils or *anterior nares* and with the upper part of the throat through the *posterior nares* or *choanae.*

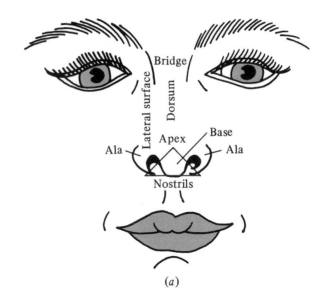

(a)

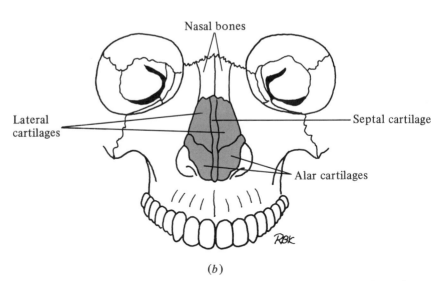

(b)

Figure 11.2 *External Nose.* (a) Front view. (b) Front view with skin and muscles removed to reveal cartilaginous and bony framework.

Roof and Floor The roof of the nasal cavity is formed by parts of the frontal, ethmoid, and sphenoid bones, while its floor is formed by the hard palate. The nasal septum partitions the cavity into two symmetrical halves called *nasal fossae.*

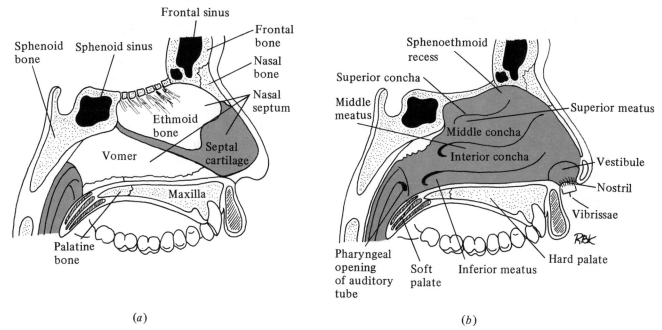

Figure 11.3 *Nasal Cavity.* (a) Medial wall of right nasal fossa. Mucous membrane is removed to reveal bony and cartilaginous components of nasal septum. (b) Lateral wall of left nasal fossa.

Walls The medial wall of each nasal fossa, which is formed by the nasal septum, has a straight smooth surface. The lateral wall, in contrast, has a sloping irregular surface. The surface irregularities are caused by three shelflike bony processes, called *nasal conchae* or *turbinates*, that project out from the lateral wall. In addition, four passages are found in the lateral wall above, between, and below the conchae.

Nasal Conchae. The three nasal conchae are named according to their position on the lateral wall: *superior concha, middle concha,* and *inferior concha.* The superior and middle conchae are part of the ethmoid bone, while the inferior concha is a separate bone.

Passages. Four air passages mark the lateral walls of each nasal fossa. The uppermost passage, called the *sphenoethmoid recess,* is situated between the roof and the superior concha. Three groove-like passages, the *superior, middle,* and *inferior meatuses,* are situated below and lateral to their corresponding conchae. The inferior meatus is the largest of the three and forms the chief air passage through the nasal fossa.

Microscopic Anatomy   The nasal fossae are lined with *nasal mucosa,* a highly vascular mucous membrane that plays a vital role in conditioning inspired air (air taken into the respiratory tract) and in detecting odors. On the basis of structural and functional differences in its epithelium, the nasal mucosa is subdivided into two types: *respiratory mucosa* and *olfactory mucosa.*

Respiratory Mucosa   Respiratory mucosa [Fig. 11.4(a)] covers the lower part of the nasal fossa. It consists of a layer of pseudostratified ciliated columnar epithelium supported by a highly vascular layer of loose connective tissue. Goblet cells are especially abundant in this type of epithelium. They secrete a continuous film of mucus over the surface of the epithelium that serves to trap fine particulate matter such as dust and bacteria. The beating of the cilia moves this mucous film back toward the throat, where it is eventually swallowed. The underlying connective tissue contains an extensive network of venous sinuses especially prominent over the middle and inferior conchae and the opposite surface of the nasal septum. Under certain conditions, such as contact with pollen, these sinuses become engorged with blood and the mucosa swells in much the same way as the erectile tissue in the genital organs. Swelling of the respiratory mucosa interferes with the passage of air through the nasal fossae and makes breathing through the nose difficult.

Olfactory Mucosa   In humans the olfactory mucosa [Fig. 11.4(b)] covers the upper part of the septum, the middle of the roof, and the superior concha of each nasal fossa. The epithelial layer contains two types of cells: *supporting cells* and *olfactory cells.* The supporting cells are pseudostratified columnar cells, while the olfactory cells are bipolar neurons modified to detect odors. The basal ends of the olfactory cells give rise to axons that form small nerve bundles *(olfactory nerves)* in the underlying connective tissue. The olfactory nerves then pass through foramina in the cribriform plate of the ethmoid bone and terminate in the olfactory bulbs of the brain. The moderately vascular connective tissue contains large *olfactory glands* whose serous secretions keep the surface of the olfactory epithelium moist and serve as a solvent for odoriferous substances.

Blood Supply   The nasal cavity receives a rich blood supply from branches of the facial, maxillary, and ophthalmic arteries. Venous drainage occurs by way of the sphenopalatine, anterior facial, and ophthalmic veins. In addition, a few veins communicate with veins draining the frontal lobes of the brain. An infection in the nasal cavity can spread to the brain by this route, often with serious consequences.

Innervation   The nasal mucosa receives general sensory innervation from the

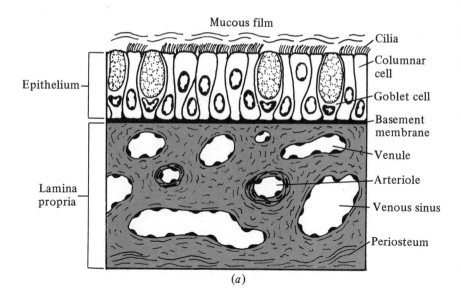

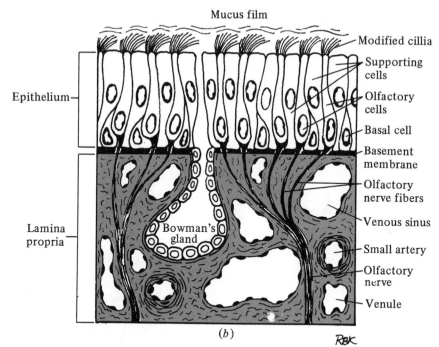

Figure 11.4 *Mucosa.* (a) Respiratory. Note that epithelium is pseudostratified ciliated columnar with goblet cells. (b) Olfactory. Note that epithelium contains olfactory cells and supporting cells. Also note presence of olfactory nerves in lamina propria.

ophthalmic and maxillary branches of the trigeminal nerve and special sensory innervation from the olfactory nerves. The mucosal glands receive parasympathetic innervation from branches of the facial nerve by way of the pterygopalatine ganglion.

Function   The major functions of the nasal cavity consist of olfaction and air conditioning.

Olfaction   The odoriferous substances are chemical in nature and are dissolved in the fluid on the surface of the olfactory epithelium. These substances stimulate the receptor portions of the olfactory cells, which send nerve impulses to the olfactory regions of the brain. The continuous stream of secretions from the olfactory glands washes away the odoriferous substances and keeps the receptors ready for new stimuli.

Air Conditioning   The nasal conchae greatly increase the surface area over which inspired air must pass as it travels through the nasal fossae. The respiratory mucosa covering the conchae warms and moistens the air that comes in contact with it; the blood in the venous sinuses brings heat to the surface of the mucosa while the mucosal glands contribute moisture in the form of serous secretions. This air conditioning helps protect the delicate alveolar tissue of the lungs from the harmful effects of very cold or very dry air. In addition, the mucous film covering the respiratory mucosa removes dust and other particulate matter from the inspired air. The advantage of nose breathing over mouth breathing is due primarily to the changes in the inspired air taking place in the nasal fossae.

PARANASAL SINUSES
Gross Anatomy   The paranasal sinuses (Fig. 11.5) are a collection of air-filled spaces in the cranial and facial bones surrounding the nasal fossae. These

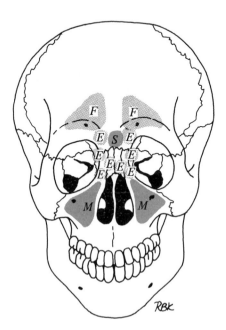

Figure 11.5 Paranasal Sinuses. Frontal aspect of skull with paranasal sinuses indicated in *gray. F,* frontal sinuses; *E,* ethmoid air cells; *S,* sphenoid sinuses; *M,* maxillary sinuses.

air spaces are named according to the bones in which they are located: the *frontal, maxillary, ethmoid,* and *sphenoid sinuses.*

Frontal Sinuses
The frontal sinuses are located in the substance of the frontal bone behind the superciliary arches. They are generally paired and are roughly pyramidal in shape. Each frontal sinus drains into the middle meatus of its corresponding nasal fossa through the frontonasal duct.

Maxillary Sinuses
The maxillary sinuses are the largest on the paranasal sinuses. Each maxillary sinus is pyramidal in shape and is located in the body of the maxilla. It drains into the middle meatus of its corresponding nasal fossa through an opening high on its medial wall. Since this opening is so far above the sinus floor, a considerable amount of fluid can collect in the maxillary sinus before it starts to drain. In fact, it may be necessary for a surgeon to make a new opening lower down in order to drain it.

Ethmoid Sinuses
The ethmoid sinuses consist of a collection of small intercommunicating air cells in the lateral masses of the ethmoid bone. Within each lateral mass the ethmoid air cells are arranged in three groups: an anterior and a middle group that open into the middle meatus and a posterior group that opens into the superior meatus of the corresponding nasal fossa.

Sphenoid Sinuses
The sphenoid sinuses are located in the body of the sphenoid bone. They are generally paired but not symmetrical and vary greatly in size and shape. Each sphenoid sinus drains into the sphenoethmoid recess of its corresponding nasal fossa.

Microscopic Anatomy
The paranasal sinuses are lined with respiratory mucosa that is continuous with the lining of the nasal fossae. This continuity makes it possible for infections in the nose to spread into the paranasal sinuses.

Function
The paranasal sinuses lighten the weight of the skull and act as resonating chambers in voice production.

The Pharynx

GROSS ANATOMY
The pharynx, or throat (Fig. 11.6), is a fibromuscular tube, about 5 inches long, that extends from the base of the skull to the esophagus and serves as a common passageway for both the respiratory and digestive tracts. Since it lies behind the nasal cavity, the oral cavity, and the larynx, it is subdivided into three corresponding regions: the *nasopharynx,* the *oropharynx,* and the *laryngopharynx.*

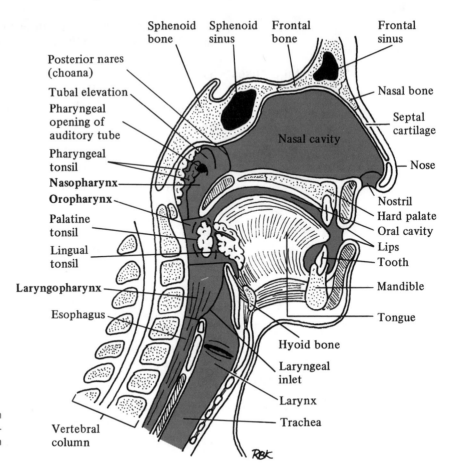

Sphenoid bone   Sphenoid sinus   Frontal bone   Frontal sinus

Posterior nares (choana)

Tubal elevation

Pharyngeal opening of auditory tube

Pharyngeal tonsil

Nasopharynx

Oropharynx

Palatine tonsil

Lingual tonsil

Laryngopharynx

Esophagus

Vertebral column

Nasal cavity

Nasal bone

Septal cartilage

Nose

Nostril

Hard palate

Oral cavity

Lips

Tooth

Mandible

Tongue

Hyoid bone

Laryngeal inlet

Larynx

Trachea

Figure 11.6 Pharynx. Sagittal section of head and neck showing nasal, pharyngeal, and laryngeal cavities. Note location of tonsils.

Nasopharynx   The nasopharynx is the upper portion of the throat. It lies posterior to the nasal cavity and extends from the base of the skull to the level of the soft palate, where it becomes continuous with the oropharynx. The posterior and lateral walls are fixed to bone, which insures that the tube always remains patent (open). The anterior wall is incomplete where it communicates with the nasal fossae through the choanae. On the lateral walls of the nasopharynx are the openings of the *auditory (Eustachian) tubes.* These are curved fibromuscular tubes that extend from the nasopharynx to the middle ear cavity on their respective sides. The auditory tube, which serves to equalize air pressure on both sides of the ear drum, provides a route for upper respiratory tract infection to spread into the middle ear and cause *otitis media* (inflammation of the mucosa of the middle ear). On the posterior wall of the nasopharynx is a collection of lymphoid tissue known as the *pharyngeal tonsil* or *adenoids* (Chapter 9). Enlargement of the adenoids may block the choanae to the point where nose breathing becomes difficult or impossible.

Oropharynx The oropharynx is the middle portion of the throat and serves both the respiratory and digestive tracts. It lies posterior to the oral cavity and extends from the level of the soft palate to the level of the hyoid bone, where it becomes continuous with the laryngopharynx. Its anterior wall is incomplete where it communicates with the oral cavity through the *fauces*.

Laryngopharynx The laryngopharynx is the lower portion of the throat and also serves both the respiratory and digestive tracts. It lies posterior to the larynx and extends from the level of the hyoid bone to the level of the cricoid cartilage, where it becomes continuous with the esophagus. Its anterior wall is incomplete where it communicates with the larynx through the laryngeal inlet.

MICROSCOPIC ANATOMY The pharyngeal tube is composed of three coats: a *mucosa*, a *fibrosa*, and a *muscularis*. The mucosa contains pseudostratified ciliated columnar epithelium in the nasopharynx and stratified squamous epithelium in the oropharynx and laryngopharynx, where the mucosa is subject to abrasion. The fibrosa, or *pharyngobasilar fascia*, is a thin but tough layer of fibrous connective tissue that forms the framework of the pharynx. It is situated between the mucosa and muscularis. The muscularis consists of a thin inner (longitudinal) layer and a thick outer (circular) layer of skeletal muscle fibers. Since these skeletal muscles are important in the act of swallowing, they will be described in Chapter 12.

BLOOD SUPPLY The pharynx receives blood from the ascending pharyngeal artery plus branches of the facial and maxillary arteries. Venous plexuses in the pharyngeal walls drain into the pharyngeal veins, which are tributaries of the internal jugular vein.

INNERVATION Most of the pharynx is innervated by branches of the glossopharyngeal and vagus nerves by way of the pharyngeal plexus.

FUNCTION The pharynx conveys air from the nasal fossae to the larynx and also conveys food and liquids from the oral cavity to the esophagus.

The Larynx

GROSS ANATOMY The larynx, or voice box (Fig. 11.7), is continuous with the pharynx above and the trachea below. It is situated in the anterior part of the neck (below the hyoid bone), and extends from vertebral level *C4* to

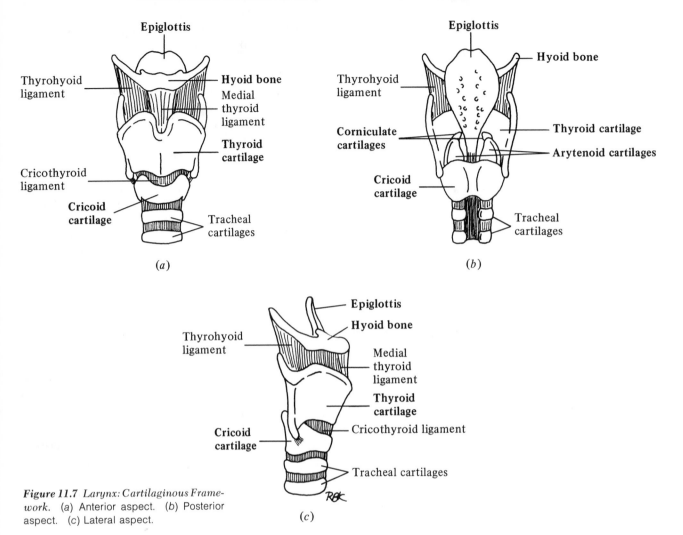

Figure 11.7 Larynx: Cartilaginous Framework. (a) Anterior aspect. (b) Posterior aspect. (c) Lateral aspect.

C6. The framework of the larynx consists of nine pieces of cartilage bound together by ligaments, membranes, and skeletal muscles.

Cartilages   The laryngeal cartilages include three large single cartilages *(thyroid, cricoid,* and *epiglottis)* plus three small paired cartilages *(arytenoid, corniculate,* and *cuneiform).*

Thyroid Cartilage   The thyroid cartilage, largest of the laryngeal cartilages, is shield-shaped. Its two laminae have fused together in the midline to form a prominence, the *Adam's apple,* that is visible beneath the skin in the anterior region of the neck. The thyroid cartilage is larger and the Adam's apple is more conspicuous in the adult male than the adult female due to the influence of testosterone, the male sex hormone.

Cricoid Cartilage	The cricoid cartilage is shaped like a signet ring with the signet portion (lamina) situated posteriorly and the band portion (arch) situated anteriorly. It is connected to the thyroid cartilage above and the first tracheal ring below.
Epiglottis	The epiglottis is a tennis-racket-shaped cartilage that forms the lid of the larynx. The handlelike portion of this cartilage is attached to the interior of the thyroid cartilage while the racketlike portion projects up behind the root of the tongue. During the act of swallowing, the epiglottis prevents food and liquids from entering the air passages by covering the laryngeal inlet.
Arytenoid Cartilages	The arytenoid cartilages are small pyramidal cartilages that sit on the upper border of the arch of the cricoid cartilage. The vocal cords are attached to the thyroid cartilage anteriorly and the arytenoid cartilages posteriorly. Movements of the arytenoid cartilages regulate the position and degree of tension of these cords.
Corniculate Cartilages	The corniculate cartilages are small cartilaginous nodules that sit on the apices of the arytenoid cartilages and sometimes fuse with them.
Cuneiform Cartilages	The cuneiform cartilages are small cartilaginous rods that are located in the membranous folds that extend between the sides of the epiglottis and the arytenoid cartilages.
Ligaments	The extrinsic ligaments of the larynx serve to connect the thyroid cartilage and epiglottis to the hyoid bone above and the cricoid cartilage to the trachea below. The intrinsic ligaments serve to connect the laryngeal cartilages to one another.
Membranes	The fibroelastic membranes of the larynx fill in gaps in the laryngeal walls. The *thyrohyoid membrane* extends between the upper border of the thyroid cartilage and the lower border of the hyoid bone, making it possible for the larynx to move with the hyoid bone during the act of swallowing. The paired *quadrangular membranes* extend from the lateral borders of the epiglottis to the arytenoid cartilages. The superior free borders of these membranes form the *aryepiglottic folds*, while their inferior free borders form the *ventricular folds* (false vocal cords). The *cricothyroid membrane* extends between the cricoid and thyroid cartilages anteriorly. Its superior free margins form the *vocal folds*.
Laryngeal Cavity	The laryngeal cavity (Fig. 11.8) lies within the framework of the laryngeal cartilages. The *laryngeal inlet* (entrance to the larynx) is bounded anteriorly by the epiglottis, laterally by the aryepiglottic

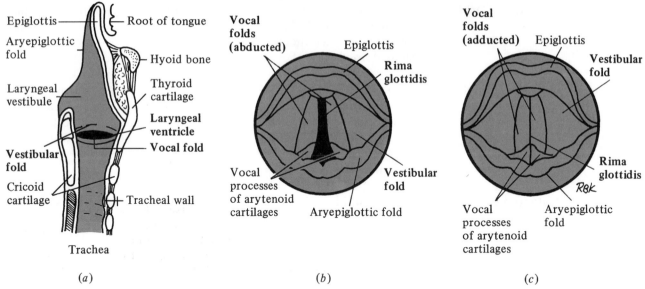

Epiglottis — Root of tongue
Aryepiglottic fold
Hyoid bone
Laryngeal vestibule
Thyroid cartilage
Laryngeal ventricle
Vocal fold
Vestibular fold
Cricoid cartilage
Tracheal wall
Trachea

(a)

Vocal folds (abducted)
Epiglottis
Rima glottidis
Vocal processes of arytenoid cartilages
Vestibular fold
Aryepiglottic fold

(b)

Vocal folds (adducted)
Epiglottis
Vestibular fold
Rima glottidis
Vocal processes of arytenoid cartilages
Aryepiglottic fold

(c)

Figure 11.8 Laryngeal Cavity. (a) Lateral aspect showing vestibular and vocal folds. (b) Superior aspect with vocal folds abducted. (c) Superior aspect with vocal folds adducted.

folds, and posteriorly by the arytenoid cartilages. Two pairs of shelf-like folds project inward from the lateral walls of the laryngeal cavity: the superior or vestibular folds, better known as the *false vocal cords,* and the inferior or vocal folds, better known as the *true vocal cords.* The true vocal cords, which contain the elastic vocal ligaments, can be moved by the rotation of the arytenoid cartilages. The true vocal cords and the opening between them form the vocal apparatus, or *glottis.* The *rima glottidis,* or glottal opening, varies in width and shape with the movements of the true vocal cords during respiration and voice production.

Muscles   The muscles acting on the larynx are divided into extrinsic and intrinsic groups.

Extrinsic Muscles   The extrinsic muscles of the larynx consist primarily of the infrahyoid muscles, which were described in Chapter 7.

Intrinsic Muscles   The intrinsic muscles of the larynx (Fig. 11.9) control the movements and degree of tension of the true vocal cords plus the width of the rima glottidis (Table 11.1). The *posterior cricoarytenoid* muscles abduct the vocal cords and widen the rima by rotating the arytenoid cartilages outwardly; the *lateral cricoarytenoid muscles* adduct them and narrow the rima by rotating the arytenoid cartilages inwardly. The

Table 11.1   Intrinsic Muscles of Larynx

Muscle	Origin	Insertion	Nerve	Action
Posterior cricoarytenoid	Lamina of cricoid cartilage	Muscular portion of arytenoid cartilage	Internal laryngeal	Abducts vocal cords and widens rima glottidis
Lateral cricoarytenoid	Arch of cricoid cartilage	Muscular portion of arytenoid cartilage	Internal laryngeal	Adducts vocal cords and narrows rima glottidis
Arytenoid Transverse	Posterior aspect of arytenoid cartilage	Posterior aspect of opposite arytenoid cartilage	Internal laryngeal	Closes rima glottidis
Oblique	Muscular process of arytenoid cartilage	Apex of opposite arytenoid cartilage	Internal laryngeal	Closes rima glottidis
Cricothyroid	Arch of cricoid cartilage	Lamina of thyroid cartilage	External laryngeal	Tenses vocal cords
Thyroarytenoid (vocalis)	Angle of thyroid cartilage	Vocal process of arytenoid cartilage	Internal laryngeal	Relaxes vocal cords

arytenoid muscle closes the rima by approximating the arytenoid cartilages. The *cricothyroid muscles* tense the vocal cords while the *thyroarytenoid muscles* relax them.

MICROSCOPIC ANATOMY The laryngeal wall consists of a framework of cartilages and fibroelastic membranes lined with a mucous membrane and covered by the intrinsic muscles of the larynx. The laryngeal mucosa contains two types of epithelium: pseudostratified ciliated columnar and stratified squamous. Stratified squamous epithelium covers the aryepiglottic folds, the epiglottis, and the vocal folds. There is also a scattering of taste buds in the mucosa, mainly on the epiglottis. The laryngeal cartilages are either hyaline or elastic. Hyaline cartilages include the thyroid, cricoid, and arytenoids. These cartilages often become calcified in adults.

BLOOD SUPPLY The larynx receives blood from the laryngeal branches of the superior and inferior thyroid arteries and is drained by the laryngeal veins, which are tributaries of the superior and inferior thyroid veins.

INNERVATION The larynx is innervated by the superior and recurrent laryngeal nerves, which are branches of the vagus nerve. The internal branch of the recurrent laryngeal nerve supplies motor innervation to all the intrinsic muscles of the larynx except the cricothyroids, which are innervated by the external branch of the superior laryngeal nerve. The

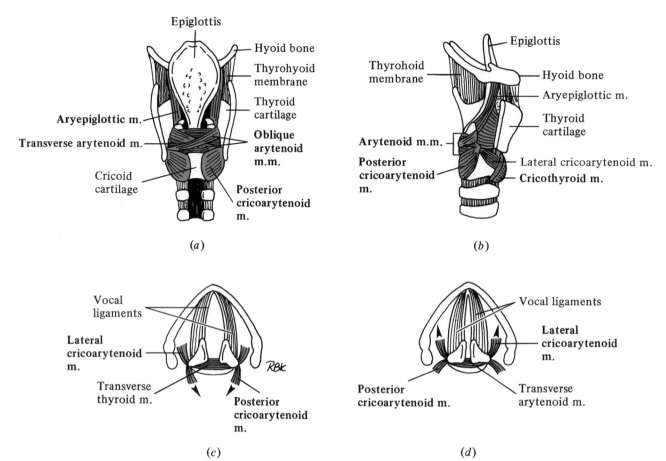

Figure 11.9 *Laryngeal Musculature.* (a) Posterior aspect. (b) Lateral aspect. (c) Superior aspect (with vocal ligaments abducted). (d) Superior aspect (with vocal ligaments adducted).

internal branch also supplies sensory innervation to the laryngeal mucosa.

FUNCTION The two primary functions of the larynx are to conduct air to and from the lower respiratory tract (trachea, bronchi, and lungs) and to produce sound. Sound or voice is produced by the vibration of the true vocal cords as intermittent bursts of expired air are forced out between them. The pitch of the voice is determined by the length and degree of tension of the vocal cords and by the size of the laryngeal cavity. Men generally have lower-pitched voices than women because their vocal cords are longer and their laryngeal cavities are larger. The quality of the voice is dependent on the resonating chambers of the upper respiratory tract: the pharynx, oral cavity, nasal fossae, and paranasal sinuses. Recall how different your own voice sounds when your sinuses are blocked by a head cold. The voice is modified by the palate, tongue, lips, and teeth, which transform the sounds produced by the vocal cords and resonators into words. *Laryngectomy,* surgical removal of the larynx, is usually performed to prevent the spread of laryngeal cancer to the surrounding structures of the head and neck. Patients who have had a laryngectomy are trained to speak without their vocal cords by one of three methods: esophageal speech, pharyngeal speech, or use of an electric voice box. Esophageal speech is produced by swallowing air and burping it up through the esophagus in controlled patterns that simulate laryngeal speech.

The Trachea

GROSS ANATOMY The trachea, or windpipe [Fig. 11.10(*a*)], is a fibroelastic tube supported by cartilaginous rings. It lies in the midline of the neck, anterior to the esophagus. From the lower border of the larynx, the trachea extends down the neck and into the thorax to vertebral level *T*5, where it divides into the right and left primary bronchi.

MICROSCOPIC ANATOMY The tracheal wall is composed of three coats: a *mucosa,* a *submucosa,* and an *adventitia* [Fig. 11.10(*b*)]. The tracheal mucosa consists of a layer of pseudostratified ciliated columnar epithelium supported by a thin layer of dense connective tissue containing numerous elastic fibers. The cilia of the tracheobronchial mucosa beat in a direction that moves the surface film of protective mucus up to the pharynx, where it is either spit out or swallowed. The submucosa consists of a thick layer of loose connective tissue that contains many seromucous glands. The adventitia contains approximately sixteen to twenty rings of hyaline cartilage bound together by a fibroelastic membrane. These rings are incomplete posteriorly, where the trachea comes in contact with the esophagus. The gap is bridged by a fibroelastic membrane and by transversely oriented smooth muscle fibers. This type of construction permits the esophagus to bulge into the posterior tracheal wall whenever food is swallowed. If the trachea were completely rigid, swallowing would be a very painful process.

BLOOD SUPPLY The trachea receives blood from the inferior thyroid arteries and is drained by veins terminating in the thyroid venous plexus.

INNERVATION The trachea is innervated by the autonomic nervous system. It receives parasympathetic fibers from the recurrent laryngeal branch of the vagus nerve and sympathetic fibers from the sympathetic trunks.

FUNCTION The trachea is the trunk of a vast respiratory tree whose branches are the bronchi. It is the first part of the lower respiratory tract to serve a purely respiratory function. A *tracheotomy* is a surgical procedure

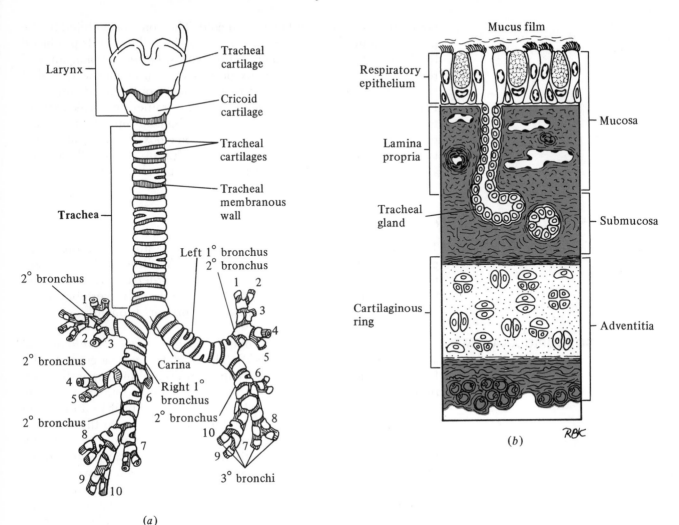

Figure 11.10 *Trachea.* (a) Gross anatomy. Anterior aspect of larynx, trachea, and bronchi. (b) Microscopic anatomy. Segment of cross section of tracheal wall showing its three coats.

in which an artificial opening, or *tracheostomy,* is made in the trachea to improve air conduction or to remove secretions. It is frequently an emergency procedure done on patients with an acute upper respiratory tract obstruction. A longitudinal incision is made through the second and third tracheal rings and a tracheostomy tube is inserted into the trachea to keep the new airway open. This tube must be kept free of secretions so that the patient does not choke to death.

The Bronchi

GROSS ANATOMY   The bronchi are fibroelastic tubes that arise from repeated division of the tracheobronchial tree. Three sets of bronchi are recognized: *primary*, *secondary*, and *tertiary*.

Primary Bronchi   The right and left primary bronchi arise from the bifurcation of the trachea at vertebral level *T5*. For most of their length they are *extrapulmonary* (outside the lungs). The right primary bronchus is shorter, wider, and more vertically directed than the left. For this reason small foreign objects, such as coins and safety pins, that are accidentally inhaled almost always lodge in the right primary bronchus or one of its branches.

Secondary Bronchi   The primary bronchi enter the substance of the lungs and give rise to the secondary bronchi, which are entirely *intrapulmonary*. The right primary bronchus divides into three secondary bronchi, while the left divides into two. Since the secondary bronchi correspond to the number of lobes in their respective lungs, they are also called *lobar bronchi*.

Tertiary Bronchi   Within the substances of the lungs the secondary bronchi give rise to the tertiary bronchi. There are ten tertiary bronchi in the right lung and also ten in the left. Since the tertiary bronchi correspond to the number of bronchiopulmonary segments in their respective lungs, they are also called *segmental bronchi*.

MICROSCOPIC ANATOMY   The microscopic anatomy of the tracheobronchial tree is illustrated in Fig. 11.11. The bronchi have the same basic structure as the trachea and are lined with the same type of epithelium. However, as the bronchi become smaller and smaller, the amount of smooth muscle and elastic fibers in their mucosa increases, while the amount of cartilage in their adventitia decreases. The spiral arrangement of smooth muscle fibers in the bronchial walls serves to shorten the tubes and constrict their lumens during muscular contraction. The dense meshwork of elastic fibers present in the walls of the intrapulmonary bronchi permits their expansion during inspiration and contraction by elastic recoil during expiration. The hyaline cartilage in the bronchial walls is present in the form of plates and plaques rather than rings. These pieces of cartilage provide support for the bronchi but do not interfere with changes in their length or diameter.

BLOOD SUPPLY   The bronchi receive blood from the bronchial arteries, which arise from the anterior surface of the aorta. Venous drainage occurs by way of the bronchial and pulmonary veins.

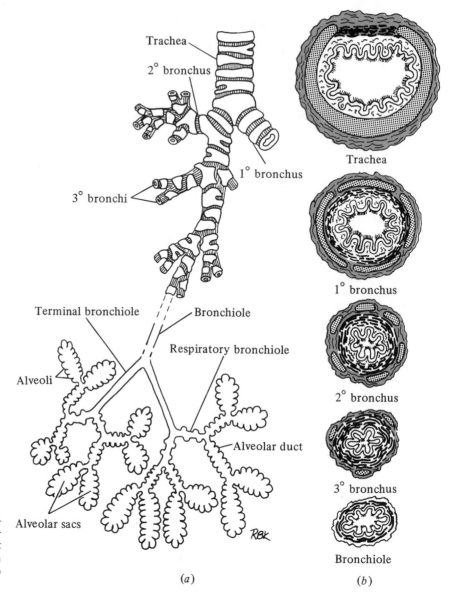

Figure 11.11 Tracheobronchial Tree.
(a) Schematic diagram. (b) Cross sections taken at various levels. Mucosa, *white;* adventitia, *gray;* cartilage, *stippled;* smooth muscle fibers, *black.* 1° = primary, 2° = secondary, 3°=tertiary.

(a)

(b)

INNERVATION The bronchi are innervated by the autonomic nervous system. They receive sympathetic fibers from the sympathetic trunks and parasympathetic fibers from the vagus nerves by way of the pulmonary plexuses. Impulses from the sympathetic fibers causes relaxation of bronchial smooth muscle *(bronchodilation);* impulses from the parasympathetic fibers cause contraction of bronchial smooth muscle *(bronchoconstriction).*

FUNCTION   The bronchi conduct air to and from the substance of the lungs.

The Lungs

GROSS ANATOMY
Location   The lungs are located in the lateral portions of the chest (thoracic) cavity. The floor of the thoracic cavity is formed by the diaphragm, the walls are formed by the rib cage and intercostal muscles, and the roof is formed by the fascia of the neck. The two lungs lie in the right and left halves of the thoracic cavity. They are separated from one another by the mediastinum and its organs, previously described in Chapter 8.

Pleura   Each lung is surrounded by a serous membrane called the *pleura* (Fig. 11.12). This membrane consists of two layers—parietal and visceral—that form a closed sac known as the pleural sac. The *parietal pleura* lines the thoracic cavity and is reflected over the structures in the mediastinum. The *visceral pleura* covers the surface of the lung. The visceral and parietal pleurae are continuous with one another at the root of the lung. The two pleural layers are separated by a potential space, called the *pleural cavity,* that normally contains a thin film of *pleural fluid.* Pleural fluid permits the lungs to move in their pleural sacs without friction during respiration.

External Structure   The lungs are large spongy organs that are generally a bluish gray color in adults. The color of the lungs, which really should be light pink, is due to inhaled dust and other pollutants present in the pulmonary lymphatics. Smokers' lungs are often badly discolored from inhaled cigarette tars and smoke.

Surfaces and Borders   Each lung is conical in shape, having a narrow *apex,* a broad *base,* and *costal* and *mediastinal surfaces.* The apex of the lung is a blunt rounded projection that extends up into the root of the neck. The base or *diaphragmatic surface* is concave. The convex costal surface conforms to the curvature of the rib cage and bears impressions made by the ribs. The concave mediastinal surface conforms to the shape of the mediastinum and bears impressions made by the heart and great vessels.

The sharp *anterior border* and the blunt *posterior border* separate the costal surface of the lung from the mediastinal surface. The *inferior border,* which is sharp laterally and blunt medially, separates the diaphragmatic surface from the costal and mediastinal surfaces.

Lobes and Fissures   Deep fissures divide the lungs into several lobes. The right lung (Fig. 11.13) is divided into three lobes (*superior, middle,* and *inferior*) by two interlobar fissures. The *horizontal* fissure separates the superior lobe from the middle lobe, while the *oblique* fissure sepa-

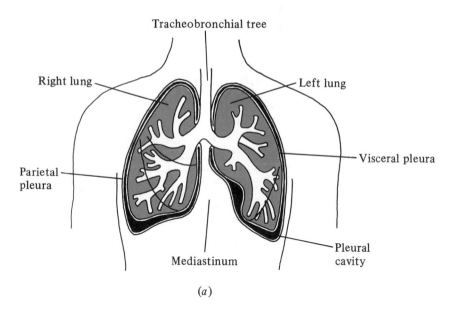

Tracheobronchial tree

Right lung

Left lung

Visceral pleura

Parietal pleura

Pleural cavity

Mediastinum

(a)

Parietal pleura

Visceral pleura

Right lung

Left lung

Root

Root

Mediastinum

Pleural cavity

(b)

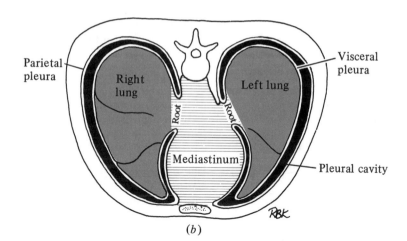

Figure 11.12   Pleura. (a) Frontal section through thorax shows lungs and pleura. (b) Transverse section through thorax shows mediastinum, lungs, and pleura. Note that pleural cavities are exaggerated since there is normally no large space between parietal and visceral pleura.

rates the superior and middle lobes from the inferior lobe. The left lung (Fig. 11.14) is divided into *superior* and *inferior* lobes by the *oblique fissure*. The superior lobe of the left lung has a small tongue-like projection, called the *lingula*, that roughly corresponds to the middle lobe of the right lung.

Internal Structure   Each pulmonary lobe contains several *bronchopulmonary segments* (Fig. 11.15). A bronchopulmonary segment is pyramidal in shape and corresponds to the area of pulmonary parenchyma supplied by a segmental (tertiary) bronchus, segmental artery, and segmental vein. Each lung contains *ten* bronchopulmonary segments. The broncho-

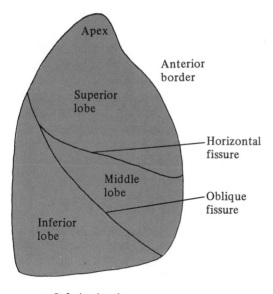

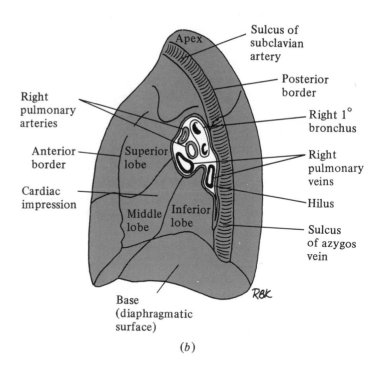

Figure 11.13 Right Lung. (a) Lateral view. (b) Medial view.

(a)

(b)

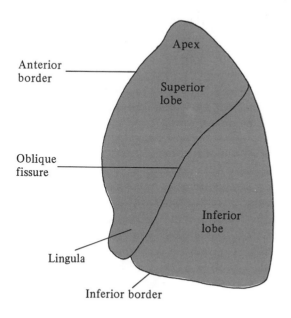

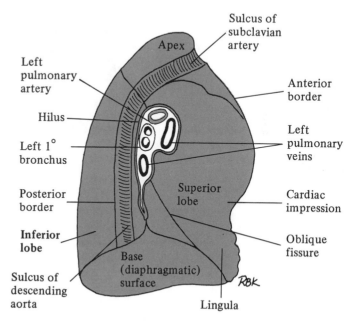

Figure 11.14 Left Lung. (a) Lateral view. (b) Medial view.

(a)

(b)

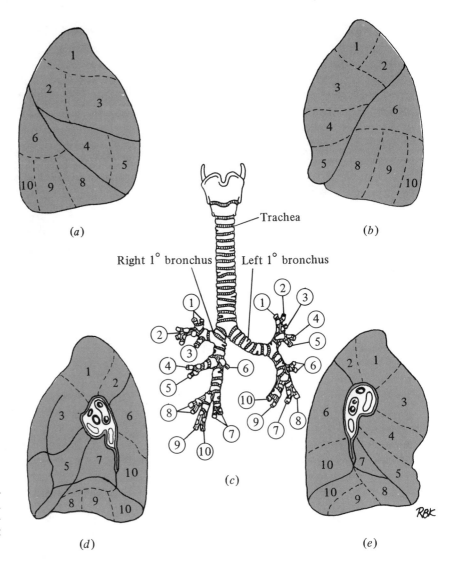

Figure 11.15 Bronchopulmonary Segments. (a) Lateral aspect of right lung. (b) Lateral aspect of left lung. (c) Tracheobronchial tree. (d) Medial aspect of right lung. (e) Medial aspect of left lung.

pulmonary segments are of clinical importance in situations where lung surgery is required. Since each bronchopulmonary segment is functionally independent, it is possible for a thoracic surgeon to remove a diseased segment with minimal disturbance to the rest of the lung.

MICROSCOPIC ANATOMY

The pulmonary parenchyma consists of the many subdivisions of the intrapulmonary bronchi supported by a small amount of fibroelastic connective tissue and supplied with abundant blood and lymphatic vessels.

Bronchioles   As the bronchi continue to divide within the substance of the lung, their branches become progressively smaller. When the branches attain a diameter of 1.0 millimeter (almost microscopic), they are called *bronchioles.* The large bronchioles undergo about twenty divisions before the tiniest conducting branches, called *terminal bronchioles,* are formed. The terminal bronchioles in turn divide to form *respiratory bronchioles,* the first structures in the bronchial tree in which gas exchange occurs.

With the exception of the respiratory bronchioles, the bronchiolar wall is composed of a highly muscular mucosa surrounded by a fibrous adventitia totally lacking in cartilage. The lack of cartilage plus the presence of a great deal of smooth muscle in their walls makes it possible for bronchioles to regulate the diameters of their lumens as do arterioles. The bronchiolar epithelium is ciliated simple columnar and is devoid of goblet cells in the terminal bronchioles. Mucus production stops at this level of the tracheobronchial tree, which helps prevent mucus from collecting below the point where it can be removed by ciliary action.

Primary Lobule   The primary lobule (Fig. 11.16) is the functional unit of the lung. It consists of a respiratory bronchiole and all its branches.

Respiratory Bronchiole   The respiratory bronchiole is a short tube, about 0.5 millimeter in diameter, with occasional alveoli (respiratory chambers) protruding out from its wall. The respiratory bronchioles are lined with a simple cuboidal epithelium almost entirely lacking in cilia and contain considerably less smooth muscle than do the conducting bronchioles.

Alveolar Ducts and Sacs   Each respiratory bronchiole gives rise to several alveolar ducts. The alveolar duct is a long thin-walled tube whose wall contains many alveoli. In turn, the alveolar duct generally gives rise to several short branches that terminate in alveolar sacs (clusters of alveoli). The alveolar duct consists of a simple squamous epithelium supported by a few connective tissue fibers. Smooth muscle fibers are found only at the entrances to the alveoli, where they act as sphincters.

Alveoli   Alveoli (Fig. 11.17) are thin-walled, cup-shaped chambers from which oxygen is exchanged for the carbon dioxide in surrounding pulmonary capillaries. The walls of the alveoli are lined with squamous epithelial cells so thin that they cannot be seen under the light microscope. In addition, the alveolar lining contains many cuboidal cells, called *septal cells,* that protrude into the alveolar lumen. The septal cells secrete *surfactant,* a lipid substance that forms a thin film over the alveolar lining. This film lowers surface tension in the alveoli, which prevents their moist surfaces from collapsing and sticking together on expiration. *Dust cells* are alveolar macrophages that ingest

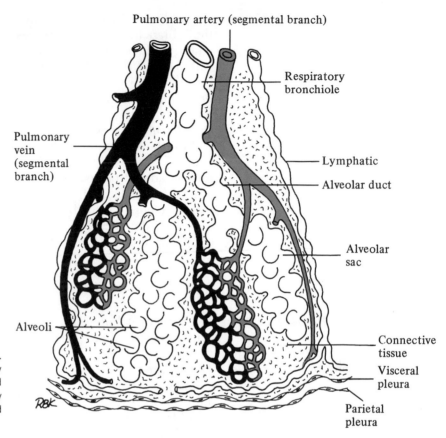

Pulmonary artery (segmental branch)

Respiratory bronchiole

Pulmonary vein (segmental branch)

Lymphatic

Alveolar duct

Alveolar sac

Alveoli

Connective tissue

Visceral pleura

Parietal pleura

Figure 11.16 *Primary Lung Lobule.* Note that segmental branch of pulmonary vein *(black)* contains oxygenated blood while segmental branch of pulmonary artery *(gray)* contains deoxygenated blood.

dust and other foreign particles that have managed to enter the alveoli.

Respiratory Membrane

The respiratory membrane, or *blood-air barrier* (Fig. 11.18), consists of all the structures through which the respiratory gases must diffuse: the alveolar epithelial cell, the basement membrane, and the capillary endothelial cell. This membrane is normally less than 1 micron thick and permits rapid exchange of respiratory gases. However, thickening of any one of these structures can cause a marked decrease in the rate at which oxygen diffuses across the membrane and in the degree of oxygenation of the blood.

BLOOD SUPPLY

The lungs receive blood from two different sources: the pulmonary arteries from the pulmonary circuit and the bronchial arteries from the systemic circuit. Each lung receives deoxygenated blood from a corresponding pulmonary artery. Branches of the pulmonary artery travel as far as the respiratory bronchioles, where they break up into extensive capillary networks that surround all the alveoli. Following respiratory gas exchange across the blood-air barrier, oxygenated blood is carried back to the heart by the pulmonary venules and veins. In addition, bronchial arteries bring oxygenated blood to the walls of the bronchi and the peribronchial connective tissue. Some of the blood supplied by the bronchial arteries drains into the bronchial veins. However, due to extensive anastomoses between the bronchial and pulmonary veins, the greater portion drains into the pulmonary veins.

INNERVATION The lungs are innervated by the autonomic nervous system. They receive sympathetic fibers from the sympathetic trunks and parasympathetic fibers from the vagus nerves by way of the pulmonary plexuses.

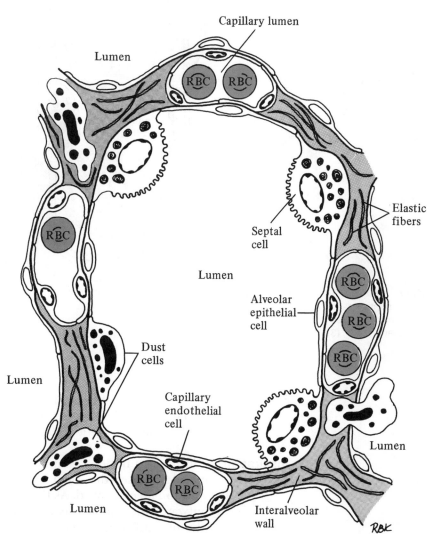

Figure 11.17 Pulmonary Alveolus. Note extreme thinness of alveolar epithelial cells. Drawing based on electron microscopic view.

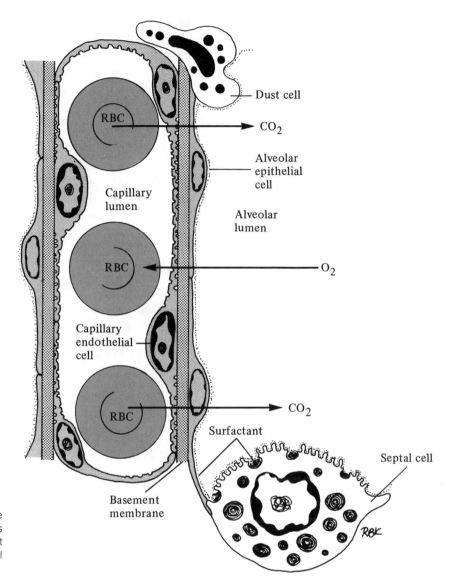

Figure 11.18 Blood-Air Barrier. Note that respiratory gases have to diffuse across (1) alveolar epithelial cell, (2) basement membrane, and (3) capillary endothelial cell.

FUNCTION   During the act of breathing, oxygen-rich air is taken into the lungs, called *inspiration;* gas exchange occurs in the alveoli; and oxygen-depleted air is expelled, called *expiration.*

Inspiration   During inspiration, contraction of the diaphragm and intercostal muscles causes the diaphragm to move downward and the thoracic wall to move outward. As the thoracic walls and parietal pleura move outward, the thoracic cavity expands in volume. Because the thin film of fluid in the pleural cavities has such a high tensile strength, the visceral pleura and lungs are pulled outward with the parietal pleura and the lungs expand. This causes a decrease in the air pressure inside the lungs (intrapulmonic pressure) below that of the air pressure outside the body (atmospheric pressure), and air rushes into the lungs until the two pressures have equalized.

Expiration Expiration is normally a passive process. When the inspiratory muscles relax, the elastic tissue of the lungs (which had been put under stretch during inspiration) recoils. As the lungs move inward, the thoracic walls follow, due to the cohesion between the two layers of pleura. The resulting decrease in the size of the lungs and thoracic cavity causes the intrapulmonic air pressure to rise above atmospheric pressure, and air rushes out of the lungs until the two pressures are equalized.

Diseases and Disorders of the Respiratory System

DISEASES AND DISORDERS OF THE NOSE AND PARANASAL SINUSES

Rhinitis. Rhinitis refers to inflammation of the nasal mucosa. It is characterized by mucosal swelling plus secretion of large amounts of seromucous fluid. Among the many agents that cause rhinitis are viruses, bacteria, dust, pollen, excessive dryness, and exposure to cold air.

Sinusitis. Sinusitis refers to inflammation of the mucosa of one or more paranasal sinuses. It generally results from mucosal extension of an infectious process in the nose during an upper respiratory tract infection. If the sinus openings are blocked by mucosal swelling, the seromucous fluid that builds up can put tremendous pressure on the sinus walls, causing headache, facial pain, and disturbances of vocalization and olfaction.

DISEASES AND DISORDERS OF THE PHARYNX

Pharyngitis. Pharyngitis (sore throat) refers to inflammation of the pharyngeal mucosa. It is a common occurrence in many upper respiratory tract infections. The mucosa frequently becomes swollen and covered with a thick coat of mucous material.

Tonsillitis. Tonsillitis refers to inflammation of the tonsils, a condition that often accompanies pharyngitis. Enlargement of the adenoids as the result of chronic tonsillitis may interfere with nose breathing.

DISEASES AND DISORDERS OF THE LARYNX

Laryngitis. Laryngitis refers to inflammation of the laryngeal mucosa. It is characterized by mucosal swelling, which causes coughing and hoarseness if the vocal cords are involved. Laryngitis occurs in acute and chronic forms. *Acute laryngitis* may be caused by viral or bacterial infections, allergies, air pollution, or overuse of the voice. Swelling of the vocal cords or epiglottis may impair breathing to the point where an emergency tracheotomy is necessary. *Chronic laryngitis* is a condition that develops from chronic irritation of the laryngeal mucosa, such as by overstraining the vocal cords, heavy smoking, or heavy drinking. The irritated mucosa may become thickened, especially over the vocal cords.

Laryngospasm. Laryngospasm refers to spasmodic contractions of the arytenoid muscle that result in closure of the rima glottidis. This is what happens when people choke to death after accidentally getting food, usually steak, caught in their laryngeal cavities. These so-called restaurant coronaries can be prevented by using the *Heimlich maneuver,* one in which the victim is given a bear hug from behind and the rescuer presses his or her fist against the victim's abdomen, just below the rib cage. This forces the diaphragm upward and compresses the air trapped in the lungs, which expels the food trapped in the larynx.

Carcinoma of the Larynx. Carcinoma of the larynx originates from the epithelium lining the laryngeal cavity and most commonly begins on an irritated vocal cord. This type of cancer occurs in middle age and is more frequent in men than women. Carcinoma of the larynx is associated with heavy smoking (two to three packs of cigarettes per day) or heavy drinking. Early detection followed by surgical removal of the larynx offers the best chance for long-term survival.

DISEASES AND DISORDERS OF THE TRACHEOBRONCHIAL TREE

Bronchitis. Bronchitis refers to inflammation of the bronchial mucosa. It occurs in acute and chronic forms. *Acute bronchitis* is generally due to the downward extension of an upper respiratory tract infection of a viral nature that is complicated by a secondary bacterial infection. It is characterized by soreness of the chest and a painful cough that eventually expels thick yellow mucus produced by the inflamed mucosa. *Chronic bronchitis* occurs in people who have had repeated respiratory tract infections or who smoke heavily. It is characterized by a chronic cough (smoker's cough) plus a thick mucous secretion that is generally coughed up in the early morning.

Asthma. Asthma is primarily a response of the bronchial mucosa to allergenic substances such as dust, pollen, feathers, animal hair, and so on. It is characterized by mucosal swelling, production of large amounts of mucus, and bronchiolar constriction (which leads to labored breathing and wheezing). Asthma attacks can be treated with *bronchodilator* drugs, such as epinephrine, which cause the constricted bronchioles to dilate.

Bronchiogenic Carcinoma. Bronchiogenic carcinoma is now the most common type of cancer in men and is rapidly increasing in women. It arises from the epithelium and glands of the bronchial mucosa and appears to be caused by the chronic irritation of heavy smoking or breathing of polluted air. The earliest symptoms of bronchiogenic carcinoma include coughing and shortness of breath. The best form of treatment is surgical removal of the cancerous portion of the lung. However, by the time this disease is diagnosed correctly, the cancer has usually metastasized or has become too extensive for surgery.

DISEASES AND DISORDERS OF THE LUNGS AND PLEURA

Pleurisy. Pleurisy refers to inflammation of the pleura, a frequent complication of lung diseases such as pneumonia or tuberculosis. Pleurisy occurs in dry or wet forms. In *dry pleurisy* the parietal and visceral membranes become swollen and rub against one another as the lungs move; this causes considerable pain. In *wet pleurisy* the pleurae secrete an excessive amount of serous fluid, which accumulates in the pleural cavities and restricts respiratory movements.

Pneumonia. Pneumonia is a general term for inflammation of the lungs. It can be caused by a wide variety of infectious, physical, or chemical agents. Several different types of pneumonia are recognized:

1. *Lobar Pneumonia.* Inflammation is localized in the alveoli. One or more segments of a lung or both lungs are affected at the same time. The affected segments become solidified by inflammatory material so that air cannot enter the alveoli. This type of pneumonia is generally caused by bacterial infection (bacterial pneumonia).
2. *Bronchopneumonia.* Inflammation is localized in or around the bronchi and bronchioles. Patches of affected tissue are scattered throughout both lungs. This type of pneumonia is generally caused by viral infection (viral pneumonia).
3. *Inhalation Pneumonia.* Inflammation of lung tissue is produced by inhalation of toxic chemicals or by accidental inhalation of food, liquids, or vomitus while unconscious.

Tuberculosis. Tuberculosis is an infectious lung disease caused by the tubercle bacillus. Lung tissue reacts to the presence of these bacteria by forming *tubercles,* localized nodules of inflammatory tissue, around them. If the infected person's resistance is high, the bacteria eventually are killed and the tubercles are converted to connective tissue scars. However, if the infected person's resistance is low, the bacteria multiply and the tubercles degenerate, leaving cavities in the lungs. Progressive destruction of lung tissue by the tubercle bacilli eventually depresses respiratory function far enough to cause death. Tuberculosis can be cured, even in a person with low resistance and advanced disease, by the use of powerful antibiotic drugs such as streptomycin and isoniazid plus rest, fresh air, and good food.

Emphysema.   Emphysema is a chronic lung disease that is rapidly becoming a major health problem. Air becomes trapped in pathologically enlarged alveoli as the result of either bronchiolar collapse or bronchiolar obstruction. In severe cases, where many bronchioles are involved, the affected person is so short of breath that he or she becomes a respiratory cripple. The development of emphysema is associated with chronic bronchitis or asthma, with heavy smoking, and with air pollution.

Dust Diseases.   The dust diseases of the lungs, called *pneumoconioses,* are the result of inhalation of industrial dusts such as silica dust, coal dust, cotton dust, or asbestos dust. These substances are extremely irritating to lung tissue. In response to the irritation, the lungs form small connective tissue scars around the dust particles. Eventually, so much functional lung tissue has been converted to scar tissue that the affected miner or mill worker becomes a respiratory cripple and is unable to work.

Metastatic Lung Cancer.   The lung is one of several body organs that act as filters for blood-borne metastatic cancer cells. Cells from malignant tumors of the stomach, breast, and prostate gland, in particular, spread to the lungs by way of the bloodstream and form secondary growths.

Suggested Readings

AVERY, MARY ELLEN, NAI-SAN WANG, and H. WILLIAM TAEUSCH, Jr. "The lung of the newborn infant," *Sci. Am., 228,* no. 4 (April 1973), pp. 74–85.

CLEMENTS, JOHN A. "Surface tension in the lungs," *Sci. Am., 207,* no. 6 (December 1962), pp. 120–130.

COMROE, JULIUS H., Jr. "The lung," *Sci. Am., 214,* no. 2 (February 1966), pp. 56–68.

FENN, WALLACE O. "The mechanism of breathing," *Sci. Am., 202,* no. 1 (January 1960), pp. 138–148.

GLUCK, LOUIS. "Pulmonary surfactant and neonatal respiratory distress," *Hosp. Pract., 6,* no. 11 (November 1971), pp. 45–56.

HAMMOND, E. CUYLER. "The effects of smoking," *Sci. Am., 207,* no. 1 (July 1962), pp. 39–51.

° MURRAY, JOHN F. *The Normal Lung.* Philadelphia: Saunders, 1976.

PARADISE, JACK L., and CHARLES D. BLUESTONE. "Toward rational indications for tonsil and adenoid surgery," *Hosp. Pract., 11,* no. 2 (February 1976), pp. 79–87.

SMITH, CLEMENT A. "The first breath," *Sci. Am., 209,* no. 4 (October 1963), pp. 23–35.

WINTER, PETER M., and EDWARD LOWENSTEIN. "Acute respiratory failure," *Sci. Am., 221,* no. 5 (November 1969), pp. 23–29.

° Advanced level.

12

The Digestive System

The digestive system is concerned with supplying nutrients to the cells of the body for their continued function, growth, repair, and replacement. The functions of the digestive system are as follows:

1. *Ingestion.* The taking of food into the superior (cephalic) end of the digestive tract.
2. *Digestion.* The conversion of food by a series of physical and chemical actions into simple molecules capable of being absorbed.
3. *Absorption.* The passage of food in its molecular state through the mucous membrane of the small intestine and into the blood or the lymph.
4. *Egestion.* The elimination of undigested residue from the inferior (caudal) end of the digestive tract.

The organs of the digestive system consist of the digestive tract and the accessory organs of digestion (Fig. 12.1). In humans the digestive tract consists of a muscular tube, approximately 30 feet long, subdivided into the *oral cavity, fauces, pharynx, esophagus, stomach, small intestine, large intestine,* and *anus.* The accessory organs of digestion lie external to the digestive tract but are connected with it by means of ducts. These organs include the *salivary glands, pancreas, liver,* and *gallbladder.*

Before we begin our discussion of the digestive tract and its accessory organs, it is necessary for us to describe several items of a general nature that are pertinent to our discussion: the wall of the digestive tube, the peritoneum, and the mesenteries.

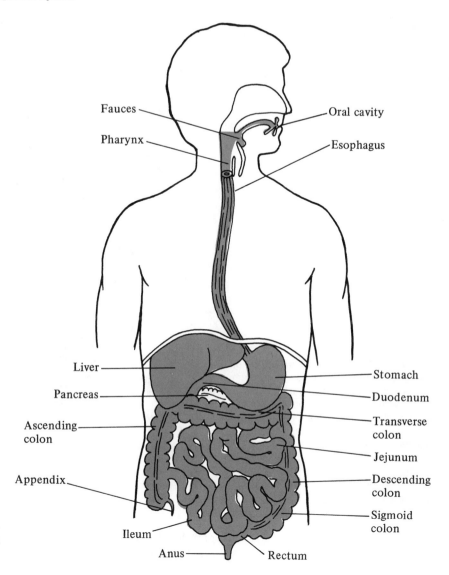

Fauces

Pharynx

Oral cavity

Esophagus

Liver

Pancreas

Ascending colon

Appendix

Ileum

Anus

Stomach

Duodenum

Transverse colon

Jejunum

Descending colon

Sigmoid colon

Rectum

Figure 12.1 Digestive System. Digestive tract and accessory organs of digestion (except for salivary glands) are shown.

The Wall of the Digestive Tube

For most of its length the wall of the digestive tube (Fig. 12.2) is composed of four coats: *mucosa, submucosa, muscularis externa*, and *adventitia/serosa.*

Mucosa The mucosa is the innermost coat of the digestive tube. The type of epithelium found in the mucosa varies with function: stratified squamous for ingestion and egestion and simple columnar for digestion and absorption. The epithelial layer is supported by a connective tissue layer, or *lamina propria*, that contains many mucosal

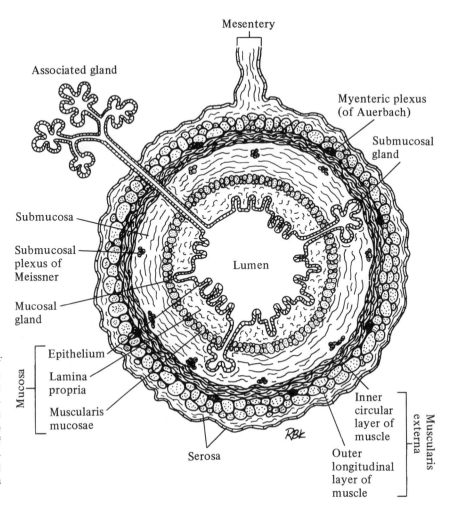

Mesentery

Associated gland

Myenteric plexus
(of Auerbach)

Submucosal
gland

Submucosa

Submucosal
plexus of
Meissner

Lumen

Mucosal
gland

Epithelium

Lamina
propria

Muscularis
mucosae

Mucosa

Inner
circular
layer of
muscle

Muscularis
externa

Outer
longitudinal
layer of
muscle

Serosa

RBK

Figure 12.2 Cross-sectional Plan of Digestive Tube. Schematic diagram shows various coats of digestive tube in cross section: mucosa, submucosa, muscularis externa, and serosa (or adventitia in some regions). Different types of digestive glands are shown emptying into lumen of digestive tube. Note that circularly disposed muscle fibers appear in longitudinal section, while longitudinally oriented fibers appear in cross section.

glands, blood vessels, lymphatics, and lymph nodules. The *muscularis mucosae*, a thin layer of smooth muscle fibers, completes this coat.

Submucosa The submucosa is a loose connective tissue coat that binds the mucous membrane to the external muscular coat. It contains submucosal glands plus many blood vessels, lymphatics, and nerve fibers. The *submucosal plexus* (of Meissner) provides autonomic nerve fibers for the innervation of muscularis mucosae plus mucosal and submucosal glands. Sympathetic impulses inhibit secretion, while parasympathetic impulses stimulate secretion.

Muscularis Externa The muscularis externa is an extensive coat of smooth muscle fibers that serves to mix food with digestive enzymes and to move food through the digestive tract. The muscle fibers are organized into inner (circular) and outer (longitudinal) layers. The *myenteric plexus* (of Auerbach), a plexus that provides autonomic nerve fibers for the innervation of muscularis externa, is located between the two muscle layers. Sympathetic impulses inhibit persistalsis, while parasympathetic impulses stimulate peristalsis.

Adventitia/Serosa   Depending on its location, the outermost coat of the digestive tube consists of either an adventitia or a serosa. Above the diaphragm the digestive tube is covered by an adventitia; below the diaphragm the digestive tube is covered for most of its length by a serosa.

Any modifications of this basic plan will be described in the section dealing with the portion of the digestive tube in which the modification occurs.

Peritoneum

The peritoneum is a serous membrane that lines the abdominal cavity and is reflected over or encloses the abdominal organs. The peritoneum lining the abdominal walls is called *parietal peritoneum,* while that covering or enclosing the abdominal organs is called *visceral peritoneum.* The abdominal organs that lie between the posterior abdominal wall and the parietal peritoneum are said to be *retroperitoneal.*

The peritoneum forms a serous sac that is closed in males but open (through the uterine tubes) in females. The space enclosed by the peritoneal sac is the *peritoneal cavity.* Under normal conditions the peritoneal cavity is a potential space that contains a thin film of serous fluid. Peritoneal fluid permits the organs in the peritoneal cavity to move upon one another and against the walls without friction. *Peritonitis* (inflammation of the peritoneum) is usually the result of an infectious process. It is a very serious condition that can be fatal if not treated promptly with antibiotics. *Ascites* refers to the accumulation of an abnormally large amount of fluid in the peritoneal cavity. It is caused by conditions such as cirrhosis of the liver, kidney and heart disease, and inflammation or tumors within the peritoneal cavity.

Mesenteries

Mesenteries and visceral ligaments are double-layered sheets of peritoneum that suspend the organs they enclose from the walls of the abdominal cavity or from other abdominal organs (Fig. 12.3). Mesenteries not only provide support but also transmit blood vessels,

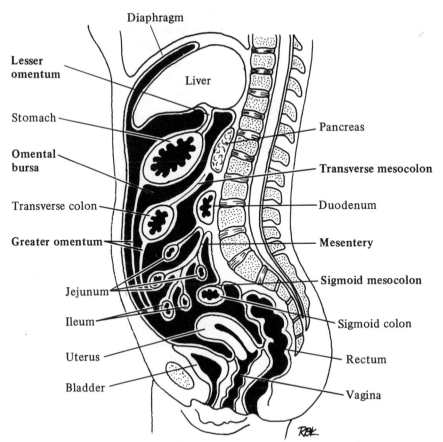

Figure 12.3 Peritoneal Reflexions. Schematic diagram shows sagittally sectioned adult female trunk. Peritoneal sac is subdivided into greater sac and lesser sac (omental bursa), which are in communication through epiploic foramen (not shown in this section). Note that greater omentum, which forms walls of lesser sac, hangs down in front of transverse colon and bulk of small intestine. Note also that pancreas and duodenum lie behind parietal peritoneum lining posterior abdominal wall and are thus retroperitoneal.

nerves, and lymphatics to the organs they suspend. The *mesentery* (of the small intestine) is a fan-shaped sheet of peritoneum that suspends the small intestine from the posterior abdominal wall. The *lesser omentum* is a mesentery that suspends the stomach from the liver. The *greater omentum* is a mesentery extending from the stomach to the transverse colon. The *omental bursa,* a saclike fatty fold of the greater omentum, hangs down like a curtain over the small intestine and serves as a shock absorber. Other mesenteries and visceral ligaments will be described as they are encountered.

Part A. **The Digestive Tract**

The digestive tract or tube extends from the lips to the anus. It is divided into several regions with different functions and structural specializations. These regions are best studied in the order in which they are traveled by ingested food.

Oral Cavity

GROSS ANATOMY The mouth or oral cavity (Fig. 12.4) is the portion of the digestive tract that extends from the lips to the fauces. It contains the teeth and the tongue, which play an important role in chewing and swallowing.

Entrance The entrace to the oral cavity, called the *rima oris,* is located in its anterior wall. The rima oris is surrounded by the *lips,* flexible muscular folds covered externally by skin and internally by mucous membrane. The muscular core of the lips is formed by the orbicularis oris, a sphincter that regulates the width of the rima oris.

Roof The roof of the oral cavity is formed by the palate. It consists of a bony anterior portion, called the *hard palate,* and a muscular posterior portion, called the *soft palate.* The hard palate is composed of the palatine processes of the maxillae in front and the horizontal processes of the palatine bones behind. The soft palate is a muscular flap covered with mucous membrane that is suspended from the back of the hard palate. A small conical structure called the *uvula* projects down into the pharynx from the free posterior border of the soft palate.

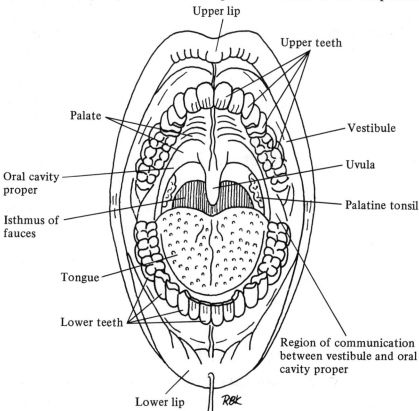

Figure 12.4 *Oral Cavity.* Front view. Lower jaw has been depressed and lips retracted for better visualization of oral cavity structures. Note that vestibule and oral cavity proper communicate behind last molar tooth.

Lateral Walls   The lateral walls of the oral cavity are formed by the *cheeks*. The cheeks are flexible muscular structures covered externally with skin and internally with mucous membrane. The muscular core of the cheeks is formed principally by the buccinator muscles, which are responsible for compressing the cheeks during mastication (chewing).

Floor   The floor of the oral cavity is formed by a mucous-membrane-covered muscular diaphragm, comprising the myohyoid muscles, and is occupied by the tongue.

Exit   The exit from the oral cavity is a narrow passageway, called the fauces, which leads into the oropharynx.

Subdivisions   The oral cavity is subdivided into two parts: a smaller outer cavity called the *vestibule* and a larger inner cavity called the *oral cavity proper*. The vestibule is a slitlike space bounded on the outside by the lips and cheeks and on the inside by the gums and teeth. The oral cavity proper is bounded in the front and on the sides by the alveolar processes of the jawbones and the teeth they contain. When the jaws are closed, the oral cavity proper communicates with the vestibule by way of the spaces behind the last molar teeth.

MICROSCOPIC ANATOMY   The oral cavity is lined with oral mucosa, which consists of a layer of stratified squamous nonkeratinizing epithelium supported by fibrous connective tissue. Where a submucosa is present, the mucosa is loosely attached to underlying structures. Many submucosal glands, called minor salivary glands, empty their secretions into the oral cavity. Their secretions plus the secretions of the major salivary glands constitute *saliva*.

FUNCTION   Food is taken into the oral cavity through the rima oris and mechanically broken down into smaller particles by the teeth. The digestion of carbohydrates is begun in the oral cavity by the the enzyme *ptyalin* (salivary amylase). Ptyalin breaks down starches into dextrins and disaccharide (double) sugars such as maltose. The tongue, teeth, and cheek muscles mix the partially digested food particles with saliva and form them into balls, or *boluses*. Food boluses are passed through the fauces and into the oropharynx during the act ,of swallowing.

Teeth and Gums

GROSS ANATOMY   The teeth are the organs of mastication. They consist of modified connective tissue papillae covered by some extremely hard substances.

Location The teeth are implanted in deep sockets in the alveolar processes of the jawbones. They are arranged in two horseshoe-shaped rows called *dental arches*.

General Description Each tooth (Fig. 12.5) consists of a *crown, neck,* and *root*. The crown is the exposed portion of the tooth. It contains the biting or grinding surface. The neck is the constricted region of the tooth between the crown and root and is located at the gumline. The root is the portion of the tooth that is embedded in the alveolar process of the jawbone.

Classification The teeth are placed in the following classes on the basis of their function and appearance:

1. *Incisors.*   Chisel-shaped teeth used for biting and chewing. The incisor has a chisel-shaped crown and a single root.
2. *Canines (Cuspids).*   Pointed teeth used for tearing. The canine has a conical crown with a single cusp (point) and a long single root.
3. *Premolars (Bicuspids).*   Moderately broad teeth used for grinding. The premolar has a crown with two cusps and one or two roots.
4. *Molars (Multicuspids).*   Broad teeth used for grinding and crushing. The molar has a crown with three to five cusps and two or three roots.

As a rule, each person has two sets of teeth: a *deciduous set* and a *permanent set*.

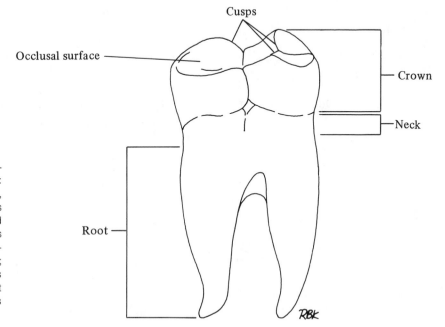

Figure 12.5 Molar Tooth: Gross Anatomy. Each tooth consists of three regions: crown, which projects above the gum; root, which is embedded in the alveolar process of jawbone; and neck, which is constricted area between crown and root. Crown has several surfaces: surface that contacts opposing teeth is known as occlusal surface; surface that faces lips or cheek is known as labial or buccal surface; and surface that touches a neighboring tooth is known as contact surface.

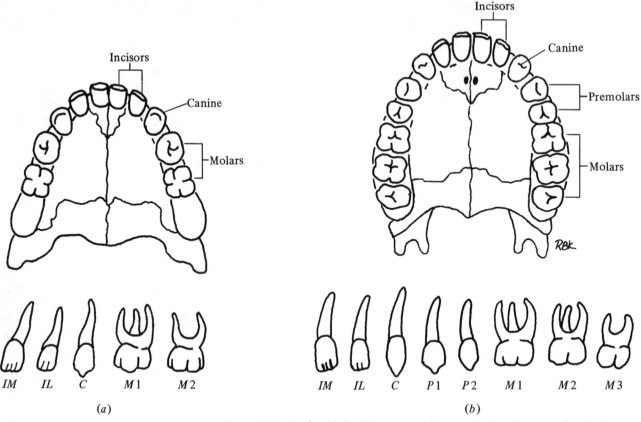

Figure 12.6 *Teeth.* (a) Deciduous teeth of upper dental arch as seen from below and from right side. *IM,* medial incisor; *IL,* lateral incisor; *C,* canine; *M1,* first molar; *M2,* second molar. (b) Permanent teeth of upper dental arch as seen from below and from the right side. *1M,* medial incisor; *IL,* lateral incisor; *P1,* first premolar; *P2,* second premolar; *M1,* first molar; *M2,* second molar; *M3,* third molar (wisdom tooth).

Deciduous Teeth   The deciduous teeth [Fig. 12.6(*a*)] are 20 in number and are the teeth of childhood. They appear in infancy and are shed between the sixth and twelfth years of life.

Permanent Teeth   The permanent teeth [Fig. 12.6(*b*)] are 32 in number and are the teeth of adulthood. They appear between the sixth and twenty-first years of life. If proper care is taken of the permanent teeth, they should last a lifetime.

MICROSCOPIC ANATOMY
Teeth   Each tooth (Fig. 12.7) consists of an outer shell of calcified tissues enclosing a central core of soft tissue called the dental pulp. It is anchored in its socket by the periodontal ligament.

Calcified Tissues   The outer shell of the tooth is composed of the three highly calcified dental tissues: *enamel, dentin,* and *cementum.*

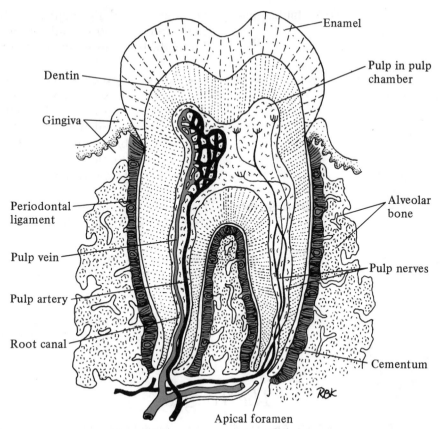

Figure 12.7 Molar Tooth: Microscopic Anatomy. Longitudinal section in position in lower jaw. Bulk of tooth consists of dentin, which surrounds root canals and pulp chamber. Crown of tooth is covered with thick layer of enamel, while root is covered with relatively thin layer of cementum. Blood vessels (shown on left side) and nerves (shown on right side) enter tooth through apical foramina in roots, pass through root canals, and terminate in pulp chamber. Note that tooth is anchored in its socket in alveolar bone by periodontal ligament.

Enamel. Enamel is a hard, brittle white tissue that covers the crown and neck of the tooth. It is the most highly calcified and hardest tissue in the human body. In its mature form enamel contains no cells and cannot regenerate if worn away.

Dentin. Dentin is a yellowish white tissue, somewhat softer than enamel, that forms the bulk of the tooth. Similar to bone, but more resilient, dentin acts as a cushion for the enamel that lies above it.

Cementum. Cementum is an unusual form of bone that contains no blood vessels. It covers the root of the tooth.

Dental Pulp   The dental pulp lies in a cavity in the dentin called the *pulp chamber*. It consists of loose connective tissue richly supplied with blood vessels and nerve fibers. The dental pulp keeps the tooth alive by providing nutrients and innervation for the calcified tissues. A *root canal* is a minute channel through the root of the tooth that transmits blood vessels and nerves from the jawbone to the pulp chamber.

Peridontal Ligament

The periodontal ligament is a fibrous connective tissue membrane that extends from the walls of the alveolar socket to the cementum of the tooth. It serves to anchor the tooth in its socket.

Gums

The gums, or *gingivae,* consist of the oral mucosa that covers the alveolar processes of the jawbones. This mucosa surrounds the necks of the teeth and is reflected into the alveolar sockets, where it merges with the peridontal membrane. The gums are the most common site of mucous membrane disease in the oral cavity. Since gingival disease may result in the loss of teeth, taking good care of the gums is an important part of oral hygiene.

BLOOD SUPPLY

The teeth are supplied by the alveolar branches of the maxillary artery and are drained by their companion veins.

INNERVATION

The teeth are innervated by the alveolar branches of the maxillary (upper jaw) and mandibular (lower jaw) divisions of the trigeminal nerve.

FUNCTION

Because of their great strength and hard occlusal surfaces, the teeth are effective organs for biting, tearing, grinding, and crushing food.

The Tongue

GROSS ANATOMY
Location

The tongue [Fig. 12.8(*a*)] is a fibromuscular organ located on the floor of the oral cavity within the curve of the body of the mandible. It is the chief organ of taste, an important organ of speech, and an accessory organ of chewing and swallowing.

Parts

The tongue consists of three parts: *apex, body,* and *root.* The apex is the free rounded tip of the tongue that lies against the lingual surface of the lower incisors. The body forms the anterior two-thirds of the tongue and is also free. The root forms the posterior one-third of the tongue and is fixed. It is connected to the hyoid bone, epiglottis, soft palate, and pharynx.

Surfaces

The tongue has inferior and superior surfaces that meet at the sides of the tongue.

Inferior Surface

The inferior surface of the tongue is concave and smooth. A midline fold of mucous membrane, called the *frenulum* of the tongue, anchors the inferior surface to the floor of the oral cavity. In some people the frenulum is so short that it limits movement of the tongue and interferes with speech.

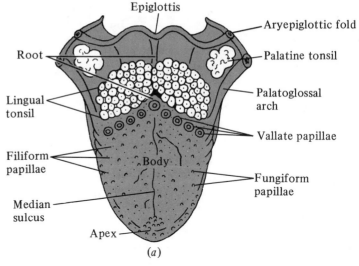

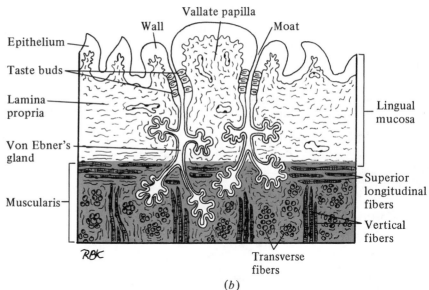

Figure 12.8 Tongue. (a) Gross anatomy. Dorsal aspect of tongue shows various types of lingual papillae and lingual tonsils. Note that V-shaped row of vallate papillae marks division between anterior two-thirds and posterior one-third tongue. Also note that each lingual tonsil consists of a number of lymphoid tissue islands located at root of tongue. (b) Microscopic anatomy. Segment of sagittal section through tongue that includes vallate papilla. Note that taste buds are located on sides of vallate papilla and its surrounding wall. Secretions from serous glands of von Ebner flow constantly over taste buds, removing food particles and dissolved substances from buds so that they can receive new taste stimuli.

Superior Surface

The superior surface, or dorsum, of the tongue is convex and rough. Its mucous membrane is thrown into numerous projections called *lingual papillae*. There are three types of lingual papillae:

1. *Filiform Papillae.* Slender conical papillae evenly distributed over the dorsum of the tongue.
2. *Fungiform Papillae.* Globular, mushroom-shaped papillae distributed on the sides and apex of the tongue and scattered irregularly over the dorsum.

3. *Vallate Papillae.* Eight to twelve large round papillae arranged in a V-shaped row that marks the junction of the body and root of the tongue.

Musculature

The tongue contains intrinsic and extrinsic skeletal muscles (Table 12.1). The intrinsic muscles lie within the substance of the tongue, while the extrinsic muscles have their origins outside the tongue.

Intrinsic Muscles

The intrinsic muscles of the tongue are named according to the planes in which their fibers run: *longitudinalis, transversus,* and *verticalis.* These muscles help alter the shape of the tongue.

Extrinsic Muscles

The extrinsic muscles of the tongue are named according to their origins and insertions: *genioglossus, hyoglossus,* and *styloglossus.* These muscles anchor the tongue to bony structures and enable it to perform many movements, such as protrusion and retraction.

MICROSCOPIC ANATOMY

The tongue is covered by a mucous membrane that is separated from the underlying skeletal muscle by a layer of fibrous connective tissue. On the root of the tongue the mucous membrane is elevated by masses of submucosal lymphoid tissue that constitute the *lingual tonsil.* On the dorsum of the tongue, the mucous membrane gives rise to the lingual papillae described previously. Each lingual papilla consists of a connective tissue core covered with stratified squamous epithelium

Table 12.1 Muscles of Tongue

Muscle	Origin	Insertion	Nerve	Action
Genioglossus	Genial tubercle of mandible	Hyoid bone and inferior surface of tongue	Hypoglossal	Assists in protrusion, retraction, and depression of tongue
Hyoglossus	Body and greater horn of hyoid bone	Side of tongue	Hypoglossal	Depresses tongue and draws sides down
Styloglossus	Styloid process of temporal bone	Side of tongue	Hypoglossal	Assists in elevation and retraction of tongue
Longitudinalis	Originates and inserts within substance of tongue		Hypoglossal	Shortens tongue
Transversus	Originates and inserts within substance of tongue		Hypoglossal	Narrows and lengthens tongue
Verticalis	Originates and inserts within substance of tongue		Hypoglossal	Flattens and broadens tongue

and richly supplied with nerves and blood vessels. Taste buds (taste receptor organs) are scattered over the mucous membrane of the tongue, especially on the sides of the vallate papillae.

BLOOD SUPPLY The tongue is supplied by the lingual artery, a branch of the external carotid artery, and is drained by the lingual vein, a tributary of the external jugular vein.

INNERVATION The tongue is innervated by branches of five cranial nerves: the trigeminal (general sensation), facial (taste on the anterior two-thirds), glossopharyngeal (taste on the posterior one-third), vagus (taste on the root near the epiglottis), and hypoglossal (tongue musculature).

FUNCTION The tongue plays an important role in chewing and swallowing. The tongue assists in the act of chewing by moving the food into position to be chewed, mixing chewed food with saliva, and helping to form the mixture of food and saliva into boluses. During the act of swallowing, the body of the tongue is raised against the hard palate, which moves the bolus to the level of the soft palate. Elevation of the root of the tongue then forces the bolus into the fauces, the next portion of the digestive tract.

The Fauces

GROSS ANATOMY The fauces is the narrow arched passageway connecting the oral cavity to the oropharynx (Fig. 12.9). It is bounded superiorly by the soft palate, laterally by the faucial pillars, and inferiorly by the root of the tongue.

Faucial Pillars The lateral walls of the fauces are marked by two pairs of vertically oriented structures known as the faucial pillars. The anterior pillars

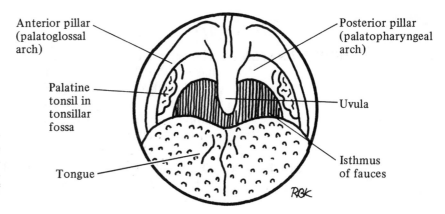

Figure 12.9 Fauces. Schematic diagram shows space between soft palate and root of tongue. Note that palatine tonsils are located in fossae between palatoglossal and palatopharyngeal arches of soft palate.

consist of folds of mucous membrane covering the *palatoglossal muscles*, which extend from the soft palate to the root of the tongue. The posterior pillars consist of folds of mucous membrane covering the *palatopharyngeal muscles*, which extend from the soft palate to the oropharynx.

Palatine Tonsils   The tonsillar fossae are triangular depressions in the lateral faucial walls between the anterior and posterior pillars. They contain ovoid masses of lymphoid tissue known as the *palatine tonsils*. When the palatine tonsils enlarge, they tend to interfere with swallowing.

MICROSCOPIC ANATOMY   The fauces is lined with a mucous membrane whose stratified squamous epithelium is continuous with that of the oral cavity and the oropharynx. This epithelium penetrates the substance of the palatine tonsils at 15 to 20 points to form the tonsillar crypts. Due to the presence of these crypts, the palatine tonsils become infected more easily than the adenoids and lingual tonsils.

BLOOD SUPPLY   The palatine tonsils receive a very rich blood supply from branches of the lingual, facial, and maxillary arteries. It is the reason why tonsillectomies tend to be very bloody operations.

INNERVATION   Most muscles of the soft palate and faucial arches are innervated by branches of the vagus and spinal accessory nerves by way of the pharyngeal plexus.

FUNCTION   During the act of swallowing, the fauces receives a bolus of food from the oral cavity and propels it into the oropharynx. When the bolus passes through the isthmus of the fauces, the palatoglossus muscles contract; this action shuts off the oral cavity behind the bolus. Contraction of the palatopharyngeal muscles causes elevation of the soft palate, which closes off the nasopharynx.

The Pharynx

GROSS ANATOMY   The pharynx or throat is a fibromuscular funnel that extends from the base of the skull to the beginning of the esophagus at vertebral level C6. The three divisions of the pharynx are the nasopharynx (entirely respiratory) the oropharynx (respiratory and digestive), and the laryngopharynx (respiratory and digestive). The nasopharynx was described in Chapter 11.

Oropharynx   The oropharynx extends from the level of the soft palate to the level of the hyoid bone. It is continuous with the nasopharynx above, with the oral cavity in front (through the fauces), and with the laryngo-pharynx below.

Laryngopharynx The laryngopharynx extends from the level of the hyoid bone to the level of the cricoid cartilage, where it becomes continuous with the esophagus. It is continuous with the oropharynx above, with the larynx in front (through the laryngeal inlet), and with the esophagus below. On either side of the laryngeal inlet there is a pocket or *piriform sinus*. When food gets wedged in the piriform sinuses, it gives a person the sensation of having something caught in his or her throat.

Muscles of the Pharynx The muscular coat of the pharyngeal wall consists of a thin inner longitudinal layer and a relatively thick, outer circular layer of skeletal muscles that are described in Table 12.2 and illustrated in Fig. 12.10.

Longitudinal Muscles Muscles of the inner longitudinal layer, which are named according to their origins and insertions, include the paired *palatopharyngeus* (which really belong to the fauces), *stylopharyngeus*, and *salpingopharyngeus muscles*. Contraction of these muscles elevates the pharynx and closes off the nasopharynx from the oropharynx.

Table 12.2 Muscles of Pharynx

Muscle	Origin	Insertion	Nerve	Action
Palatopharyngeus	Soft palate	Thyroid cartilage and musculature of pharynx	Pharyngeal plexus (accessory and vagus nerves)	Elevates pharynx and helps close nasopharynx
Stylopharyngeus	Styloid process of temporal bone	Thyroid cartilage and musculature of pharynx	Glossopharyngeal	Elevates pharynx; expands sides of pharynx
Salpingopharyngeus	Auditory tube	Musculature of pharynx	Pharyngeal plexus	Elevates pharynx and opens auditory tube
Superior Constrictor	Medial pterygoid plate of sphenoid bone, pterygomandibular ligament, and mandible	Pharyngeal tubercle of occipital bone and pharyngeal raphe	Pharyngeal plexus	Constricts upper pharynx
Middle Constrictor	Greater and lesser horns of hyoid bone	Pharyngeal raphe	Pharyngeal plexus	Constricts middle pharynx
Inferior Constrictor	Sides of thyroid and cricoid cartilages	Pharyngeal raphe	Pharyngeal plexus	Constricts lower pharynx

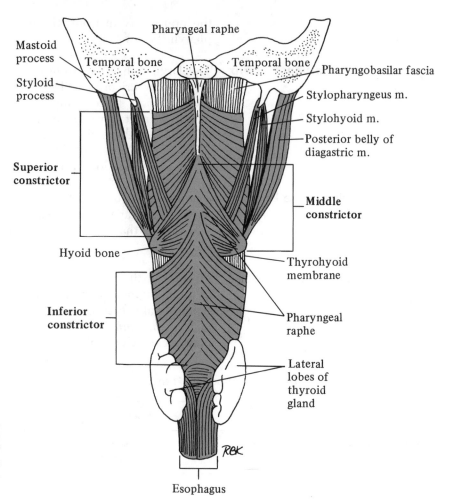

Figure 12.10 Pharyngeal Musculature. Posterior aspect of pharynx. Pharyngeal raphe is connective tissue in midline of posterior aspect of pharynx that unites right and left muscular elements of pharyngeal constrictors.

Circular Muscles   Muscles of the outer circular layer, which are named according to their location and function, include *superior, middle,* and *inferior constructors.* Contraction of the pharyngeal constructors forces the food bolus into the esophagus.

MICROSCOPIC ANATOMY   As described in Chapter 11, the pharyngeal wall consists of mucosal, fibrous, and muscular coats. The mucosal lining of the oropharynx and laryngopharynx contains stratified squamous epithelium, which withstands the abrasion of food boluses. The skeletal muscles are restricted to the oral and laryngeal portions of the pharynx. Their fiber direction — inner layer longitudinal and outer layer circular — is the reverse of that found in the remainder of the digestive tract.

BLOOD SUPPLY   The pharynx is supplied by the ascending pharyngeal artery plus branches of the facial and maxillary arteries. It is drained by the pharyngeal veins, which are tributaries of the internal jugular vein.

INNERVATION   The stylopharyngeus muscle is innervated by the glossopharyngeal nerve. The remainder of the pharyngeal muscles are innervated by branches of the glossopharyngeal and vagus nerves by way of the pharyngeal plexus.

FUNCTION   During the act of swallowing, the pharynx is elevated and expanded to receive the bolus of food from the fauces. Once the bolus has been received, the elevator muscles relax, the pharynx descends, and the bolus is grasped by the constrictor muscles. Sequential contraction of the constrictors conveys the bolus to the upper esophagus, where peristaltic activity begins.

The Esophagus

GROSS ANATOMY   The esophagus [Fig. 12.11(a)] is a flattened fibromuscular tube, about 10 inches long, that lies posterior to the trachea for most of its length. The esophagus begins at the level of the cricoid cartilage (vertebral level $C6$), descends through the neck and posterior part of the thoracic cavity, passes through the diaphragm, and enters the abdominal cavity. At vertebral level $T10$ it becomes continuous with the stomach.

MICROSCOPIC ANATOMY   The esophageal wall [Fig. 12.11(b)] is the first part of the digestive tube to conform to the basic plan. The mucous membrane contains a thick layer of stratified squamous epithelium that is able to withstand the friction of food boluses passing down the tube. Mucosal and submucosal glands secrete abundant mucus to lubricate the surface of the epithelium, which facilitates the passage of boluses. The muscular coat is a mixture of skeletal and smooth muscle fibers: the upper third contains only skeletal muscle fibers, the middle third contains both smooth and skeletal muscle fibers, and the lower third contains only smooth muscle fibers. Just proximal to its union with the stomach wall, the muscular coat of the esophagus thickens slightly to form the *cardiac sphincter,* a structure that regulates the passage of food into the stomach. The cervical and thoracic portions of the esophagus are covered with an adventitia, while the abdominal portion is covered with a serosa.

BLOOD SUPPLY   The esophagus is supplied by esophageal branches from the thoracic aorta, plus the inferior thyroid, left gastric, and left inferior phrenic arteries. It is drained by esophageal tributaries of the inferior thyroid, azygos, hemiazygos, and gastric (coronary) veins. Communication be-

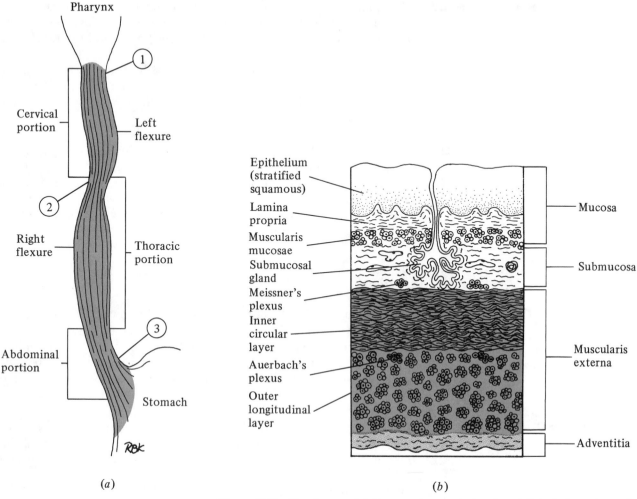

(a)

(b)

Figure 12.11 *Esophagus.* (a) Gross anatomy, anterior aspect. Note that esophagus is constricted at three points: (1) at its orgin, (2) at bifurcation of trachea, and (3) at esophageal hiatus of diaphragm. These constricted areas are vulnerable to trauma and may be perforated by swallowed foreign objects. (b) Microscopic anatomy. Segment of cross section through distal one-third of esophagus. Note that epithelium lining esophagus is nonkeratinized stratified squamous. Also note that muscularis mucosae and muscularis externa contain only smooth muscle fibers in distal one-third of esophagus.

tween the esophageal and gastric veins provides an alternative route for blood from the stomach to return to the heart if the portal vein is obstructed.

INNERVATION The esophagus receives sympathetic fibers from the sympathetic trunks and parasympathetic from the vagus nerve by way of the esophageal plexus.

FUNCTION Once a bolus of food enters the esophagus, peristaltic contraction waves pass the bolus down the esophageal tube until it reaches the cardiac sphincter. Under normal conditions the cardiac sphincter relaxes as a peristaltic wave approaches and permits the food bolus to enter the stomach.

The Stomach

GROSS ANATOMY The stomach (Fig. 12.12) is the dilated, J-shaped portion of the digestive tract. It lies in the upper left quadrant of the abdomen, just below the diaphragm and to the left of the midline of the body. The stomach is continuous with the esophagus above and the small intestine below.

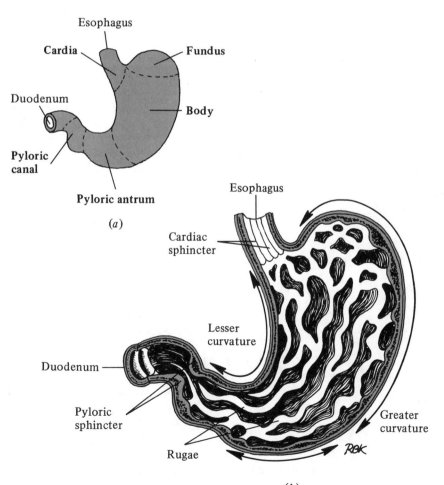

Figure 12.12 Stomach: Gross Anatomy. (a) External, anterior aspect. (b) Internal, anterior aspect. Stomach is cut open to permit visualization of its interior. Note appearance of gastric rugae, mucosal folds that run chiefly in longitudinal direction. Note also location and appearance of cardiac and pyloric sphincters.

Surfaces and Borders The stomach has two surfaces: an anterosuperior surface and a posterolateral surface. Its concave right border, or *lesser curvature*, is continuous with the right side of the esophagus, while its convex left border, or *greater curvature*, is continuous with the left side of the esophagus.

Divisions The stomach has four divisions: *cardia, fundus, body,* and *pyloris*. The cardia is the short constricted region adjacent to the gastroesophageal junction. The fundus is the saclike region that bulges above and to the left of the gastroesophageal junction. The body is the expanded central portion that lies between the fundus and the pyloris. The pyloris is the funnel-shaped terminal portion of the stomach. It is subdivided into the relatively wide pyloric antrum and the narrow pyloric canal, which is continuous with the duodenum. The pyloric opening is guarded by a well-developed muscular thickening, called the *pyloric sphincter,* that regulates the passage of stomach contents into the duodenum.

MICROSCOPIC ANATOMY The stomach wall [Fig. 12.13(a)] is lined with a relatively thick mucous membrane that contains simple columnar epithelium. This epithelium covers the luminal surface and extends into depressions called *gastric pits,* where it becomes continuous with the epithelium of the *gastric glands.* The gastric glands located in the fundus and body of the stomach are branched tubular mucosal glands [Fig. 12.13(b)]. They secrete *gastric juice,* a mixture of mucus, digestive enzymes, and hydrochloric acid. Mucosal glands in the cardia and pylorus secrete mucus, which helps protect the stomach and the portions of the digestive tract immediately above and below the stomach from the acidity of the gastric juice. The mucous membrane is loosely attached to the underlying muscular coat by the submucosa. When the muscular coat contracts or when the stomach is empty, this arrangement permits the gastric mucosa to be thrown into folds called *rugae.* The muscular coat of the stomach contains three rather than two layers of smooth muscle fibers, which helps the stomach to function as a churn. The outer coat of the stomach consists of a serosa composed of visceral peritoneum.

BLOOD SUPPLY The stomach is supplied by direct or indirect branches of the celiac trunk: the right and left gastric arteries (lesser curvature), the right and left gastroepiploic arteries (greater curvature), and the short gastric arteries (greater curvature). It is drained by direct or indirect tributaries of the portal vein that have the same names and distribution as their companion arteries.

INNERVATION The stomach receives sympathetic fibers by way of the celiac plexus and parasympathetic fibers by way of the vagus nerves.

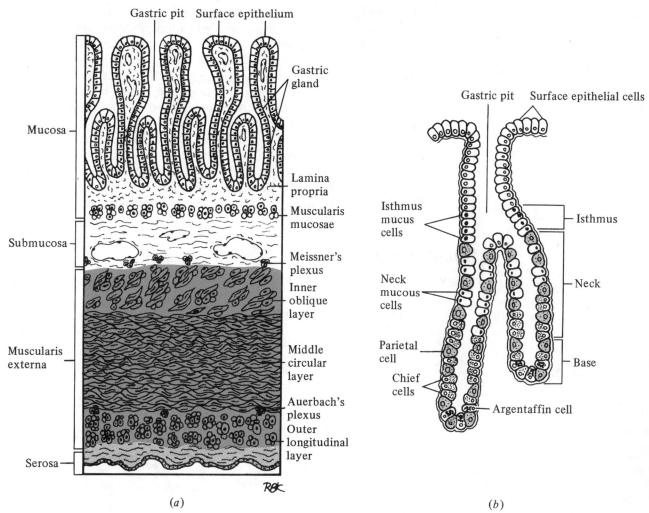

Figure 12.13 *Stomach.* (a) Microscopic anatomy. Segment of cross section through stomach wall. Note that muscularis externa contains oblique, circular, and longitudinal layers of smooth muscle fibers. (b) Gastric gland. Isthmus and neck mucous cells secrete mucus, parietal cells secrete HCl, chief (zymogenic) cells secrete digestive enzymes and intrinsic factor, and argentaffin cells produce and store serotonin.

FUNCTION   The stomach receives boluses of food from the esophagus and subjects them to mechanical and chemical action. The muscular coat of the stomach wall functions as a churn, mechanically breaking up large bits of food into smaller particles and mixing them with gastric juice to form a thick liquid called *chyme*. Food is broken down still further by the action of the gastric digestive enzymes, primarily *pepsins* and *rennin*. Hydrochloric acid provides the optimum pH for these enzymes to function. The pepsin enzymes break down meat and vegetable proteins into smaller molecules called peptones and proteoses; the enzyme rennin (which may be absent in humans) clots milk proteins. The activity of salivary amylase is inhibited by the acid gastric juice. Secretion of gastric juice is stimulated by parasympathetic impulses and by *gastrin,* a hormone secreted by the pyloric mucosa. When gastric digestion is completed, chyme is released into the first part of the small intestine. With the exception of water and alcohol, no absorption takes place in the stomach. Another function of the stomach, which is really its only essential function, is the secretion of intrinsic factor. Intrinsic factor promotes the absorption of Vitamin B_{12} from the ileum. If the stomach is removed, B_{12} shots must be given to compensate for the loss of intestinal B_{12}.

The Small Intestine

GROSS ANATOMY   The small intestine (Fig. 12.14) is the portion of the digestive tract where digestion is completed and absorption of nutrients occurs. It is about 21 feet long and is continuous with the stomach at the pyloris and the large intestine at the ileocecal junction. Most of the small intestine is relatively mobile and is suspended from the posterior

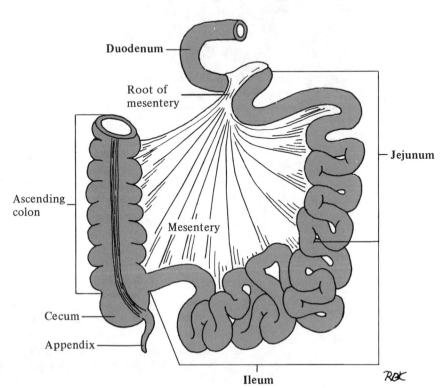

Figure 12.14 Small Intestine: Gross Anatomy. Regional subdivisions of small intestine are indicated.

abdominal wall by a fan-shaped fold of peritoneum called the *mesentery*. The attached portion of the mesentery is narrow, while the portion that encloses the small intestine is wide. This arrangement throws the small intestine into a series of regular loops and lessens the danger of it becoming kinked or twisted. Blood vessels, nerves, and lymphatics travel in the mesentery to reach the wall of the small intestine.

The small intestine is divided into three portions of unequal length: the *duodenum, jejunum,* and *ileum*.

Duodenum

The duodenum (Fig. 12.15) is the first and shortest part of the small intestine. It is about 1 foot long and extends from the pyloris to the duodenojejunal flexure. This part of the small intestine is horseshoe-shaped, with the open end of the horseshoe facing left and the head of the pancreas lying in the curve of the horseshoe. Except for a short portion adjacent to the pyloris, the duodenum is retroperitoneal. The duodenum receives secretions from the pancreas and gall bladder through a common duct, the *hepatopancreatic ampulla* (of Vater), which pierces the duodenal wall a few inches below the pyloris. These secretions are essential for completion of the digestive process.

Jejunum

The jejunum is the second part of the small intestine and is approximately 8 feet long. It begins at the duodenojejunal junction, just to the left of vertebra *L2*, and is located in the lower left quadrant of the abdominal cavity. It is hard to tell exactly where the jejunum becomes

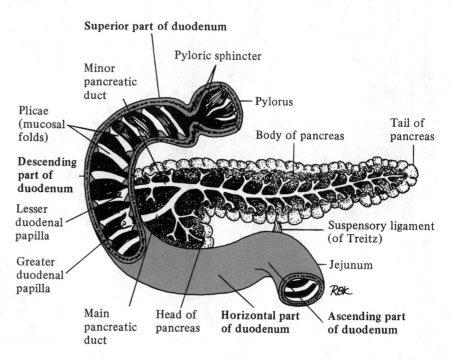

Figure 12.15 *Duodenum: Gross Anatomy.* Relationship of duodenum and pancreas is shown. Pancreas is partially dissected to expose its duct system. Part of duodenal wall is cut away to reveal openings of major and minor pancreatic ducts into duodenum. Note that duodenum has four parts: superior, descending, horizontal, and ascending.

continuous with the ileum, since the differences between these two parts of the small intestine are not that great. Generally speaking, the jejunum is wider, more vascular, and has a thicker wall than the ileum.

Ileum The ileum is the third and longest part of the small intestine. It is approximately 12 feet long and is located in the lower right quadrant

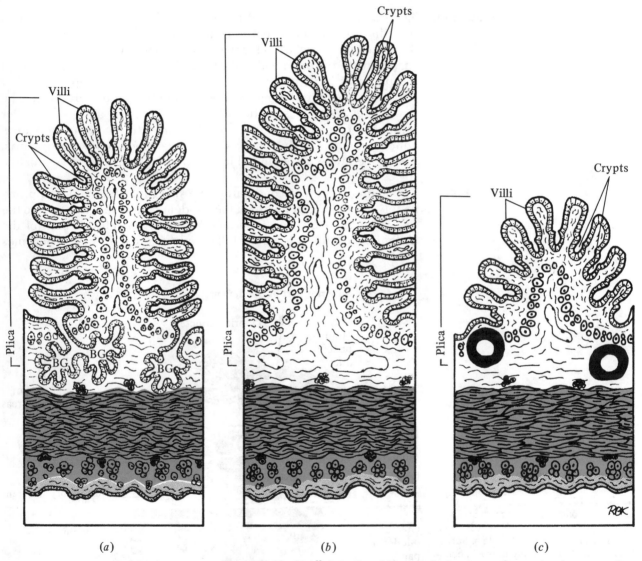

(a)   *(b)*   *(c)*

Figure 12.16 Small Intestine: Microscopic Anatomy. Segments of cross sections through three different regions of small intestine. (a) Duodenum. (b) Jejunum (c) Ileum. Compare height of plicae (mucosal folds) and shape of villi in different regions. Note presence of Brunner's glands *(BG)* in the submucosa of the duodenum.

of the abdomen and in the pelvis. In the right iliac fossa, it joins the first portion of the large intestine.

MICROSCOPIC ANATOMY

The wall of the small intestine (Fig. 12.16) contains certain structural modifications to facilitate the absorption of nutrients. The mucosa and underlying submucosa form *plicae circulares*, circular shelflike folds that project into the lumen of the small intestine. These folds, which reach their greatest height in the jejunum, help to increase the absorptive surface of the small intestine without increasing its length. *Intestinal villi* (Fig. 12.17) are tiny fingerlike projections of the mucosa that also help to increase the absorptive surface area. The connective tissue core of each villus contains a lymph capillary (lacteal) and many blood capillaries to receive the nutrients absorbed through its epithelial covering.

The intestinal mucosa contains a layer of simple columnar epithelium consisting of mucus-secreting goblet cells and intestinal absorptive cells. Each intestinal absorptive cell (Fig. 12.18) has so many microvilli protruding from its apical surface that they are visible under

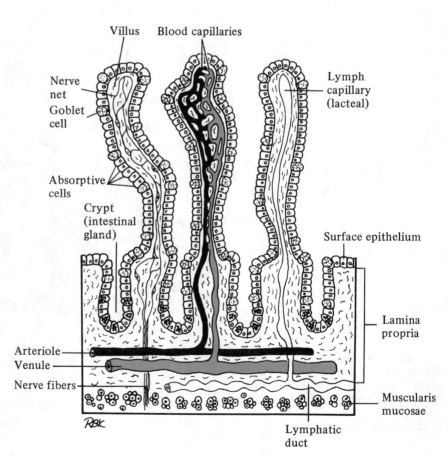

Figure 12.17 Intestinal Mucosa. Schematic diagram shows relationship between intestinal villi and glands (crypts of Lieberkuhn). Nerves, capillaries, and lacteal illustrated in different villi.

the light microscope as a *striate border*. The microvilli help increase the surface area of the cell for the absorption of nutrients.

Peyer's patches (aggregated lymph nodules) are found in the mucous membrane of the small intestine, mainly in the ileum. They always occur on the side of the intestine opposite the attachment of the mesentery.

Two types of glands are found in the small intestine: mucosal and submucosal. The mucosal glands, called *intestinal glands* or *crypts of Lieberkuhn,* are simple tubular glands that secrete *intestinal juice.* Intestinal juice contains a mixture of mucus and enzymes that help complete the digestive process. The submucosal glands, called the *duodenal glands* (of Brunner), are compound tubular glands located in

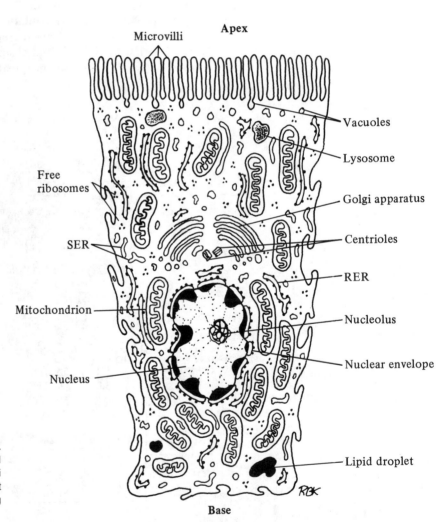

Figure 12.18 Intestinal Absorptive Cell. Note dense array of microvilli on apical (free) surface of cell. These microvilli increase surface area of cell in contact with food in intestinal lumen. Drawing based on electron microscopic view.

the duodenum. They are most numerous near the gastroduodenal junction, where their mucous secretion protects the mucosa from the acidity of entering chyme.

The small intestine has both an adventitia and a serosa. The duodenum, which is retroperitoneal, has an adventitia. The jejunum and ileum, which are enclosed in the free border of the mesentery, have a serosa.

BLOOD SUPPLY

The duodenum is supplied by the superior and inferior pancreatoduodenal arteries, while the jejunum and ileum are supplied by intestinal branches of the superior mesenteric artery. The branches of the superior mesenteric artery anastomose with one another in the mesentery to form arterial arcades. Arterial branches to the jejunum form only one or two arcades, while branches to the distal part of the ileum may form up to five sets of arcades before giving off terminal branches to the intestinal wall.

The veins draining the small intestine unite to form the superior mesenteric vein, one of the major tributaries of the portal vein.

INNERVATION

The small intestine receives sympathetic fibers from the sympathetic trunks by way of the celiac plexus and parasympathetic fibers from the vagus nerves by way of the superior mesenteric plexus.

FUNCTION

Two functions occur in the small intestine: digestion and absorption.

Digestion

The process of digestion, begun in the oral cavity and continued in the stomach, is completed in the small intestine. The bulk of intestinal digestion occurs in the duodenum. Intestinal digestion is brought about by the bile salts plus the various enzymes present in the pancreatic juices and on the surface of intestinal absorptive cells.

Bile Salts

Bile salts are produced in the liver, stored in the gall bladder, and delivered to the duodenum by way of the common bile duct. Bile salts emulsify fats so that lipases (fat-digesting enzymes) can attack them more readily.

Pancreatic Juice

Pancreatic juice is produced in the exocrine pancreas and delivered to the duodenum by way of the major and minor pancreatic ducts. Pancreatic juice contains the enzymes *trypsin, chymotrypsin, carboxypeptidase, pancreatic amylase,* and *pancreatic lipase.* Trypsin and chymotrypsin break down proteoses and peptones into small polypeptides. Carboxypeptidase cleaves amino acids off of polypeptides. Pancreatic amylase acts just like salivary amylase. Pancreatic lipase breaks down emulsified fats into glycerides, glycerol, and fatty acids. Pancreatic juice also contains sodium bicarbonate, which makes it highly alkaline. The alkalinity of the pancreatic juice counteracts the acidity of the chyme and provides the optimum pH for the pancreatic and intestinal enzymes to function.

Intestinal Juice

Intestinal juice is produced by the intestinal glands and released into the lumen of the small intestine. It consists mainly of watery mucus and salts. The digestive enzymes, *aminopeptidases, disaccharidases,* and *intestinal lipase,* are contained in the membranes of sloughed off intestinal absorptive cells and not in the juice itself as once thought. Aminopeptidases cleave amino acids off of polypeptides. Disaccharidases break down disaccharides such as sucrose (cane sugar) into monosaccharides (single sugars) such as glucose (blood sugar). Intestinal lipase has the same action as pancreatic lipase.

Absorption

Absorption is the process whereby the end products of digestion (amino acids, monosaccharides, glycerides, glycerol, and fatty acids) are transported across the intestinal epithelium and into the blood and lymph capillaries in the underlying connective tissue. Structures providing for a large intestinal absorptive area include the plicae circulares, the intestinal villi, and the microvilli on the absorptive cells. The bulk of nutrient absorption occurs in the jejunum, where the plicae are tall and branched and the villi are long and slender.

The Large Intestine

GROSS ANATOMY

The large intestine (Fig. 12.19) is the terminal portion of the digestive tract. It is the portion in which undigested residue is converted into feces and expelled from the body. Approximately 5 feet long, the large intestine is continuous with the small intestine at the ileocecal junction and with the exterior of the body at the anus. The large intestine is subdivided into the following parts: the *cecum* and *appendix, colon, rectum,* and *anal canal.*

Cecum and Appendix

The cecum [Fig. 12.20(*a*)] is the first part of the large intestine. It is a large blind pouch, about 2 inches long, that lies in the right iliac fossa and becomes continuous with the ascending colon at the level of the ileocecal junction. A slitlike valve, the *ileocecal valve,* guards the opening of the ileum into the cecum. This valve prevents the backflow of fecal matter into the small intestine.

The appendix [Fig. 12.20(*b*)] is a slender blind tube that arises from the apex of the cecum. It averages about $3\frac{1}{2}$ inches in length and is most frequently found behind the cecum or lying over the pelvic brim. The appendix is entirely covered by peritoneum and fixed in position by a triangular extension of the mesentery. The wall of the appendix contains a large number of lymph nodules. Because it contains so

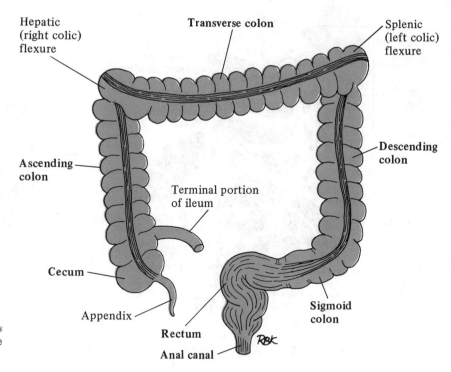

Hepatic (right colic) flexure

Transverse colon

Splenic (left colic) flexure

Ascending colon

Descending colon

Terminal portion of ileum

Cecum

Appendix

Rectum

Sigmoid colon

Anal canal

Figure 12.19 Large Intestine: Gross Anatomy. Regional subdivisions of large intestine are indicated.

much lymphoid tissue, many anatomists consider the appendix to be a lymphoid organ like the tonsils. The lumen of the appendix extends the whole length of the tube and communicates with the cecum distal to the ileocecal opening. Since the appendix is in communication with the cecum, fecal matter can accumulate in the appendix and cannot be expelled by peristalsis because the tube is blind-ended. Trapped fecal material frequently causes *appendicitis* (inflammation of the appendix). Appendicitis is a potentially dangerous condition, for rupture of an inflamed appendix can lead to *peritonitis*. For this reason an inflamed appendix is usually removed surgically before it has the chance to rupture.

Colon The colon (Fig. 12.21) forms the longest part of the large intestine. It is subdivided into four parts: the *ascending, transverse, descending,* and *sigmoid colon.*

Ascending Colon The ascending colon is approximately 6 inches long. It begins at the level of the ileocecal valve and ascends through the right lumbar and hypochondriac regions of the abdomen to the undersurface of the liver. There it makes a sharp bend to the left as the *right colic (hepatic) flexure* and becomes continuous with the transverse colon. Its poste-

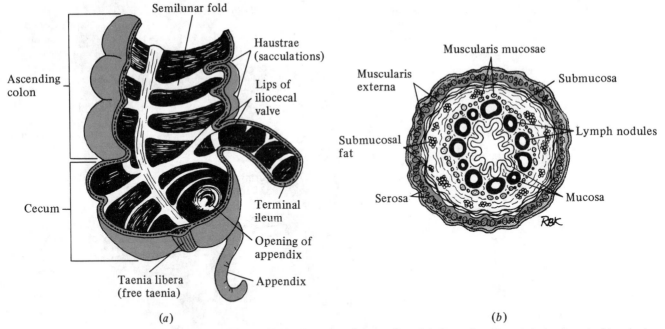

Figure 12.20 *Cecum and Appendix.* (a) *Gross Anatomy.* Anterior aspect of terminal ileum, appendix, cecum, and lower ascending colon. Anterior walls of these structures (excluding appendix) are cut away to expose ileocecal junction and opening of appendix. (b) *Appendix: Microscopic Anatomy.* Cross section through appendix wall. Note large number of lymph nodules in mucosa of appendix.

rior surface is attached to the posterior abdominal wall, so it has no mesentery.

Transverse Colon   The transverse colon is approximately 20 inches long. From the undersurface of the liver, it arches across the abdominal cavity to the undersurface of the spleen in the left hypochondriac region. There it makes a sharp downward bend as the *left colic (splenic) flexure* and becomes continuous with the descending colon. The transverse colon is completely covered by peritoneum and is suspended by a mesentery, the *transverse mesocolon,* that is connected to the inferior border of the pancreas. The hepatic and splenic flexures of the colon are firmly attached to the posterior abdominal wall, which helps prevent the transverse colon from sagging or kinking.

Descending Colon   The descending colon is approximately 12 inches long. It descends through the left hypochondriac and lumbar regions of the abdomen to the left iliac fossa then turns medially to become continuous with the sigmoid colon at the pelvic brim. Since the descending colon is attached to the posterior abdominal wall, it has no mesentery.

Sigmoid Colon   The sigmoid colon is approximately 16 inches long. It extends from the pelvic brim to the rectocecal junction in the pelvis. Within the pelvic cavity it makes an S-shaped curve, hence the name sigmoid. The sigmoid colon is completely covered by peritoneum and is attached to the posterior pelvic wall by an extensive mesentery, the *mesosigmoid.*

Rectum   The rectum (Fig. 12.22) is a 6-inch-long portion of the large intestine that lies in the pelvic cavity. It is continuous with the sigmoid colon

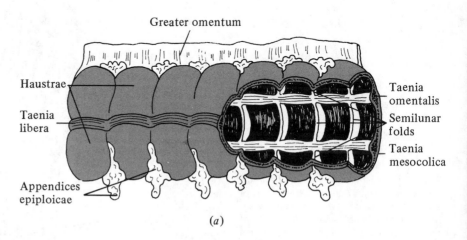

(a)

Figure 12.21   Colon. (a) Gross anatomy. Segment of transverse colon with part of anterior wall cut away to permit visualization of internal structures. Note structures that characterize colon: haustrae (sacculations), semilunar mucosal folds, taenia coli (longitudinal muscle bands), and appendices epiploica (fat-filled sacs). (b) Microscopic anatomy. Segment of cross section through wall of colon. Note that colon has no plicae and no villi. Mucosa does contain intestinal glands (crypts of Lieberkuhn), which have large number of goblet cells. Also note that smooth muscle fibers of outer layer of muscularis externa are grouped into three separate longitudinal bundles, one of which is shown in this section.

(b)

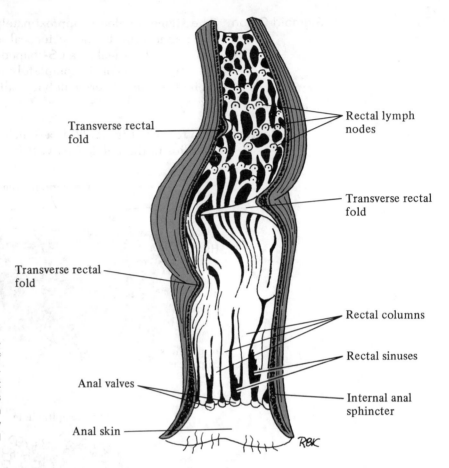

Transverse rectal fold

Rectal lymph nodes

Transverse rectal fold

Transverse rectal fold

Rectal columns

Rectal sinuses

Anal valves

Internal anal sphincter

Anal skin

Figure 12.22 Rectum and Anal Canal: Gross Anatomy. Anterior aspect. Part of anterior wall is cut away to permit visualization of internal structures. Note that mucosa of anal canal is thrown into series of longitudinal folds (anal columns). Each anal column generally contains an artery and vein. Between anal columns are small pockets called anal sinuses.

above and the anal canal below. Just above its union with the anal canal, the rectum becomes dilated. Since the rectum is attached to the posterior abdominal wall, it has no mesentery.

Anal Canal   The anal canal (Fig. 12.22) is the short terminal portion of the large intestine. It is approximately 1½ inches long and extends from the rectoanal junction to the *anus* (anal opening). The anal canal has no peritoneal covering; it is supported by the levator ani muscles and is surrounded by the *internal* and *external anal sphincters.* The internal sphincter is a terminal thickening of the smooth muscle of the intestinal wall and is under involuntary control. The external sphincter consists of skeletal muscle surrounding the anus and is under voluntary control. The mucous membrane of the anal canal is thrown into approximately eight longitudinal folds. These mucosal folds enclose a plexus of veins and are known as the *rectal columns.* When the rectal columns become swollen and inflamed, they are called *hemorrhoids* or *piles*. The rectal columns are separated by furrows called rectal sinuses, whose distal ends unite with one another to form the pocketlike anal valves.

MICROSCOPIC ANATOMY

The wall of the large intestine differs markedly in structure from that of the small intestine. Plicae circulares are absent and the mucous membrane contains no villi. The large intestine is lined with a simple columnar epithelium that contains many goblet cells and produces a great deal of mucus. In the anal canal the epithelium changes from simple columnar to stratified squamous, which is better able to withstand the abrasion of fecal material. Numerous intestinal glands are present, but they produce mucus rather than digestive enzymes. Another important structural difference is found in the muscular coat. The outer longitudinal layer is incomplete and consists of three ribbonlike bands, called *taenia coli,* that extend from the root of the appendix to the rectum. Since the taenia are shorter than the inner circular layer of smooth muscle, they create sacculations *(haustrae)* in the intestinal wall. Certain parts of the large intestine do not have a complete serosal coat because they are attached to the posterior abdominal or pelvic walls. Small fat-filled peritoneal sacs called *appendices epiploicae* hang down from the serosa of the large intestine except in the regions of the cecum and rectum. These structures provide an external means for identifying the colon.

BLOOD SUPPLY

The cecum, appendix, ascending colon, and proximal half of the transverse colon are supplied by branches of the superior mesenteric artery; the distal half of the transverse colon plus the descending and sigmoid portions of the colon are supplied by branches of the inferior mesenteric artery. The rectum and anal canal are supplied by the superior, middle, and inferior rectal (hemorrhoidal) arteries.

Tributaries of the superior mesenteric vein drain the portions of the colon supplied by the superior mesenteric artery, while tributaries of the inferior mesenteric vein drain those portions supplied by the inferior mesenteric artery. The rectum and anal canal are drained by the rectal veins, which are companions of the rectal arteries. Anastomotic communications between the superior rectal vein, which drains into the portal system, and the middle and inferior rectal veins, which drain into the systemic system, provide an alternative route for blood from the intestines to return to the heart if the portal vein is obstructed.

INNERVATION

Except for the external anal sphincter, the large intestine is innervated by the autonomic nervous system. It receives sympathetic fibers by way of the superior and inferior mesenteric plexuses and parasympathetic fibers from the sacral portion of the spinal cord by way of the pelvic splanchnic nerves.

FUNCTION The major functions of the large intestine are the absorption of water from chyme to form feces and the storage of fecal matter until it can be expelled from the body. Minor functions include the formation of Vitamin K and vitamins of the B complex by intestinal bacteria.

About 500 milliliters of fluid chyme pass from the small intestine into the large intestine each day. Among other things, it contains water, electrolytes, and the undigested residue of food. As the chyme passes through the proximal half of the colon, most of the water and electrolytes are absorbed through the mucosa. Mixing movements, caused by contractions of the muscular coat, promote absorption by bringing the chyme in contact with the mucosa. Progressive dehydration of chyme as it passes through the colon reduces its fluid content to about 100 milliliters and converts it into a semisolid mass called *feces*. Feces contain approximately 75 percent water and 25 percent solid matter. The solid matter includes undigested fiber (such as cellulose), dead intestinal bacteria, fat, bile pigments, and dead

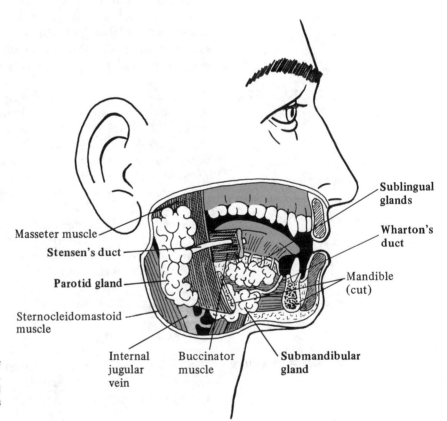

Figure 12.23 Salivary Glands: Gross Anatomy. Right side of face with cheek, part of mandible, and part of floor of oral cavity cut away to expose salivary glands and their ducts.

epithelial cells. The water content of the feces depends a great deal on colonic motility: a sluggish colon causes increased water absorption and hard feces *(constipation),* while a hyperactive colon causes decreased water absorption and loose feces *(diarrhea).* The amount of feces formed depends on the type and amount of food ingested. Ingestion of a lot of fiber, such as bran, increases the amount of the feces and decreases the time the feces spend in the colon.

Fecal matter is stored in the sigmoid colon until mass movements of the colonic musculature propel the feces into the rectum. As the rectum becomes distended with feces, the stretch receptors in the rectal wall activate the defecation reflex. Motor impulses stimulate peristaltic contractions in the descending and sigmoid portions of the colon, which force the feces toward the anus, plus relaxation of the anal sphincters, which permits expulsion of feces from the body.

Part B. Accessory Organs of Digestion

Several organs lie external to the digestive tract but are connected to it by ducts. These organs either produce or store products that aid the process of digestion. Accessory digestive organs include the salivary glands, the pancreas, the liver, and the gallbladder.

The Salivary Glands

The major salivary glands consist of three pairs of compound acinar or tubuloacinar glands associated with the mandible: *parotid, submandibular,* and *sublingual.* The secretions of these glands are released into the oral cavity, where they contribute to the formation of saliva.

PAROTID GLANDS
Gross Anatomy

The parotid glands are the largest of the major salivary glands. Each parotid gland (Fig. 12.23) lies below and in front of the external ear. It wraps around the posterior border of the ramus of the mandible and partially covers the masseter muscle. The main excretory duct (Stensen's duct) runs forward across the masseter muscle, pierces the buccinator muscle, and opens into the vestibule of the oral cavity opposite the second upper molar tooth.

Microscopic Anatomy

The secretion of the parotid gland is entirely serous (watery). It is produced in serous acini and released into an elaborate duct system that joins the main excretory duct [Fig. 12.24(*a*)]. The secretory acini and their ducts are embedded in a loose connective tissue stroma and enclosed in a fibrous capsule whose inward projections subdivide the gland into lobes and lobules.

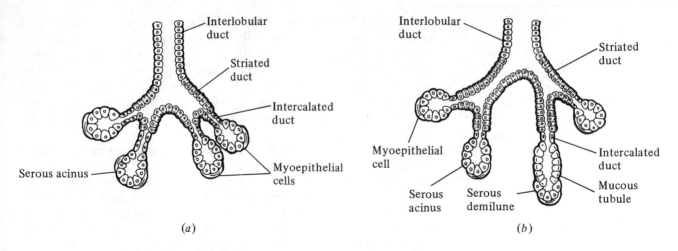

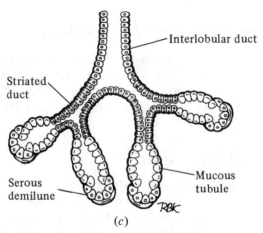

Figure 12.24 Salivary Glands: Microscopic Anatomy. (a) Parotid gland. Secretory units consist of serous acini; intercalated ducts are present. Myoepithelial cells, when stimulated, squeeze secretions out of secretory units. (b) Submandibular gland. Secretory units consist of serous acini plus mucous tubules with serous demilunes; intercalated ducts present but much-reduced. (c) Sublingual gland. Secretory units consist of mucous tubules with serous demilunes; intercalated ducts are absent.

Blood Supply   The parotid gland is supplied by branches of the external carotid artery, which is embedded in the gland, and is drained by tributaries of the external jugular vein.

Innervation   All the major salivary glands are innervated by the autonomic nervous system. The parotid gland receives sympathetic fibers from the superior cervical ganglion by way of the external carotid plexus and parasympathetic fibers from the otic ganglion by way of the mandibular division of the trigeminal nerve. Sympathetic impulses inhibit secretion, while parasympathetic impulses stimulate copious production of a thin serous secretion.

SUBMANDIBULAR GLANDS
Gross Anatomy

The submandibular glands are the second largest of the major salivary glands. Each submandibular gland is located in the floor of the oral cavity beneath the body of the mandible. The main excretory duct (Wharton's duct) passes forward along the root of the tongue and opens into the oral cavity on the side of the frenulum of the tongue.

Microscopic Anatomy

The secretion of the submandibular gland is part serous and part mucus. It is produced in serous acini and mucous tubules and released into an elaborate duct system that joins the main excretory duct [Fig. 12.24(b)]. The glandular parenchyma is enclosed in a thin capsule of fibrous connective tissue.

Blood Supply

The submandibular gland is supplied by branches of the facial and lingual arteries and drained by tributaries of the corresponding veins.

Innervation

The submandibular gland receives sympathetic fibers from the superior cervical ganglion and parasympathetic fibers from the submandibular ganglion by way of the mandibular division of the trigeminal nerve. Sympathetic impulses stimulate production of a thick mucous secretion, while parasympathetic impulses stimulate production of a thin serous secretion.

SUBLINGUAL GLANDS
Gross Anatomy

The sublingual glands are the smallest of the major salivary glands. Each sublingual gland consists of a mass of glandular tissue that lies on the floor of the oral cavity lateral to the duct of the submandibular gland. Each glandular mass has its own excretory duct, which opens onto a mucosal fold at the side of the frenulum of the tongue.

Microscopic Anatomy

The secretion of the sublingual gland is mostly mucus. It is produced primarily in mucous tubules capped with serous-secreting cells (serous demilunes) and is released into a simplified duct system terminating in an excretory duct [Fig. 12.24(c)]. These glandular masses are embedded in the floor of the oral cavity and lack a distinct capsule.

Blood Supply

The blood supply of the sublingual gland is the same as that of the submandibular gland.

Innervation

The innervation of the sublingual gland is the same as that of the submandibular gland. Parasympathetic impulses stimulate production of a thick mucous secretion.

The Pancreas

The pancreas is a large digestive and endocrine gland located in the abdominal cavity. Because of its structural similarity to the parotid gland, many anatomists consider it to be a displaced salivary gland.

GROSS ANATOMY
Location

The pancreas [Fig. 12.25(a)] is a retroperitoneal organ, about 6 inches long, that extends across the posterior abdominal wall in the epigastric and left hypochondriac regions. It lies behind the stomach and in front of the inferior vena cava, aorta, and left kidney.

Divisions

The pancreas is a soft yellowish lobulated gland better known to butchers as the "sweetbread." It is divided into four parts: *head, neck, body,* and *tail.* The head is the widest part of the gland; it lies in the curve of the duodenum. The neck is the constricted part to the left of the head. The body is the long pyramidal part of the gland that lies

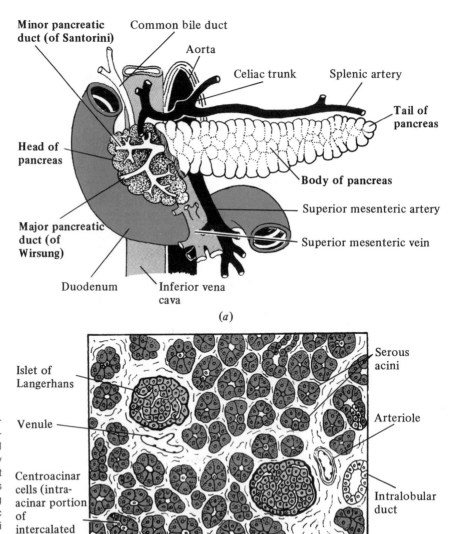

Figure 12.25 Pancreas. (a) Gross anatomy. Diagram shows relationship of pancreas, duodenum, and major abdominal blood vessels. Part of pancreas is cut away and its excretory ducts exposed. Note that superior mesenteric artery and vein pass between body of pancreas and ascending portion of duodenum. (b) Microscopic anatomy. Section shows pancreatic acini (exocrine) and pancreatic islets (endocrine).

behind the stomach. The tail or left extremity is the narrow tapering part of the gland that touches the medial border of the spleen.

Ducts Two excretory ducts drain the pancreas. The *major pancreatic duct* (of Wirsung) extends the length of the pancreas from tail to neck. It emerges from the neck and unites with the common bile duct to form the hepatopancreatic ampulla (of Vater). The ampulla of Vater pierces the duodenal wall and opens into the lumen of the descending portion of the duodenum. The *minor pancreatic duct* (of Santorini) drains the head of the pancreas. It either unites with the major duct or opens separately into the duodenal lumen.

MICROSCOPIC ANATOMY The pancreas [Fig. 12.25(*b*)] is a solid glandular organ that contains both endocrine and exocrine tissue. The endocrine pancreas consists of scattered clusters of hormone-secreting epithelial cells called *pancreatic islets* (of Langerhans). The exocrine pancreas is a compound acinar gland whose serous acini, which resemble those seen in the parotid gland, secrete pancreatic juice. Pancreatic juice is released into a system of ducts that unite with either the major or minor excretory ducts. The pancreatic parenchyma is enclosed in an ex-

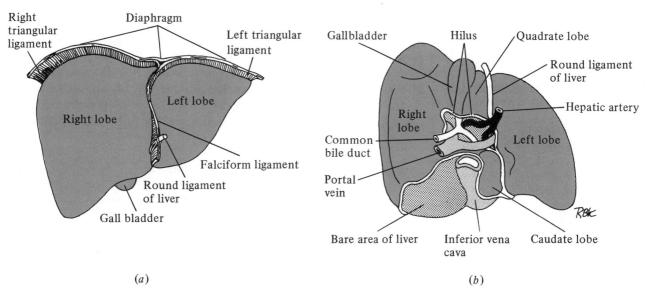

(a) *(b)*

Figure 12.26 Liver. (*a*) Gross anatomy, diaphragmatic surface. Note that falciform ligament separates large right lobe from smaller left lobe. Within falciform ligament is fibrous cord (round ligament of liver) that is remnant of umbilical vein. (*b*) Gross anatomy, visceral surface. Note that inferior vena cava separates caudate lobe from right lobe and that gallbladder, portal vein, hepatic artery, and common bile duct separate quadrate lobe from right lobe. Also note location of base area of liver (area where liver is not covered with visceral peritoneum).

tremely thin connective tissue capsule whose inward projections subdivide the gland into lobes and lobules.

BLOOD SUPPLY The pancreas receives a rich blood supply from branches of the splenic artery plus the pancreatoduodenal branches of both the hepatic and superior mesenteric arteries. It is drained by tributaries of the splenic and superior mesenteric veins.

INNERVATION The pancreas is innervated by the autonomic nervous system. It receives sympathetic fibers from the sympathetic trunks and parasympathetic fibers from the vagus nerves by way of the celiac plexus. Parasympathetic impulses stimulate secretion of pancreatic juice; however, hormonal stimulation of the pancreas produces more pancreatic juice than neural stimulation.

FUNCTION The pancreas produces both endocrine and exocrine secretions. The endocrine secretions, which will be described in Chapter 13, are important in regulating carbohydrate metabolism. The exocrine secretion contains a mixture of enzymes that are important in completing intestinal digestion. The entry of acidic chyme into the duodenum stimulates the intestinal mucosa to produce two hormones, *secretin* and *cholecystokinin-pancreozymin,* that stimulate the pancreas to secrete copious amounts of enzyme-rich pancreatic juice.

The Liver

GROSS ANATOMY The liver, which weighs about $3\frac{1}{2}$ pounds in the adult, is the largest gland in the human body. A healthy liver has a deep reddish brown color, due to its great vascularity, and is firm but not hard in texture.

Location The liver lies in the right hypochondriac and epigastric regions of the abdomen and is in contact with the undersurface of the diaphragm.

Surfaces The liver has two surfaces: *diaphragmatic* and *visceral.*

Diaphragmatic Surface The diaphragmatic surface of the liver [Fig. 12.26(a)] is convex and fits the curvature of the diaphragm. A large part of the diaphragmatic surface is in direct contact with the undersurface of the diaphragm. Since this part is not covered by peritoneum, it is referred to as the "bare area" of the liver.

Visceral Surface The visceral surface of the liver [Fig. 12.26(b)] is concave and marked by impressions of neighboring organs such as the right kidney, transverse colon, stomach, and duodenum. It is almost completely covered by peritoneum except where the gallbladder is attached to the liver and where the porta hepatis (the hilus of the liver) is located.

Lobes The liver consists of four partially defined lobes. The *right lobe* is the largest of the four. It is separated from the much smaller *left lobe* by the falciform ligament on the diaphragmatic surface and the longitudinal fissure on the visceral surface. The *caudate* and *quadrate lobes* are located on the visceral surface of the liver. They are separated from one another by the *porta hepatis*. The porta hepatis is a deep transverse fissure on the visceral surface of the liver that transmits the hepatic artery, portal vein, bile ducts, nerves, and lymphatics. The caudate lobe is posterior to the porta and lies between the groove for the inferior vena cava and the fissure for the ligamentum venosum (the remnant of the ductus venosus). The quadrate lobe is anterior to the porta and lies between the groove for the gallbladder and the fissure for the ligamentum teres (the remnant of the umbilical vein).

Ligaments The liver is attached to the undersurface of the diaphragm and to neighboring organs by peritoneal reflections (ligaments). The *coronary* and *triangular ligaments* are peritoneal reflections from the undersurface of the diaphragm to the diaphragmatic surface of the liver. These ligaments help attach the liver to the diaphragm and enclose the bare area of the liver. The *falciform ligament* attaches the liver to the anterior abdominal wall.

MICROSCOPIC ANATOMY The liver is a compound tubular gland whose parenchyma is enclosed in a thin connective tissue capsule called *Glisson's capsule*. The connective tissue of the capsule enters the substance of the liver at the porta hepatis along with the blood vessels. It follows the branches of the blood vessels throughout the substance of the liver, partitioning the hepatic parenchyma into lobules. These lobules are poorly demarcated in humans but are well demarcated in animals such as the pig.

The *hepatic lobule* (Fig. 12.27) is the functional unit of the liver. It is roughly hexagonal (six-sided) in shape and separated from adjacent lobules by connective tissue. Where several lobules join, three small vessels can be seen: a portal venule, a hepatic arteriole, and a bile ductule. These structures and the connective tissue that encloses them constitute a *portal canal.* In the center of each lobule is a central vein. Radiating inward from the periphery of the lobule to the central vein are a number of hepatic sinusoids, thin-walled vessels that transport blood from the portal canals to the central vein. The sinusoids are surrounded by cords of hepatocytes (liver cells). The hepatocyte (Fig. 12.28) is a large many-sided cell whose location permits it to remove nutrients from and add substances to sinusoidal blood.

BILE DUCTS Bile is produced in the liver, stored in the gallbladder, and utilized in the small intestine. A system of ducts transports bile from its site of production to the sites of storage and utilization.

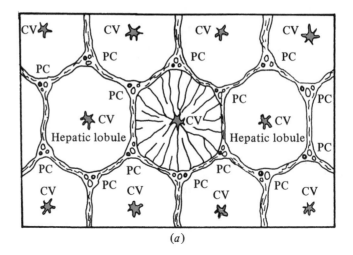

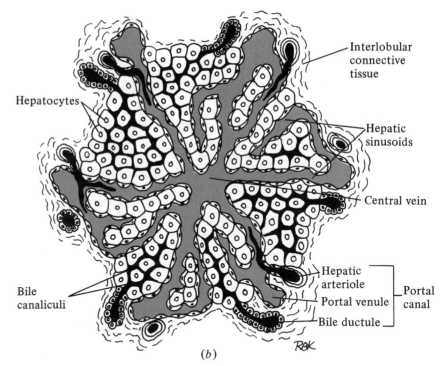

Figure 12.27 *Liver: Microscopic Anatomy.* (a) Schematic representation of segment of liver tissue shows territories of hepatic lobules, which are separated from one another by interlobular connective tissue and portal canals *(PC)*. Each hepatic lobule contains a central vein *(CV)* into which blood drains (from vessels in portal canals). (b) Schematic representation of single hepatic lobule. Note that blood flows centripetally (from vessels in portal canal to central vein), while newly synthesized bile flows centrifugally (from bile canaliculi in lobule to bile ductule in portal canal).

Intrahepatic Bile Ducts   Bile is released from hepatocytes into *bile capillaries*, tiny channels formed by the cell membranes of adjacent hepatocytes in a cord. The bile capillaries unite with *bile ductules* in the portal canals, and these unite with a series of larger and larger ducts that ultimately form the *right* and *left hepatic ducts*. These ducts, which drain the right and left liver lobes, emerge from the substance of the liver at the porta hepatis.

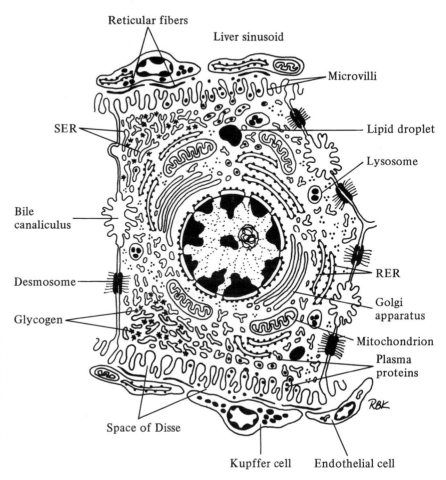

Reticular fibers

Liver sinusoid

Microvilli

SER

Lipid droplet

Lysosome

Bile
canaliculus

RER

Desmosome

Golgi
apparatus

Glycogen

Mitochondrion

Plasma
proteins

Figure 12.28 Hepatocyte. Schematic representation. Note that newly synthesized plasma proteins are released into space of Disse (space between hepatocyte and liver sinusoid) and pass through gaps in sinusoidal wall to enter peripheral circulation. Also note presence of Kupffer cells (fixed macrophages) in sinusoidal wall. Drawing based on electron microscopic view.

Space of Disse

Kupffer cell   Endothelial cell

Extrahepatic Bile Ducts   The extrahepatic bile duct system is illustrated in Fig. 12.29. After emerging from the liver, the right and left hepatic ducts unite to form the *common hepatic duct.* The common hepatic duct extends downward and unites with the *cystic duct,* the duct that transports bile to and from the gallbladder, to form the *common bile duct.* The common bile duct extends down to the level of the duodenum, where it unites with the major pancreatic duct to form the ampulla of Vater.

BLOOD SUPPLY   The liver is another organ that receives blood from two sources. It receives oxygenated blood from the right and left hepatic arteries and nutrient-rich deoxygenated blood from the portal vein. These vessels branch repeatedly in the substance of the liver until their smallest branches reach the periphery of the hepatic lobules and discharge their blood into the sinusoids. The radially oriented sinusoids converge on the center of each lobule, where they unite with a central vein.

The central veins form the first part of the venous drainage system

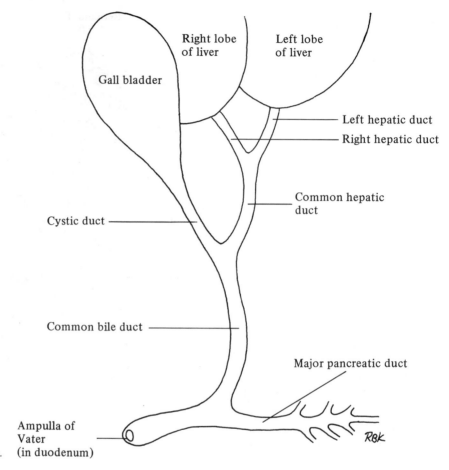

Figure 12.29 Extrahepatic Bile Ducts.

of the liver. At the base of each lobule, the central vein unites with a sublobular vein. The sublobular veins in turn unite with larger and larger veins that unite at last with the hepatic veins. Three hepatic veins converge on and enter the inferior vena cava at the posterior aspect of the liver.

INNERVATION   The liver is innervated by the autonomic nervous system. It receives sympathetic fibers from the sympathetic trunks and parasympathetic fibers from the left vagus nerve. There is no evidence of any neural regulation of bile secretion.

FUNCTION   The liver performs a large number of functions, some of which are listed next.

1. *Synthesis of Bile.*   Bile is a yellowish brown fluid containing bile salts, bile pigments, cholesterol, and water. Bile salts are utilized in the small intestine for the emulsification of fats.

2. *Formation and Storage of Glycogen.* Glucose is removed from the bloodstream, converted into glycogen, stored, reconverted back to glucose, and released into the bloodstream when needed by the body.

3. *Synthesis of Blood-clotting Factors.* Prothrombin and other blood-clotting factors are synthesized by the liver.

4. *Synthesis of Plasma Proteins.* With the exception of some of the gamma globulins, the plasma proteins are synthesized by the liver.

5. *Urea Formation.* The ammonia formed from the cellular breakdown of amino acids is converted to urea in the liver. Without this conversion the concentration of ammonia in the blood would rapidly reach the level where coma and death occur.

6. *Detoxification.* The liver detoxifies certain drugs, metabolic by-products, and other toxic substances and excretes them with the bile.

The Gallbladder

GROSS ANATOMY

The gallbladder [Fig. 12.30(*a*)] is a hollow muscular sac that lies in a groove on the visceral surface of the liver. It is divided into three regions: *fundus, body,* and *neck.* The fundus is directed forward, downward, and to the right. It usually projects slightly beyond the anterior margin of the liver. The body and neck are directed backward, upward, and to the left. The neck is connected to the cystic duct, which unites with the common hepatic duct.

MICROSCOPIC ANATOMY

The wall of the gallbladder [Fig. 12.30(*b*)] is composed of three coats: *mucosa, muscularis,* and *serosa/adventitia.* The mucosa contains a simple columnar epithelium. The muscularis consists of three ill-defined layers of smooth muscle fibers. The portion of the gallbladder in contact with the liver has an adventitia, while its free surface has a serosa for an outer coat.

BLOOD SUPPLY

The gallbladder receives blood from the cystic artery, a branch of the common hepatic artery. It is drained by the cystic vein, a tributary of the portal vein.

INNERVATION

The gallbladder is innervated by the autonomic nervous system. It receives sympathetic fibers from the sympathetic trunks and parasympathetic fibers from the vagus nerves by way of the hepatic plexus.

FUNCTION

The gallbladder stores bile until it is needed in the duodenum. Within the gallbladder, water is absorbed from the bile through the mucosa, which concentrates the bile about fivefold. Bile concentration makes it possible for a maximum amount of bile to be stored in a minimum amount of space. Under abnormal conditions cholesterol may precipitate out of the concentrated bile and form *gallstones*.

The release of bile from the gallbladder is dependent on the secretion of an intestinal hormone. When fat-rich chyme enters the duodenum, the duodenal mucosa secretes *cholecystokinin*. Cholecystokinin stimulates the gallbladder to contract and release bile into the common bile duct.

Diseases and Disorders of the Digestive System

DISEASES AND DISORDERS OF THE ORAL CAVITY

Inflammation. *Stomatitis* refers to inflammation of the oral mucosa, while *gingivitis* refers specifically to inflammation of the gums (gingivae).

Dental Caries. Dental caries, or tooth decay, refers to erosion of the calcified tissues of a tooth. It results from a number of factors including a carbohydrate-rich diet, acid-producing bacteria in the oral cavity, and improper tooth care. When decay progresses into the pulp chamber, the pulp becomes vulnerable to infection. Infection of the pulp causes destruction of the tooth's nerve supply and ultimate death of the tooth. Tooth decay is a significant cause of tooth loss in young adults.

Pyorrhea. Pyorrhea is a pus-producing inflammatory disease of the tooth socket that causes destruction of alveolar bone and loosening of the teeth. It is a significant cause of tooth loss in middle-aged adults.

DISEASES AND DISORDERS OF THE ESOPHAGUS

Esophagitis. Esophagitis ("heartburn") refers to inflammation of the esophageal mucosa. Causes include bacterial infection, ingestion of corrosive substances, or the backflow of gastric juice into the esophagus.

Esophageal Stricture. Inflammation of the esophagus may be followed by the formation of scar tissue that narrows the esophageal lumen. A common cause of esophageal stricture is the swallowing of a corrosive substance such as lye. Most corrosive substances are swallowed by small children, who are unaware of their harmful effects.

Cardiospasm. Cardiospasm is a condition in which the cardiac sphincter fails to relax as a bolus of food approaches the stomach. As a result, passage of food into the stomach is slowed down or completely blocked.

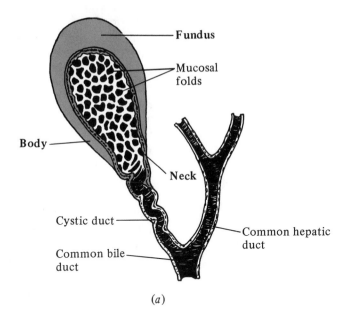

Fundus

Mucosal folds

Body

Neck

Cystic duct

Common hepatic duct

Common bile duct

(a)

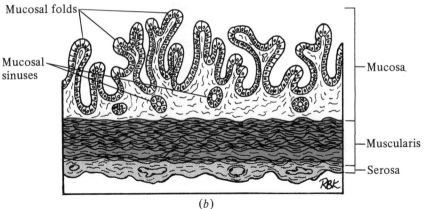

Mucosal folds

Mucosal sinuses

Mucosa

Muscularis

Serosa

(b)

Figure 12.30 Gallbladder. (a) Gross anatomy. Gallbladder is separated from liver and portion of its wall plus walls of related ducts are cut away to permit visualization of internal structures. Note presence of mucosal folds in gallbladder. (b) Microscopic anatomy. Segment of section through gallbladder wall shows its various coats. On its superior surface gallbladder wall would have adventitia instead of serosa because it is not covered by visceral peritoneum.

Hiatus Hernia. Hiatus hernia refers to protrusion of abdominal viscera through a defect in the esophageal hiatus of the diaphragm. This condition is either congenital or acquired and occurs most frequently in older or obese people. People with hiatus hernia experience epigastric discomfort and indigestion, especially if they lie down after eating.

DISEASES AND DISORDERS OF THE STOMACH

Gastritis. Gastritis refers to inflammation of the gastric mucosa. It can result from the action of irritating food or liquids (especially alcohol), from excessive production of gastric juice, or from the action of microorganisms or their toxins.

Gastric Ulcer.   A gastric ulcer is a peptic ulcer that occurs in the stomach, usually in the region of the pyloris. Peptic ulcers are defects in the mucous membrane of the esophagus, stomach, or duodenum produced by the action of gastric juice on tissue with low resistance to ulceration. A peptic ulcer may perforate through the wall of the digestive tube into the peritoneal cavity, causing severe hemorrhage and peritonitis. Peptic ulcers are most common in people suffering from chronic anxiety or tension. People with peptic ulcers are commonly treated with antacids, a bland diet, and drugs that reduce gastric motility and secretion.

Gastric Carcinoma.   Carcinoma of the stomach is one of the leading causes of cancer deaths in the United States. It appears to be caused largely by environmental factors, and food additives are now under suspicion. Also, people with gastric atrophy or long-standing pernicious anemia have a greater risk of developing gastric carcinoma. The lesion tends to arise over a wide area of mucosal surface and either constricts the lumen or forms a single large cancerous mass that impinges on the lumen. Metatases to the liver, lungs, brain, bones, and ovaries are common. Gastric carcinomas produce minimal symptoms of discomfort and most are inoperable by the time of discovery.

Pyloric Stenosis.   Pyloric stenosis refers to a narrowing of the pyloric canal. In some instances it is due to a congenital thickening of the pyloric sphincter. *Pylorospasm,* spastic contraction of the pyloric sphincter, has the same end result as pyloric stenosis. Both conditions prevent or retard the exit of chyme from the stomach.

DISEASES AND DISORDERS OF THE INTESTINES

Enteritis.   Enteritis refers to inflammation of the intestinal mucosa. Causes of enteritis include bacterial or viral infection, food poisoning, ingestion of irritating foods or liquids (especially alcohol), overeating, or emotional stress.

Duodenal Ulcer.   A duodenal ulcer is a peptic ulcer that occurs in the duodenum. Most duodenal ulcers occur in the first 2 inches of the duodenum. Duodenal ulcers are more common than gastric ulcers and constitute a major health problem.

Hernia.   Hernia refers to the protrusion of abdominal viscera, usually the small intestine, outside of the normal confines of the abdominal cavity. Inguinal, diaphragmatic, femoral, and umbilical hernias are most common. A hernia may become strangulated and gangrenous when there is tight constriction of the neck of the hernial sac.

Diverticulosis.   Diverticulosis refers to the formation of pouchlike projections in the intestinal wall, especially in the sigmoid colon, due to weak spots in the muscular coat. *Diverticulitis* (inflammation of the diverticula) can occur if fiber or fecal matter lodges in these pockets. If the inflamed diverticula rupture, peritonitis generally results.

Ulcerative Colitis. Ulcerative colitis is an inflammatory disease of autoimmune origin that is characterized by ulceration and scarring of the colonic wall. It is primarily a disease of young adults and tends to be chronic in nature.

Carcinoma of the Large Intestine. Carcinoma of the large intestine is the second most common cause of cancer deaths in the United States. It has been attributed recently to a diet that is low in fiber and high in beef products, which stimulate the production of cancer-causing chemicals by the colonic bacteria. Carcinoma of the large intestine is derived from the intestinal mucosa and occurs most frequently in the rectum and sigmoid colon. Pathologists recognize constricting and obstructing types. The standard form of treatment is surgical removal of the diseased portion of the intestine followed by a *colostomy*. A colostomy is an artificial opening made between the colon and the anterior abdominal wall for the purpose of emptying the colon, especially when the rectum has been removed.

DISEASES AND DISORDERS OF THE SALIVARY GLANDS

Mumps. Mumps is a viral infection that localizes in the parotid glands, and often in the pancreas or testes. Swelling of the infected parotid glands is quite painful but not serious. Destruction of the pancreatic islets or testicular cells is serious.

Tumors. The salivary glands develop both benign and malignant tumors. Surgical removal of a parotid gland tumor is complicated by the fact that the facial nerve and its major branches pass through the substance of the parotid gland.

DISEASES AND DISORDERS OF THE PANCREAS

Pancreatitis. Pancreatitis is a serious condition characterized by inflammation of the pancreatic parenchyma. It is complicated by the escape of pancreatic enzymes into the surrounding tissues. Repeated attacks of pancreatitis may cause pancreatic insufficiency or destroy the organ entirely.

Adenocarcinoma. Adenocarcinoma of the pancreas is a malignant tumor that arises from the glandular epithelium of the exocrine pancreas. It is more common in the head than the body or the tail. The cure rate for this particular cancer is very low.

DISEASES AND DISORDERS OF THE LIVER

Hepatitis. Hepatitis refers to inflammation of the liver. It may be caused by exposure to toxic agents such as carbon tetrachloride (an ingredient of cleaning fluid) or infectious agents such as viruses. The virus that causes serum hepatitis is transmitted by transfusions of infected blood and constitutes a serious problem in many hospitals.

Cirrhosis.   Cirrhosis of the liver is a chronic disease characterized by atrophy of the liver lobules plus hypertrophy of the perilobular connective tissue. Portal hypertension frequently accompanies cirrhosis due to obstruction of blood flow through the substance of the liver. Cirrhosis may be the result of chronic alcoholism, infectious disease, or the retention of bile in the liver.

Hepatomas.   Primary hepatomas (liver carcinomas) are relatively rare malignant tumors that arise from the parenchyma of the liver. They often arise in a cirrhotic liver. Secondary hepatomas are malignant tumors from almost any organ in the body that metastasize to the liver. Because of the liver's vascularity and filter function, secondary hepatomas are very common.

DISEASES AND DISORDERS OF THE GALLBLADDER

Cholelithiasis.   Cholelithiasis refers to the formation of gallstones. Infection of the gallbladder is a frequent cause of gallstone formation. Gallstones may pass into the common bile duct and cause obstruction. Prolonged obstruction of the common bile duct causes retention of bile in the liver, which leads to cirrhosis.

Cholecystitis.   Cholecystitis refers to inflammation of the mucosa of the gallbladder. It is usually associated with the presence of gallstones and may result in life-threatening complications. When surgeons perform a *cholecystectomy* for cholecystitis or cholelithiasis, they remove the gallbladder and as much of the cystic duct as possible. They must be very careful to leave the common bile duct intact; otherwise bile cannot pass into the duodenum and is retained in the liver.

Suggested Readings

BOWEN, W. H. "Prospects for the prevention of dental caries," *Hosp. Pract.*, 9, no. 5 (May 1974), pp. 163–168.

CONNELL, ALASTAIR M. "Dietary fiber and diverticular disease," *Hosp. Pract.*, 11, no. 3 (March 1976), pp. 119–124.

DAVENPORT, HORACE W. "Why the stomach does not digest itself," *Sci. Am.*, 226, no. 1 (January 1972), pp. 87–93.

JAMES, ARTHUR G. "Surgical treatment of parotid tumors," *Hosp. Pract.*, 8, no. 3 (March 1973), pp. 94–101.

KAPPAS, ATTALLAH, AND ALVITO P. ALVARES. "How the liver metabolizes foreign substances," *Sci. Am.*, 232, no. 6 (June 1975), pp. 22–31.

LIEBER, CHARLES S. "The metabolism of alcohol," *Sci. Am.*, 234, no. 3 (March 1976), pp. 25–33.

NETTER, FRANK H. *The Ciba Collection of Medical Illustrations.* Vol. 3: *Digestive System.* Part I: *Upper Digestive Tract.* Summit, N.J.: Ciba, 1959.

_____. *The Ciba Collection of Medical Illustrations.* Vol. 3: *Digestive System.* Part II: *Lower Digestive Tract.* Summit, N.J.: Ciba, 1962.

_____. *The Ciba Collection of Medical Illustrations.* Vol. 3: *Digestive System.* Part III: *Liver, Biliary Tract, and Pancreas.* Summit, N.J.: Ciba, 1964.

NEURATH, H. "Protein digesting enzymes," *Sci. Am., 211,* no. 6 (December 1964), pp. 68–79.

13

The Endocrine System

The endocrine system consists of scattered glands and cell clusters whose secretions are essential for normal body function. Like exocrine glands, most endocrine glands develop from some sort of epithelial tissue; unlike exocrine glands, they have no ducts to carry away their secretions. Therefore, endocrine glands must release their secretions directly into the bloodstream. For this reason endocrine glands have a rich blood supply, and every secreting cell lies next to a capillary or sinusoid.

The secretion of an endocrine gland is called a *hormone*. A hormone is a chemical messenger that regulates the rate at which cellular metabolic processes take place. It is transported by the bloodstream to its site of action, which is known as a *target organ*. Often the target organ is far distant from the cells producing the hormone.

Hormones play three major roles in the human body:

1. *Integration of Body Functions.* Hormones permit different groups of tissues to act together in response to internal or external stimuli.
2. *Promotion of Homeostasis.* Hormones play a vital role in maintaining the constancy of the internal environment. When there is an excess or deficiency of a particular hormone, the internal environment is disturbed; a specific endocrine disease may result if the condition is severe and chronic.
3. *Growth Regulation.* Hormones regulate the rate and type of growth of the body's tissues.

The major endocrine glands of the human body include the *pituitary gland, ovaries, testes, placenta, adrenal glands, thyroid gland, parathyroid glands, pancreas,* and *pineal gland* (Fig. 13.1 and Table 13.1). The hypothalamus (part of the brain) can function as an endocrine gland although anatomists do not classify it as one.

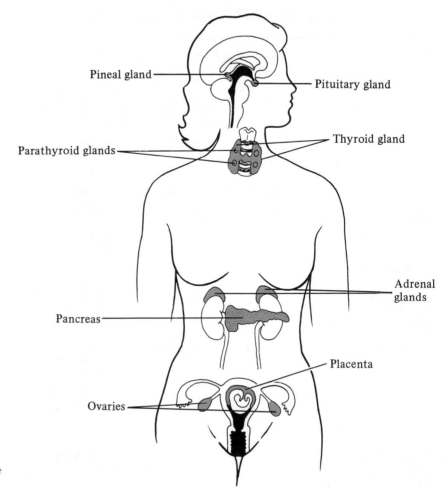

Pineal gland

Pituitary gland

Parathyroid glands

Thyroid gland

Adrenal glands

Pancreas

Placenta

Ovaries

Figure 13.1 *Location of Major Endocrine Organs in Adult Female.*

The Pituitary Gland (Hypophysis)

The pituitary gland, or *hypophysis* (Fig. 13.2), is a pea-sized organ attached by a stalk to the base of the brain just posterior to the optic chiasma (the region where the optic nerves cross). It lies directly below the hypothalamus, with which it has important anatomical and functional connections, in the sella turcica (a depression in the sphenoid bone). The pituitary consists of two parts, the *anterior pituitary* and the *posterior pituitary*, which have different embryological origins and different functions.

ANTERIOR PITUITARY
Anatomy

The anterior lobe, or *adenohypophysis*, forms 75 percent of the pituitary gland. It is of epithelial origin, being derived from the roof of the embryonic oral cavity. The anterior pituitary is divided into three

Table 13.1 Major Hormones of Human Body

Hormone	Origin	Regulated by	Function	Hyposecretion	Hypersecretion
Growth hormone (GH)	Anterior pituitary (pars distalis)	GHRH, somatostatin, general nutritional status	Elevates blood glucose; promotes bone, muscle, and visceral growth	Child, dwarfism	Child, gigantism; adult, acromegaly
Prolactin (PRL)	Anterior pituitary (pars distalis)	PIF and PRF	Stimulates milk production and secretion	Lack of milk secretion	Milk secretion without pregnancy, amenorrhea
Follicle-stimulating hormone (FSH)	Anterior pituitary (pars distalis)	FSHRH; negative estrogen feedback	Stimulates growth of ovarian follicle and estrogen secretion; stimulates sperm production	Sterility	Girl, precocious puberty; woman, hyperestrinism
Luteinizing hormone (LH)/interstitial cell-stimulating hormone (ICSH)	Anterior pituitary (pars distalis)	LHRH; negative progesterone or testosterone feedback	Triggers ovulation; stimulates progesterone and testosterone secretion	Sterility	Woman, sterility; man, hyperandrogenism
Thyrotropin (TSH)	Anterior pituitary (pars distalis)	TRF; negative thyroxin feedback	Stimulates thyroxin and triodothyronine secretion	Hypothyroidism	Hyperthyroidism
Adrenocorticotropin (ACTH)	Anterior pituitary (pars distalis)	CRF; negative cortisol feedback	Stimulates cortisol and sex steroid secretion	Low blood pressure and susceptibility to stress	Cushing's syndrome
Melanocyte-stimulating hormone (MSH)	Anterior (pars intermedia)	MRF and MIF	Undetermined in humans	Undetermined in humans	Hyperpigmentation of skin
Oxytocin (OT)	Hypothalamus (supraoptic and paraventricular nuclei)	Neural reflex	Strong uterine contractions, milk letdown from mammary glands	Weak uterine contractions, inability to lactate	No known effect
Vasopressin or antidiuretic hormone (ADH)	Hypothalamus (supraoptic and paraventricular nuclei)	Osmotic pressure of blood	Causes reabsorption of water from urine	Diabetes insipidus (water loss)	Water intoxication
Estrogens	Ovaries (follicles, corpora lutea)	FSHRH/FSH secretion; negative estrogen feedback	Cause female secondary sex characteristics; stimulate growth of uterine lining and mammary gland ducts	Girl, primary amenorrhea; woman, secondary amenorrhea	Girl, precocious puberty; woman, hyperestrinism
Progesterone	Ovaries (corpora lutea)	LHRH/LH secretion; negative progesterone feedback	Stimulates secretory activity of uterine lining and mammary gland acini	Sterility	Sterility

Table 13.1 *(Continued)*

Hormone	Origin	Regulated by	Function	Hyposecretion	Hypersecretion
Testosterone	Testes	LHRH/LH (ICSH) secretion; negative testosterone feedback	Causes male secondary sex characteristics; stimulates development of male reproductive organs	Boy, eunuchism; man, regression of secondary sex characteristics	Boy, precocious puberty; man, hyperandrogenism
Chorionic gonadotropin (CG)	Placenta (fetal chorion)	Placental estrogen and progesterone production	Maintains corpus luteum of pregnancy	Early spontaneous abortion	Late spontaneous abortion
Aldosterone	Adrenal cortex (zona glomerulosa)	Renin-angiotensin system, blood sodium and potassium levels	Regulates blood pressure and sodium-potassium balance	Low blood pressure	Conn's syndrome (high blood pressure)
Cortisol	Adrenal cortex (zona fasciculata and reticularis)	CRF/ACTH secretion; negative cortisol feedback	Elevates blood–glucose levels; has antistress, antiallergy, antiinflammatory actions	Low blood pressure, low resistance to stress	Cushing's syndrome (high blood pressure)
Adrenal sex steroids	Adrenal cortex (zona fasciculata and reticularis)	CRF/ACTH secretion; negative cortisol feedback	Replace gonadal steroids in middle age	No known effect	Adrenogenital syndrome (precocious puberty or masculinization)
Epinephrine	Adrenal medulla	Sympathetic nervous system	Elevates blood–glucose levels, increases heart and metabolic rate	*	High blood pressure
Norepinephrine	Adrenal medulla	Sympathetic nervous system	Vasoconstriction	*	High blood pressure
Thyroxin, triiodothyronine	Thyroid gland (follicular cells)	TRF/TSH secretion; negative thyroxin feedback	Skeletal growth, maturation of CNS, elevation of BMR	Child, cretinism; adult, myxedema	Graves' disease (thyrotoxicosis)
Calcitonin	Thyroid gland (parafollicular cells)	Blood–calcium levels; parathormone secretion	Lowers blood–calcium levels	Elevation of blood–calcium levels	Depression of blood–calcium levels; osteopetrosis
Parathormone	Parathyroid glands	Blood–calcium; calcitonin secretion	Raises blood–calcium levels	Depression of blood–calcium levels	Elevation of blood–calcium levels; cystic lesions of bone
Insulin	Pancreatic islets (beta cells)	Blood–glucose levels	Lowers blood–glucose levels	Diabetes mellitus (hyperglycemia)	Hyperinsulinism (hypoglycemia)

Table 13.1 *(Continued)*

Hormone	Origin	Regulated by	Function	Hyposecretion	Hypersecretion
Glucagon	Pancreatic islets (alpha cells)	Blood–glucose levels	Raises blood–glucose levels	Hypoglycemia	Hyperglycemia
Gastrin	Pancreatic islets (**delta cells**) (?)	Food entering stomach	Secretion of gastric juice	*	Gastric hyperacidity; peptic ulcers
Melatonin	Pineal gland	CNS, light	Inhibits gonadal function (?)	Precocious puberty (?)	Delayed puberty (?)

* There are other sources of this hormone.

parts: a large part, the *pars distalis;* a cranial part, the *pars tuberalis,* which wraps around the infundibulum (neural stalk); and an intermediate part, the *pars intermedia,* situated between the pars distalis and the posterior pituitary.

Pars Distalis The pars distalis produces the greatest number of anterior pituitary hormones. These include *growth hormone (GH), prolactin (PRL),*

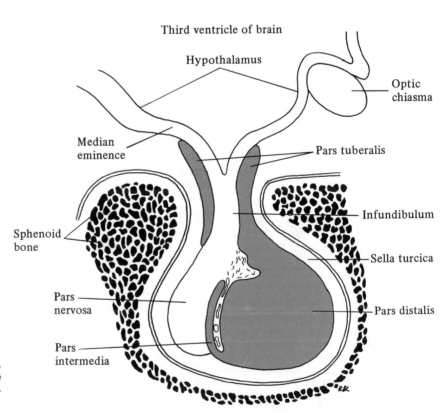

Figure 13.2 *Pituitary Gland.* Adenohypophysis *(gray)* includes pars tuberalis, pars distalis, and pars intermedia, while neurohypophysis *(white)* includes infundibulum and pars nervosa.

follicle-stimulating hormone (FSH), luteinizing hormone (LH or ICSH), thyrotropin (TSH), and *adrenocorticotropin (ACTH).*

The cells of the pars distalis are epithelioid (like epithelial cells in appearance). They occur in cords and clusters that are in close contact with a vast network of sinusoidal capillaries. These cells are classified according to the staining reactions of their secretory granules. *Chromophobes* are cells that have little or no affinity for histological stains. They include what are probably nonsecreting reserve cells and cells that probably secrete ACTH. *Acidophils* are cells whose large secretory granules have an affinity for acidic stains. They secrete GH and PRL. *Basophils* are cells whose small secretory granules have an affinity for basic stains. These cells secrete FSH, LH, and TSH.

Pars Intermedia
The cells of the pars intermedia in humans form a discontinuous layer between the pars distalis and the posterior pituitary and are similar in appearance to the basophils of the pars distalis. These cells secrete *melanocyte stimulating hormone (MSH).*

Pars Tuberalis
The pars tuberalis, which surrounds the infundibulum like a sleeve, is highly vascularized. Its cells are organized into long cords that run parallel to the blood vessels. Though these basophilic cells have the appearance of secretory activity, no known hormone has been attributed to them as yet.

Blood Supply
The anterior pituitary receives arterial blood from the right and left superior hypophyseal arteries and venous blood from the hypophyseal portal veins. The hypophyseal portal veins arise from a capillary bed in the hypothalamus and infundibulum and empty into the sinusoids of the anterior pituitary. This *hypophyseal portal system* (Fig. 13.3) plays an important role in pituitary function because it transports regulatory substances from the hypothalamus to the anterior lobe. The anterior pituitary is drained by the lateral hypophyseal veins, which are tributaries of the dural venous sinuses.

Function
The anterior pituitary secretes seven proteinaceous or polypeptide hormones. Many of these hormones are *trophic hormones* (hormones that regulate other endocrine glands). The hypothalamus secretes a number of regulatory substances, called *releasing factors (hormones)* and *inhibiting factors (hormones),* which are transported to the anterior pituitary by the hypophyseal portal veins. These hypothalamic hormones stimulate or inhibit the release of anterior pituitary hormones.

The relationships between the hypothalamus, anterior pituitary, and a hypothetical target organ are illustrated in Fig. 13.4. The releasing factor secreted by the hypothalamus stimulates the anterior pituitary to secrete its trophic hormone. The trophic hormone in turn

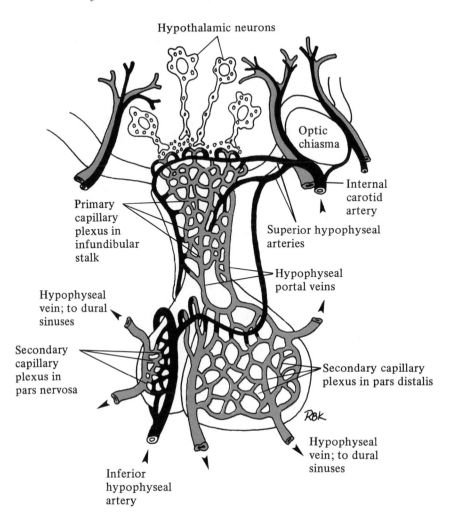

Hypothalamic neurons

Optic chiasma

Internal carotid artery

Primary capillary plexus in infundibular stalk

Superior hypophyseal arteries

Hypophyseal portal veins

Hypophyseal vein; to dural sinuses

Secondary capillary plexus in pars nervosa

Secondary capillary plexus in pars distalis

Inferior hypophyseal artery

Hypophyseal vein; to dural sinuses

Figure 13.3 Blood Supply of Pituitary Gland. Note that hypophyseal portal system of vessels carries secretory products of hypothalamic neurons to adenohypophysis.

stimulates the target organ to produce and secrete its hormones. These hormones, besides affecting peripheral tissues, regulate the secretory activity of the hypothalamus and anterior pituitary by a *negative feedback* mechanism. In negative feedback regulation, a high output of the system (high target organ hormone levels) causes a decrease in the input (low releasing factor and trophic hormone levels).

The anterior pituitary hormones (Fig. 13.5) are discussed next. For more detailed information you are referred to standard endocrinology and physiology texts.

Growth Hormone (GH). Growth hormone, also called somatotropin (STH), is responsible for normal body growth. In particular, it promotes the growth of long bones, skeletal muscles, and viscera. GH secretion has been shown to exert a specific growth effect on the epiphyseal cartilages of long bones, which causes the bones to increase in length. It also plays an important role in the metabolism of proteins, carbohydrates, and fats and appears to potentiate the action of other hormones. The pancreatic hormone insulin increases the effectiveness of GH in promoting growth. Secretion of GH is regulated by the GH releasing factor (GRF or GHRH) and somatostatin (GIF or SRIF) plus the general nutritional state of the body.

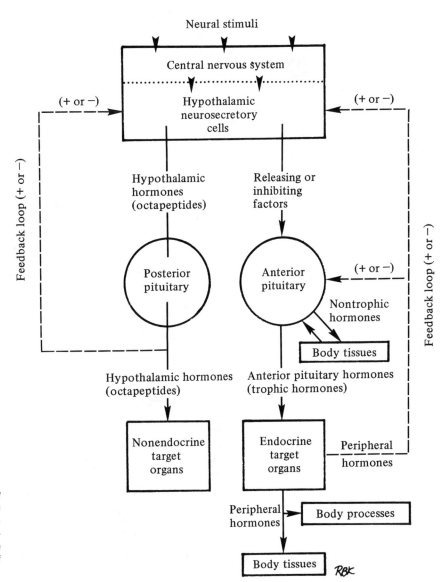

Figure 13.4 Regulation of Pituitary Hormone Secretion. Schematic diagram shows relationships between hypothalamus, pituitary, and target organs. Note that posterior pituitary (neurohypophysis) is site of storage and subsequent release of hypothalamic hormones.

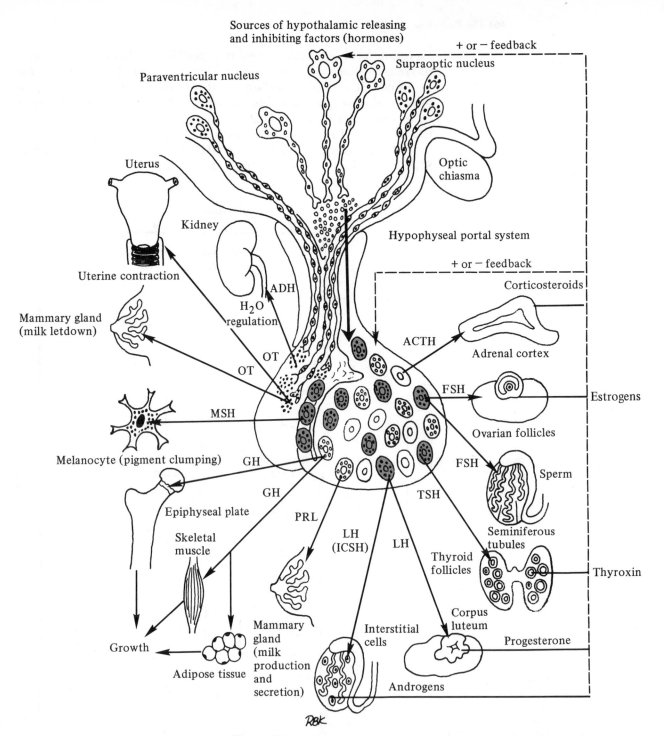

Figure 13.5 Pituitary Function. Schematic diagram shows effects of pituitary hormones on their target organs. Note that hormones produced by target organs of certain anterior pituitary hormones can act on pituitary or hypothalamus (by positive or negative feedback) to regulate their secretion. Anterior pituitary cells are indicated as follows: basophils, *gray with black granules;* acidophils, *white with white granules;* chromophobes, *white with no granules.*

Prolactin (PRL). Prolactin, also called lactogenic hormone (LTH), exerts its effect primarily on the mammary glands. In the human female it stimulates the mammary glands to produce milk and maintains milk production in association with other hormones (estrogen and progesterone) after birth. In some species, but not in the human female, it acts on the corpus luteum of the ovary. The function of PRL in males, if any, has not been determined. PRL secretion is regulated by the PRL inhibiting factor (PIF), which is secreted at all times except during pregnancy and lactation, and the PRL releasing factor (PRF).

Follicle-stimulating Hormone (FSH). The follicle-stimulating hormone is a *gonadotropin* (a hormone that exerts its effects on the gonads). In the female FSH acts on the ovary to stimulate ovarian follicular growth and estrogen secretion; in the male it acts on the testis to stimulate spermatogenesis. FSH secretion is regulated by the FSH releasing factor (FRF or FSHRH) plus blood estrogen levels in the female and blood androgen levels in the male.

Luteinizing Hormone (LH). The luteinizing hormone, formerly known as interstitial cell-stimulating hormone (ICSH) in the male, is also a gonadotropin. In the female, LH triggers ovulation and transforms the ovarian follicle into a corpus luteum ("yellow body"). It also stimulates the corpus luteum to secrete progesterone. In the male, LH stimulates the interstitial cells of the testis to secrete testosterone and other androgens. LH secretion is regulated by the LH releasing hormone (LHRH) plus blood progesterone levels in the female and blood testosterone levels in the male.

Thyrotropin (TSH). Thyrotropin, also known as thyroid-stimulating hormone, stimulates thyroid follicular cells to secrete thyroxin. TSH secretion is regulated by the TSH releasing hormone (TRH) plus blood thyroxin levels.

Adrenocorticotropic Hormone (ACTH). The adrenocorticotropic hormone, also called corticotropin, stimulates the adrenal cortex to secrete glucocorticoids and sex steroids. ACTH secretion is regulated by the ACTH releasing factor (CRF) plus blood corticosteroid levels.

Melanocyte-stimulating Hormone (MSH). The melanocyte-stimulating hormone stimulates the dispersion of melanin granules in the melanocytes of amphibians, which causes darkening of the skin. MSH secretion is regulated by the MSH releasing factor (MRF) and the MSH inhibiting factor (MIF). The function of MSH in humans is not well understood. The human body has two forms of MSH, both of which are identical to portions of the ACTH molecule. Humans suffering from degeneration of the adrenal cortex develop hyperpigmentation of the skin, which apparently is due to the release of excess ACTH and MSH from the anterior pituitary.

POSTERIOR PITUITARY
Anatomy

The posterior lobe, or *neurohypophysis,* forms 25 percent of the pituitary gland. It is the neural part of the pituitary, being formed from the floor of the embryonic brain. The posterior pituitary is divided into two parts: the *pars nervosa* and the *infundibulum* (neural stalk). The pars nervosa forms the bulk of the posterior pituitary, while the infundibulum connects the pars nervosa with the median eminence of the hypothalamus.

Pars Nervosa

The pars nervosa consists mainly of the unmyelinated axons of secretory neurons, whose cell bodies are located in hypothalamic nuclei. The axons of these neurons converge in the median eminence to form the *hypothalamohypophyseal tract.* They then traverse the infundibulum and terminate in close proximity to the capillaries in the pars nervosa. Two hypothalamic hormones, *oxytocin* and *vasopressin,* are produced in these nerve cell bodies. The hormones are transported down their axons in granular form, stored in their axon terminals, and released into the general circulation as needed. The accumulations of secretory granules in the axon terminals, which are visible under the light microscope, are called *Herring bodies.* The pars nervosa also contains connective tissue cells and a type of glial cell called a *pituicyte.* Pituicytes were once mistakenly thought to produce oxytocin and vasopressin. Thus the posterior pituitary is merely a storage depot and not an endocrine gland, since its secretions are produced elsewhere.

Infundibulum

The infundibulum is the direct link between the hypothalamus and the pituitary gland. It transmits the hypophyseal portal veins to the pars distalis and intermedia and the hypothalamohypophyseal nerve tract to the pars nervosa.

Blood Supply

The posterior pituitary is supplied by the right and left inferior hypophyseal arteries and is drained by the lateral hypophyseal veins.

Function

The posterior pituitary is the site of storage and release of *oxytocin* and *vasopressin* (Fig. 13.5).

Oxytocin (OT). Oxytocin is produced by nerve cell bodies in the *paraventricular* and *supraoptic nuclei* of the hypothalamus (nuclei situated alongside the third ventricle and above the optic chiasma of the brain). It stimulates contraction of the smooth muscle of the uterine wall during sexual intercourse and childbirth. Pitocin®, a commercial oxytocin preparation, is used to induce labor and to assist women in labor whose uterine contractions are weak. Oxytocin also stimulates contraction of the myoepithelial cells in the mammary glands, which action is responsible for the release of milk when a mother breast-feeds her baby.

Vasopressin (VP). Vasopressin is also produced by nerve cell bodies in the paraventricular and supraoptic nuclei of the hypothalamus. It stimulates contraction of smooth muscle fibers in small blood vessels, which action elevates the blood pressure (pressor effect). Vasopressin also helps conserve body water by promoting the reabsorption of water from the distal convoluted tubules and collecting tubules of the kidneys. Since the water-conserving effect of vasopressin is physiologically more important than the pressor effect, vasopressin is more commonly known as the *antidiuretic hormone (ADH)*. Vasopressin secretion is regulated by the osmotic pressure of blood passing through the hypothalamus.

The Gonads

The gonads (sex glands) are the sources of hormones that regulate human sexual development and reproduction. They include the ovaries and testes.

OVARIES
Anatomy

The ovaries are the female gonads. They are situated in depressions on the lateral pelvic walls, behind and on either side of the uterus. The ovary (Fig. 13.6) is a gland whose parenchyma consists of an inner core (medulla) surrounded by a peripheral zone (cortex). Embedded in the connective tissue of the cortex are *ovarian follicles* of various

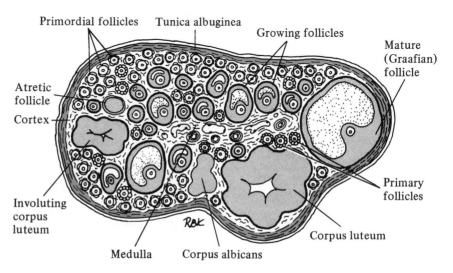

Figure 13.6 Ovary.

sizes and stages of development, each consisting of an *oocyte* (immature egg cell) surrounded by follicular cells of epithelial origin. The cortex also contains transformed ovarian follicles called *corpora lutea.*

Function Approximately every 28 days, from puberty (when the ovaries begin to cycle) to menopause (when the ovaries cease to cycle), one or more ovarian follicles mature under the stimulus of the hypophyseal hormones FSH and LH. The maturing ovarian follicle secretes a class of female sex hormones called *estrogens.* A large burst of LH at midcycle triggers ovulation (the expulsion of the oocyte from the follicle) and transforms the empty follicle into a corpus luteum. The corpus luteum secretes estrogens plus a second female sex hormone called *progesterone.* In addition to estrogens and progesterone, ovarian tissues also secrete androgens and relaxin.

Estrogens. *Estradiol* and its metabolic derivatives, *estriol* and *estrone,* are steroid hormones collectively known as estrogens. Estrogens are responsible for the development of the female reproductive organs and of female secondary sex characteristics (mammary gland enlargement plus female distribution of fat and body hair). During the preovulatory portion of the menstrual cycle (see Chapter 14), estrogens stimulate proliferation of the uterine lining and the ducts of the mammary gland. Estrogens are also important for the maintenance of pregnancy. Estrogen secretion is regulated by FRF/FSH and LHRH/LH secretion and by negative feedback involving blood estrogen levels. This relationship is illustrated in Fig. 13.7.

Progesterone. Progesterone is a steroid hormone that prepares the female body for pregnancy and, with estrogens, helps maintain pregnancy. During the postovulatory portion of the menstrual cycle, progesterone stimulates secretory activity in the uterine lining in preparation for implantation of a fertilized ovum. If implantation fails to occur, a portion of the uterine lining is sloughed off and discharged during menstruation. Progesterone also stimulates growth and differentiation of the secretory acini of the mammary glands in preparation for lactation. Progesterone secretion is regulated by LHRH/LH secretion and by negative feedback involving blood progesterone levels. This relationship is also illustrated in Fig. 13.7.

Androgens. Androgens are male sex hormones that are also found in small quantities in women. They are produced by the ovary and also by the adrenal cortex. Though present in a low level in women, androgens, particularly adrenal androgens, are thought to be very important in stimulating general body growth and maintaining the female sexual drive.

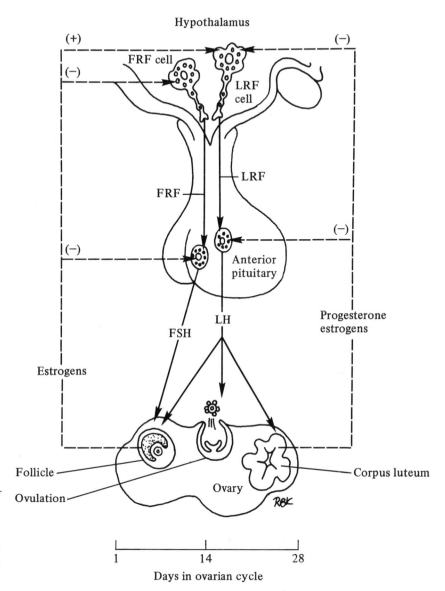

Hypothalamus

(+)   (−)

(−)

FRF cell

LRF cell

LRF

FRF

(−)

(−)

Anterior pituitary

Progesterone estrogens

LH

Estrogens

FSH

Follicle

Ovulation

Corpus luteum

Ovary

RBK

1   14   28

Days in ovarian cycle

Figure 13.7 Hormonal Regulation of Ovary. Schematic diagram shows relationships among hypothalamus, anterior pituitary, and ovarian structures in feedback mechanism regulating secretion of hormones during ovarian cycle. Positive feedback (+); negative feedback (−).

Relaxin. Relaxin is a protein hormone secreted by the corpus luteum during the last trimester of pregnancy. The precise role of this hormone in pregnancy in humans is still poorly understood, but it is known to inhibit uterine motility. It also causes softening and dilation of the cervix (neck of the uterus) in preparation for delivery. In some animals (and probably in humans) it causes relaxation of the pubic symphysis and the pelvic ligaments, which permits easier passage of the fetal head through the birth canal.

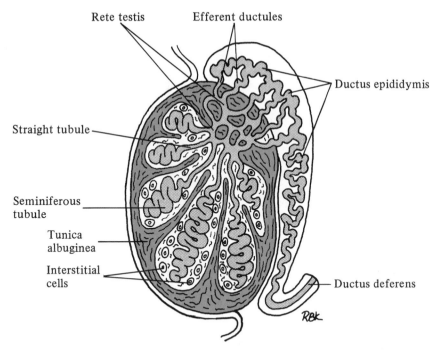

Figure 13.8   Testis.

TESTES
Anatomy

The testes are the male gonads. They lie outside the abdominopelvic cavity in an integumentary pouch called the scrotum. The testis (Fig. 13.8) is a solid glandular organ whose parenchyma consists of epithelial tubules called *seminiferous tubules.* The seminiferous tubules are supported by a loose connective tissue stroma. Within the connective tissue stroma are clusters of secretory cells called interstitial cells (of Leydig).

Function

Under the stimulus of FSH, the seminiferous tubules produce sperm (male sex cells). The Leydig cells secrete androgens under the stimulus of LH (ICSH). Testicular tissues also secrete estrogens.

Androgens.   Androgens are the male sex hormones and, like the female sex hormones, are steroids. The principal androgen is called *testosterone.* Testosterone is responsible for both the embryolgical development and the maintenance of the male reproductive organs, for male secondary sex characteristics (muscular build, low-pitched voice, plus male distribution of fat and body hair), and for the male sex drive. Testosterone secretion is regulated by LHRH/LH (ICSH) secretion and by negative feedback involving blood testosterone levels. This relationship is illustrated in Fig. 13.9.

Estrogens.   A small amount of estrogens are produced by the testis. Both the interstitial cells and the epithelium of the seminiferous

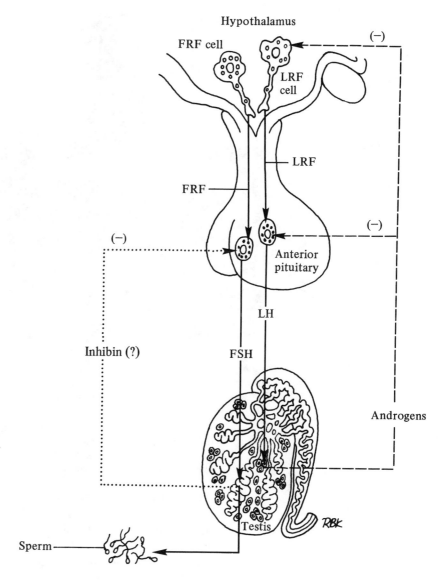

Figure 13.9 Hormonal Regulation of Testis. Schematic diagram shows relationships among hypothalamus, anterior pituitary, and testicular structures in feedback mechanism regulating sperm formation and androgen secretion. Ill-defined substance called inhibin appears to play role in regulating secretion of FSH. Negative feedback (−).

tubules appear to secrete estrogens. The function of estrogens in the male is unknown as yet.

The Placenta

ANATOMY   The placenta (Fig. 13.10) is a spongy vascular organ that grows on the wall of the uterus during pregnancy and provides oxygen and nutrients for the fetus. The placenta is a composite organ, consisting of fetal and maternal tissue. The superficial portion, or *chorion,* is of fetal origin.

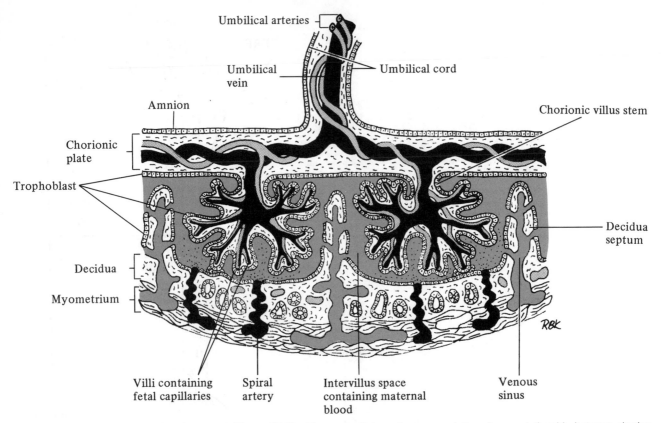

Umbilical arteries

Umbilical vein

Umbilical cord

Amnion

Chorionic villus stem

Chorionic plate

Trophoblast

Decidua septum

Decidua

Myometrium

Villi containing fetal capillaries

Spiral artery

Intervillus space containing maternal blood

Venous sinus

Figure 13.10 Placenta. Schematic representation shows relationship between chorion (fetal portion) and decidua (maternal portion). Note that umbilical vein *(black)* carries oxygenated blood from placenta to fetus, while umbilical arteries *(gray)* carry deoxygenated blood from fetus to placenta.

The chorion gives rise to many treelike outgrowths called *chorionic villi.* Each villus has a mesenchymal core containing fetal capillaries and is covered with trophoblast. The deep portion, or *decidua basalis,* is of maternal origin. The decidua basalis represents the highly modified lining of the uterus. The spaces between the chorionic villi and the decidua basalis are filled with maternal blood rich in oxygen and nutrients. Respiratory gases and metabolites diffuse across the thin placental barrier that separates the fetal and maternal bloodstreams.

FUNCTION The placenta transfers oxygen, water, electrolytes, nutrients, vitamins, hormones, antibodies, and some drugs from the maternal blood to the fetal blood. It also transfers carbon dioxide, water, hormones, and waste products from the fetal blood to the maternal blood. In addition to serving as a transfer organ, the placenta is a temporary endocrine organ that secretes hormones important in maintaining pregnancy. These hormones include, among others, *chorionic gonado-*

tropin, placental steroids, and *placental lactogen.* All these hormones appear to be secreted by the trophoblast cells of the chorion.

Chorionic Gonadotropin (CG). Chorionic gonadotropin is a protein hormone secreted by the placenta that stimulates the corpus luteum, especially during the first trimester (three months) of pregnancy. It is similar in function, though not in chemical structure, to LH. CG prevents regression of the corpus luteum, stimulating it to produce the large amounts of estrogen and progesterone needed to maintain pregnancy. It also inhibits pituitary secretion of FSH and LH, eliminating further follicular maturation, ovulation, and menstrual cycles for the duration of pregnancy. CG secretion begins shortly after implantation and reaches a peak about 60 to 80 days after the end of the last menstrual cycle. The CG excreted in the mother's urine in early pregnancy forms the basis for most pregnancy tests. CG secretion declines rapidly after reaching its peak, so that by the end of the first trimester it has reached a low level that remains constant for the remainder of the pregnancy. The decline in CG secretion occurs just when the placenta itself begins to secrete large amounts of estrogens and progesterone.

Placental Steroids. Estrogens and progesterone are secreted in large quantities by the placenta during the second and third trimesters of pregnancy. The placenta thus comes to replace the corpus luteum as the major steroid-secreting structure in the latter part of pregnancy. In some pregnant women placental steroid secretion remains abnormally low. Unless treated with large amounts of exogenous (outside) estrogens or progestogens, these women undergo spontaneous abortion.

Placental Lactogen. Placental lactogen is a protein hormone secreted by the placenta whose effects are similar to those of growth hormone and prolactin. It is thought that this hormone maintains the mother in a positive protein balance and stabilizes her blood glucose at levels high enough to meet the demands of the growing fetus.

Adrenal Gland

GROSS ANATOMY The adrenal glands [Fig. 13.11(*a*)] are paired organs situated near the upper poles of the kidneys and embedded in the perirenal fat. They are about $1\frac{1}{4}$ to 2 inches long, $\frac{3}{4}$ inch wide, and $\frac{3}{4}$ to $1\frac{1}{2}$ inches thick. The right adrenal is roughly pyramidal in shape, while the left adrenal is crescent-shaped and somewhat larger than the right. Each adrenal has anterior, posterior, and renal (basal) surfaces. A hilus is present on the anterior surface; this is where the adrenal vein emerges from the substance of the gland.

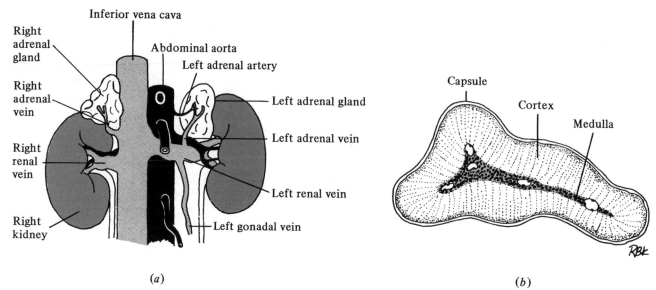

(a)

(b)

Figure 13.11   Adrenal Gland. (a) Gross anatomy. Note that adrenal glands sit on superior poles of kidneys. (b) Microscopic anatomy. Section shows its two major subdivisions: cortex *(white)* and medulla *(gray)*.

MICROSCOPIC ANATOMY   Each adrenal gland [Fig. 13.11(b)] consists of a highly vascular parenchyma enclosed by a fibrous connective tissue capsule. The parenchyma consists of two concentric layers: the *adrenal cortex,* a yellowish peripheral layer, and the *adrenal medulla,* a pinkish central layer. Each layer has a separate embryological origin and secretes distinctly different hormones. In fact, the adrenal gland can be considered to be two morphologically and functionally distinct organs that became united during embryonic development.

Adrenal Cortex   The adrenal cortex (Fig. 13.12) consists of epithelial cells that are derived from embryonic mesoderm and secrete steroid hormones. Due to differences in the arrangements of the cells, three distinct layers can be distinguished. The outermost and thinnest layer is called the *zona glomerulosa.* It consists of cells arranged in round clusters surrounded by capillaries. The middle and widest layer is called the *zona fasciculata.* It consists of cells arranged in radiating columns separated by sinusoids. The innermost layer, which lies next to the medulla, is called the *zona reticularis.* It consists of cords of cells arranged in an irregular network.

Adrenal Medulla   The adrenal medulla is derived from neural crest cells that became incorporated in the developing cortex. The principal cell type is the *chromaffin cell,* so named because its secretory granules turn brown when exposed to chromium salts. The ovoid chromaffin cells are ar-

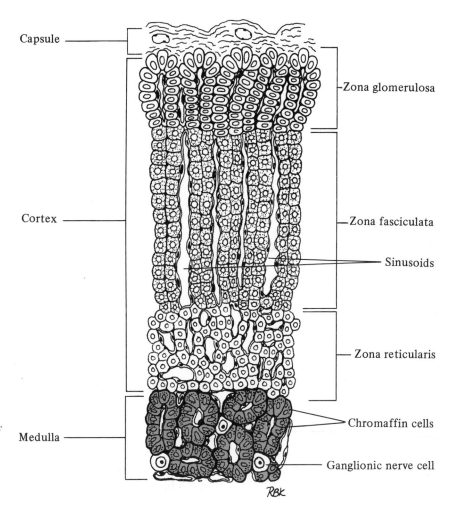

Capsule

Cortex

Medulla

Zona glomerulosa

Zona fasciculata

Sinusoids

Zona reticularis

Chromaffin cells

Ganglionic nerve cell

RBK

Figure 13.12 Cellular Organization of Adrenal Cortex and Medulla. Highly magnified section through adrenal cortex and medulla shows arrangement and appearance of various cell types.

ranged in irregular clusters around the capillaries and venules of adrenal medulla and are surrounded by numerous sympathetic nerve fibers. They secrete hormones, called *catecholamines,* that mimic the action of the sympathetic nervous system. Scattered here and there among the chromaffin cells are a few autonomic ganglion cells, which occur singly or in small groups. These are regarded as sympathetic postganglionic neurons (see Chapter 15).

BLOOD SUPPLY The adrenal glands receive a rich blood supply from the superior, middle, and inferior adrenal arteries. The middle adrenal arteries are direct branches of the abdominal aorta. These glands are drained by the adrenal veins; the right adrenal vein is a tributary of the inferior vena cava, while the left is a tributary of the left renal vein.

INNERVATION   The adrenal gland is innervated by the sympathetic division of the autonomic nervous system. Preganglionic sympathetic fibers arising from nerve cell bodies in the lower thoracic and lumbar segments of the spinal cord pass through the sympathetic trunks (see Chapter 15) and reach the adrenal gland by way of the splanchnic nerves. These fibers penetrate the adrenal medulla and terminate on the chromaffin cells. The chromaffin cells, which are highly modified sympathetic postganglionic neurons, secrete catecholamines under the stimulus of sympathetic nerve impulses.

FUNCTION   Both the cortex and medulla of the adrenal gland secrete hormones. Loss of the hormones of the adrenal cortex is incompatible with life while loss of those of the adrenal medulla is not.

Hormones of the Adrenal Cortex   The hormones of the adrenal cortex are known as corticosteroids. The corticosteroids fall into three categories: *mineralocorticoids, glucocorticoids,* and *sex steroids.*

Mineralocorticoids.   Mineralocorticoids are hormones that help regulate salt and fluid balance in the body. They are secreted by the cells of the zona glomerulosa. *Aldosterone,* the principal mineralocorticoid, promotes the reabsorption of sodium and the excretion of potassium. It acts mainly on the distal tubules of the kidney as well as the intestinal mucosa and ducts of the salivary and sweat glands. The action of aldosterone maintains the body's electrolytic balance, which is essential for normal blood volume and pressure. Aldosterone secretion is regulated primarily by the following mechanisms: (a) the renin-angiotensin system (see Chapter 10), (b) the direct effect of decreased blood-sodium concentration and increased blood-potassium concentration on the cells of the adrenal cortex, and (c) marked changes in the extracellular fluid volume.

Glucocorticoids.   Glucocorticoids are hormones that regulate the intermediary metabolism of carbohydrates, proteins, and lipids. They are secreted by the cells of the zona fasciculata and reticularis under the stimulus of ACTH. *Cortisol,* the principal glucocorticoid, promotes an increase in blood levels of glucose, amino acids, and fatty acids. These substances are used to provide energy, to increase body resistance, and to aid in tissue repair in times of stress or injury. The glucocorticoids also suppress the immune and inflammatory responses. *Cortisone,* another glucocorticoid, is used in the treatment of arthritis and certain allergic conditions. *Prednisone,* still another glucocorticoid, is used to prevent homograft rejection. The secretion of glucocorticoids is regulated by CRF/ACTH secretion and by negative feedback mechanisms involving blood glucocorticoid levels (Fig. 13.13).

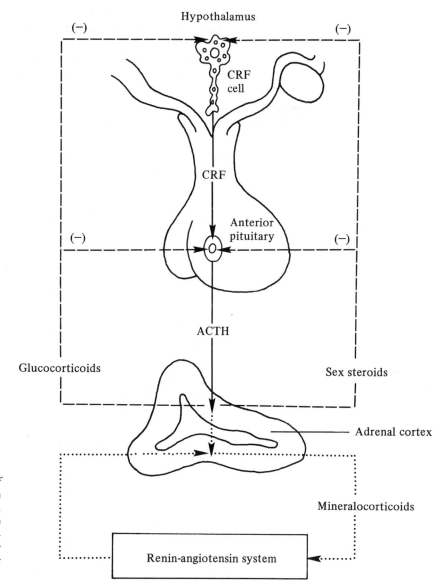

Figure 13.13 Hormonal Regulation of Adrenal Gland. Schematic diagram shows relationships among hypothalamus, anterior pituitary, and adrenal cortex. Note that secretion of mineralocorticoids is regulated by renin-angiotensin system rather than by trophic hormone from anterior pituitary. Negative feedback (−).

Sex Steroids. The adrenal sex steroids are sex hormones produced in both sexes in addition to those produced by the gonads. They are secreted by the cells of the zona fasciculata and reticularis under the stimulus of ACTH. These sex hormones consist primarily of androgens plus small amounts of estrogen and progestogens. Under normal conditions they have little effect on sexual development. For example, *dehydroepiandrosterone,* the only adrenal sex hormone secreted in physiologically significant amounts, has a masculinizing effect but is much less potent than testosterone. It is interesting to note that as the gonads decline in function with age, the adrenal cortex becomes an important source of sex hormones. The adrenal glands thus become the "gonads of middle age."

Hormones of the Adrenal Medulla

The hormones of the adrenal medulla are known as *catecholamines*. The two catecholamines secreted by the chromaffin cells are *epinephrine* (adrenalin) and *norepinephrine* (noradrenalin). These two hormones resemble one another chemically and have many overlapping functions.

Epinephrine.   Epinephrine (adrenalin) is a hormone whose effects are similar to stimulation by the sympathetic division of the autonomic nervous system (see Chapter 15). It is secreted only by the chromaffin cells of the adrenal medulla. Epinephrine is essential to the "fight-or-flight" response: it increases the rate and force of the heartbeat, causes constriction of skin and visceral arterioles, elevates the blood pressure, dilates the bronchioles, causes piloerection (hairs stand on end), dilates the pupils, inhibits gastrointestinal motility, and elevates blood-glucose levels. Secretion of epinephrine is mediated through the sympathetic nervous system in response to stressful situations, and its chief function is to reinforce and prolong the effects of sympathetic stimulation. Epinephrine is used clinically as an emergency heart stimulant, as a bronchodilator, and to counteract severe allergic reactions.

Norepinephrine.   Norepinephrine (noradrenalin) shares many of the functions of epinephrine. However, it differs from epinephrine in that it does not elevate blood-glucose levels nor does it cause bronchodilation. Also, it is a more potent vasoconstrictor than epinephrine because it constricts the arterioles of skeletal muscles as well as those of the skin and viscera. Norepinephrine is secreted by sympathetic postganglionic nerve endings all over the body as well as by the chromaffin cells of the medulla. Its secretion is mediated through the sympathetic nervous system in response to shock or low blood pressure. Because it is such a powerful vasoconstrictor, norepinephrine is used clinically to counteract shock and elevate the blood pressure.

The Thyroid Gland

GROSS ANATOMY

The thyroid gland develops as an outgrowth from the anterior (ventral) wall of the embryonic pharynx. In the adult [Fig. 13.14(a)] it is a butterfly-shaped gland situated in the anterior part of the neck alongside the larynx and trachea. The thyroid gland consists of two lateral lobes

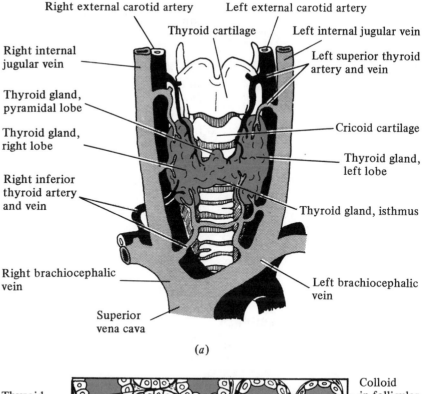

Right external carotid artery Left external carotid artery

Thyroid cartilage

Right internal jugular vein Left internal jugular vein

Left superior thyroid artery and vein

Thyroid gland, pyramidal lobe

Thyroid gland, right lobe Cricoid cartilage

Right inferior thyroid artery and vein Thyroid gland, left lobe

Thyroid gland, isthmus

Right brachiocephalic vein Left brachiocephalic vein

Superior vena cava

(a)

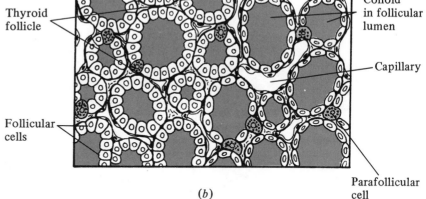

Thyroid follicle Colloid in follicular lumen

Capillary

Follicular cells

Parafollicular cell

(b)

Figure 13.14 Thyroid Gland. (a) Gross anatomy. Anterior aspect of neck, with infrahyoid and sternocleidomastoid muscles removed. Note relationship of thyroid gland to larynx and trachea. Note also that thyroid gland receives abundant blood supply from deep vessels of neck. (b) Microscopic anatomy. Note presence of colloid in follicular lumens. This is storage form of thyroid hormones.

connected by an isthmus. The lateral lobes are pyramidal in shape and extend from the middle of the thyroid cartilage to the level of the fifth tracheal ring (about 2 inches). Each lobe has medial and lateral surfaces and anterior and posterior borders. The lateral surface is convex and is covered by skin, fascia, and portions of the sternocleidomastoid and infrahyoid muscles. The medial surface is concave and is molded by neighboring structures and organs. The isthmus is a transverse strip of thyroid tissue that extends between the lateral lobes at the level of the second and third tracheal rings. A pyramidal lobe sometimes arises from the isthmus and extends to the level of the hyoid bone.

MICROSCOPIC ANATOMY

The thyroid gland is enclosed in a thin fibrous connective tissue capsule whose inward projections subdivide the parenchyma into many irregular lobules. Each lobule [Fig. 13.14(*b*)] contains a number of spherical epithelial structures, called *thyroid follicles,* that are embedded in a loose connective tissue stroma and surrounded by capillaries. The thyroid follicle consists of a single layer of epithelial cells arranged in the form of a hollow ball. The lumen of the follicle contains *colloid,* the storage form of the follicular hormones. The thyroid follicles vary in structure according to the region of the gland and functional activity: an inactive follicle contains a large mass of colloid surrounded by a cuboidal or squamous epithelium; an active follicle contains a small mass of eroded-looking colloid surrounded by a columnar epithelium. In addition, small nests of epithelial cells, called *parafollicular cells* or *C-cells* (clear cells), can be found in the stromal connective tissue between the thyroid follicles.

BLOOD SUPPLY

The thyroid gland receives a rich blood supply from the superior and inferior thyroid arteries and is drained by the superior, middle, and inferior thyroid veins.

FUNCTION

The thyroid gland contains two different populations of cells, *follicular cells* and *parafollicular cells,* each of which has a different embryological origin and secretes different hormones.

Follicular Cell Hormones

The follicular cells are derived from a median outpocketing of the anterior wall of the embryonic pharynx. These cells secrete two amino acid hormones, *thyroxin* and *triiodothyronine,* that regulate the basal metabolic rate of the body. Both of these hormones are iodinated: thyroxin (T_4) contains four iodine atoms, while triiodothyronine (T_3) contains three iodine atoms. T_4 and T_3 are linked to a protein polysaccharide substance called *thyroglobulin* and stored as colloid in the follicular lumen. When stimulated by TSH, the follicular cells remove colloid from the lumen, detach the hormones from thyroglobulin enzymatically, and release T_4 and T_3 into peripheral circulation. It is important to note that though T_4 constitutes 95 percent of the circulating thyroid hormones, T_3 is metabolically more active. The major effect of T_4 and T_3 is to increase the basal metabolic rate by speeding up metabolic activities in most of the body's tissues. These hormones are responsible for an individual's energy (or lack of it), for growth and development (especially of the skeletal and nervous systems), for the functioning of the nervous system, for the texture of the skin, and for the luster of the hair. Secretion of the follicular hormones is regulated

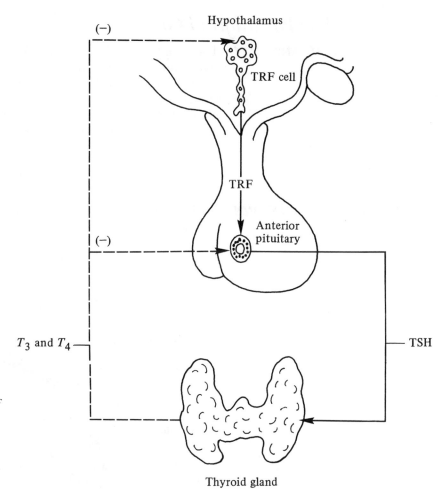

Hypothalamus

TRF cell

TRF

Anterior pituitary

$(-)$

$(-)$

T_3 and T_4 — TSH

Thyroid gland

Figure 13.15 Hormonal Regulation of Thyroid Gland. Schematic diagram shows relationships among hypothalamus, anterior pituitary, and thyroid gland. Remember that parafollicular cells are not regulated by trophic hormone from anterior pituitary. Negative feedback $(-)$.

by TRF/TSH secretion, by the availability of dietary iodine, and by a negative feedback mechanism involving blood-thyroxin levels (Fig. 13.15).

Parafollicular Cell Hormone

The parafollicular cells are derived from the fifth pharyngeal pouches (outpocketings of the lateral walls of the embryonic pharynx) and are incorporated into the developing thyroid gland. These cells secrete *calcitonin,* a polypeptide hormone that lowers blood-calcium levels. Calcitonin acts primarily on bone. It decreases the rate of bone resorption plus the rate of calcium absorption from the intestines, which effectively reduces the blood-calcium level. Secretion of calcitonin is regulated by blood-calcium levels and by the counteracting effect of the parathyroid hormone.

The Parathyroid Glands

GROSS ANATOMY   The parathyroid glands [Fig. 13.16(a)] are four small oval bodies derived from the third and fourth pharyngeal pouches (outpocketings from the lateral walls of the embryonic pharynx). They are usually situated on the posterior borders of the lateral lobes of the thyroid gland, between the parenchyma and the capsule. The parathyroid glands are arranged in superior and inferior pairs; the superior pair lie at the level of the lower border of the cricoid cartilage, while the inferior pair may lie near the basal end of the lateral lobes or even at some distance below the thyroid gland.

MICROSCOPIC ANATOMY   Each parathyroid gland [Fig. 13.16(b)] consists of a pea-sized mass of parenchyma enclosed by a thin connective tissue capsule. The parathyroid parenchyma contains two cell types, *chief cells* and *oxyphilic cells*, which are arranged in anastomosing cords or follicles. The chief

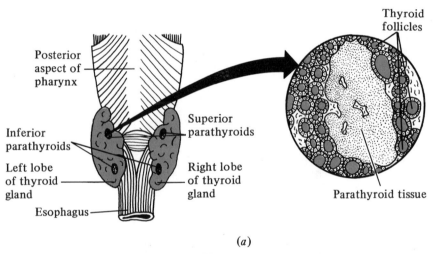

(a)

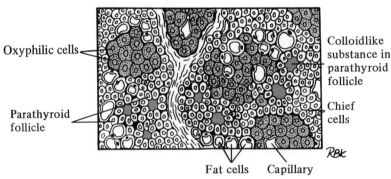

(b)

Figure 13.16 Parathyroid Glands. (a) Gross anatomy. Posterior aspect of pharynx shows relationship of parathyroid glands to lobes of thyroid gland. *Inset* shows parathyroid tissue embedded in thyroid tissue. (b) Microscopic anatomy. Section shows chief cells *(white)* and oxyphilic cells *(gray)*. Note that some of chief cells form follicles.

cells, which are more numerous than the oxyphilic cells, secrete the parathyroid hormone. The oxyphilic cells, which are larger than the chief cells, increase in number with the age of the individual. Their function is unknown; it has been suggested that they might be worn-out chief cells.

BLOOD SUPPLY   The superior parathyroid glands are supplied by the superior thyroid arteries, while the inferior thyroid glands are supplied by the inferior thyroid arteries. The glands are drained by the corresponding thyroid veins.

FUNCTION   The chief cells of the parathyroid glands secrete a protein hormone called *parathormone (PTH)*. PTH is of great importance in regulating calcium and phosphate balance and is therefore essential for life. It elevates the blood-calcium level by stimulating the resorption of bone and maintains the body's electrolytic balance by stimulating the renal reabsorption of calcium and excretion of phosphate. It also promotes the intestinal absorption of dietary calcium. Secretion of PTH is regulated by blood-calcium levels, the availability of dietary calcium, and the counteracting effect of calcitonin.

The Endocrine Pancreas

GROSS ANATOMY   The pancreas [Fig. 13.17(*a*)] is a large flattened digestive gland that lies in the abdominal cavity between the duodenum and spleen. Its gross appearance has been described in Chapter 12.

MICROSCOPIC ANATOMY   The endocrine portion of the pancreas [Fig. 13.17(*b*)] consists of approximately one million scattered spherical clusters of cells known as the *pancreatic islets* (of Langerhans). Each islet is surrounded by a reticular fiber capsule that separates it from exocrine pancreatic tissue. The cells forming the pancreatic islet are arranged in cords or clusters separated by pancreatic capillaries. Three different cell types have been recognized in the islets on the basis of the staining reactions of their secretory granules: *alpha cells* (20 to 30 percent), *beta cells* (60 to 80 percent), and *delta cells* (2 to 8 percent).

FUNCTION   At present, three different hormones are associated with the pancreatic islet cells: *insulin, glucagon,* and *gastrin.*

Insulin.   Insulin is a protein hormone secreted by the beta cells of the pancreatic islets and is essential for life. It regulates the rate at which the body utilizes carbohydrates and the amount of glucose present in the blood. Insulin increases the uptake and cellular utilization of glucose, decreases the blood-glucose level, and promotes the conversion

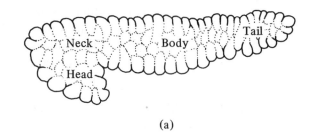

(a)

Figure 13.17 Pancreas. (a) Gross anatomy. Anterior aspect. (b) Microscopic anatomy. Section of endocrine pancreas shows islet of Langerhans surrounded by exocrine pancreatic acini. Islet cells are indicated as follows: beta cells, *gray with granules;* alpha cells, *white with granules;* delta cells, *white with no granules.*

Pancreatic acini (exocrine pancreas)

Islet of Langerhans

Alpha cells

Delta cell

Capillary

Beta cells

(b)

of glucose to glycogen in liver and muscle cells. Insulin secretion is regulated by blood-glucose levels and the counteracting effect of glucagon. GH, ACTH, and epinephrine indirectly influence insulin secretion, since they all can cause elevation of blood-glucose levels.

Glucagon. Glucagon is a protein hormone secreted by the alpha cells of the pancreatic islets. Its action is opposite to that of insulin. Glucagon elevates the blood-glucose level by stimulating the conversion of liver glycogen to glucose. Glucagon secretion is regulated by blood-glucose levels and the counteracting effect of insulin.

Gastrin. Gastrin is a protein hormone thought by some to be secreted by the delta cells of the pancreatic islets as well as by the pyloric mucosa. Gastrin stimulates the gastric glands to secrete hydrochloric acid. Its secretion is regulated by the presence of food in the stomach.

The Pineal Gland

GROSS ANATOMY The pineal gland or body [Fig. 13.18(*a*)] develops as an outgrowth from the roof of the embryonic brain. In the adult it is a small cone-

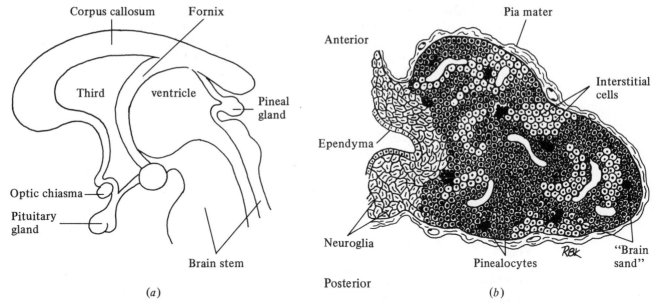

(a)

(b)

Figure 13.18 Pineal Gland. (a) Gross anatomy. Sagittal section through brain shows relationship of pineal gland to roof of diencephalon and third ventricle. Note also position of pituitary gland. (b) Microscopic anatomy. Sagittal section through pineal gland shows location of pinealocytes, interstitial cells, and calcified bodies ("brain sand"). These calcified bodies are used as reference point in skull X rays because they stand out very clearly in skull films.

shaped gland, about $\frac{1}{4}$ inch long, that is attached by a stalk to the roof of the diencephalon (a part of the brain).

MICROSCOPIC ANATOMY The pineal parenchyma [Fig. 13.18(*b*)] is enclosed in a capsule continuous with the pia mater (innermost connective tissue covering of the brain). Inward projections from the capsule subdivide the pineal parenchyma into irregular lobules. There are two main cell types present in the pineal parenchyma: *pinealocytes* and *interstitial cells*. Pinealocytes are epithelioid cells that make up the bulk of the parenchyma. They are thought to secrete the pineal hormone(s). The interstitial cells are considered by some to be neuroglia. The pineal gland reaches its maximum development about the seventh year of life and then starts to undergo involution. After puberty the glandular elements are gradually replaced by connective tissue. The human pineal gland also contains laminated calcified bodies called *brain sand*. The amount of brain sand increases with the age of the individual, but its significance is unknown.

BLOOD SUPPLY The pineal gland is supplied by the posterior cerebral arteries and is drained by veins that are tributaries of the dural venous sinuses.

FUNCTION The endocrine function of the pineal gland, especially in humans, is still not very clear, although it is being studied intensively. The pineal has been demonstrated to produce a hormone called *melatonin.* In amphibians melatonin causes clumping of melanin granules in the melanocytes, which results in lightening of the skin. It does not have the same effect on mammalian melanocytes. In higher animals and possibly in humans, melatonin is thought to have an antigonadotropic effect; that is, it inhibits the release of the pituitary gonadotropins, preventing early onset of puberty. Melatonin is also thought to exert an influence on light-induced biological cycles. In addition to melatonin, the pineal gland appears to produce a group of hormonelike substances that impose *circadian (24-hour) rhythms* on the secretory activities of other endocrine glands.

Diseases and Disorders of the Endocrine System

DISEASES AND DISORDERS OF THE PITUITARY GLAND

Hypofunction of Anterior Pituitary. Hypofunction of the anterior pituitary, called *Simmonds' disease,* is due to the destruction of the anterior pituitary by tumors or necrosis following severe blood loss (especially during childbirth). Destruction or necrosis of the glandular tissue is followed by loss of the anterior pituitary hormones and atrophy of their target organs. The clinical symptoms are primarily those of sex hormone, thyroid hormone, and corticosteroid hormone hyposecretion. Simmonds' disease is treated by giving the patient commercially produced hormones (androgens or estrogens, thyroxin, and hydrocortisone) to replace those that he or she is lacking. Growth hormone may be given to young patients to prevent premature closure of the epiphyseal plates.

Hyperfunction of the Anterior Pituitary. Hyperfunction of the anterior pituitary is almost always caused by a benign acidophil tumor and is expressed as hypersecretion of growth hormone. As these tumors enlarge they may impair vision or cause blindness by pressing on the optic nerves, which cross just above and in front of the pituitary gland. Hypersecretion of growth hormone before the epiphyseal plates close produces *gigantism,* a proportional increase in size (think of the 8-foot-tall circus giant). Hypersecretion of growth hormone after the epiphyseal plates have closed produces *acromegaly.* This disease is characterized by the progressive growth of soft tissues and thickening of bone, especially membrane bones of the skull and the long bones of the hands and feet. There is conspicuous enlargement of the supraorbital ridges, cheeks, lower jaw, nose, lips, hands, and feet.

Treatment for both gigantism and acromegaly consists of irradiation or surgical removal of the tumor.

Hypofunction of the Posterior Pituitary.   Hypofunction of the posterior pituitary is usually caused by a hypothalamic or pituitary tumor and is expressed as hyposecretion of ADH. ADH hyposecretion produces *diabetes insipidus* (water diabetes). Diabetes insipidus is characterized by the production of a large volume of very dilute urine, since the kidney tubules are unable to reabsorb water from the urine. There is no glucose in the urine, and blood-glucose levels are normal; this distinguishes diabetes insipidus from diabetes mellitus (sugar diabetes). Treatment consists of destruction or removal of the tumor plus ADH replacement therapy with Pitressin®, a commercial posterior pituitary extract.

DISEASES AND DISORDERS OF THE GONADS

Ovarian Hypofunction.   Ovarian hypofunction is usually expressed as estrogen deficiency. Estrogen deficiency before puberty causes *primary amenorrhea* (failure to develop menstrual periods and female secondary sex characteristics); estrogen deficiency after puberty causes *secondary amenorrhea* (cessation of menstrual periods and regression of female secondary sex characteristics). The causes of ovarian hypofunction range from hypothalamic or pituitary dysfunction (which interfers with FSH and LH secretion) to ovarian disease or absence of the ovaries. Treatment usually consists of FSH replacement therapy (Pergonal®) if hypothalamic or pituitary function is disturbed or estrogen replacement therapy (Premarin®) if ovarian function is absent, disturbed, or lost.

Ovarian Hyperfunction.   Ovarian hyperfunction is usually expressed as estrogen excess (hyperestrinism). Hyperestrinism in a young girl causes precocious puberty. Hyperestrinism in an adult woman may cause uterine and breast tumors or heavy menstrual bleeding. In a postmenopausal woman, hyperestrinism stimulates the return of menstrual bleeding. The usual cause of hyperestrinism is an estrogen-secreting ovarian tumor, which is treated by *ovariectomy* (surgical removal of the ovary).

Testicular Hypofunction.   Testicular hypofunction is expressed as testosterone deficiency. Testosterone deficiency at puberty causes *eunuchism* (failure to develop male secondary sex characteristics); testosterone deficiency after puberty causes regression of male secondary sex characteristics. The causes of testicular hypofunction range from hypothalamic or pituitary dysfunction (which interferes with LH secretion) to testicular injury or disease. Treatment for hypothalamic or pituitary dysfunction or for undescended testes consists of administering a commercial chorionic gonadotropin preparation, which has LH-like activity. If testicular function is lost or disturbed, a commercial testosterone preparation is administered.

Testicular Hyperfunction. Testicular hyperfunction is usually caused by some type of hypothalamic, pituitary, or testicular tumor and is expressed as testosterone excess. Testosterone excess in a young boy causes precocious puberty; in an adult male it causes an exaggeration of male secondary sex characteristics. Treatment for testicular hyperfunction consists of irradiation destruction or surgical removal of the responsible tumor.

DISEASES AND DISORDERS OF THE ADRENAL GLANDS
Cortical Hypofunction

Chronic adrenocortical hypofunction, called *Addison's disease,* is characterized by low blood pressure, fatigue, and gastrointestinal upsets (mineralocorticoid deficiency) plus weakness and weight loss (glucocorticoid deficiency), and hyperpigmentation of skin. Addison's disease is caused by destruction or degeneration of the adrenal cortex. Combined mineralocorticoid (deoxycorticosterone) and glucocorticoid (hydrocortisone) replacement therapy makes it possible for patients with Addison's disease to live a relatively normal life. However, they are unable to react to stressful situations and may go into acute vascular collapse (Addisonian crisis) when under stress.

Cortical Hyperfunction

Adrenocortical hyperfunction is usually expressed as mineralocorticoid excess (*Conn's syndrome*), glucocorticoid excess (*Cushing's syndrome*), or adrenal androgen excess (*adrenogenital syndrome*).

Conn's Syndrome. Conn's syndrome, or *primary hyperaldosteronism,* is a minor but important cause of high blood pressure in humans. It is characterized by high blood pressure plus high blood-sodium and low blood-potassium levels. Conn's syndrome is usually caused by a benign tumor of zona glomerulosa cells and is treated by surgical removal of the tumor.

Cushing's Syndrome. Cushing's syndrome is characterized by moon face, truncal obesity (large abdomen and thin extremities), buffalo hump, purple striae (on the abdomen), excess facial and body hair, and high blood pressure. Cushing's syndrome can be caused by ACTH-secreting pituitary tumors as well as adrenocortical tumors. Treatment consists of surgical removal of the responsible tumor. In some instances bilateral adrenalectomy is necessary, and the patient must then be maintained on corticosteroid replacement therapy. Cushing's syndrome can also be produced as a side effect of glucocorticoid administration in the treatment of diseases such as leukemia or homograft rejection.

Adrenogenital Syndrome. The adrenogenital syndrome is characterized by *virilization* (masculinization). Adrenal androgen excess causes precocious puberty in young males and the appearance of male secondary sex characteristics in females. In its congenital form this syndrome is caused by a defect in the corticosteroid biosynthetic pathway. The defect leads to increased synthesis of adrenal androgens and decreased synthesis of glucocorticoids. When caused by a biochemical defect, the adrenogenital syndrome is treated by exogenous glucocorticoid suppression of the adrenal cortex.

Medullary Hyperfunction

Medullary hyperfunction is expressed as catecholamine excess and is characterized by intermittent episodes of hypertension, palpitations, sweating, and nervousness. It is caused by a *pheochromocytoma,* a catecholamine-secreting medullary tumor, and is treated by surgical removal of the tumor.

DISEASES AND DISORDERS OF THE THYROID GLAND
Hypofunction of the Thyroid Gland

Goiter.   Simple goiter is characterized by enlargement of the thyroid gland and is expressed as mild thyroxin deficiency. It is caused by lack of dietary iodine. Iodine deficiency stimulates the thyroid to produce an excessive amount of iodine-poor colloid, which remains stored in the thyroid follicles. Goiter is common in certain parts of the world, such as the American Midwest and Switzerland, where the soil and the food grown on it are poor in iodine. Simple goiter is treated by increasing dietary iodine intake through use of iodized salt or iodine tablets.

Cretinism.   Severe hypothyroidism in a child, called *cretinism,* is expressed as marked thyroxin deficiency and is characterized by physical and mental retardation. Causes of cretinism range from TSH insufficiency to congenital absence of the thyroid gland. The earlier thyroxin replacement therapy is instituted, the better the child's chances are for normal physical and mental development.

Myxedema.   Severe hypothyroidism in an adult, called *myxedema,* is expressed as marked thyroxin deficiency and is characterized by physical and mental slowdown plus the accumulation of a mucoid fluid in the superficial fascia. Causes of myxedema range from TSH insufficiency to atrophy or surgical removal of the thyroid gland. Treatment consists of thyroxin replacement therapy.

Hyperfunction of the Thyroid Gland

Severe hyperthyroidism, called *Graves' disease,* is expressed as thyroxin excess and is characterized by nervousness and heat intolerance, often accompanied by *exophthalmos* (bulging of the eyeballs). Graves' disease now appears to be an autoimmune disease in which the thyroid gland is stimulated by an abnromal antibody called long-acting thyroid stimulator (LATS). It is generally treated by surgical or chemical reduction of thyroxin secretion. For example, the hypersecreting tissue can be destroyed by the administration of radioactive iodine, which concentrates in the thyroid gland.

DISEASES AND DISORDERS OF THE PARATHYROID GLANDS

Hypofunction of the Parathyroid Glands. *Hypoparathyroidism* is expressed as parathormone deficiency and is characterized by low blood-calcium levels, tetany (muscle spasms), and convulsions. The convulsions may terminate in death. Hypoparathyroidism is most commonly seen after intentional or accidental surgical removal of the parathyroid glands. It is treated by diet therapy (high calcium, low phosphate) plus long-term administration of calcium salts and vitamin D.

Hyperfunction of the Parathyroid Glands. *Hyperparathyroidism* is expressed as parathormone excess and is characterized by excessive urinary excretion of calcium and phosphate, which leads to the formation of kidney stones, and bone lesions. Hyperparathyroidism is caused by parathyroid hyperplasia or parathormone-secreting tumors. It is treated by surgical removal of the parathyroids followed by diet therapy plus calcium and vitamin D replacement therapy.

DISEASES AND DISORDERS OF THE PANCREATIC ISLETS

Hypofunction of the Pancreatic Islets. Hypofunction of the pancreatic islets is usually expressed as hypoinsulinism, which causes *diabetes mellitus* (sugar diabetes). Diabetes mellitus is characterized by hyperglycemia (high blood-glucose levels), glycosuria (excretion of glucose in the urine), polyuria (production of a large volume of urine), and polydipsia (excessive thirst). Diabetes mellitus is primarily an inherited metabolic defect that results in beta cell failure. It may be triggered by obesity, emotional stress, or viral infections of the pancreas. Mild diabetes (adult onset type) is generally treated by diet therapy and oral antidiabetic agents, which stimulate insulin secretion by surviving beta cells. Severe diabetes (juvenile onset type) is treated by diet therapy plus insulin replacement therapy. Diabetic coma, which may terminate in death, is a frequent complication of untreated diabetes.

Hyperfunction of the Pancreatic Islets. Hyperfunction of the pancreatic islets is usually expressed as hyperinsulinism and is characterized by hypoglycemia (low blood-glucose levels), hunger, weakness, and convulsions. Hyperinsulinism is usually caused by a benign beta cell tumor and is treated by surgical removal of the tumor.

Suggested Readings

ASCHOFF, JUERGEN. "Circadian systems in man and their implications," *Hosp. Pract., 11,* no. 5 (May 1976), pp. 51–57.

BARZEL, URIEL S. "Parathyroid hormone and the buffering function of bone," *Hosp. Pract., 6,* no. 9 (September 1971), pp. 131–138.

DAVIDSON, ERIC H. "Hormones and genes," *Sci. Am., 212,* no. 6 (June 1965), pp. 36–45.

FRIESEN, HENRY G. "Prolactin: Its physiological role and therapeutic potential," *Hosp. Pract.*, 7, no. 9 (September 1972), pp. 123–130.

GILLIE, R. BRUCE. "Endemic goiter," *Sci. Am.*, 224, no. 6 (June 1971), pp. 92–101.

GUILLEMIN, ROGER, "Hypothalamic hormones: Releasing and inhibiting factors," *Hosp. Pract.*, 8, no. 11 (November 1973), pp. 111–120.

————, and ROGER BURGUS. "The hormones of the hypothalamus," *Sci. Am.*, 227, no. 5 (November 1972), pp. 24–33.

HOLLANDER, CHARLES S. "Newer aspects of hyperthyroidism," *Hosp. Pract.*, 7, no. 5 (May 1972), pp. 87–96.

JAFFEE, ROBERT B. "Endocrine interactions and the placenta," *Hosp. Pract.*, 6, no. 12 (December 1971), pp. 71–82.

° KASHGARIAN, MICHAEL, and GERARD N. BURROW. *The Endocrine Glands.* Baltimore: Williams & Wilkins, 1974.

KOPIN, IRWIN J. "Catecholamines, adrenal hormones, and stress," *Hosp. Pract.*, 11, no. 3 (March 1976), pp. 49–55.

KRIEGER, DOROTHY T. "The hypothalamus and neuroendocrinology," *Hosp. Pract.*, 6, no. 9 (September 1971), pp. 87–99.

LERNER, AARON B. "Hormones and skin color," *Sci. Am.*, 205, no. 1 (July 1961), pp. 98–108.

LEVINE, SEYMOUR. "Sex differences in the brain," *Sci. Am.*, 214, no. 4 (April 1966), pp. 84–90.

LI, CHOH HAO. "The ACTH molecule," *Sci. Am.*, 209, no. 1 (July 1963), pp. 46–53.

McEWEN, BRUCE S. "Interactions between hormones and nerve tissue," *Sci. Am.*, 235, no. 1 (July 1976), pp. 48–58.

McKUSICK, VICTOR A., and DAVID L. RIMOIN. "General Tom Thumb and other midgets," *Sci. Am.*, 217, no. 1 (July 1967), pp. 102–110.

MONIF, GILLES R. G. "Can diabetes mellitus result from an infectious disease?" *Hosp. Pract.*, 8, no. 3 (March 1973), pp. 124–130.

NETTER, FRANK H. *The Ciba Collection of Medical Illustrations.* Vol. 4: *The Endocrine System and Selected Metabolic Diseases.* Summit, N.J.: Ciba, 1965.

RAISZ, LAURENCE G. "Calcium homeostasis and bone metabolism," *Hosp. Pract.*, 5, no. 5 (May 1970), pp. 74–83.

RASMUSSEN, HOWARD, and MAURICE M. PECHET, "Calcitonin," *Sci. Am.*, 223, no. 4 (October 1970), pp. 42–50.

RIMOIN, DAVID L. "Genetic defects of growth hormone," *Hosp. Pract.*, 6, no. 2 (February 1971), pp. 113–124.

°° ————, and R. NEIL SCHIMKE. *Genetic Disorders of the Endocrine Glands.* St. Louis: Mosby, 1971.

° TURNER, C. DONNELL, and JOSEPH T. BAGNARA. *General Endocrinology.* 6th ed. Philadelphia: Saunders, 1976.

WEITZMAN, ELLIOT D. "Biologic rhythms and hormone secretion patterns," *Hosp. Pract.*, 11, no. 8 (August 1976), pp. 79–86.

WILKINS, LAWSON. "The thyroid gland," *Sci. Am.*, 202, no. 3 (March 1960), pp. 119–129.

WURTMAN, RICHARD J. "The effects of light on the human body," *Sci. Am.*, 233, no. 1 (July 1975), pp. 68–77.

———, and JULIUS AXELROD. "The pineal gland," *Sci. Am., 213,* no. 1 (July 1965), pp. 50–60.

ZUCKER, IRVING. "Light, behavior, and biologic rhythms," *Hosp. Pract., 11,* no. 10 (October 1976), pp. 83–91.

° Advanced level.

°° Highly advanced level.

14

The Reproductive System

In higher animals, including humans, reproduction is *sexual*. Individuals are either male or female and produce male or female sex cells, which must unite to form a new individual and perpetuate the species. Though the male and female reproductive systems differ markedly in sexually mature individuals, they both contain the following types of genital organs:

1. *Gonads (Sex Glands).* Glands that manufacture the sex cells and secrete the sex hormones.
2. *Genital Ducts.* Ducts that transport the sex cells from the site of production to the site of fertilization.
3. *Accessory Glands.* Exocrine glands whose secretions are essential for sexual intercourse.
4. *External Genitalia.* Perineal organs that are directly or indirectly involved in sexual intercourse.

This chapter is divided into three parts: Part A deals with the male reproductive system, Part B deals with the female reproductive system, and Part C deals with human reproduction and its regulation.

Part A. The Male Reproductive System

The male reproductive system (Fig. 14.1) consists of the testes, male genital ducts and accessory glands, and male external genitalia.

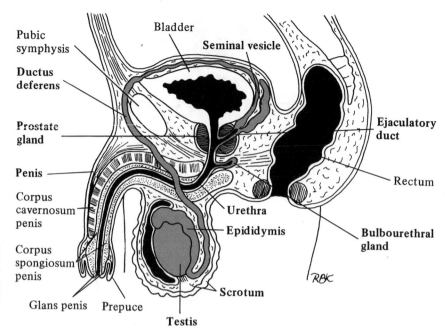

Figure 14.1 *Male Reproductive System.* Sagittal section through male pelvis and external genitalia showing testis, male genital ducts, and accessory organs. Course of ductus deferens is indicated though it would not normally appear in such a section.

Labels on figure: Pubic symphysis, Ductus deferens, Prostate gland, Penis, Corpus cavernosum penis, Corpus spongiosum penis, Glans penis, Prepuce, Bladder, Seminal vesicle, Urethra, Epididymis, Scrotum, Testis, Ejaculatory duct, Rectum, Bulbourethral gland, RBK

The Testes

The testes (male gonads) produce *sperm* (the male sex cells) and secrete *androgens* (the male sex hormones).

GROSS ANATOMY The testes develop in the abdominal cavity but descend into the *scrotum,* an integumentary pouch outside the abdominal cavity, about two months before birth. Sometimes one or both testes fail to descend and remain in the abdominal cavity. This condition is known as *cryptorchidism* (hidden testis). Since the temperature in the abdominal cavity is too high for normal sperm development, it is important for cryptorchidism to be corrected surgically before puberty (when sperm production begins).

Each adult testis (Fig. 14.2) is an ovoid glandular organ about 1.8 inches long and 1 inch wide. It has medial and lateral surfaces, anterior and posterior borders, and superior and inferior poles. The testis is suspended in its scrotal sac at the end of a long vascular stalk (the spermatic cord), with its lateral surface facing upward and backward and its medial surface facing downward and forward.

The testis has four coverings, which it acquires from the abdominal wall as it descends into the scrotum: the external spermatic fascia, which is continuous with the fascia of the external oblique muscle; the middle spermatic fascia *(cremaster layer),* which is continuous with the muscle fibers and fascia of the internal oblique muscle; the internal spermatic fascia, which is continuous with the fascia of the

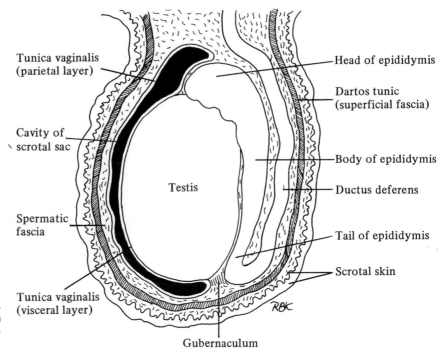

Tunica vaginalis (parietal layer)

Cavity of scrotal sac

Testis

Spermatic fascia

Tunica vaginalis (visceral layer)

Head of epididymis

Dartos tunic (superficial fascia)

Body of epididymis

Ductus deferens

Tail of epididymis

Scrotal skin

Gubernaculum

Figure 14.2 Testis: Gross Anatomy. Schematic representation of adult testis in scrotum. Scrotal wall is sectioned to reveal its component layers.

transversus abdominis muscle; and the *tunica vaginalis,* which is continuous with the peritoneum. The tunica vaginalis lines the scrotal wall and is reflected over the testis and epididymis to form the *scrotal sac.* The scrotal sac has a potential space containing a thin film of serous fluid that provides for frictionless movement of the testis.

MICROSCOPIC ANATOMY

The testis is a compound tubular gland enclosed in a thick fibrous connective tissue capsule called the *tunica albuginea.* The tunica albuginea is greatly thickened along the posterior border of the testis. This thickened region is called the *mediastinum testis.* Connective tissue septa project inward from the mediastinum to divide the parenchyma into numerous wedge-shaped testicular lobules (Fig. 14.3).

Each testicular lobule contains one to four coiled *seminiferous tubules.* The seminiferous tubules are supported by a highly vascular loose connective tissue stroma containing clusters of interstitial cells (of Leydig). Each seminiferous tubule is lined with a complex stratified epithelium consisting of spermatogenic cells and supporting cells (Fig. 14.4). The spermatogenic cells are stacked four to eight layers deep between the periphery and the lumen of the tubule. They represent different stages in the continuous process of male sex cell formation. The cells closest to the periphery of the tubule are germ cells called *spermatogonia.* The cells closest to the lumen of the

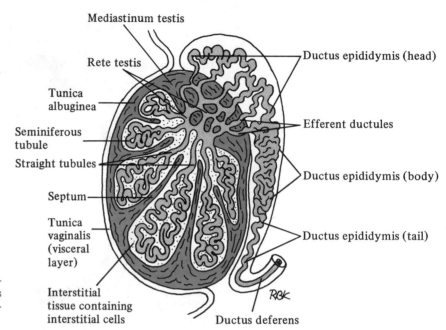

Mediastinum testis

Rete testis

Tunica albuginea

Seminiferous tubule

Straight tubules

Septum

Tunica vaginalis (visceral layer)

Interstitial tissue containing interstitial cells

Ductus epididymis (head)

Efferent ductules

Ductus epididymis (body)

Ductus epididymis (tail)

Ductus deferens

Figure 14.3 Testis: Microscopic Anatomy. Sagittal section through adult testis shows seminiferous tubules plus intra- and extratesticular ducts.

tubule are maturing sex cells or *spermatozoa* (sperm). The supporting cells, called *Sertoli cells,* extend the full thickness of the seminiferous tubule. These wedge-shaped cells are thought to provide support and nutrients for the spermatogenic cells.

BLOOD SUPPLY   The testes are supplied by the testicular arteries, which are direct branches of the abdominal aorta. They are drained by tributaries of the testicular veins, which form a dense network in the spermatic cord called the *pampiniform plexus.* The right testicular vein is a tributary of the inferior vena cava, while the left is a tributary of the left renal vein.

INNERVATION   The testes receive autonomic nerve fibers by way of the pelvic and testicular plexuses.

FUNCTION   The testes have two major functions: sperm formation and androgen secretion.

Sperm Formation   Sperm can be considered to be the secretory products of the seminiferous tubules. The term *spermatogenesis* is used to describe the entire sequence of events in sperm formation (Fig. 14.5). Spermatogenesis begins at puberty and continues until death. For descriptive purposes spermatogenesis is divided into three phases: *spermatocytogenesis, meiosis,* and *spermiogenesis.* Spermatocytogenesis refers to the proliferation of spermatogonia by mitotic division. A newly

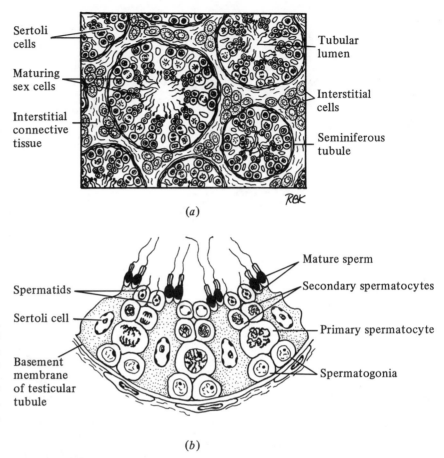

Sertoli cells

Maturing sex cells

Interstitial connective tissue

Tubular lumen

Interstitial cells

Seminiferous tubule

(a)

RBK

Spermatids

Sertoli cell

Basement membrane of testicular tubule

Mature sperm

Secondary spermatocytes

Primary spermatocyte

Spermatogonia

(b)

Figure 14.4 Testicular Tubules: Microscopic Anatomy. (a) Cross section of testicular tissue shows several seminiferous tubules. Note that interstitial cells are located in connective tissue surrounding tubules. (b) Highly magnified segment of seminiferous tubule in cross section showing Sertoli (supporting) cells and male sex cells in various stages of spermatogenesis.

formed spermatogonium may either continue to undergo mitotic division or differentiate into a primary spermatocyte, a cell capable of undergoing meiotic division. During meiosis each primary spermatocyte divides to form two *secondary spermatocytes* (meiosis I), and each secondary spermatocyte divides to form two *spermatids* (meiosis II). Thus each primary spermatocyte that enters meiosis gives rise to a cluster of four spermatids. Each spermatid has half the number of chromosomes as the spermatogonium or the primary spermatocyte. During spermiogenesis (Fig. 14.6) spermatids undergo a series of cytological transformations that turn them into sperm. The mature sperm is a highly modified cell that has lost most of its cytoplasm and gained a flagellum.

Spermatogenesis in humans is a slow cyclic process that takes about 64 days from start to finish. It occurs in a wavelike fashion throughout the seminiferous tubule so that the spermatogenic cells in each region of the tubule are in a different phase of spermatogenesis. Temperature plays an important role in the regulation of spermatogenesis.

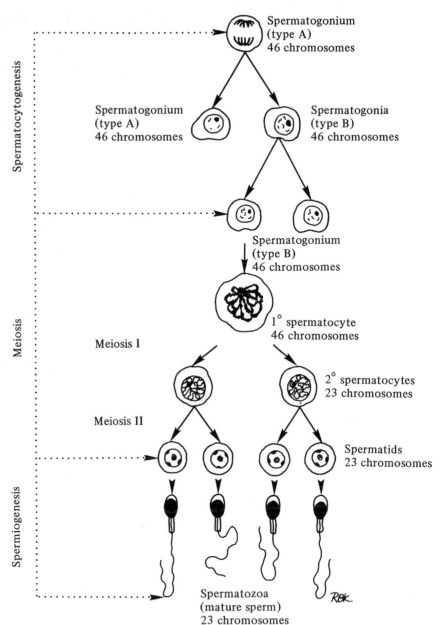

Figure 14.5 Spermatogenesis. Schematic representation of events occurring during sperm formation. Note that type A spermatogonia serve as stem cells and that type B spermatogonia enter meiosis. Also note that first meiotic division (meiosis I) is reductional and second meiotic division (meiosis II) is equational with respect to chromosome number.

Temperatures at or above 98.6°F (normal body temperature) inhibit spermatogenesis; this is why undescended testes in an adult male produce no sperm. A hot bath or tight pants may adversely affect spermatogenesis by elevating the testicular temperature. Spermatogenesis is also adversely affected by malnutrition, alcoholism, use of certain drugs, and administration of female sex hormones. The endocrine regulation of spermatogenesis has been discussed in Chap-

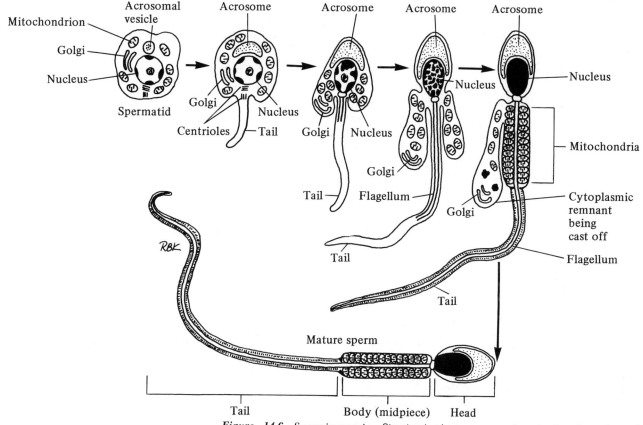

Figure 14.6 Spermiogenesis. Structural changes occurring in transformation of spermatid into mature sperm.

ter 13. X-irradiation of the testes causes irreversible sterility through destruction of the spermatogonia.

Androgen Secretion Androgens (male sex hormones) are produced and secreted by the Leydig cells under the stimulus of LH from the pituitary. Testosterone, the principal androgen secreted by the testes, is essential for sperm formation and for the differentiation and maintenance of the male genitalia.

Genital Ducts and Accessory Glands

INTRATESTICULAR DUCTS
Anatomy

In the posterior region of the testis, at the apex of each lobule, the seminiferous tubules unite with short straight epithelial tubules called *tubuli recti*, which represent the first part of the excretory duct system. The tubuli recti in turn unite with a network of anastomosing epithelial channels in the mediastinum called the *rete testis*. Approximately ten to fifteen *efferent ductules,* short epithelial tubules that arise from the upper part of the rete testis, emerge from the substance of the testis at the mediastinum and unite with the epididymis.

Function The intratesticular ducts serve to conduct sperm from the seminiferous tubules to the epididymis, the first part of the extratesticular duct system. The sperm are suspended in a slow-moving fluid continuously secreted by the seminiferous epithelium.

EPIDIDYMIS
Gross Anatomy The epididymis is a long highly coiled tube that consists of a *head, body,* and *tail.* The head, which receives the efferent ductules, caps the superior pole of the testis. The body extends down the posterior border of the testis. The tail, located at the inferior pole of the testis, becomes continuous with the ductus deferens. The epididymis is covered by the visceral layer of the tunica vaginalis, which helps bind this duct to the testis.

Microscopic Anatomy The ductus epididymis is lined with a pseudostratified columnar epithelium that appears to have both secretory and phagocytic functions. It also reabsorbs most of the testicular fluid. External to the epithelium is a coat of smooth muscle fibers that increases in thickness from head to tail. Contractions of the muscular coat propel sperm down through the epididymis in slow peristaltic waves.

Function The ductus epididymis, especially the tail, is the site of accumulation and storage of sperm. It may take up to six weeks for sperm to travel the entire length of the epididymis. As sperm pass through the epididymis, they undergo functional maturation; that is, they become motile and capable of fertilizing an ovum.

DUCTUS DEFERENS
AND SPERMATIC CORD
Gross Anatomy The ductus, or *vas,* deferens [Fig. 14.7(*a*)], is a long fibromuscular tube that extends from the tail of the epididymis to the prostatic urethra. It ascends the posterior border of the testis medial to the epididymis and is incorporated into the spermatic cord. As part of the spermatic cord, it emerges from the scrotum and passes through the inguinal canal. At the internal inguinal ring, it leaves the spermatic cord and takes a roundabout route through the abdominopelvic cavity to the posterior surface of the bladder. On the posterior surface of the bladder, it widens to form a dilated region called the *ampulla.* The distal portion of the ampulla unites with the duct of the seminal vesicle to form the *ejaculatory duct.* The ejaculatory duct extends obliquely through the substance of the prostate gland and opens onto the posterior wall of the prostatic urethra.

The spermatic cord [Fig. 14.7(*b*)] extends from the tail of the epididymis to the internal inguinal ring. In addition to the ductus deferens, it contains the testicular and deferential arteries, the

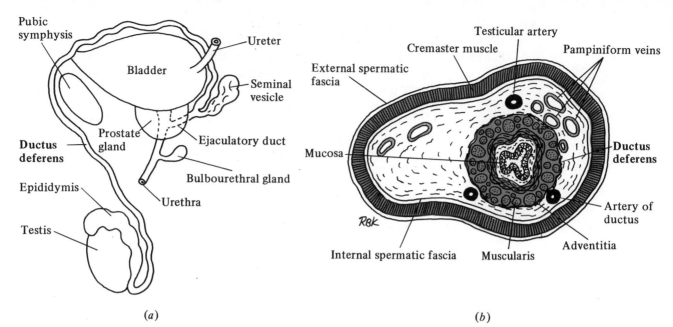

Figure 14.7 *Ductus Deferens.* (a) Gross anatomy. Course of ductus deferens as seen from lateral aspect of bladder and male reproductive organs. (b) Microscopic anatomy. Cross section through spermatic cord, with various coats of ductus wall and spermatic cord indicated.

pampiniform plexus of veins, and the nerves and lymphatics of the testis and epididymis. These structures are bound together by connective tissue and enclosed by the same fascial layers that cover the testis.

Microscopic Anatomy

For most of its length the ductus deferens is a firm cylindrical tube with a relatively narrow lumen and a thick wall. The wall of the ductus deferens consists of thee coats: *mucosa, muscularis,* and *adventitia.* The mucosa, which contains pseudostratified columnar epithelium, is thrown into longitudinal folds when the ductus is contracted. It is surrounded by a thick coat of spirally arranged smooth muscle fibers whose contractions cause the expulsion of sperm during ejaculation. Enclosing the muscularis is an adventitial coat of fibrous connective tissue. The ejaculatory duct loses the adventitia when it enters the prostate gland.

Blood Supply

The ductus deferens is supplied by the deferential artery, a branch of the superior vesicle artery, and is drained by tributaries of the pampiniform plexus, the vesicle veins, and the prostatic plexus.

Innervation

The ductus deferens is innervated by the autonomic nervous system. It receives sympathetic and parasympathetic fibers by way of the pelvic plexus.

Function Under the stimulus of parasympathetic impulses during the ejaculatory reflex, forceful rhythmic contractions of the paired deferential ducts propel stored sperm into the ejaculatory ducts, from which they are expelled into the prostatic urethra. *Vasectomy* (cutting and tying off the cut ends of the deferential ducts) is a relatively simple operation performed on men to prevent viable sperm from reaching the ejaculatory ducts. Today vasectomy is the most effective way of regulating male fertility.

SEMINAL VESICLES
Gross Anatomy The seminal vesicles (Fig. 14.8) are elongated saccular glands that are outgrowths of the deferential ducts. They are located above the base of the prostate gland, between the posterior surface of the bladder and the anterior surface of the rectum. Each seminal vesicle lies lateral to its companion deferential duct.

Microscopic Anatomy The seminal vesicles consist of two highly coiled tubes bound by fibrous connective tissue to form a pair of saccular organs. The wall of the seminal vesicle is similar in construction to that of the ductus deferens. However, its mucosa is thrown into a series of branching folds that subdivide the lumen into numerous pouches. The mucosa contains a pseudostratified columnar epithelium whose secretory cells are highly sensitive to testosterone. The muscularis of the seminal vesicle is much thinner than that of the ductus deferens and its adventitia is rich in elastic fibers.

Blood Supply The seminal vesicles are supplied by branches of the middle and in-

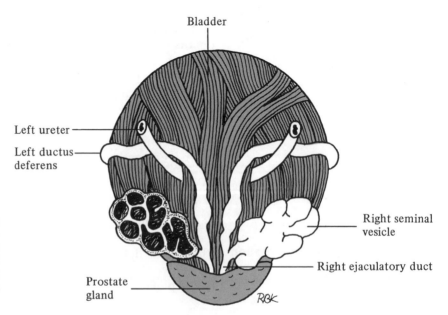

Figure 14.8 Seminal Vesicles. Posterior aspect of bladder showing seminal vesicles in normal position. Note relationship of seminal vesicles to ureters, deferential ducts, and prostate gland. Left seminal vesicle is sectioned to show pouches created by mucosal folds.

Bladder

Left ureter

Left ductus deferens

Right seminal vesicle

Right ejaculatory duct

Prostate gland

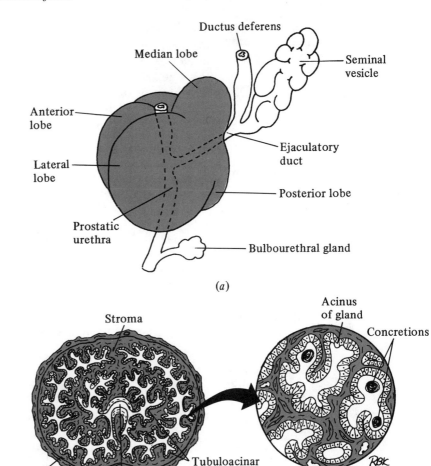

Figure 14.9 *Prostate Gland.* (a) Gross anatomy. Lateral aspect showing relative positions of male genital ducts. Note that ejaculatory duct penetrates substance of prostate gland to unite with prostatic urethra. (b) Microscopic anatomy. Cross section through prostate gland shows relationship of tubuloacinar glands and their ducts to urethra. Microscopic appearance of glandular elements is shown in *inset*.

ferior vesicle arteries and by branches of the middle rectal artery. They are drained by tributaries of the corresponding veins.

Innervation

The seminal vesicles are innervated by the autonomic nervous system. They receive sympathetic and parasympathetic fibers by way of the inferior hypogastric plexus.

Function

Under the stimulus of testosterone, the seminal vesicles secrete a thick yellowish alkaline fluid that forms the bulk of the semen. It is rich in vitamin C and fructose, substances that are important for the nutrition and motility of the sperm. Thus seminal fluid serves as a transport medium for sperm and also provides them with an energy source. Parasympathetic impulses cause contraction of the seminal vesicles and the release of seminal fluid into the ejaculatory ducts.

PROSTATE GLAND
Gross Anatomy

The prostate gland [Fig. 14.9(a)] is the largest accessory gland of the male reproductive system. It lies beneath the urinary bladder and encircles the proximal portion of the urethra. This gland, which is about the size of a horse chestnut, has a base, an apex, and four surfaces. The base of the prostate gland is applied to the neck of the bladder while its apex rests on the urogenital diaphragm. The ejaculatory ducts enter the substance of the prostate gland on its posterior surface, near the base.

Microscopic Anatomy

The parenchyma of the prostate gland consists of 30 to 50 small branched tubuloacinar glands whose ducts open into the prostatic urethra. There are several groups of glands present, and they are arranged in concentric layers around the urethra [Fig. 14.9(b)]. The glandular elements are embedded in a dense connective tissue stroma containing many smooth muscle and elastic fibers. The entire gland is enclosed by a fibromuscular capsule whose inward projections partition the parenchyma into lobes.

Blood Supply

The prostate gland is supplied by branches of the inferior vesicle and middle rectal arteries and is drained by small veins that empty into the prostatic plexus.

Innervation

The prostate gland is innervated by the autonomic nervous system. It receives sympathetic and parasympathetic fibers by way of the pelvic and prostatic plexuses.

Function

Under the stimulus of testosterone, the prostate gland secretes a thin milky slightly acidic fluid. During the ejaculation reflex, parasympathetic impulses cause contractions of the prostate gland. These contractions expel prostatic fluid into the urethra, where it is added to the semen. In most middle-aged men the glandular units begin to hypertrophy. This benign prostatic hypertrophy can cause partial or total obstruction of the urethra. In addition, adenocarcinoma of the prostate gland, the most common malignant tumor of old men, also begins in the glandular units.

BULBOURETHRAL GLANDS
Gross Anatomy

The bulbourethral glands, also known as Cowper's glands, are two small pea-shaped glands located in the urogenital diaphragm behind and lateral to the membranous urethra.

Microscopic Anatomy

Each bulbourethral gland is a compound tubuloacinar gland whose secretory units produce mucus. The glandular parenchyma is enclosed in a fibromuscular capsule whose inward projections partition the parenchyma into lobules. Each gland has a single excretory duct that pierces the urogenital diaphragm and opens into the penile portion of the urethra.

Function During the ejaculation reflex, the bulbourethral glands release their mucous secretion into the penile urethra, where it serves as a lubricant.

The Male Perineum

The perineum is the diamond-shaped region between the thighs that contains the urethral and anal openings. It is bounded anteriorly by the pubic arch and arcuate ligaments, posteriorly by the tip of the coccyx, and laterally by the inferior rami of the pubis and ischium. The anterior half of this diamond is called the *urogenital triangle*,

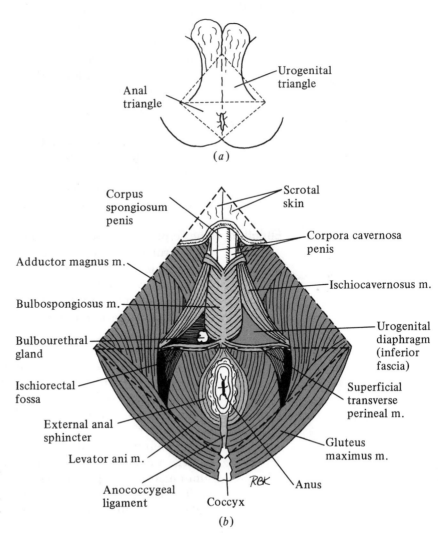

Figure 14.10 Male Perineum. (a) External aspect with urogenital and anal triangles indicated. (b) Schematic representation of muscles of superficial perineal compartment.

while the posterior half is called the *anal triangle*. The superficial compartment of the male perineum is shown in Fig. 14.10. Important structures include the crura and bulb of the penis, the ischiocavernosus and bulbospongiosus muscles, the external anal sphincter, and the superficial transverse perineal muscle.

External Genitalia

The male external genitalia, located in the perineal area, include the *scrotum* and the *penis*.

SCROTUM
Gross Anatomy

The scrotum is an integumentary pouch that contains the testes and their ducts. It is suspended from the perineum directly behind the penis and a short distance in front of the anus. A connective tissue septum partitions the scrotum into two lateral compartments, one for each testis. The scrotum is not symmetrical; the left side of the scrotum hangs somewhat lower than the right due to the greater length of the left spermatic cord.

Microscopic Anatomy

The wall of the scrotum is composed of skin and subcutaneous fascia. The scrotal skin is thin, pigmented, and usually wrinkled. It is covered with coarse hairs and contains well-developed sebaceous glands. Beneath the skin lies the subcutaneous fascia, or *dartos tunic*, a connective tissue layer that contains many smooth muscle fibers. The scrotal wall is lined by the parietal layer of the tunica vaginalis.

Function

The scrotum acts as a heat-regulating container that attempts to keep the spermatogenic cells in the testes at the right temperature to produce viable sperm. The smooth muscle fibers in the dartos tunic cause the scrotal wall to contract and draw the testes closer to the body in a cold environment and to relax and draw the testes further away from the body in a warm environment.

PENIS
Gross Anatomy

The penis [Fig. 14.11(a)] is the male copulatory organ. When not sexually stimulated, it hangs down in front of the scrotum. The shaft of the penis contains three cylindrical bodies of erectile tissue. Two of the cylindrical bodies, the *corpora cavernosa,* lie side by side on the dorsum of the penis, while the third, the *corpus spongiosum,* lies ventral to the other two. The distal portions of the corpora cavernosa are fused; their proximal tendinous portions, the *crura* (roots) of the penis, diverge and are attached to the ischiopubic rami. The corpus spongiosum encloses the penile urethra. Its expanded proximal portion, the *bulb* of the penis, lies between the crura of the corpora cavernosa. The distal end of the corpus spongiosum expands into an acorn-shaped

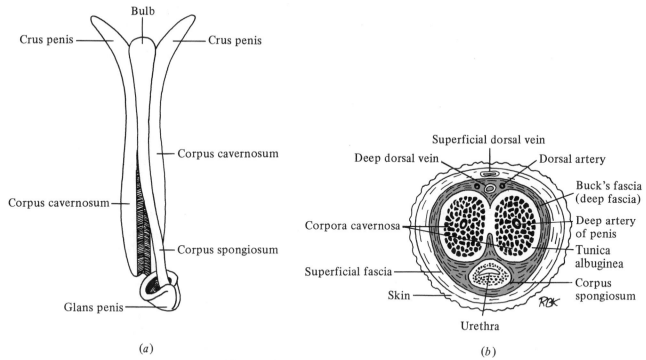

Figure 14.11 Penis. (a) Gross anatomy. Penis as seen from below; skin and fascial coverings are removed to expose cavernous bodies. Bulb and crura of penis are proximal; glans is distal. *(b)* Microscopic anatomy. Cross section through penis shows cavernous bodies, fascial layers, and blood vessels.

structure, the *glans penis,* that projects beyond and caps the corpora cavernosa. On the tip of the glans is the slitlike external opening of the urethra. The *ischiocavernosus muscles* enclose the crura of the corpora cavernosa, while the *bulbospongiosus muscles* enclose the bulb of the penis. These voluntary muscles play an important role in erection and ejaculation.

Microscopic Anatomy The penis contains three cylindrical bodies of erectile tissue bound together by fascia and covered by thin skin [Fig. 14.11(*b*)]. Each erectile body contains numerous venous sinuses supported by fibromuscular trabaculae and enclosed in a thick connective tissue capsule called the *tunica albuginea.* The erectile bodies in turn are surrounded by superficial and deep fascial layers and supported by a suspensory ligament that extends from the pubic symphysis to the dorsum of the penis. The penis is covered with thin hairless skin that forms a loose fold over the glans called the *prepuce* (foreskin). Surgical removal of the foreskin, for hygienic or religious purposes, is called *circumcision.*

Blood Supply   The erectile bodies of the penis are supplied by superficial (dorsal) and deep penile arteries, which are branches of the internal pudendal arteries, and are drained by superficial (dorsal) and deep penile veins, which are tributaries of the external and internal pudendal veins, respectively.

Innervation   The penis has a dual innervation. It receives voluntary nerve fibers by way of the pudendal nerves and autonomic fibers by way of the pelvic plexuses (sympathetic) and pelvic nerves (parasympathetic).

Function   The male sexual act consists of erection of the normally flaccid penis and ejaculation of semen.

Erection   Erection is triggered by psychic stimuli or direct stimulation of the penis. Parasympathetic impulses from spinal cord centers cause the penile arteries to dilate, which allows blood to enter the venous sinuses in the erectile bodies under high pressure. This plus the contraction of the bulbospongiosus and ischiocavernosus muscles prevents venous drainage and causes the erectile bodies to expand against their resistant capsules. The resulting engorgement of the penis with blood makes it rigid enough to penetrate the vagina of the female during sexual intercourse.

Ejaculation   When sexual stimulation has become sufficiently intense, sympathetic impulses from spinal cord centers cause peristaltic contractions of the epididymis and vas deferens, which expel sperm into the urethra. Simultaneous contractions of the seminal vesicles and prostate gland expel seminal and prostatic fluid along with the sperm. The fluid and sperm mix with mucus from the bulbourethral glands to form semen. At the peak of sexual excitement, rhythmic contractions of the perineal muscles cause ejaculation (male orgasm), in which semen is expelled from the penile urethra.

From 2 to 4 cubic centimeters of semen containing approximately 300 million sperm are expelled at each ejaculation. Following ejaculation, the venous sinuses in the erectile bodies are drained off and the penis resumes its flaccid state.

Diseases and Disorders of the Male Reproductive System

INGUINAL HERNIA   The inguinal canal is the canal in the anterior abdominal wall through which the spermatic cord passes. Normally the region surrounding the inguinal canal is fairly well reinforced with connective tissue. However, the inguinal canal constitutes a weak spot in the anterior abdominal wall and is the most frequent site of herniation in the male.

Direct inguinal hernia refers to herniation of a loop of small intestine through a defect in the abdominal wall surrounding the canal. *Indirect inguinal hernia* refers to herniation of a loop of small intestine into the scrotum through the inguinal canal.

VENERAL DISEASE　Venereal diseases are diseases spread by sexual intercourse. The two most common venereal diseases are gonorrhea and syphilis, closely followed by genital herpes virus infection.

Gonorrhea.　Gonorrhea is an infection of the genitourinary mucosa caused by gonococcal bacteria. In the male this disease is characterized by discharge of a whitish fluid or pus from the urethra plus a burning sensation during urination. The infection usually spreads up the urethra and genital ducts to involve the prostate gland and testes. If the testes become infected, sterility may result. Penicillin is usually the drug of choice for treating gonorrhea.

Syphilis.　Syphilis is an infection caused by a spirochete, *Treponema pallidum,* that causes widespread damage to the body if untreated. The first symptom of this disease is a skin lesion, called a *chancre,* at the site of infection. In the male the chancre is most commonly found on or near the glans penis. Several months after the chancre disappears, the infected male usually develops penile ulcerations plus wartlike lesions on his skin and mucous membranes. His testes are invaded and become inflamed. If, however, he is treated with penicillin in these early stages, he is readily cured. In the late stage of this disease, the microorganisms spread throughout the body, attacking the cardiovascular and central nervous systems in particular.

DISEASES AND DISORDERS OF THE TESTIS　*Cryptorchidism.*　Cryptorchidism is a condition in which one or both testes fail to complete their descent into the scrotum. In some cases this is due to a short spermatic cord or a tight inguinal canal. If cryptorchidism is not discovered and corrected by the time of puberty, the undescended testis atrophies and is much more likely to develop tumors than a normal scrotal testis.

Orchitis.　Orchitis refers to inflammation of the testis. It is characterized by a painful swelling of the affected organ. Infection by the microorganisms that cause gonorrhea, syphilis, tuberculosis, and mumps are the most common causes of orchitis in the adult male. If the orchitis is severe enough to destroy Leydig cells, sterility may result.

Testicular Tumors.　The testis develops a wide variety of tumors, most of which arise from the seminiferous epithelium and are highly malignant. Germ cell tumors of the testis are the most common form of cancer in men between the ages of 25 and 35. The two major types of germ cell tumors are the *seminoma* and the *teratoma.* Seminomas arise from and sometimes resemble spermatogenic cells. Teratomas, which arise from primordial germ cells, contain a wide variety of adult and fetal tissue types. The tissues within the teratoma are quite disorganized and frequently undergo malignant transformation. The remainder of the testicular tumors arise from Sertoli or Leydig cells and are almost always benign. However, they usually secrete steroid hormone in sufficient quantities to cause endocrine disorders.

DISEASES AND DISORDERS OF THE PROSTATE GLAND

Prostatitis. Prostatitis refers to inflammation of the prostate gland. The acute variety is usually caused by extension of a urethral or bladder infection into the prostate gland. Prostatitis commonly follows urethral catheterization or cystocopy.

Benign Prostatic Hypertrophy. This condition is very common in men over the age of 50. It is probably caused by the relative hyperestrinism that develops as androgen production decreases. The adrenal-derived estrogens stimulate hypertrophic changes in the prostate gland, which cause urethral constriction, urine retention, and kidney infections. If kidney infections occur frequently, the prostate gland should be removed surgically in order to prevent serious kidney damage.

Prostatic Tumors. Prostatic tumors occur quite frequently in men over the age of 50. The cause is unknown but is probably related to changes in the androgen level with age. Adenocarcinoma of the prostate is the most common type of prostatic tumor and is the third most common cause of death in the adult male. Like benign prostatic hypertrophy, it is characterized by urinary obstruction and kidney infections. In many instances metastases, especially to the skeletal system, have occurred by the time the disease is discovered. Prostatic cancer is commonly treated with estrogen therapy or orchiectomy (surgical removal of the testes).

DISEASES AND DISORDERS OF THE PENIS

Congenital Malformations. Ectopic locations of the urethral opening are the most common congenital malformations of the male genitalia. *Hypospadias* is characterized by an abnormal urethral opening on the ventral surface of the penile shaft. *Epispadias,* which is less common, is characterized by an abnormal urethral opening on the dorsal surface of the penile shaft. Urinary tract infections are common in infants with these malformations.

Phimosis. Phimosis is a condition in which the foreskin of the penis becomes so tight that it cannot be retracted over the glans. If forcibly retracted, the tight foreskin may cause painful swelling of the glans and urethral construction. This condition is best treated by circumcision.

Part B. **The Female Reproductive System**

The female reproductive system (Fig. 14.12) consists of the ovaries, female genital ducts and accessory glands, female external genitalia, and the mammary glands.

The Ovaries

The ovaries (female gonads) produce *ova* (the female sex cells) and secrete *estrogens* and *progesterone* (the female sex hormones).

GROSS ANATOMY The ovaries develop in the abdominal cavity but descend into the pelvis about two months before birth. Each adult ovary [Fig. 14.13(*a*)] is an almond-shaped glandular organ about $1\frac{1}{2}$ inches long and $\frac{3}{4}$ inch wide. It is located in the ovarian fossa on the lateral pelvic wall,

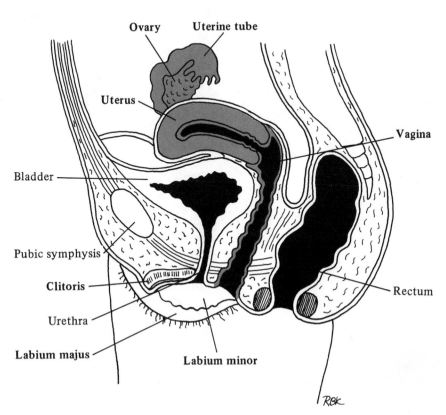

Figure 14.12 *Female Reproductive System*. Sagittal section through female pelvis and external genitalia shows ovary and female genital ducts.

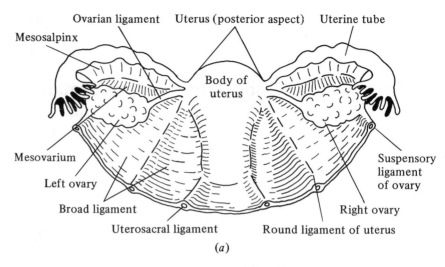

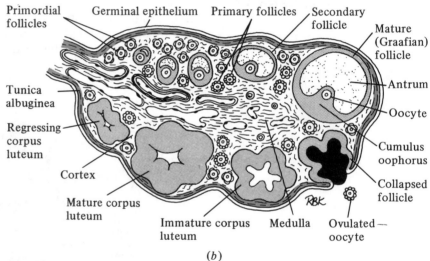

Figure 14.13   Ovary. (a) With female internal genitalia. Posterior aspect of uterus and uterine tubes shows relationship of ovaries to these structures. (b) Microscopic anatomy. Coronal section with stages in ovarian cycle illustrated schematically in clockwise fashion from primordial follicles to regressing corpus luteum.

which is bounded by the external iliac vessels, the lateral umbilical ligament, and the ureter. The ovary has medial and lateral surfaces, anterior (mesovarian) and posterior (free) borders and superior (tubal) and inferior (uterine) poles. The ovary is fixed in position by the *suspensory ligament,* which runs from the iliac vessels to the tubal pole, the *ovarian ligament,* which runs from the side of the uterus to the uterine pole, and the *mesovarium,* a mesentery that suspends the ovary from the back of the broad ligament. The broad ligament is a peritoneal reflection that covers the uterine tubes and uterus.

MICROSCOPIC ANATOMY   The ovary [Fig. 14.13(b)] is a solid glandular organ whose parenchyma is enclosed in a thick fibrous connective tissue capsule called the *tunica albuginea*. The tunica albuginea is in turn covered by the *germinal epithelium*, a thin glistening layer of peritoneum that corresponds to the tunica vaginalis of the testis. The term "germinal epithelium" is a poor one, since this epithelium does not give rise to germ cells. The ovarian parenchyma is divided into poorly demarcated cortical and medullary regions. The *medulla* (inner zone) is a core of loose connective tissue containing elastic and smooth muscle fibers plus numerous blood vessels. The *cortex* (outer zone) consists of a loose connective tissue stroma in which *ovarian follicles* predominate.

Three different types of ovarian follicles (Fig. 14.14) can be distinguished on the basis of their morphology: *primordial follicles*, *growing follicles*, and *mature* or *Graafian follicles*.

Primordial Follicle The primordial follicle consists of an *oocyte* (immature egg cell) surrounded by a single layer of epithelial cells called *follicular* or *granulosa cells*. Primordial follicles are the only ones present in the ovary before puberty.

Growing Follicle Growing follicles are found in the ovary between *menarche* (start of the menstrual periods) and *menopause* (cessation of the menstrual periods). At the beginning of each ovarian cycle (Fig. 14.15), one or more primordial follicles are transformed into actively growing *primary follicles* under the stimulus of FSH. As the oocyte begins to enlarge, the granulosa cells surrounding it multiply to form a stratified epithelium. The connective tissue cells surrounding the granulosa layer form a sheath, or *theca*, for the growing follicle. The internal thecal cells differentiate into secretory cells and begin to secrete a fluid rich in estrogenic hormones. As the follicle continues to grow, this fluid begins to accumulate between the granulosa cells. The primary follicle becomes a *secondary follicle* when it develops a crescent-shaped cavity, or *antrum*, to contain the increasing volume of follicular fluid. The oocyte becomes incorporated in a mound of granulosa cells, the *cumulus oophorus*, which protrudes into the interior of the antrum.

Mature Follicle By 10 to 14 days after the beginning of the ovarian cycle, the follicle attains maximum growth and is called a mature (Graafian) follicle. The mature follicle is a large transparent vesicular structure that bulges from the free surface of the ovary. A surge of LH from the anterior pituitary causes the mature follicle to swell and rupture, ejecting the oocyte from the follicle along with the follicular fluid. This process is known as *ovulation*. After ovulation, LH stimulates the transformation of the collapsed follicle into a *corpus luteum*, which secretes estrogens and progesterone. If conception and implantation do not occur, the corpus luteum begins to degenerate about 12 days after ovulation.

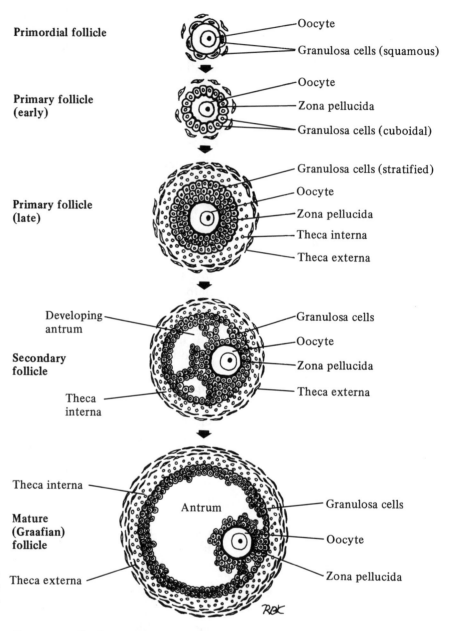

Primordial follicle

Oocyte
Granulosa cells (squamous)

Primary follicle (early)

Oocyte
Zona pellucida
Granulosa cells (cuboidal)

Primary follicle (late)

Granulosa cells (stratified)
Oocyte
Zona pellucida
Theca interna
Theca externa

Secondary follicle

Developing antrum
Theca interna

Granulosa cells
Oocyte
Zona pellucida
Theca externa

Mature (Graafian) follicle

Theca interna
Theca externa
Antrum

Granulosa cells
Oocyte
Zona pellucida

Figure 14.14 Development of Ovarian Follicles. Schematic representation shows progressive stages from primordial follicle to mature (Graafian) follicle.

It eventually becomes a *corpus albicans* (a connective tissue scar). Occasionally two or more follicles mature and release their oocytes, which usually leads to multiple births.

BLOOD SUPPLY The ovaries are supplied by the ovarian arteries, which are direct branches of the abdominal aorta, plus anastomotic branches of the uterine arteries. They are drained by tributaries of the ovarian veins,

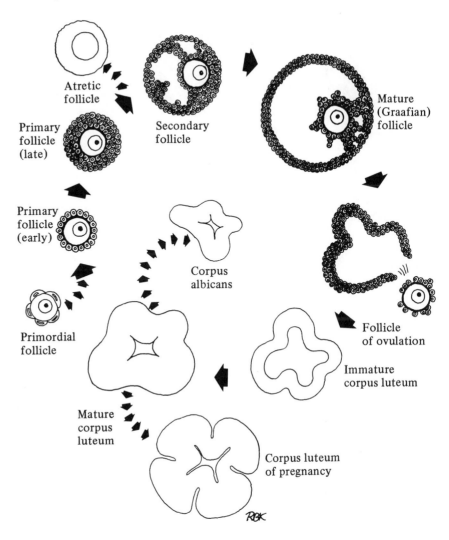

Figure 14.15   Ovarian Cycle.

which form a dense network in the mesovarium called the *pampini-form plexus*. The right ovarian vein is a tributary of the inferior vena cava, while the left ovarian vein is a tributary of the left renal vein.

INNERVATION   The ovaries receive autonomic nerve fibers by way of the pelvic, hypogastric, and ovarian plexuses.

FUNCTION   The ovaries have two functions: formation of ova plus estrogen and progesterone secretion.

Ova Formation   The term *oogenesis* is used to describe the entire sequence of events in the production of egg cells or ova (Fig. 14.16). Sex cell formation in the female is quite different from that in the male. Oogenesis be-

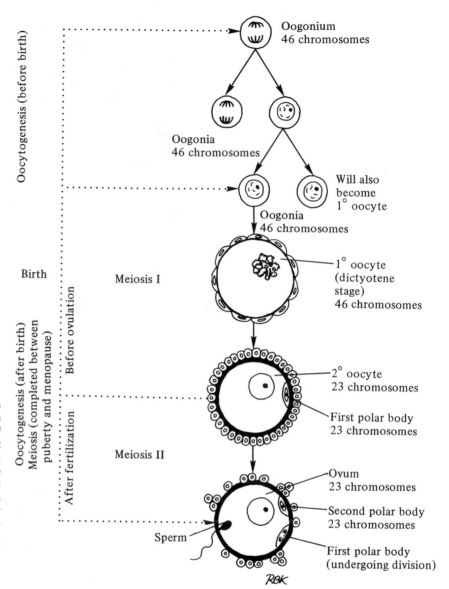

Figure 14.16 *Oogenesis.* All oogonia convert to oocytes before birth; no stem cells are left in ovary after birth. At time of birth all 1° oocytes are arrested in first meiotic division (dictyotene stage); this division may be completed in any cycle from puberty to menopause. Second meiotic division does not take place unless oocyte is fertilized. Note that for every oogonium that enters meiosis, one functional cell (ovum) and two to three nonfunctional cells (polar bodies) are produced.

gins before birth and is not completed until fertilization occurs. *Oogonia,* the female germ cells, proliferate rapidly by mitotic division in the fetal ovary. By the time of birth, all the oogonia have transformed into *primary oocytes* and entered the first meiotic division. It is interesting to note that of the one million primary oocytes present in the ovaries at birth, only about four hundred fifty will be ovulated during a woman's reproductive life and the remainder will degenerate. At birth the primary oocytes are arrested in late prophase of the first meiotic division and enter the *dictyotene stage* (a resting stage). They remain in the dictyotene stage until stimulated by FSH to complete development. Some primary oocytes may remain in the dictyotene stage for 30 to 40 years. This is one explanation why the percentage of children born with chromosomal defects increases with maternal age.

In the course of a woman's reproductive life, which extends from menarche to menopause, one of several growing follicles matures and its oocyte is ovulated approximately every 28 days. Under the stimulus of the preovulatory LH surge, the primary oocyte completes the first meiotic division just before ovulation. Two daughter cells of unequal size, the large *secondary oocyte* and the small *first polar body,* are produced by this division. The secondary oocyte, which receives the bulk of the nutrient-rich cytoplasm, immediately enters the second meiotic division. If a sperm succeeds in penetrating the secondary oocyte, it completes the second meiotic division. This division also produces a large cell, the *ovum,* and a small cell, the *second polar body.* Thus every primary oocyte that enters meiosis produces one functional sex cell; in contrast, every primary spermatocyte produces four functional sex cells. If fertilization does not occur, the secondary oocyte degenerates without completing the second meiotic division. The life span of the secondary oocyte is about 24 hours.

Estrogen and Progesterone Secretion

Estrogenic hormones are secreted by the internal thecal cells of the actively growing ovarian follicle under the stimulus of FSH. After ovulation, the granulosa and internal thecal cells of the collapsed follicle are transformed into *lutein* cells under the stimulus of LH. The lutein cells secrete progesterone plus large amounts of estrogens. If pregnancy does not occur, the LH stimulus ceases about 10 to 14 days after ovulation and the corpus luteum stops producing its hormones.

Genital Ducts and Accessory Glands

The female genital ducts consist of a series of tubes that facilitate fertilization (the uterine tubes), provide a place for a new human being to develop (the uterus), and serve as a birth canal (the vagina).

UTERINE (FALLOPIAN) TUBES
Gross Anatomy

The uterine or Fallopian tubes are paired musculomembranous tubes that transport the oocyte or embryo to the uterus. Each tube [Fig. 14.17(*a*)] lies in the *mesosalpinx,* the free anterior border of the broad ligament. It is about 4 to 5 inches long and extends from the lateral pelvic wall, where it is in close proximity to its respective ovary, to the superolateral angle of the uterus. For descriptive purposes the uterine tube is divided into three portions:

1. *Infundibulum.* The flared lateral portion of the trumpet-shaped uterine tube. Its ostium (open end) is surrounded by a number of

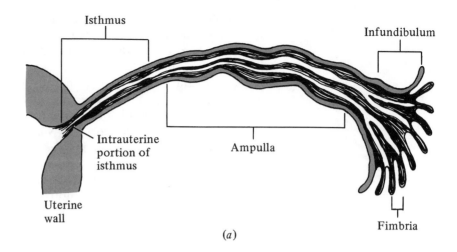

(a)

Figure 14.17 *Uterine Tube.* (a) Gross anatomy. Uterine tube and adjacent segment of uterus are sectioned to show folding of tubal mucosa. Note that fimbria are fingerlike appendages of infundibulum, the funnel-shaped lateral portion of uterine tube. (b) Microscopic anatomy. Cross section shows three coats of tubal wall. Layers of smooth muscle in muscularis are actually not as well demarcated as indicated in figure.

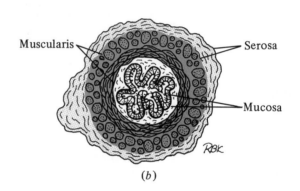

(b)

fingerlike processes called *fimbria* that sweep over the surface of the ovary and guide the ovulated oocyte into the tube.

2. *Ampulla.* The long dilated middle portion of the uterine tube that curves over the ovary.

3. *Isthmus.* The short constricted medial portion of the uterine tube that enters the body of the uterus.

Microscopic Anatomy

The wall of the uterine tube consists of three coats: *mucosa, muscularis,* and *serosa* [Fig. 14.17(b)]. The thick mucosa is thrown into branched longitudinal folds. The simple columnar epithelium lining the mucosa contains both ciliated and secretory cells. Its mucous secretion forms a surface film that is moved toward the uterus by the beating cilia. The tubular mucous film provides transportation and possibly some nutritive material for the oocyte-embryo. The muscularis consists of two ill-defined layers of smooth muscle fibers capable of undergoing rhythmic peristaltic contractions. The serosa consists of a layer of visceral peritoneum continuous with the mesosalpinx.

Blood Supply The uterine tube is supplied by branches of the ovarian and uterine arteries and is drained by tributaries of the corresponding veins.

Innervation The uterine tube receives sympathetic nerve fibers by way of the pelvic and ovarian plexuses and parasympathetic nerve fibers by way of the pelvic nerves.

Function The fimbria of the uterine tube hover over the ovary at the time of ovulation and appear to draw the ovulated oocyte into the ostium. The uterine tube provides the appropriate environment for fertilization, which usually occurs in the infundibulum. The unfertilized oocyte or embryo is propelled toward the uterus by the current created by the beating of the mucosal cilia and the peristaltic contractions of the muscular coat. The rhythmic peristaltic contractions of the muscullaris, starting at the infundibulum and leading toward the uterine cavity, are stimulated by parasympathetic impulses and influenced by estrogen secretion.

UTERUS
Gross Anatomy The uterus, or womb (Fig. 14.18), is a hollow muscular organ that lies in the pelvic cavity between the bladder and the rectum. In women who have never borne children, it is shaped like a pear and is about 3 inches long, 1½ inches wide, and 1 inch from front to back.

Parts of the Uterus For descriptive purposes the uterus is divided into the following parts:

1. *Fundus.* The rounded upper end of the uterus above the entry of the uterine tubes.
2. *Body.* The tapering portion of the uterus that extends from the fundus to the cervix. It has a flattened anterior surface apposed to the bladder, a convex posterior surface, and slightly convex lateral margins that receive the uterine tubes, the round ligaments of the uterus, and the ovarian ligaments.
3. *Cervix.* The constricted neck of the uterus. Its distal end projects into the proximal portion of the vagina. The distal end of the cervix contains the *uterine os,* a small opening through which the cavity of the uterus communicates with the vaginal lumen.

Supports of the Uterus In addition to the levator ani muscles, four pairs of ligaments provide support for the uterus and help prevent uterine prolapse (abnormal protrusion of the uterus into the vagina):

1. *Broad Ligaments.* Transverse folds of peritoneum extending from the lateral margins of the uterus that anchor the body to the lateral pelvic walls.
2. *Cardinal Ligaments.* Condensations of fascia surrounding the uterine and vaginal blood vessels that anchor the cervix and upper

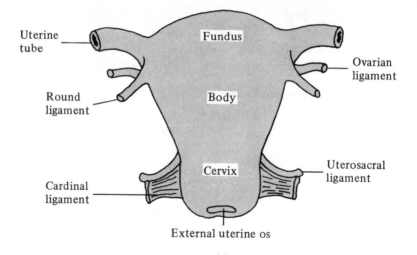

Uterine tube

Fundus

Ovarian ligament

Round ligament

Body

Cervix

Uterosacral ligament

Cardinal ligament

External uterine os

(a)

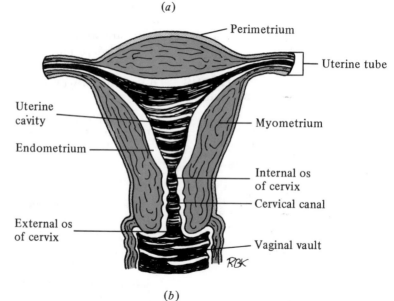

Perimetrium

Uterine tube

Uterine cavity

Myometrium

Endometrium

Internal os of cervix

Cervical canal

External os of cervix

Vaginal vault

(b)

Figure 14.18 Uterus: Gross Anatomy. (a) External aspect of uterus with ligaments indicated. Broad ligament is removed for better visualization. (b) Internal aspect of uterus and upper part of vagina. Uterus is sectioned in coronal plane to show shape of uterine cavity and thickness of uterine wall.

portion of the vagina to the lateral pelvic walls. They are the chief supports of the uterus and may require surgical repair after childbirth.

3. *Round Ligaments.* Fibromuscular cords that extend from the superolateral margins of the uterus to the labia majora by way of the inguinal canals.

4. *Uterosacral Ligaments.* Flattened fibromuscular bands extending from the lateral margins of the upper cervix that anchor the cervix to the sacrum.

Positions of the Uterus   The uterus is a rather mobile organ that can bend forward or backward on the fulcrum formed by the cardinal ligaments. The uterus thus has

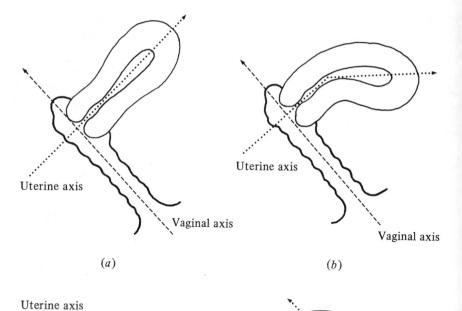

(a)

(b)

Figure 14.19 *Positions of Uterus.* Schematic representation shows normal and abnormal positions of uterus. Uterus and vagina are shown in sagittal section, with uterine and vaginal axes indicated. (*a*) Anteversion. Normal position in which uterus is pointed forward. (*b*) Anteflexion. Abnormal position in which uterus is bent forward at cervix. (*c*) Retroversion. Abnormal position in which uterus is pointed backward. (*d*) Retroflexion. Abnormal position in which uterus is bent backward at cervix.

(c)

(d)

four basic positions: *anteverted, anteflexed, retroverted,* and *retroflexed* (Fig. 14.19). In the most common (anteverted) position, the uterus and vagina are at an angle of about 90° to each other. When the individual is standing and the bladder and rectum are empty, the body of the uterus is nearly horizontal.

Microscopic Anatomy The uterine wall (Fig. 14.20) consists of three coats: *mucosa, muscularis,* and *serosa.*

The uterine mucosa, or *endometrium,* which lines the uterine

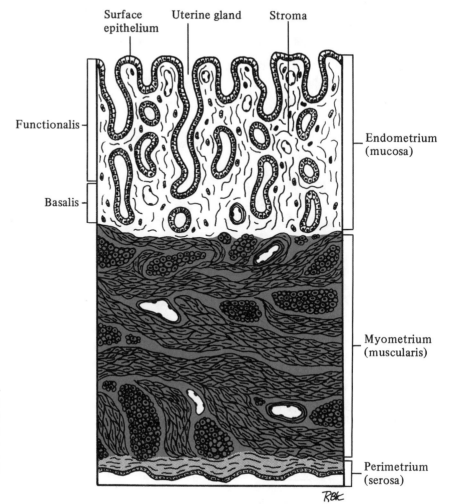

Surface epithelium   Uterine gland   Stroma

Functionalis

Basalis

Endometrium (mucosa)

Myometrium (muscularis)

Perimetrium (serosa)

Figure 14.20 *Uterus: Microscopic Anatomy.* Segment of cross section through uterine wall showing its three coats. Uterine endometrium consists of epithelial lining and connective tissue stroma which supports uterine glands. Thickness of endometrium varies with stages of menstrual cycle.

cavity, contains a layer of simple ciliated columnar epithelium that changes to stratified squamous near the uterine os. The epithelium is supported by a highly vascular stroma of mesenchymal connective tissue. In the fundus and body, the endometrium consists of two portions: a thick superficial layer, or *functionalis,* which is sloughed off during menstruation, and a thin deep layer, or *basalis,* which is retained during menstruation and contributes to the regeneration of the functionalis. In addition, long tubular glycogen-secreting glands occur in the fundus and body; these are the *uterine glands.* In the cervix the endometrium is thinner and thrown into a number of oblique folds. It is not shed during menstruation. Numerous large-branched, mucus-secreting glands occur in the cervix; these are the *cervical mucus glands.*

The muscularis, or *myometrium,* is the massive muscular coat of the uterine wall. It contains three ill-defined layers of interlaced smooth muscle fibers. During pregnancy these muscle fibers increase in both size and number, which allows the uterus to expand and accommodate the growing fetus. After pregnancy there is destruction of some muscle fibers and regression of others, so that the uterus returns almost to its prepregnancy dimensions.

The serosa, or *perimetrium,* is a peritoneal reflexion, continuous with the broad ligaments, that covers the entire posterior surface of the uterus and the upper part of the anterior surface.

Blood Supply The uterus receives blood from the uterine arteries, which are branches of the internal iliac arteries, and from the ovarian arteries, which are branches of the abdominal aorta. It is drained by tributaries of the uterine and ovarian veins.

Innervation The uterus receives sympathetic and parasympathetic nerve fibers by way of the hypogastric and ovarian plexuses.

Function The uterus is an organ that is specialized to receive a fertilized ovum and provide it with protection and nourishment as it undergoes embryonic and fetal development. In preparation for a possible pregnancy, the uterine endometrium is highly responsive to cyclic changes in the levels of the ovarian hormones.

Endometrial (Menstrual) Cycle During a woman's reproductive life the endometrium undergoes cyclic structural changes approximately every 28 days that prepare it to receive a fertilized ovum and cause it to slough off if implantation does not occur. The endometrial or menstrual cycle (Fig. 14.21), which is correlated with the ovarian cycle through the ovarian hormones, is divided into the following three phases:

1. *The Menstrual Phase* (Days 1 to 4). This phase is chosen as the start of the cycle because it is the most outwardly visible of the three phases. It is characterized by the sloughing off of necrotic fragments of the functionalis and the discharge *(menstruation)* of bloody fluid and necrotic tissue from the vagina.
2. *The Proliferative Phase* (Days 5 to 14). This phase occurs under the stimulus of estrogens secreted by the growing ovarian follicle(s). It is characterized by repair and rapid growth of the endometrium. The raw endometrial surface is reepithelialized by cells derived from the bases of the uterine glands (days 5 to 6). This is followed by rapid regrowth of the functionalis. The uterine glands become tall and wavy, and the functionalis is invaded by coiled arteries (days 7 to 14).
3. *The Secretory Phase* (Days 15 to 28). This phase follows ovulation

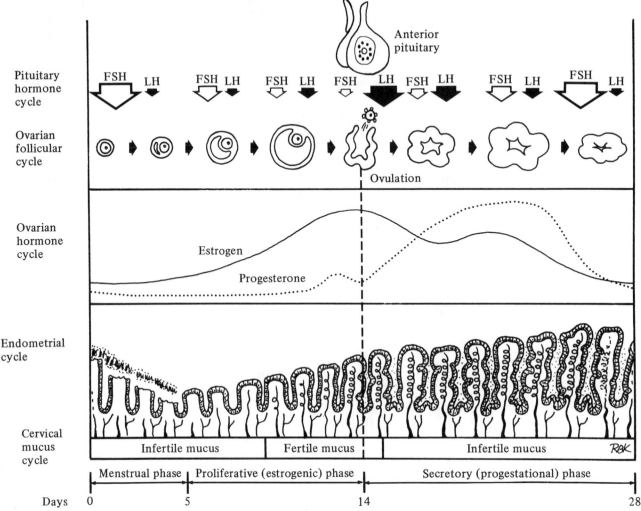

Figure 14.21 Menstrual Cycle. Schematic representation of relationships among pituitary hormone cycle, ovarian follicular cycle, ovarian hormone cycle, endometrial cycle, and cervical mucus cycle.

and corpus luteum formation. It occurs under the stimulus of estrogens and progesterone secreted by the corpus luteum. The secretory phase is characterized by an accumulation of glycogen-rich fluid in highly coiled uterine glands, an increase in tissue fluid in the endometrial stroma, and further growth of the coiled arteries (days 15 to 25). The endometrium now provides a suitable environment for the implantation of a fertilized ovum. If implantation does not occur, the corpus luteum regresses and estrogen and progesterone levels drop precipitously. Withdrawal of hormonal stimulus causes constriction of the coiled arteries and ischemia of the functionalis (days 26 to 28). Necrosis of the functionalis occurs, and the necrotic tissue begins to slough. Blood oozes out of the damaged arteries. The resulting menstrual flow marks the beginning of a new cycle.

Cervical Mucous Cycle

The activity of the cervical mucous glands is also correlated with the ovarian cycle through the ovarian hormones. Estrogen stimulation causes the cervical mucous glands to secrete *fertile mucus*, a thin clear nonviscous mucus that has the consistency of raw egg white. Fertile mucus is most abundant and its characteristics most pronounced just before ovulation. It promotes sperm penetration of the uterus and may provide the sperm with nourishment. Progesterone stimulation causes the cervical mucous glands to secrete *infertile mucus*, a thick opaque viscous mucus that in essence forms a "vaginal plug." Infertile mucus acts as a physical barrier to sperm penetration and may also prevent bacteria within the vagina from entering the uterus. Changes in the cervical mucus can be used to identify the time of ovulation and as a means to avoid or achieve pregnancy.

VAGINA
Gross Anatomy

The vagina [Fig. 14.22(*a*)] is the female copulatory organ. It is a distensible fibromuscular tube, situated posterior to the bladder and anterior to the rectum, that extends from the uterus to the vaginal vestibule. The vaginal opening is partially blocked by a membranous partition in the young girl. This partition, called the *hymen*, occurs in several basic types [Fig. 14.22(*b*)]. It is either torn or stretched during the first sexual intercourse.

Microscopic Anatomy

The vaginal wall consists of a *mucosa*, a *muscularis*, and an *adventitia*. The mucosa contains a stratified squamous epithelium whose superficial cells become cornified (slightly keratinized) during the peak of estrogen secretion. Transverse rugae (ridges) are normally present in the vaginal mucosa; they are smoothed out when the vagina is distended. The muscularis consists of interlacing bundles of smooth muscle fibers that permit distension of the vagina during sexual intercourse and childbirth. A fibrous adventitial coat completes the vaginal wall.

Blood Supply

The vagina is supplied by the vaginal artery, which anastomoses with branches of the uterine, inferior vesicle, and middle rectal arteries. It is surrounded by a venous plexus that drains into the internal iliac veins.

Innervation

The vagina has a dual innervation. It receives autonomic nerve fibers from the hypogastric plexus and voluntary nerve fibers from the pudendal nerve.

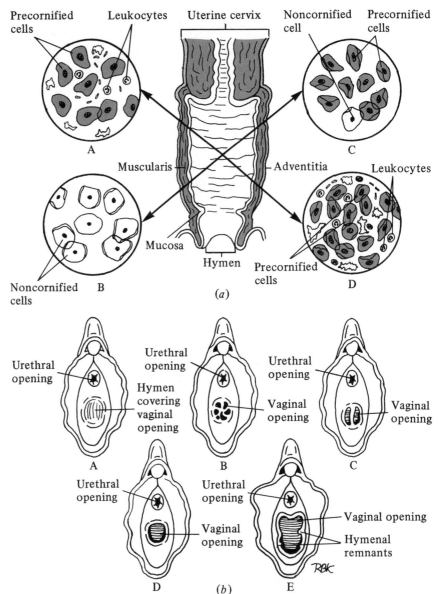

Figure 14.22 Anatomy of Vagina. (a) Frontal section through uterine cervix and vagina showing three coats of vaginal wall. Changes in vaginal cytology (epithelial cells) indicated in *insets: A,* early proliferative phase; *B,* late proliferative phase; *C,* midsecretory phase; *D,* late secretory phase.

Hymenal Types. (b) Vaginal vestibule. Some hymenal variations in virginal state and hymenal remnants following childbirth are shown: *A,* imperforate hymen; *B,* cribriform hymen; *C,* septate hymen; *D,* annular hymen; *E,* hymen following childbirth.

Function The vagina serves as a distensible sheath for the penis during sexual intercourse, forms the membranous portion of the birth canal, and conveys menstrual blood from the uterus to the exterior of the body. Cytological examination of cells detached from the vaginal mucosa *(Pap smear)* provides information on ovarian hormone balance and permits early detection of some types of genital cancer.

VESTIBULAR GLANDS
Anatomy

The vestibular or *Bartholin's glands* are a pair of small round glands comparable to the bulbourethral glands in the male. They are situated in the vaginal vestibule on either side of the vaginal opening. Their ducts open into the groove between the hymen and the labia minora.

Function

The vestibular glands produce a mucous secretion that acts as a lubricant during sexual intercourse.

PARAURETHRAL GLANDS
Anatomy

The paraurethral or *Skene's glands* are a group of small glands that surround the urethral opening. They are comparable to the periurethral part of the prostate gland in the male.

Function

The paraurethral glands secrete a small amount of mucus that also acts as a lubricant during sexual intercourse.

The Female Perineum

The female perineum is comparable to the male perineum. The anal triangles are the same in both sexes; only the urogenital triangles differ. In the female both the urethra and the vagina open into the urogenital triangle. Structures that would occupy the midline of the urogenital triangle in the male are located on either side of the urethral and vaginal openings in the female. The superficial compartment of the female perineum is shown in Fig. 14.23. Important structures include the crura of the clitoris, the vaginal bulbs (comparable to the bulb of the penis), the vestibular glands, the ischiocavernosus and bulbospongiosus muscles, the perineal body, and the superficial transverse perineal muscles. The female perineum is a very important structure in childbirth.

External Genitalia

The female external genitalia consist of the *mons pubis*, the paired *labia majora* and *minora*, and the *clitoris*. These organs (Fig. 14.24), collectively known as the *vulva* or *pudenda*, are located in the perineal area.

MONS PUBIS
Anatomy

The mons pubis is a skin-covered mound of fatty tissue located in front of the pubic symphysis. After puberty it becomes covered with coarse pubic hairs.

Function

The mons pubis acts as a shock absorber during sexual intercourse.

LABIA MAJORA
Anatomy

The labia majora ("major lips") are two fat-filled rounded folds of skin that are continuous with the mons pubis anteriorly and are

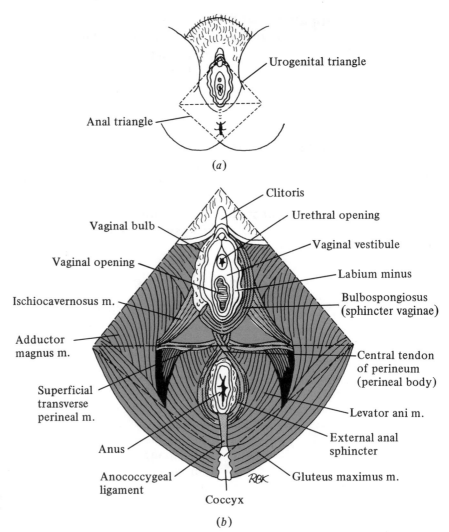

Figure 14.23 *Female Perineum.* (a) External aspects with urogenital and anal triangles indicated. (b) Skin and superficial fascia are removed to expose superficial perineal musculature. Compare female superficial perineal musculature to that of male in Fig. 14.10. Note that bulbospongiosus muscle, which surrounds bulb of penis in male, acts as vaginal sphincter in female.

united by a commissure just in front of the anus. The internal surface is hairless while the external surface is covered with pubic hairs. Both surfaces contain numerous sebaceous and sweat glands.

Function   The labia majora form the lateral boundaries of the pudendal cleft and enclose the labia minora. They are comparable to the scrotum in the male.

LABIA MINORA
Anatomy   The labia minora ("minor lips") are two thin folds of hairless skin that lie medial to the labia majora. Anteriorly these folds split to form the prepuce of the clitoris; posteriorly they are connected by the fourchette, a transverse skin fold situated behind the vaginal opening.

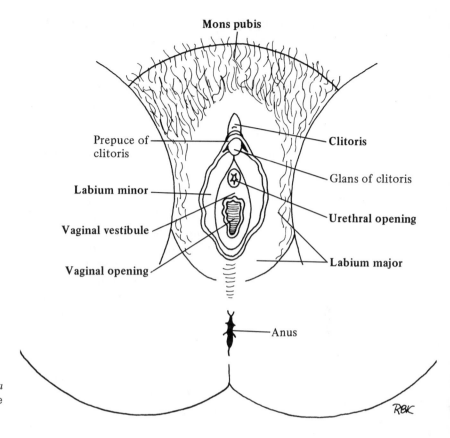

Figure 14.24 *Female External Genitalia (Vulva).* Thighs are spread to expose female perineum and external genitalia.

Function The labia minora enclose the *vaginal vestibule,* a shallow sinus within the pudendal cleft that corresponds to much of the male penile urethra. Opening into the vestibule are the urethra, the vagina, and the ducts of the paraurethral and vestibular glands.

CLITORIS
Gross Anatomy The clitoris is the female counterpart of the penis. It differs from the penis in that it is very short, it does not contain a corpus spongiosum, and it is not traversed by the female urethra. The shaft of the clitoris contains two small erectile bodies, *corpora cavernosa,* that end in a rudimentary *glans clitoridis.* Near the crura of the corpora cavernosa are two elongated masses of erectile tissue called the *vaginal bulbs.* The vaginal bulbs, which correspond to the bulb of the penis, surround the vaginal opening. Both the *ischiocavernosa muscles,* which enclose the clitoral crura, and the *bulbospongiosus (sphincter vaginae) muscles,* which enclose the vaginal bulbs, play an important role in tightening the vagina during sexual intercourse.

Microscopic Anatomy The clitoris consists of two cylindrical bodies of erectile tissue enclosed in a dense connective tissue membrane and covered with thin hairless skin. The skin forms a prepuce for the glans clitoridis.

Blood Supply   The erectile bodies of the clitoris are supplied by the superficial and deep clitoral arteries and are drained by veins that empty into the pudendal plexus.

Innervation   The clitoris has a dual innervation. It receives voluntary nerve fibers by way of the pudendal nerves and autonomic nerve fibers by way of the pelvic plexus.

Function   The female sexual act consists of erection and lubrication followed by climax.

Erection and Lubrication   Erection in the female is triggered by psychic stimuli or direct stimulation of the clitoris. Parasympathetic impulses from spinal cord centers cause vascular congestion and erection of the clitoris, vaginal bulbs, and outer third of the vagina. This serves to lengthen the vagina and tighten the vaginal tissue around the penis during sexual intercourse. At the same time, a clear slippery fluid derived from blood plasma oozes from the vaginal walls. This fluid plus the mucus secreted by the vestibular and paraurethral glands act as lubricants for the penis. Without vaginal lubrication, sexual intercourse is painful for both sexes.

Climax   At the peak of sexual excitement, the female perineal muscles undergo a series of rhythmic contractions known as climax (female orgasm). These contractions plus contractions of the genital ducts are thought to help transport sperm through the female genital ducts.

Mammary Glands

GROSS ANATOMY   The mammary glands, or breasts (Fig. 14.25), are accessory female reproductive organs that secrete milk. The breasts are located in the anterior region of the thorax, one on either side of the midline. Each breast is a conical mass of skin-covered glandular tissue that extends from the lateral margin of the sternum to the anterior border of the axilla. It lies superficial to the pectoralis major muscle and is attached to the clavipectoral fascia by strands of fibrous connective tissue. A conical pigmented papilla, or *nipple*, projects from the surface of the breast. It is surrounded by a circular area of pigmented skin called the *areola*. The size and shape of the breasts vary not only from woman to woman but also with the age, sex hormone levels, and nutritional state of each woman.

MICROSCOPIC ANATOMY   The breast contains 15 to 20 radially arranged compound acinar glands. Each gland is drained by its own excretory or *lactiferous*

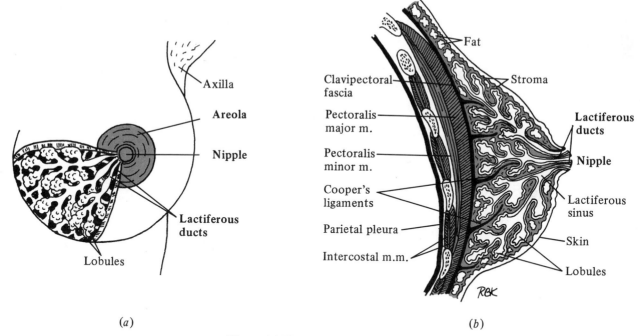

(a)

(b)

Figure 14.25 Mammary Gland. (a) Front view. Wedge-shaped segment of skin is removed to expose underlying glandular tissue. (b) Sagittal section through breast and underlying chest wall shows arrangement of glandular tissue and pectoral musculature.

duct, which opens independently on the surface of the nipple. The glands and their ducts are embedded in fat and separated from one another by fibrous connective tissue septa that partition the breast into lobes and lobules.

BLOOD SUPPLY   The breast is supplied by thoracic branches of the internal thoracic, axillary, and intercostal arteries. It is drained by tributaries of the internal thoracic and axillary veins.

INNERVATION   The breast is innervated by anterior and lateral branches of the fourth, fifth, and sixth thoracic nerves.

FUNCTION   The function of the breast is to produce and secrete *milk.* This is known as *lactation.* Milk is a fluid that provides nutrients and antibodies for the nursing child. The secretion of milk is hormonally controlled. During pregnancy, estrogens cause proliferation and branching of the duct system, while progesterone causes growth of the secretory acini. Following childbirth prolactin stimulates the sex-hormone-primed glands to secrete milk. The transport of milk from the glands to the nipple is triggered by the milk ejection reflex. During breast-feeding the sucking of the child stimulates sensory receptors in the nipple. Impulses from these sensory receptors cause the release of oxytocin from the neurohypophysis. Oxytocin in turn causes contraction of the myoepithelial cells surrounding the secretory units and milk is ejected into the duct system. Breast-feeding is beneficial to the mother as well as the child. When a woman is breast-feeding totally, FSH secretion is inhibited and ovulation usually does not occur. For this reason women in many cultures breast-feed their children for one to two years in order to space their pregnancies.

Diseases and Disorders of the Female Reproductive System

VENEREAL DISEASE

Gonorrhea. In the female the early stage of gonorrhea is characterized by inflammation of the urethra, vagina, and associated mucous glands and is almost painless. In its later stage the infection travels up the uterus to the uterine tubes, where it causes painful inflammation and scarring severe enough to produce sterility. Extension of the infection into the pelvic cavity can cause peritonitis. As in the male, gonorrheal infections usually respond to penicillin therapy. However, penicillin-resistant strains of gonococci appear to be emerging in various parts of the world, and treatment in the future may be more expensive and complicated.

Syphilis. The primary lesion of syphilis is usually undetected in the female and the disease, untreated, is spread to unsuspecting sex partners. More important than the damage the spirochetes do to the body of an untreated woman is the damage they can do to her unborn child. The spirochetes can cross the placental barrier and attack fetal tissues, causing a variety of congenital malformations.

DISEASES AND DISORDERS OF THE OVARY

Tumors are the most common form of ovarian disease. They can be divided into cystic and solid forms.

Cystic Tumors. *Cystadenomas* are benign tumors, containing serous or mucinous fluid, that can grow to large size. The serous cystadenoma is potentially malignant and often gives rise to the *cystadenocarcinoma*, which is the commonest form of ovarian cancer. *Dermoid cysts* are cystic teratomas (tumors of germ cell origin) that contain an assortment of structures primarily of ectodermal origin such as teeth and hair.

Solid Tumors. Embryonic ovarian tumors are relatively rare solid tumors that arise from sex-differentiated mesenchyme and are potentially malignant. The feminizing types, such as granulosa and theca cell tumors, secrete large amounts of estrogens; the masculinizing types, such as androblastomas, secrete large amounts of androgens. Other solid ovarian tumors include fibromas, teratomas, dysgerminomas, and secondary carcinomas. Secondary ovarian carcinomas represent metastases from adjacent pelvic organs or from the gastrointestinal tract.

DISEASES AND DISORDERS OF THE UTERINE TUBES

The only common diseases of the uterine tubes are *salpingitis* (inflammation of the uterine tube) and *tubal pregnancy.* Salpingitis is usually of gonorrheal origin. A residual effect of any form of salpingitis is tubal stenosis. Tubal stenosis (narrowing of the uterine tube) results either in infertility or tubal pregnancy (abnormal implantation of the embryo in the uterine tube).

DISEASES AND DISORDERS OF THE UTERUS

The most common diseases of the uterus are uterine fibroids, carcinoma of the cervix, and carcinoma of the corpus (body) of the uterus.

Uterine Fibroids. Uterine fibroids, or *leiomyomas,* are benign tumors that arise from the smooth muscle fibers of the myometrium in response to prolonged or abnormal estrogen stimulation. As women pass the age of 30, they generally develop one or more of these tumors. In some women fibroids distort the uterine cavity to such a degree that implantation is prevented. *Myomyectomy* (surgical removal of fibroids) can restore fertility to many of these women. Sometimes fibroids become so large that they cause pressure on adjacent organs or uterine hemorrage. It then may be necessary to surgically remove the entire uterus *(total hysterectomy).* Smaller fibroids regress spontaneously at the time of menopause.

Carcinoma of the Cervix. Carcinoma of the cervix is the most common of all female pelvic malignancies. It arises from the cervical endometrium and reaches its peak incidence in women between the ages of 40 to 50. Carcinoma of the cervix occurs more frequently in women who have had early sexual exposure and have borne children than in virgins or childless women. This type of uterine cancer extends from the cervix to the vagina and then to the pelvic walls, bladder, and rectum. Once the pelvis is involved, the prognosis is very poor. Early detection of carcinoma of the cervix through use of the *Pap (Papanicolaou) smear,* which involves microscopic examination of detached lining cells of the uterus and upper vagina, has saved the lives of many women.

Carcinoma of the Corpus (Endometrial Adenocarcinoma). Carcinoma of the corpus is almost as common as carcinoma of the cervix. It arises from the endometrial glands in the body of the uterus and reaches its peak incidence in women between the ages of 50 to 60. Endometrial adenocarcinoma is related to hyperestrinism or to prolonged estrogen therapy and occurs more frequently in women who have not borne children. This type of cancer slowly penetrates the uterine wall and extends into the pelvis. If caught before it penetrates the serosa, it is curable.

MENSTRUAL DISORDERS

Amenorrhea. Amenorrhea refers to the absence of menstrual flow. *Primary amenorrhea* is the term used to indicate that a woman has never had a menstrual period. This condition is generally due to congenital absence of the ovaries or lack of FSH. *Secondary amenorrhea* is the term used to indicate that a woman has had menstrual periods but they have stopped prematurely. The causes include hormone imbalances, pregnancy, poor health, or marked emotional disturbances.

Menorrhagia. Menorrhagia refers to excessive flow during a normal menstrual period. Common clinical disorders associated with menorrhagia include blood-clotting deficiencies, uterine fibroids, endometrial polyps, and sex hormone imbalances (hyperestrinism).

Metorrhagia. Metorrhagia refers to noncyclic bleeding from the uterus. It is usually an indication of endometrial adenocarcinoma.

ENDOMETRIOSIS

Endometriosis refers to the ectopic occurrence and growth of endometrial tissue. Extrauterine sites include the ovaries, uterine tubes, uterosacral ligaments, bladder, and rectum. The most common symptom is pelvic pain, which reaches a peak at the onset of menstruation. Extensive endometriosis can cause sterility. Exogenous sex hormone therapy may retard the progress of this disease, but the only way to stop it is by surgical removal of the internal genitalia or radiation destruction of the ovaries.

DISEASES AND DISORDERS OF THE VAGINA

Vaginitis. Vaginitis (inflammation of the vagina) may be caused by a wide variety of bacterial, fungal, and protozoan infections. It is characterized by a whitish discharge accompanied by itching. Common causative organisms include a protozoan called *Trichomonas vaginalis* and a fungus called *Candida albicans*.

Carcinoma. Carcinoma of the vagina is relatively rare, especially in young women. However, it has been observed recently that the teenage daughters of women who received diethylstilbesterol (a synthetic nonsteroidal estrogen) to prevent spontaneous abortion have a higher incidence of vaginal adenocarcinoma than the rest of the female population.

DISEASES AND DISORDERS OF THE MAMMARY GLANDS
Inflammation

Mastitis (inflammation of the breast) is primarily a disease of nursing mothers and is caused by a staphylococcal or streptococcal infection.

Tumors   *Benign Tumors.*   Benign breast tumors are relatively common. The *fibroadenoma* is a very common, slow-growing, estrogen-induced tumor of the glandular tissue and surrounding fibrous connective tissue. Its peak ocurrence is in women between the ages of 25 to 35.

Malignant Tumors.   Carcinoma of the breast is the leading cause of death in women. Its peak incidence is in women between the ages of 40 to 50. Some of the causative factors include age, hereditary predisposition, and a history of hormone imbalance (hyperestrinism or low progesterone levels). Early detection and treatment is essential for survival, as carcinoma of the breast metastasizes early and rapidly. Common sites of metastasis are the axillary lymph nodes, bone, brain, and liver. Treatment consists of *mastectomy* (surgical removal of the breast) generally accompanied by removal of the axillary lymph nodes and pectoralis muscles. Radiation therapy may be given before or after surgery. The ovaries are often removed surgically or destroyed by radiation to inhibit the growth of estrogen-dependent metastases. Chemotherapy may be employed following mastectomy, especially if there is evidence of metastases.

Part C. Conception, Pregnancy, and Childbirth

Conception, pregnancy, and childbirth are a series of events that result in the production of a new human life in the body of a woman and its subsequent expulsion into the world.

Conception and Pregnancy

An average pregnancy lasts approximately 266 days or $9\frac{1}{2}$ lunar months (28 days each) from the time of conception. This prenatal period is divided into the *period of the ovum* (first week), the *period of the embryo* (second to eighth weeks), and the *period of the fetus* (ninth to thirty-eighth weeks). Most doctors prefer to divide the prenatal period into three *trimesters,* a trimester being a period of approximately three calendar months. In describing the events occurring during pregnancy, we will first present the changes taking place in the baby and then the responses occurring in the mother.

THE BABY   *Period of the Ovum (First Week).*   The first week of life begins with
First Trimester   conception and ends with implantation (Fig. 14.26). Conception (fertilization) occurs and life begins when the nucleus of a sperm unites with the nucleus of an ovum in the distal portion of one of the mother's uterine tubes. At this time the diploid chromosome number (46) is restored and the sex of the baby is determined. If an X-bearing sperm fertilizes an X-bearing ovum, a female will result; if a Y-bearing sperm fertilizes an X-bearing ovum, a male will result.

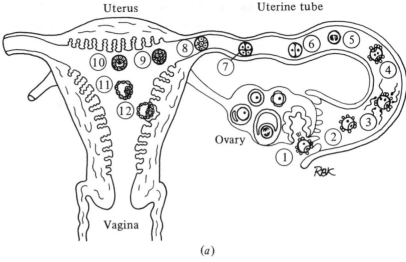

Uterus   Uterine tube

Ovary

Vagina

(*a*)

Figure 14.26 Fertilization to Implantation. (*a*) Schematic representation of events in first week of human life. (1) Ovulated oocyte; (2) oocyte in infundibulum of uterine tube; (3) fertilization; (4) DNA replication by male and female pronuclei; (5) first mitotic division of fertilized ovum; (6) two-cell stage; (7) four-cell stage; (8) morula (early); (9) morula (advanced) reaching uterine cavity; (10) early blastocyst; (11) advanced blastocyst; (12) blastocyst implanting in endometrium. (*b*) Enlarged drawings of some selected early embryonic stages.

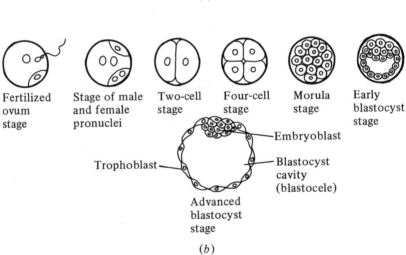

Fertilized ovum stage

Stage of male and female pronuclei

Two-cell stage

Four-cell stage

Morula stage

Early blastocyst stage

Trophoblast

Embryoblast

Blastocyst cavity (blastocele)

Advanced blastocyst stage

(*b*)

As the fertilized ovum completes its leisurely trip through the uterine tube, it undergoes a series of mitotic divisions called *cleavage.* When the fertilized ovum reaches the uterine cavity on the fourth day, it consists of a solid ball of 16 cells and is called a *morula* (because it resembles a mulberry). A fluid-filled cavity develops, converting the morula into a *blastocyst.* An inner cell mass, or *embryoblast,* which projects into the cavity of the blastocyst, will give rise to the embryofetus. The wall of the blastocyst, called the *trophoblast,* will form the extraembryonic membranes. The zona pellucida (the thick membrane surrounding the ovum and blastocyst) disappears, permitting the sticky trophoblast cells to adhere to the endometrium. On about the seventh day (twenty-first day of the menstrual cycle), the blastocyst penetrates and embeds itself into the endometrium. This process is called *implantation*. After implantation the endometrium is called the *decidua*. The portion of endometrium beneath the blastocyst is known as the *decidua basalis*, while the portion reflected over the blastocyst is known as the *decidua capsularis*. The remainder of the endometrium is known as the *decidua parietalis*.

The Embryonic Period (Second to Eighth Weeks). In the second to third weeks, the baby, now called an *embryo* (Fig. 14.27), acquires the three primary germ layers described in Chapter 3. From these germ layers—ectoderm, mesoderm, and endoderm—come all the body's tissues and organ systems. During the third to eighth weeks, the rudiments of the musculoskeletal, nervous, cardiovascular, and digestive systems make their appearance. If the embryo is exposed to teratogenic (monster-forming) agents such as viruses, drugs, or X rays during this critical period of development, major congenital malformations may occur.

Early Fetal Period (Third Month). The beginning of the third month marks the start of the fetal period. There are no striking changes as the embryo becomes a fetus. However, the change in name signifies

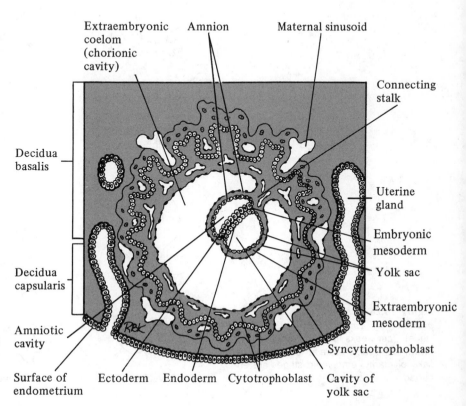

Figure 14.27 Early Human Embryo. Represented at beginning of third week of development. Note that blastocyst is completely embedded in uterine endometrium and that three germ layers have been established.

a change in the baby's appearance from a fishlike creature to one that is recognizably human (Fig. 14.28). During this period the organ systems laid down during the embryonic period undergo growth and differentiation. Very few new organs or structures appear during the fetal period. For this reason the fetus is much less susceptible to the effects of teratogenic agents. The developing muscles and nerves are hooked up early in the fetal period, enabling the fetus to make reflex movements. These movements are gross at first but become more refined as the nervous system matures.

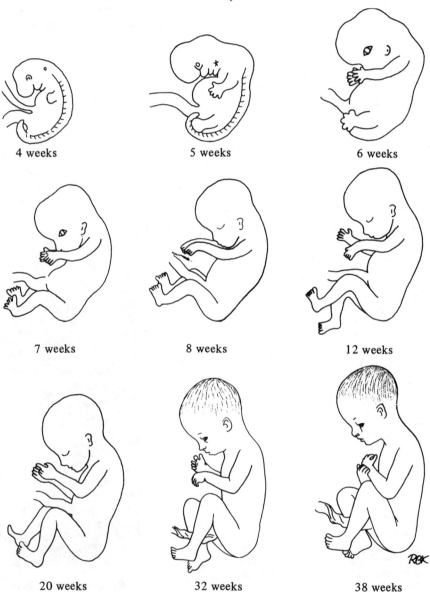

Figure 14.28 Changes in External Appearance of Embryo-Fetus. Embryonic stages (four weeks to seven weeks) are not drawn to the same scale as fetal stages; for sake of clarity, embryonic stages are enlarged.

4 weeks

5 weeks

6 weeks

7 weeks

8 weeks

12 weeks

20 weeks

32 weeks

38 weeks

Development of the Fetal Membranes

As the embryo-fetus develops within the uterine cavity, so do the membranes that provide for its support, nourishment, and protection.

Placenta

The placenta is a compound membranous organ composed of a fetal and a maternal portion. Placental formation begins when the embryonic trophoblast develops radiating fingerlike projections called *villi* that penetrate the decidua. While breaking down decidual tissue for nutrients, the villi create a series of cavities called *intervillous spaces*. Blood from eroded decidual arteries fills the intervillous spaces, forming a complex series of blood lakes that bathe the villi in maternal blood. Once the trophoblast is invaded by mesoderm and develops capillaries, it is known as the *chorion* and the villi as *chorionic villi*. Only the villi closest to the maternal blood supply in the decidua basalis continue to increase in size and complexity. This part of the chorion (the chorion frondosum or shaggy chorion) becomes the fetal portion of the placenta, while the decidua basalis becomes the maternal portion of the placenta (Fig. 14.29).

The mature placenta (Fig. 14.30) is a rounded, flattened, meaty organ about 6 to 7 inches in diameter and $1\frac{1}{4}$ inches thick. It weighs about $\frac{1}{2}$ to $\frac{3}{4}$ pound and is connected to the body of the fetus by the *umbilical cord*. The umbilical cord, the fetal lifeline, contains two arteries and a vein. The two umbilical arteries carry deoxygenated waste-laden blood from the fetus to the placenta. In the placenta carbon dioxide and waste products diffuse from the villous capillaries into the maternal blood, while oxygen and nutrients diffuse from the maternal blood into the villous capillaries. The umbilical vein then returns oxygenated nutrient-laden blood from the placenta to the fetus.

The placenta has two major functions: it acts as a transfer organ for respiratory gases and metabolic products and it produces the hormones necessary to maintain pregnancy. It thus serves as the fetus's lungs, kidneys, digestive tract, and as an endocrine gland. Normally there is no direct exchange of blood between the fetal and maternal portions of the placenta. These blood supplies are separated by a thin diffusion barrier composed of fetal vascular endothelium and a single layer of trophoblast. Sometimes this barrier breaks down late in pregnancy, and an exchange of fetal and maternal blood occurs. Such an exchange may be detrimental to the fetus.

Amnion

The amnion is a thin tough bloodless membrane derived from the trophoblast and reinforced by mesoderm. It forms a protective sac around the fetus that eventually comes in contact with the inner surface of the chorion. The amniotic sac ("bag of waters") is filled with a clear watery fluid that acts as a shock absorber for the fetus and helps it maintain a constant body temperature. A technique called *amnio-*

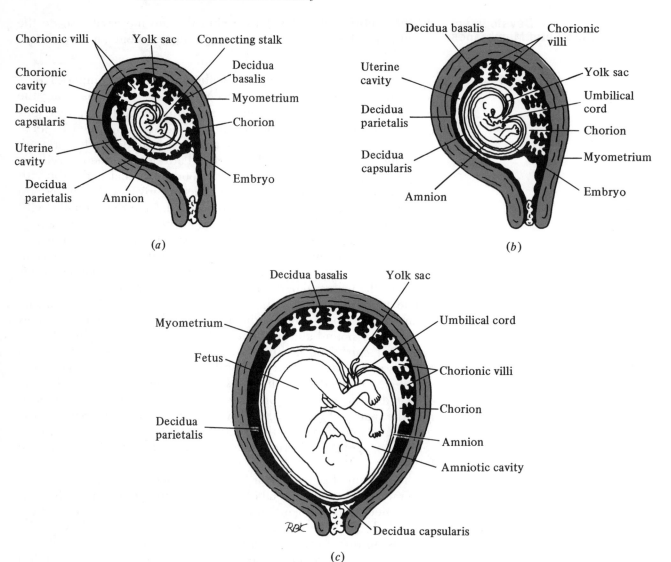

Figure 14.29 *Developmental Changes in Fetal Membranes.* (a) Four weeks. Chorionic villi cover entire surface of chorion. Yolk sac is present in chorionic cavity between amnion and chorion. (b) Six weeks. Chorionic villi on embryonic pole of chorion continue to grow, giving rise to *chorion frondosum* (shaggy chorion). Those on abembryonic pole (opposite pole of chorion) have degenerated, giving rise to *chorion laeve* (smooth chorion). Yolk sac is still present but is relatively small in comparison to size of embryo. (c) Twenty weeks. Amnion and chorion have fused. Uterine cavity has been obliterated by fusion of chorion laeve with decidua parietalis. Yolk sac has become very small.

centesis (transabdominal penetration of the amniotic sac and withdrawal of amniotic fluid) makes it possible for doctors to obtain detached fetal cells for study in cases of suspected genetic disorders.

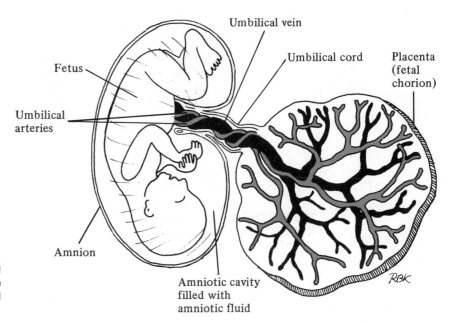

Figure 14.30 Fetal Membranes at Term. Note that umbilical vein carries oxygenated blood *(black)* from placenta to fetus, while umbilical arteries carry deoxygenated blood *(gray)* from fetus to placenta.

Second Trimester (Fourth to Sixth Months)

During the second trimester the fetus grows in length (crown to rump) from 6 to 12 inches and increases in weight from 4 ounces to $1\frac{1}{2}$ pounds. The fetus is now suspended like a swimmer in a quart of amniotic fluid that protects it from the jolts and jars to which its world is subjected. The skin of the fetus develops a covering of fine hair called *lanugo,* which will be gone at birth. In addition, it is covered with *vernix caseosa,* a mixture of sebum and dead epithelial cells. Vernix caseosa waterproofs the skin of the fetus the way grease waterproofs the skin of the long-distance swimmer. Kicking and turning movements occur more frequently as the active fetus swims in its watery environment. However active it may be, the second-trimester fetus is unable to breathe, nurse, or maintain its body temperature on its own. It rarely survives if born prematurely, even when placed in an incubator.

Third Trimester (Seventh to Ninth Months)

The third trimester is significant in that the fetus can survive outside the uterine environment if born prematurely. Chances for survival are about 10 percent in the seventh month, 70 percent in the eighth months, and approach 100 percent as the fetus nears term. Most "preemies" need the support of an incubator to maintain breathing and body temperature. They frequently develop the respiratory distress syndrome because their lungs are too immature to produce surfactant (Chapter 11).

During the third trimester the fetus grows in length (crown to rump) from 12 to 15 inches and increases in weight from $1\frac{1}{2}$ to about 6 to 8 pounds. Its body contours fill out as a large amount of fat is deposited in the subcutaneous fascia. This fat deposit provides the fetus with a nutritional reserve should it be born prematurely. As the fetus increases in size, it runs out of space to move around in the amniotic sac. Therefore, fetal movements become restricted and less frequent as the fetus approaches term.

In the ninth month the fetus usually assumes a head-down position for its trip through the birth canal. This change in position is called *turning*. About two to three weeks before the estimated date of delivery, signs of impending placental failure appear. It is thought that placental failure plays an important role in terminating a normal pregnancy.

THE MOTHER
First Trimester

For most women a missed menstrual period is the first indication of pregnancy. Because the length of time from ovulation to menstruation is approximately 14 days, a woman is entering her third week of pregnancy before she even suspects she might be pregnant. And she may not be certain that she is pregnant until her second missed period. Laboratory pregnancy tests, such as Gravindex®, are based on the immunological detection of CG (chronionic gonadrotropin) in the pregnant woman's urine. These tests are not 100 percent accurate, but considered with other signs and symptoms it is possible in most cases for a doctor to tell a woman in the first trimester whether or not she is pregnant.

Since there are no absolute indications of pregnancy in the critical early weeks of life, the mother may be exposed to infectious disease, x-irradiation, or harmful drugs without being aware of her condition. As soon as she knows she is pregnant, a woman should not take so much as an aspirin without her doctor's permission.

During the first trimester the mother's breasts become swollen and tender under the stimulus of prolactin and the sex hormones. She may even develop *morning sickness* (nausea and vomiting upon arising). Morning sickness has a physiological basis in that the mother's body is not yet adapted to the high levels of sex hormones, especially estrogens, that are present in pregnancy. High levels of the sex hormones also cause bladder irritability, so the mother has to urinate frequently.

Second Trimester

The mother's body undergoes many changes (Fig. 14.31) as the baby begins its growth spurt. Her uterus ascends into the abdominal cavity, and her waist grows thick as her abdomen expands to accommodate the enlarging uterus. The mother may be gaining a pound a week at this point. She usually develops striae (stretch marks) on her abdomen because her skin is overstretched. The mother's circulatory system undergoes changes to accommodate placental circulation. Her total blood volume increases due to an increase in the number of red blood cells needed to carry oxygen to the fetus. The mother may become severely anemic at this point and require extra dietary iron. About

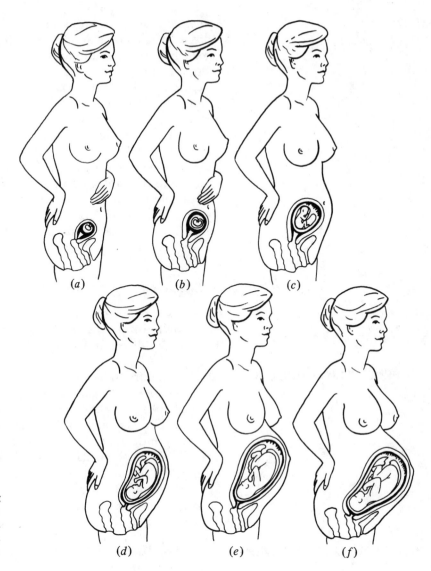

Figure 14.31   Changes in Mother During Pregnancy. Note changes in breast size and size and position of pregnant uterus. (a) 6 weeks. (b) 12 weeks. (c) 16 weeks. (d) 24 weeks. (e) 32 weeks. (f) 38 weeks.

the sixteenth week of pregnancy, the mother begins to feel fetal movements; this is known as *quickening* or "feeling life."

Third Trimester   The third trimester is a period of physical discomfort for the mother. She continues to gain weight, though not as much as in the second trimester. The enlarged uterus now presses on her diaphragm, causing shortness of breath. Displacement of her stomach and duodenum by the uterine fundus causes heartburn and indigestion. Increased frequency of urination results from pressure on her bladder. Constipation is common, caused by pressure upon and displacement of her intestines. Increased pressure on her pelvic veins leads to the formation of hemorrhoids and varicose leg veins. The mother experiences backaches as she attempts to counterbalance the forward-placed weight of the baby and as her pelvic joints relax. Sodium and water retention in late pregnancy may cause the mother to have ankle edema (swollen ankles).

About two to three weeks before the onset of labor, the fetal head (or other presenting part) descends into the pelvic cavity. The mother experiences a sensation of *"lightening"* from the reduction of pressure on her abdominal organs. *Effacement* (shortening of the cervix) begins when the presenting part of the baby presses against the cervix. When labor is imminent, the mother often has a warning sign or *"show,"* a plug of blood-tinged mucus expelled from her cervical canal. This is an indication that the cervix has begun to dilate.

Childbirth (Labor)

Labor is a series of processes by which the products of conception are expelled from the mother's body. It is divided into three stages that vary in length from woman to woman.

FIRST STAGE (DILATION)

During the first stage of labor, the cervix dilates so that the baby can pass into the birth canal. This stage extends from the beginning of regular uterine contractions until the cervix is fully dilated (about 4 inches). The first stage of labor lasts on the average of 15 hours for a first baby and about 8 hours for subsequent births. However, this stage may be longer than 24 hours or less than 1 hour depending on the strength of the uterine contractions, the ability of the cervix to efface and dilate, and the presentation and size of the baby. If uterine contractions are very weak, the mother usually receives injections of Pitocin® (oxytocin) to increase their strength.

SECOND STAGE (DELIVERY)

The second stage of labor begins once the cervix is fully dilated and ends with the complete birth of the baby. During this stage, which may last from a few minutes to several hours, the baby is pushed through the birth canal by rhythmic involuntary contractions of the uterus assisted by voluntary contractions of the mother's abdominal muscles. The amniotic sac ruptures at the beginning of this stage if it has not done so previously. In most instances the head of the baby presents first, but occasionally the buttocks present first (breech presentation). When the presenting part appears at the vaginal opening, the obstetrician generally performs an *episiotomy.* An episiotomy is a perineal incision to widen the vaginal opening. It is made to permit easier passage for the baby and to prevent perineal tearing or damage to the anal sphincters. Once the head or buttocks have passed through the vaginal opening, the rest of the body follows readily. After delivery of the baby, the umbilical cord is tied off and severed close to the baby's body.

THIRD STAGE (AFTERBIRTH) The third stage of labor consists of the delivery of the placenta or afterbirth. Within about 30 minutes after delivery of the baby, several strong uterine contractions cause the separation of the placenta from the uterine wall and its subsequent expulsion through the birth canal. Continued uterine contractions following placental separation compress the endometrium and prevent hemorrhaging from open blood vessels.

Family Planning

Various methods have been devised to regulate human fertility by either preventing or promoting conception.

STERILIZATION Sterilization is a permanent means of preventing conception. In men it is achieved by vasectomy (cutting and tying off the vasa deferentia), while in women it is achieved by tubal ligation (tying off and cauterizing the uterine tubes).

ORAL CONTRACEPTIVES Oral contraceptives, or birth control pills, contain hormones that prevent ovulation or make the endometrium unsuitable for implantation. There are basically three types of birth control pills in use today:

1. *Combination Pill.* Consists of a mixture of estrogen and progesterone that is taken for the first 20 days of the menstrual cycle. The estrogen inhibits FSH secretion, while the progestrone causes secretion of infertile cervical mucus and formation of an inhospitable endometrium. This is most effective type of birth control pill.
2. *Sequential Pill.* Consists of pure estrogen that is taken for the first 15 days of the menstrual cycle followed by a mixture of estrogen and progesterone taken for the following 5 days. The sequential pill has been used mainly by women who are sensitive to progesterone. It is less effective than the combination pill because it lacks the continuous progestational effect. The sequential pill is being withdrawn from the market because it appears to cause adenocarcinoma of the endometrium.
3. *"Mini" Pill.* Consists of low doses of pure progesterone taken throughout the entire menstrual cycle. This is the least effective of the birth control pills because it lacks the ovulation suppression effect of the other two types. However, it is the pill of choice for women who are sensitive to estrogen.

While birth control pills are highly effective in preventing conception, there is a definite element of risk involved in taking them. The side effects of taking high doses of female sex hormones include development of blood clots, strokes, heart attacks, cancer, and permanent sterility. It is also interesting to note that there are no oral contraceptives for men to date.

IUD The IUD (intrauterine device) is a small plastic or metal object inserted into the uterus, where it acts as a foreign body. It appears to keep the endometrium in a state of irritation such that implantation is discouraged. If implantation does occur, the endometrium usually cannot support the embryo and it is aborted. Some unpleasant side effects of the IUD are a heavy menstrual flow, severe cramping, perforation of the uterine wall, and septic abortion.

MECHANICAL BARRIERS Mechanical barriers prevent sperm from coming in contact with the oocyte. The *diaphragm* is a soft rubber cup with a flexible metal rim that fits over the cervix. The *condom* is a thin flexible rubber sheath that fits over the erect penis. When used with a spermicidal jelly, they are both highly effective in preventing conception. However, care must be taken to see that the rubber in these devices does not deteriorate.

NATURAL METHODS Natural methods of family planning are those that employ no drugs or mechanical devices. All natural methods are based on identification of the fertile period of the menstrual cycle and require abstinence from genital contact during that time.

Rhythm Method The rhythm method consists of calculating the approximate length of the fertile period. Since sperm have a survival rate of four to five days and oocytes have a survival rate of 12 to 24 hours, the five days before and the day after ovulation constitute the fertile period. In the rhythm method a few days are added to each end of the period for extra safety. The approximate length of the fertile period is calculated as follows:

1. Keep a record of the length of all menstrual cycles for 12 consecutive months.
2. Subtract 18 from the length of the shortest cycle to determine the first fertile day; for example, $25 - 18 =$ day 7.
3. Subtract 11 from the length of the longest cycle to determine the last fertile day; for example, $28 - 11 =$ day 17.

The rhythm method can be very effective if a woman has regular menstrual cycles. If she has irregular cycles, the length of the fertile period can be miscalculated.

Temperature Method The temperature method is based on the fact that a woman's basal body temperature (BBT) rises at least 0.4°F after ovulation and remains elevated until one to two days before the start of the next menstrual period. The BBT must remain elevated for three consecutive days before the thermal shift is considered postovulatory. This method enables a couple trying to avoid pregnancy to recognize the infertile postovulatory period and use it for intercourse. The disadvantage of this method is that fever, alcohol consumption, and emotional upset can produce a false BBT rise.

Ovulation Method   The ovulation method is based on the identification of the fertile period by changes in the pattern of cervical mucus secretion. Several days prior to ovulation, rising estrogen levels cause the cervical mucus to become thin, clear, and stretchy (fertile pattern) and the woman experiences a sensation of vaginal wetness. Fertile mucus symptoms reach their peak just before ovulation occurs. About 12 hours after ovulation, rising progesterone levels cause the cervical mucus to become thick and opaque (infertile pattern) and the woman experiences a sensation of vaginal dryness. The fertile period is considered to extend from the beginning of the preovulatory wet days until three days after the peak day. The great advantage of this method is that it can be used by women with irregular cycles, such as premenopausal women and nursing mothers.

Diseases and Disorders of Pregnancy

TOXEMIA   Toxemia of pregnancy is a poorly understood disease characterized by the following symptoms: water retention, protein in the urine, and high blood pressure. With the addition of convulsions and coma, the disease is termed *eclampsia.* The cause of toxemia is unknown. Predisposing factors include cardiovascular disease, kidney disease, and malnutrition; all presumably lead to disordered placental metabolism. Toxemia of pregnancy is almost always cured when the baby is born.

ECTOPIC PREGNANCY   Ectopic pregnancy is a condition in which the embryo implants outside the uterine cavity. The most common extrauterine site is the uterine tube. Other sites include the ovary, cervix, broad ligament, and peritoneal cavity. Ectopic pregnancies rarely survive to term because extrauterine sites are not adaptable to pregnancy. They are usually terminated by rupture into the peritoneal cavity and intraperitoneal hemorrhage during the first trimester. Without surgical intervention, a ruptured ectopic pregnancy may cause massive hemorrhage and death.

ABORTION   Abortion refers to any interruption of pregnancy before the baby is viable (at about 23 weeks).

Spontaneous Abortion.   A spontaneous abortion, commonly called a *miscarriage,* arises from some abnormality of the baby or the mother and usually results in the expulsion of a dead baby from the uterus. About 50 to 60 percent of spontaneous abortions are due to developmental defects of the baby. Another 15 percent are due to maternal factors such as trauma, infection, poor diet, or endocrine imbalances. Causes of the remaining 25 to 35 percent are unknown.

Induced Abortion. An induced abortion is a procedure performed to remove a live baby from the uterus. A *therapeutic abortion* is one that is performed to save a mother's life or mental health. Today some therapeutic abortions are performed when the baby has a serious or fatal hereditary disease. The presence of the disease must be previously confirmed in fetal cells obtained by amniocentesis.

Suggested Readings

BEER, ALAN E., and R. E. BILLINGHAM. "The embryo as a transplant," *Sci. Am., 230,* no. 4 (April 1974), pp. 36–46.

BEHRMAN, SAMUEL JAN. "Which 'pill' to choose?" *Hosp. Pract., 4,* no. 5 (May 1969), pp. 34–39.

BILLINGS, JOHN. *Natural Family Planning: The Ovulation Method.* 3d American ed. Collegeville, Minn.: Liturgical Press, 1975 (paperback).

BOSTON WOMEN'S HEALTH BOOK COLLECTIVE. *Our Bodies, Ourselves.* 2d ed. New York: Simon & Schuster, 1976.

COOPER, LOUIS Z. "German measles," *Sci. Am., 215,* no. 1 (July 1966), pp. 30–37.

FRIEDMAN, THEODORE. "Prenatal diagnosis of genetic disease," *Sci. Am., 225,* no. 5 (November 1971), pp. 34–42.

KATCHADOURIAN, HERANT A., and DONALD T. LUNDE. *Fundamentals of Human Sexuality.* New York: Holt, Rinehart and Winston, 1972.

MACLEOD, JOHN. "The parameters of male fertility," *Hosp. Pract., 8,* no. 12 (December 1973), pp. 43–52.

MASTROIANNI, LUIGI. "Fertilization and the tubal environment," *Hosp. Pract., 7,* no. 3 (March 1972), pp. 113–119.

NETTER, FRANK H. *The Ciba Collection of Medical Illustrations.* Vol. 2: *Reproductive System.* Summit, N.J.: Ciba, 1965.

NILSSON, LENNART, AXEL INGELMAN-SUNDBERG, and CLAES WIRSEN. *A Child Is Born.* New York: Dell, 1974.

°PAGE, ERNEST W., CLAUDE A. VILLEE, and DOROTHY B. VILLEE. *Human Reproduction.* 2d ed. Philadelphia: Saunders, 1976.

RUGH, R., and L. B. SHETTLES. *From Conception to Birth.* New York: Harper & Row, 1972.

SEGAL, SHELDON. "The physiology of human reproduction," *Sci. Am., 231,* no. 3 (September 1974), pp. 52–62.

SEVER, JOHN L. "Viral teratogens: A status report," *Hosp. Pract., 5,* no. 4 (April 1970), pp. 75–83.

SWYER, G. I. M. "Investigation and treatment of ovulatory failure," *Hosp. Pract.*, *4*, no. 4 (April 1969), pp. 57–67.

VANDE WIELE, RAYMOND L. "Treatment of infertility due to ovulatory failure," *Hosp. Pract.*, 7, no. 10 (October 1972), pp. 119–126.

URICCHIO, WILLIAM A., and MARY KAY WILLIAMS, eds. *Proceedings of a Research Conference on Natural Family Planning.* Washington, D.C.: The Human Life Foundation, 1973.

° Advanced level.

15

The Nervous System

The nervous system is the final system of the human body that we will consider. It has both anatomical and functional divisions. Anatomically, the nervous system is divided into the *central nervous system,* which consists of the brain and spinal cord, and the *peripheral nervous system,* which consists of the cranial and spinal nerves. Functionally, the nervous system is divided into the *voluntary nervous system,* which includes *somatic* sensory and motor nerves, and the involuntary or *autonomic nervous system,* which includes *visceral* sensory and motor nerves.

The nervous system can be compared to electrical equipment with which we are all familiar. The central nervous system is like a giant telephone switchboard that is linked to all parts of the body by peripheral nerves, which act like telephone cables. Peripheral nerves carry messages in the form of *impulses* (very small electrical discharges) to the central nervous system concerning changes in the body's internal and external environments. They also carry messages from the central nervous system to the skeletal muscles, visceral organs, and glands to enable the body to adjust to these changes.

The human nervous system is more compact and more sophisticated than the finest computer produced to date. The human "mind," which is the summation of the highest functions of the central nervous system, is capable of things no computer can duplicate: foresight, judgment, emotion, intuition, creativity, altruism, and the exercise of free will. After all, a computer is merely the product of our minds.

This chapter is divided into five parts. Part A deals with the cellular organization of the nervous system, Part B deals with the central nervous system, Part C deals with the peripheral nervous system, Part D deals with the autonomic nervous system, and Part E deals with the special sense organs.

Part A. Cellular Organization of the Nervous System

The nervous system is composed of nerve cells called *neurons* and supporting cells called *neuroglia.* Neurons are specialized for conduction of electrical impulses and, in some cases, for secretion. Neuroglia are specialized to provide physical and metabolic support for the neurons.

Neurons

Neurons are the functional units of the nervous system. These highly specialized cells come in a variety of shapes and sizes (Fig. 15.1). No matter what their size or shape, most neurons consist of three parts: the *cell body, dendrites,* and an *axon* (Fig. 15.2).

CELL BODY The nerve cell body, also called the *perikaryon,* is the part of the neuron that contains the nucleus and cytoplasm exclusive of the processes. A collection of nerve cell bodies outside the central nervous system is called a *ganglion;* inside the central nervous system it is called a *center* or *nucleus* (not to be confused with the cell nucleus).

Cytoplasm *Organelles.*   All neurons contain an abundance of free ribosomes or rough endoplasmic reticulum (RER). Large parallel stacks of RER and free ribosomes in neuronal cytoplasm are called *Nissl bodies.* They synthesize structural proteins and proteins for transport. Mitochondria, similar to those seen in other cell types, are present, as well as a Golgi apparatus and lysosomes. Fine filaments, called *neurofibrils* when seen at the light microscopic level, form an extensive network in the cell body and its processes; they appear to provide an internal scaffolding for the cell. Microtubules, called *neurotubules,* are also found in the cell body and its processes; they appear to be transporting metabolites as well as providing support.

Inclusions.   Cytoplasmic inclusions include *melanin,* a brown-black pigment that is prominent in certain neurons of the central nervous system, and *lipofuscin,* a light brown pigment that accumulates in neurons with age. Glycogen granules and lipid droplets, which repsent stored energy sources, are also found in many neurons.

Nucleus Most neurons have a large round nucleus containing diffuse chromatin and one or more prominent nucleoli.

DENDRITES Dendrites are processes that carry impulses *toward* the nerve cell body. They vary greatly in appearance with the type of neuron and

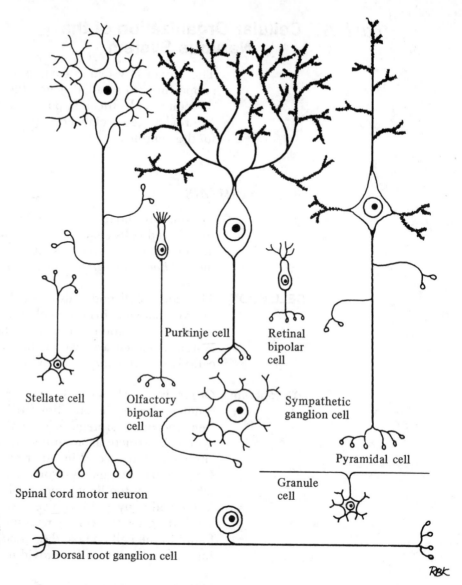

Figure 15.1 *Types of Neurons.* Neurons vary greatly in size, shape, and branching of processes.

Stellate cell

Olfactory bipolar cell

Purkinje cell

Retinal bipolar cell

Sympathetic ganglion cell

Pyramidal cell

Granule cell

Spinal cord motor neuron

Dorsal root ganglion cell

may be modified to form special sensory receptors. Most neurons have numerous dendrites, which greatly increase the receptive area of the cell. Dendrites have the same cytoplasmic organelles as the cell body except that they are devoid of Golgi apparatus.

AXON   Axons are processes that carry impulses *away* from the nerve cell body. Each neuron has only one axon, which may be quite long. The axon gives off collateral branches along its course plus terminal branches that end in small knobs called *terminal boutons*, or end

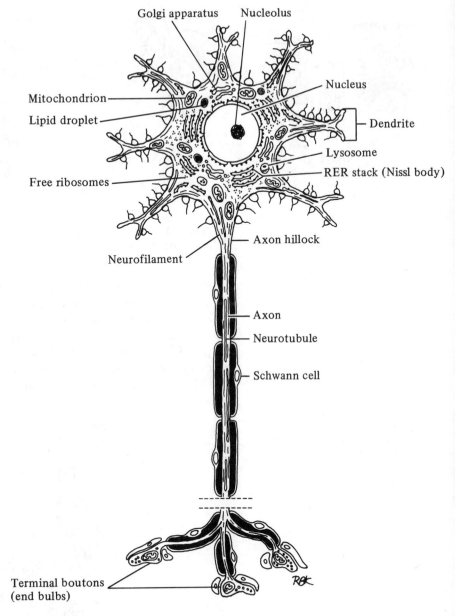

Figure 15.2 Typical Neuron. Schematic diagram of spinal cord motor neuron showing organelles and inclusions.

bulbs, through which they make contact with other neurons, glandular cells, or muscle fibers.

SYNAPSES   A nerve impulse reaches its destination by traveling over a chain of neurons. At various junction points the terminal boutons of one neuron come into contact with the dendrities, cell body, or axon (Fig. 15.3) of

Spine synapse

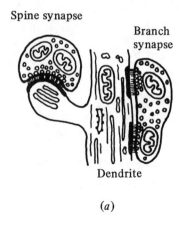

Branch synapse

Dendrite

(a)

Synapse

Cell body (soma)

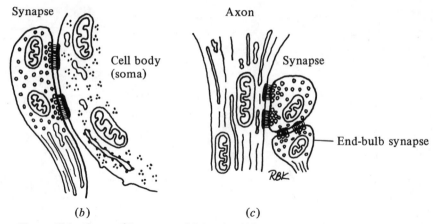

Axon

Synapse

End-bulb synapse

(b)

(c)

Figure 15.3   Types of Synapses.   (a) Axodendritic synapses.   (*b*) Axosomatic synapse. (*c*) Axoaxonic synapses.

the next neuron in the chain. These junction points are called *synapses* (synaptic junctions). A submicroscopic cleft, called the *synaptic cleft,* exists between the two neurons in a synapse; therefore, the neurons are not actually in direct contact. At synapses, influences pass between neurons by way of chemical transmitters or by electrical coupling, usually in one direction. In a typical chemical synapse (Fig. 15.4), the terminal bouton (presynaptic ending) contains *synaptic*

Terminal bouton (end bulb)

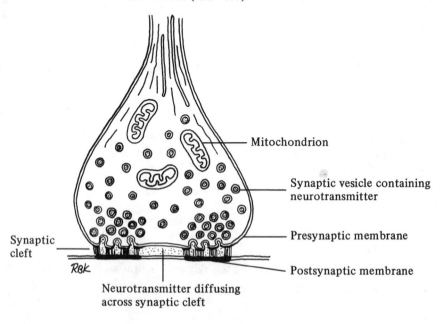

Mitochondrion

Synaptic vesicle containing neurotransmitter

Presynaptic membrane

Postsynaptic membrane

Synaptic cleft

Neurotransmitter diffusing across synaptic cleft

Figure 15.4 Morphologic Structure of Chemical Synapse. Drawing based on electron microscopic view.

vesicles, tiny sacs filled with a *neurotransmitter* substance (a chemical responsible for transmission of the nerve impulse).

CLASSIFICATION OF NEURONS   Neurons are classified on the basis of either structure or function. You should be familiar with both classifications in order to understand where the neurons are located and what they are doing.

Structural Classification   *Unipolar Neurons.*   Unipolar neurons [Fig. 15.5(*a*)] have a single axonlike process to which the cell body is attached by a stalk. Unipolar neurons are always sensory in function and their cell bodies are located in the sensory ganglia of cranial and spinal nerves.

Bipolar Neurons.   Bipolar neurons [Fig, 15-5(*b*)] have two processes, an axon and a dendrite, with the cell body located between them. Bipolar neurons are associated with special sensory functions and are found in the olfactory mucosa, the retina, and the vestibular and cochlear ganglia.

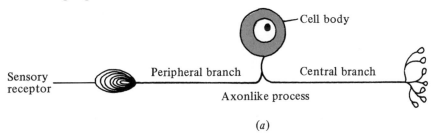

(*a*)

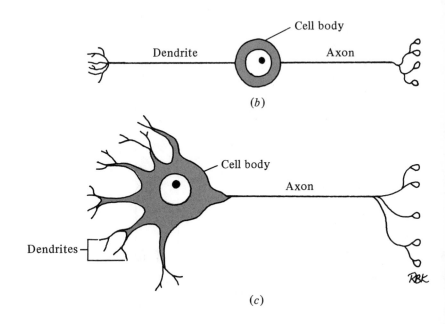

(*b*)

(*c*)

Figure 15.5   *Neurons.*   (a) Unipolar.   (b) Bipolar.   (c) Multipolar.

Multipolar Neurons. Multipolar neurons [Fig. 15.5(c)] have one axon plus two or more dendrites. Multipolar neurons are the predominant cell type in the central nervous system and are also found in autonomic ganglia.

Functional Classification

Afferent Neurons. Afferent or sensory neurons carry impulses *toward* the central nervous system. The dendritic terminal of an afferent neuron capable of receiving a sensory stimulus is called a

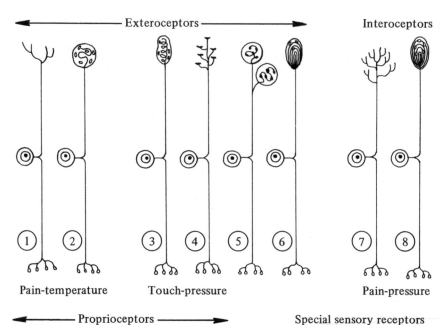

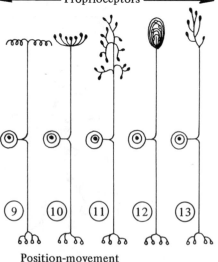
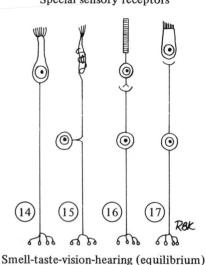

Figure 15.6 Types of Sensory Receptors. (1) Bare nerve ending—pain and warmth; (2) end bulb of Krause—cold(?); (3) Meissner's corpuscle—touch; (4) Merkel's disc—touch; (5) Golgi-Mazzoni corpuscle—pressure; (6) Pacinian corpuscle—pressure; (7) bare nerve ending—visceral pain; (8) Pacinian corpuscle—visceral fullness; (9) annulospiral ending—muscle stretch; (10) Golgi tendon receptor—tendon stretch; (11) flower-spray ending—muscle stretch; (12) Pacinian corpuscle—proprioception; (13) joint receptor—proprioception; (14) olfactory cell—smell; (15) taste cell and unipolar neuron—taste; (16) retinal photoreceptor and bipolar neuron—vision; (17) hair cell and bipolar neuron—hearing or equilibrium.

sensory receptor. Sensory receptors (Fig. 15.6) are divided into the following categories:

1. *Exteroceptors.* Sensory receptors receiving stimuli from the body surface. These stimuli include pain, touch, warmth, cold, and pressure.
2. *Interoceptors.* Sensory receptors receiving stimuli from the internal organs. These stimuli include pain, pressure, and muscle stretch.
3. *Proprioceptors.* Sensory receptors receiving stimuli from skeletal muscles, tendons, and joints. These stimuli include muscle and tendon stretch and changes in joint position.
4. *Special Sensory Receptors.* Sensory receptors receiving sensory stimuli from the organs of smell, taste, vision, equilibrium, and hearing.

Efferent Neurons. Efferent or motor neurons carry impulses *away from* the central nervous system. There are two types of motor neurons (Fig. 15.7):

1. *Somatic Motor Neurons.* Neurons that innervate voluntary (skeletal) muscle.
   a. *Upper Motor Neurons.* Motor neurons whose cell bodies are located in the cerebral or cerebellar cortex.
   b. *Lower Motor Neurons.* Motor neurons whose cell bodies are located in the brain stem or spinal cord.
2. *Visceral Motor Neurons.* Neurons that innervate involuntary (smooth and cardiac) muscle and glands.
   a. *Preganglionic Neurons.* Motor neurons whose cell bodies are located in the brain stem or spinal cord.
   b. *Postganglionic Neurons.* Motor neurons whose cell bodies are located in autonomic ganglia.

Interneurons. Interneurons are intermediary neurons in the central nervous system. They can either link afferent neurons directly with efferent neurons or form complex functional chains called *circuits*.

Supporting Cells

NEUROGLIA: CNS
Astrocytes

Astrocytes [Fig. 15.8(*a*)] are star-shaped cells with long radiating processes. Many of their processes have terminal expansions called *vascular end feet,* which contact capillary walls. Astrocytes thus appear to transport metabolites from capillaries to neurons as well as to form the structural framework of the CNS. *Fibrous astrocytes* have long thin processes that branch infrequently and contain many neurofibrils. They are found predominantly in white matter. *Protoplasmic*

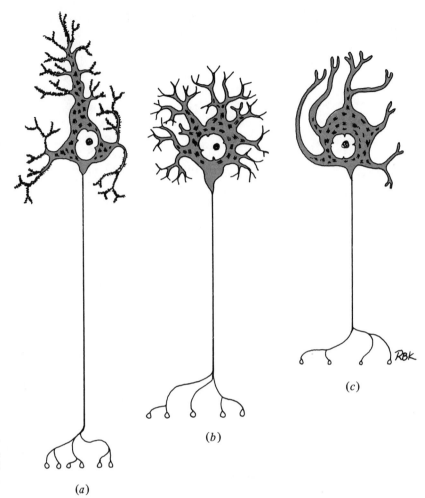

Figure 15.7 *Types of Motor Neurons.*
(a) Upper motor neuron, pyramidal cell.
(b) Lower motor neuron, anterior horn cell
(spinal cord). (c) Visceral motor neuron,
sympathetic ganglion cell.

astrocytes have numerous thick processes with many branches and contain few neurofibrils. They are found predominantly in gray matter.

Oligodendrocytes   Oligondendrocytes [Fig. 15.8(b)] are glial cells that form and maintain the myelin sheaths of nerve fibers (axons) in the central nervous system. They are much smaller than astrocytes and have fewer and shorter processes. A single astrocyte can myelinate many nerve fibers.

Ependymal Cells   Ependymal cells [Fig. 15.8(c)] are cuboidal to columnar cells, sometimes ciliated, that line the ventricular system of the brain and the central canal of the spinal cord. They form the epithelial covering of the choroid plexuses (capillaries that protrude into the ventricles).

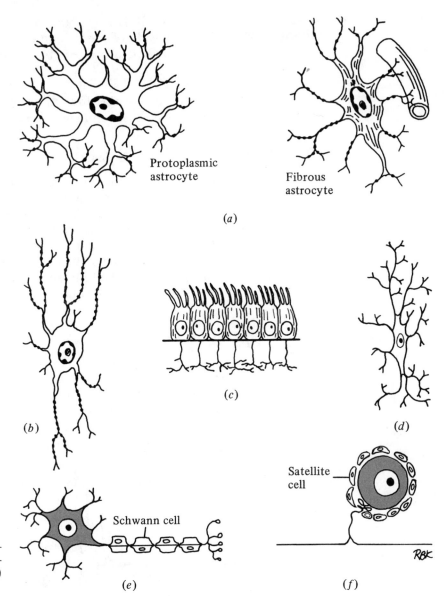

Figure 15.8 Supporting Cells. (a) Astrocytes. (b) Oligodendrocyte. (c) Ependymal cells. (d) Microgliocyte. (e) Schwann cells. (f) Satellite cells.

Microglia Microglia [Fig. 15.8(d)] are small cells with short branched processes that are scattered throughout the central nervous system. These cells are the phagocytes of the central nervous system and are functionally similar to tissue macrophages. Microglia, unlike other glial cells, are of mesodermal rather than ectodermal origin.

NEUROGLIA: PNS Schwann Cells	Schwann cells [Fig. 15.8(*e*)] are small cells that form and maintain the myelin sheaths of nerve fibers in the peripheral nervous system. Many Schwann cells are needed to myelinate a single peripheral nerve fiber.
Satellite Cells	Satellite cells [Fig. 15.8(*f*)] are small flattened cells that form protective capsules around nerve cell bodies in peripheral ganglia.

Part B. The Central Nervous System

The central nervous system is derived from an embryonic ectodermal structure called the *neural tube.* The neural tube is formed during the third week of life by the invagination and fusion of a raised plate of ectoderm on the dorsal surface of the embryo (Fig. 15.9). Growth and development of the neural tube is greater at the cephalic (head) end, which becomes the brain, than at the caudal (tail) end, which becomes the spinal cord, The lumen of the neural tube gives rise to the ventricles (cavities) of the brain and the central canal of the spinal cord.

The Spinal Cord

The spinal cord is the least complex and best understood portion of the central nervous system, which makes it the ideal structure with which to begin our discussion. The spinal cord has two fairly distinct functions: the transmission of information to and from the brain and

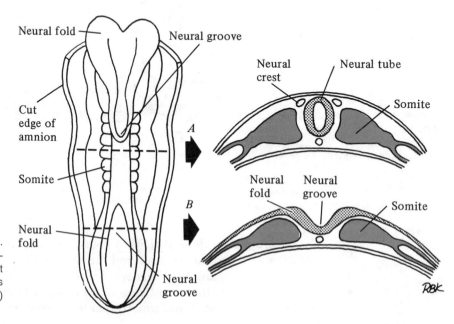

Figure 15.9 Neural Tube Formation. Transverse sections taken through three-week-old human embryo at two different levels: section *A* (more cephalic) shows closed neural tube; section *B* (more caudal) shows still-open neural tube.

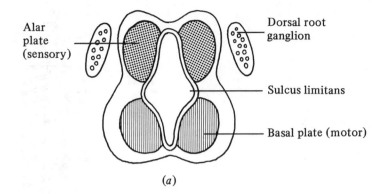

Alar plate (sensory)

Dorsal root ganglion

Sulcus limitans

Basal plate (motor)

(a)

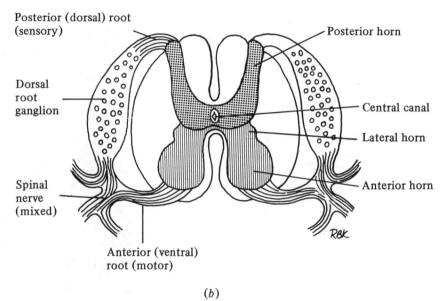

Posterior (dorsal) root (sensory)

Dorsal root ganglion

Spinal nerve (mixed)

Posterior horn

Central canal

Lateral horn

Anterior horn

Anterior (ventral) root (motor)

(b)

Figure 15.10 Development of Spinal Cord. Schematic diagrams show two successive stages in development. Note that alar plates give rise to sensory (posterior) horns of spinal cord, while basal plates give rise to motor (lateral and anterior) horns.

the processing of sensory information and initiation of stereotyped (patterned) motor movements called reflexes at the segmental level.

DEVELOPMENT The spinal cord is the most primitive and least specialized part of the embryonic neural tube. It still retains a segmental pattern reminiscent of the nervous systems of lower animals. As the caudal portion of the neural tube begins to form the spinal cord, its lateral walls thicken [Fig. 15.10(a)] by increased cell proliferation. Each lateral thickening is separated into dorsal and ventral portions by a longitudinal furrow called the *sulcus limitans*. The two dorsal thickenings, or *alar plates*, contain association neurons that synapse with incoming sensory neurons. The two ventral thickenings, or *basal plates*, contain motor neurons. The alar plates give rise to the dorsal (posterior) horns of the spinal cord, while the basal plates give rise to the lateral and ventral

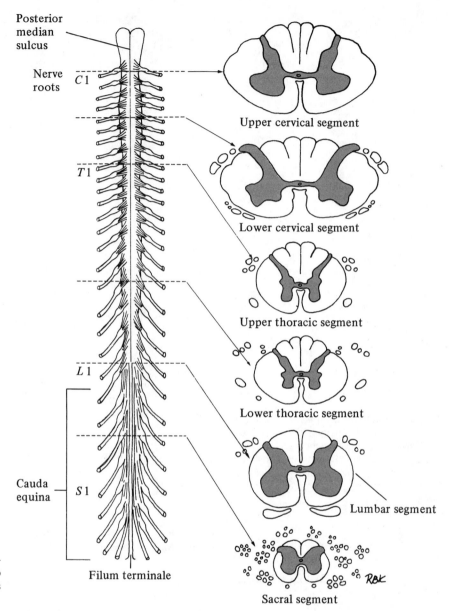

Posterior
median
sulcus

Nerve
roots   *C* 1

T 1

L 1

Cauda
equina   *S* 1

Filum terminale

Upper cervical segment

Lower cervical segment

Upper thoracic segment

Lower thoracic segment

Lumbar segment

Sacral segment

Figure 15.11 Adult Spinal Cord. Posterior aspect shows spinal nerve roots. Also indicated are segments taken from various levels of spinal cord.

(anterior) horns [Fig. 15.10(*b*)]. The lumen of the neural tube is reduced to a narrow slit that becomes the central canal of the spinal cord.

EXTERNAL ANATOMY   The adult spinal cord (Fig. 15.11) is a cylindrical tube of neural tissue, approximately 18 inches long, that occupies the upper two-thirds of the vertebral canal. The spinal cord begins at the foramen magnum and ends in a cone-shaped structure, the *conus medullaris,* at about vertebral level *L2.* The *filum terminale,* a strand of neuroglial tissue, extends down from the conus to attach the spinal cord to the coccyx. The spinal cord is not of uniform diameter. There are two conspicuous enlargements: one in the cervical region of the cord that contains the neurons innervating the upper extremities and one in the lumbar region of the cord that contains the neurons innervating the lower extremities. Running the full length of the spinal cord are two longitudinal grooves that divide the cord into right and left halves. The *posterior median sulcus* is a deep slitlike groove that separates the dorsal (posterior) halves of the cord. The *anterior median sulcus* is a deep wide groove that separates the ventral (anterior) halves of the cord. The right and left halves are connected above and below the central canal by commissural fibers.

The spinal cord consists of 31 segments: 8 *cervical* segments, 12 *thoracic* segments, 5 *lumbar* segments, 5 *sacral* segments, and 1 *coccygeal* segment. Each spinal cord segment gives rise to a pair of corresponding spinal nerves. Due to differential growth rates, the spinal cord ends up being much shorter than the vertebral column. Therefore, the cervical nerves run horizontally, but the lower nerves run more and more obliquely until their course is almost vertical (Fig. 15.12). The lumbar and sacral nerve roots travel quite a distance before they can emerge from the vertebral canal at their proper level. These long nerve roots form a structure called the *cauda equina* (horse's tail) around the filum terminale.

INTERNAL ANATOMY When the spinal cord is examined in cross section (Fig. 15.13), it has an oval shape. It is partially divided into right and left halves by the posterior and anterior median sulci. The substance of the spinal cord consists of an H-shaped area of gray matter surrounded by a shell of white matter.

Gray Matter The gray matter is composed of nerve cell bodies plus unmyelinated and lightly myelinated nerve fibers. The limbs of the H consist of the two *dorsal (posterior) horns* and the two *ventral (anterior) horns.* The small paired *lateral horns* project out from the sides of the H in the thoracic and upper lumbar segments of the cord. The crossbar of the H is formed by the *posterior* and *anterior gray commissures.* Most of the cell bodies in the spinal cord gray matter are organized into nuclei. The posterior horns contain sensory association nuclei, the anterior horns contain somatic motor nuclei, and the lateral horns contain visceral motor nuclei [Fig. 15.14(a)].

White Matter The white matter of the spinal cord lies superficial to the gray matter. It consists of myelinated nerve fibers plus glial cells. Various grooves partition the white matter in each half of the cord into *dorsal (posterior), lateral,* and *ventral (anterior) funiculi.* Each funiculus (column) contains several *tracts.* A tract consists of a bundle of nerve

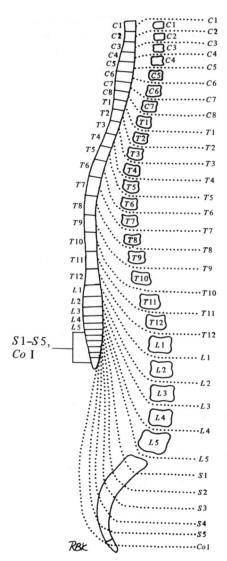

Figure 15.12 *Relationship of Spinal Cord Segments and Spinal Nerves to Vertebrae.* Spinal nerves are indicated by *dotted lines*. Note that spinal nerves C1 to C8 emerge above their corresponding vertebrae.

fibers that transmit the same type of nerve impulse and go to the same area of the central nervous system. Ascending tracts transmit sensory impulses up the spinal cord, while descending tracts transmit motor impulses down the spinal cord [Fig. 15.14(*b*)].

The Brain

The brain is the highly developed, complex portion of the central nervous system. Its simple tubular structure has become greatly distorted in the adult as the result of differential growth rates and the foldings that have occurred along its longitudinal axis during embryonic development.

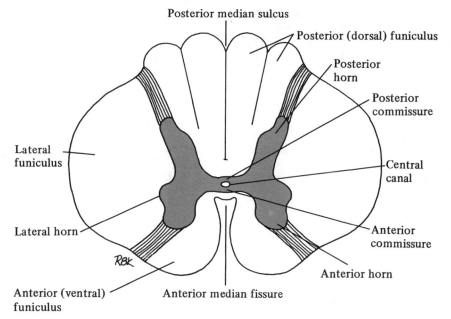

Figure 15.13 *Cross Section of Spinal Cord.* Midthoracic level.

Posterior median sulcus

Posterior (dorsal) funiculus

Posterior horn

Posterior commissure

Central canal

Anterior commissure

Anterior horn

Lateral funiculus

Lateral horn

Anterior (ventral) funiculus

Anterior median fissure

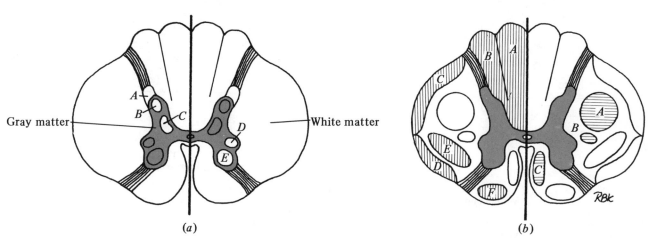

Gray matter

White matter

(a)

(b)

Figure 15.14 *Spinal Cord.* (a) Nuclei. Left side contains sensory nuclei; *A*, substantia gelatinosa; *B*, nucelus proprius; *C*, nucleus dorsalis. Right side contains motor nuclei; *D*, intermediolateral nucleus (visceral motor); *E*, lateral motor nucelus (somatic motor). (b) Tracts. Left side contains ascending (sensory) tracts; *A*, fasciculus gracilis; *B*, fasciculus cuneatus; *C*, posterior spinocerebellar tract; *D*, anterior spinocerebellar tract; *E*, lateral spinothalamic tract; *F*, anterior spinothalamic tract. Right side contains descending (motor) tracts; *A*, lateral corticospinal tract; *B*, rubrospinal tract; *C*, anterior corticospinal tract.

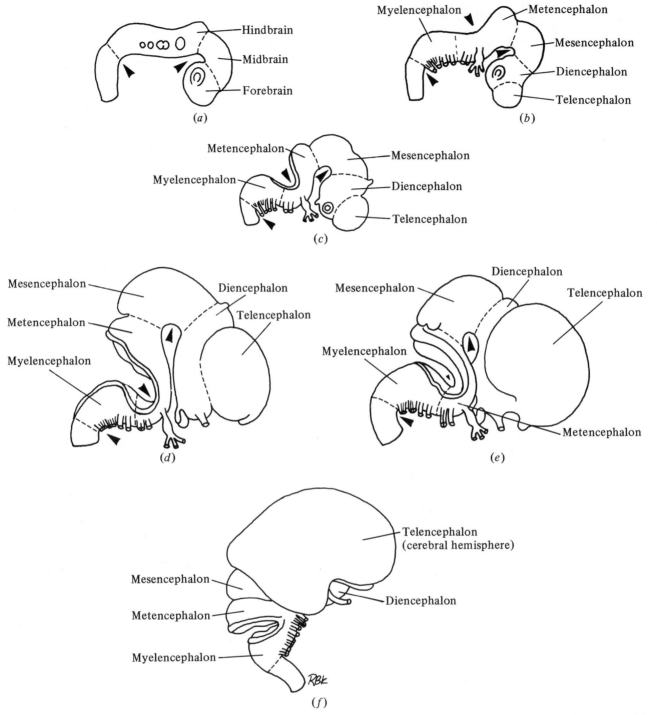

Figure 15.15 *Development of Brain.* (a) 4-week-old embryo. *Arrows* indicate cranial and cervical flexures. (b) 5-week-old embryo. *Arrows* indicate cranial, pontine, and cervical flexures. (c) 6-week-old embryo. (d) 7-week-old embryo. (e) 8-week-old embryo. (f) 3-month-old fetus. Note rapid and disproportional growth of cerebral hemisphere.

DEVELOPMENT As the cephalic end of the neural tube begins to form the brain, three swellings, or primary brain vesicles, appear. These primary vesicles represent the *forebrain, midbrain,* and *hindbrain* [Fig. 15.15(*a*)]. At the same time, the rapid growth of the neural tube in a confined space forces it to fold on itself. An anterior fold, or cephalic flexure, bends the forebrain downward, while a posterior fold, or cervical flexure, bends part of the hindbrain and spinal cord downward.

The primary vesicles give rise to secondary brain vesicles. The anterior part of the forebrain gives rise to the *telencephalon,* which is represented by large lateral outgrowths called *cerebral hemispheres.* The posterior part of the forebrain gives rise to the *diencephalon,* which comes to lie between the cerebral hemispheres. The *midbrain,* or *mesencephalon,* remains relatively unchanged and retains its basic tubular structure. The anterior part of the hindbrain gives rise to the *pons,* or *metencephalon.* The lateral walls of the pons give rise to the cerebellum. The posterior part of the hindbrain gives rise to the *medulla,* or *myelencephalon* [Fig. 15.15(*b*)]. A third fold, the pontine flexure, occurs between the pons and the medulla. This flexure bends the pons back upon the medulla and forces the cerebral hemispheres upward [Fig. 15.15(*c*)] so that they eventually appear to sit on top of the brain stem (the midbrain, pons, and medulla). With the appearance of the secondary brain vesicles and formation of the pontine flexure, the basic structures of the brain and their relationships to one another are established. Despite these changes, certain features that are typical of the spinal cord can be recognized in the brain vesicles. For instance, distinct alar and basal plates can be recognized in the lateral walls of the brain vesicles up to the diencephalon. The alar plates give rise to the sensory nuclei of the brain stem, while the basal plates give rise to the motor nuclei.

DIVISIONS The adult brain is a large, soft, ovoid organ that weighs about 3 pounds and is enclosed in the cranial cavity. It is continuous with the spinal cord at the foramen magnum. For descriptive purposes the brain is divided into three major regions (Fig. 15.16): the *brain stem, cerebellum,* and *cerebrum.*

Brain Stem The brain stem is the stemlike portion of the brain through which pass all the sensory and motor nerve fibers connecting the spinal cord with the higher brain centers. It is subdivided into the *medulla, pons,* and *midbrain* (Fig. 15.17).

Medulla. The medulla (bulb) is the most inferior and least modified portion of the brain stem. It is a cone-shaped structure whose apex is continuous with the spinal cord below and whose broad base is continuous with the pons above. The lower half of the fourth ventricle of the brain is located in the medulla. It is surrounded by medullary

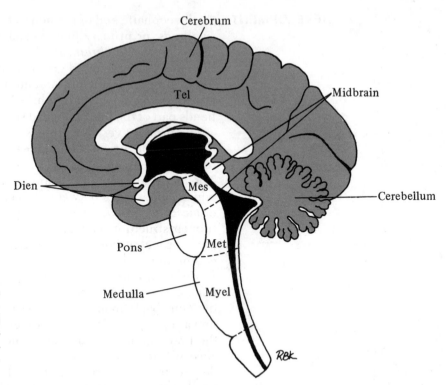

Figure 15.16 *Adult Brain: Sagittal Section.* Brain stem consists of mesencephalon (Mes), metencephalon (Met), and myelencephalon (Myel). The cerebrum consists of telencephalon (Tel), which comprises cerebral hemispheres, and diencephalon (Dien).

gray and white matter and roofed over by the posterior medullary vellum, which gives rise to the choroid plexus of the fourth ventricle. Most of the medullary gray matter is broken up into discrete nuclei and centers. Medullary nuclei include sensory relay nuclei plus the nuclei of origin of the ninth, tenth, eleventh, and twelfth cranial nerves. In addition, the medulla contains essential centers such as the *respiratory centers,* which regulate the respiratory rate, the *cardiac centers,* which regulate the heart rate, the *vasomotor centers,* which regulate the blood pressure, the *swallowing centers,* and the *vomiting centers.* Running through the medullary gray matter, as well as that of the pons and midbrain, is the *reticular formation.* The reticular formation consists of a diffuse collection of neurons that receive and integrate information from various parts of the brain. The medullary centers mentioned above are actually collections of reticular formation neurons. The medullary white matter consists of ascending and descending fiber tracts that run between the spinal cord and more cephalic brain structures. Many of these fiber tracts *decussate* (cross over) from one side of the brain to the other in the medulla.

Pons. The pons (bridge) is a portion of the brain stem situated between the medulla and the midbrain. The upper half of the fourth ventricle is located in the pons. It is surrounded by pontine gray and white matter and roofed over by the anterior medullary vellum, which

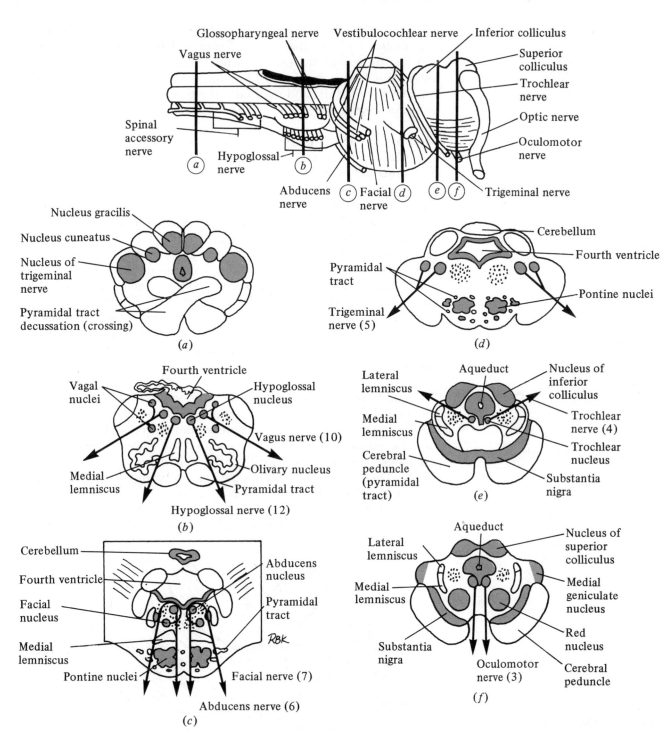

Figure 15.17 *Sections Through Brain Stem.* (a) lower medulla; (b) upper medulla; (c) lower pons; (d) upper pons; (e) lower midbrain; (f) upper midbrain. Reticular formation indicated by stippling.

contributes to the choroid plexus of the fourth ventricle. The pontine gray matter includes the nuclei of origin of the fifth, sixth, seventh, and eighth cranial nerves. The white matter of the pons consists of longitudinal and transverse fiber bundles. The longitudinal fibers connect the medulla and midbrain; the transverse fibers connect the pons and cerebellum.

Midbrain. The midbrain is the short segment of the brain stem situated between the pons and the diencephalon. It contains the *cerebral aqueduct* (of Sylvius), a narrow canal that connects the third ventricle in the diencephalon with the fourth ventricle in the pons and medulla. Four small masses of gray matter form the tectum (roof) of the midbrain. They consist of the two *superior colliculi,* which are visual reflex centers, and the two *inferior colliculi,* which are auditory reflex centers. The tegmentum (central portion of the midbrain) contains ascending fiber tracts plus the nuclei of origin of the third and fourth cranial nerves. The red nuclei and the substantia nigra, important extrapyramidal motor nuclei, are also located in the tegmentum. The base of the midbrain contains the *cerebral peduncles.* The peduncles consist mainly of descending fiber tracts that connect the cerebral hemispheres with more distal portions of the central nervous system.

Cerebellum   The cerebellum (little brain) is an outgrowth of the lateral walls of the pons. It lies above and behind the medulla and pons and occupies the posterior cranial fossa. The cerebellum is an ovoid organ consisting of a constricted medial portion called the *vermis* (worm) and two expanded lateral portions called the *cerebellar hemispheres* (Fig. 15.18). The cerebellum is connected to the brain stem by three pairs of fiber bundles called *cerebellar peduncles.*

The substance of the cerebellum (Fig. 15.19) is divided into a cortex, medulla, and cerebellar nuclei. The cortex consists of a super-

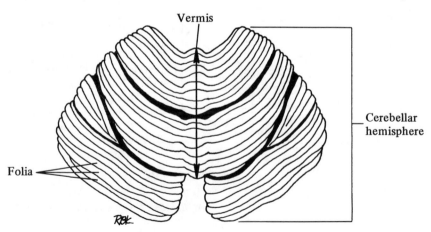

Figure 15.18 Cerebellum: External Anatomy. Cerebellum as seen from above.

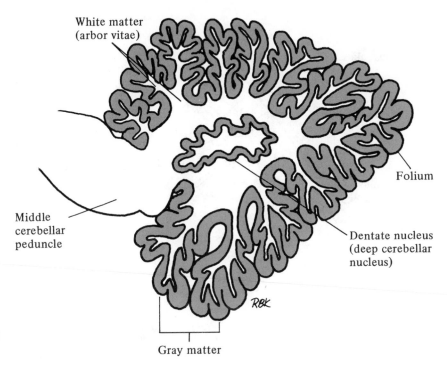

White matter
(arbor vitae)

Folium

Middle
cerebellar
peduncle

Dentate nucleus
(deep cerebellar
nucleus)

Gray matter

Figure 15.19 Cerebellum: Internal Anatomy. Parasagittal section.

ficial layer of gray matter, which is thrown into transverse folds called *folia*. The folia serve to increase the surface area of the cortex. The medulla, which lies deep to the cortex, consists of white matter. In sagittal section it resembles a tree with many branches, which is why it was named *arbor vitae* ("tree of life"). Four pairs of cerebellar nuclei are embedded deep in the white matter. These nuclei relay messages between the brain stem and the cerebellar cortex.

The cerebellum is concerned with the maintenance of equilibrium, with the regulation of muscle tone, and with the coordination of voluntary muscle movements. When the cerebellum is damaged by accident or disease, equilibrium is lost, gait (walking) is unsteady, muscle tone is reduced, and voluntary movements are inaccurate and jerky.

Cerebrum The diencephalon and the telencephalon constitute the cerebrum. The diencephalon consists of the central core of the cerebrum, while the telencephalon consists of the cerebral hemispheres.

Diencephalon. The diencephalon (Fig. 15.20) is situated between the midbrain posteriorly and the cerebral hemispheres laterally. It has three subdivisions, described below, that surround the slitlike third ventricle.

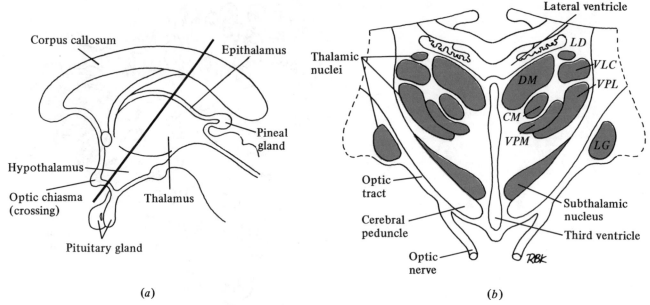

Figure 15.20 *Diencephalon.* (*a*) Sagittal section. (*b*) Coronal section. Thalamic nuclei seen at this level include the following: dorsomedial nucleus, *DM;* centromedian nucleus, *CM;* lateral dorsal nucleus, *LD;* ventral lateral nucleus, *VLC;* ventral posterolateral nucleus, *VPL;* ventral posteromedial nucleus, *VPM;* and lateral geniculate nucleus, *LG.*

The *epithalamus* (above the thalamus) forms the roof of the third ventricle. It gives rise to the *pineal gland,* whose proposed endocrine function was described in Chapter 13.

The *thalamus* (inner chamber) forms the upper walls of the third ventricle. The dorsal part of the thalamus consists of two ovoid masses of gray matter lying on either side of the third ventricle. Each mass contains sensory relay nuclei for the general senses, taste, vision, and hearing plus association nuclei for emotional and autonomic functions. The ventral part of the thalamus, or *subthalamus,* contains motor relay nuclei. Also, the reticular formation, red nuclei, and substantia nigra extend from the midbrain into the subthalamus.

The *hypothalamus* (below the thalamus) forms the lower walls and floor of the third ventricle. It is connected to the pituitary gland by the infundibulum. The hypothalamus is the major cerebral center for integrative activities of the autonomic nervous system. The hypothalamus contains a collection of nuclei and centers that are sensitive to changes in the internal environment and regulate body functions to maintain homeostasis. The hormones oxytocin and ADH as well as the releasing and inhibiting factors are produced in hypothalamic nuclei. In addition, the hypothalamus contains the *feeding center,* which regulates the desire to eat, the *satiety center,* which regulates the amount of food eaten, the *water regulation center,* which regulates water intake and ADH production, and the *temperature center,* which regulates body heat. The hypothalamus is also involved in emotional functions through input from the limbic system.

Telencephalon. The telencephalon is the most prominent part of the brain. It consists of the right and left cerebral hemispheres. The cerebral hemispheres are partially separated by a deep cleft called the *longitudinal fissure* and are connected to one another by bundles of commissural fibers. Each hemisphere has frontal, temporal, and occipital poles plus lateral, medial, and inferior surfaces. Deep furrows called *fissures* plus imaginary lines divide each hemisphere into lobes, which are named according to the skull bones they contact: *frontal, parietal, temporal,* and *occipital.* Shallow furrows called *sulci* (plural of sulcus) divide each lobe into a number of convolutions

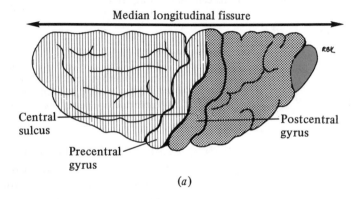

(a)

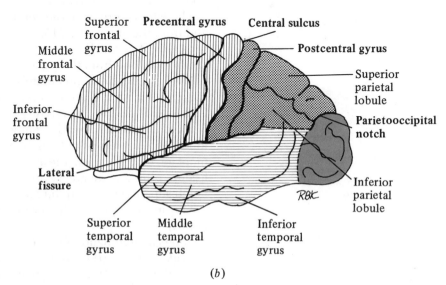

(b)

Figure 15.21 Cerebral Hemisphere. Frontal lobe, *vertical lines*; temporal lobe, *horizontal lines*; parietal lobe, *stipple*; occipital lobe, *solid gray.* (a) Dorsal aspect of left cerebral hemisphere. (b) Lateral aspect of left cerebral hemisphere.

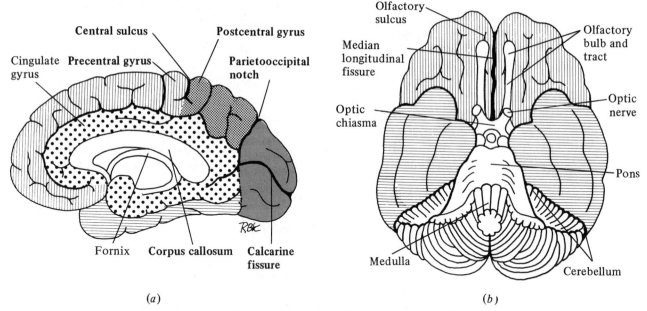

(a)

(b)

Figure 15.22 *Cerebral Hemisphere.* (a) Medial aspect, right hemisphere. Frontal lobe, *vertical lines*; temporal lobe, *horizontal lines*; parietal lobe, *small stipple*; cingulate gyrus, *large stipple*; occipital lobe, *solid gray*. (b) Inferior aspect of brain. Frontal lobe, *vertical lines*; temporal lobe, *horizontal lines*.

called *gyri* (plural of gyrus). While most of the fissures, sulci, and gyri are named, we will mention only those that serve as major anatomical landmarks. For more detailed information you should consult a textbook or atlas of neuroanatomy.

The lateral surface of the cerebral hemisphere is shown from above and from the side in Fig. 15.21. The *central sulcus* (fissure of Rolando) runs almost vertically down the middle of the hemisphere, separating the frontal lobe from the parietal lobe. A major sensory area, the *postcentral gyrus,* lies immediately behind the central sulcus, while a major motor area, the *precentral gyrus,* lies immediately in front of it. The *lateral fissure* (of Sylvius) runs diagonally down the side of the hemisphere, separating the temporal lobe from the frontal and parietal lobes. When the lips of the lateral fissure are separated, a buried segment of cortex called the *insula* can be seen. The parietooccipital fissure plus an imaginary line connecting it to the parietooccipital notch separate the occipital lobe from the parietal and temporal lobes.

The most important feature of the medial surface of the cerebral hemisphere [Fig. 15.22(a)] is the *corpus callosum.* The corpus callosum is a large band of white commissural fibers that curves over the *lateral ventricle* (the lumen of the cerebral hemisphere). The *cingulate sulcus* is a long furrow that runs parallel to the corpus callosum.

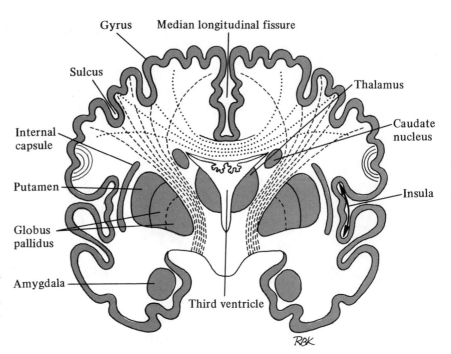

Gyrus Median longitudinal fissure

Sulcus

Thalamus

Internal capsule

Caudate nucleus

Putamen

Insula

Globus pallidus

Amygdala

Third ventricle

Figure 15.23 *Coronal Section Through Cerebrum.* Projection fibers (internal capsule), *dashed lines*; commissural fibers (corpus callosum), *dotted lines*; association fibers, *solid lines*.

It demarcates the *cingulate gyrus*, a cortical area that is part of the limbic system. The *calcarine sulcus* (better known as the calcarine fissure) is a short furrow that begins at the occipital pole and arches across the occipital lobe to terminate in the parietooccipital fissure. Most of the cortical area concerned with vision is located on either side of the calcarine fissure.

The inferior surface of the cerebral hemisphere [Fig. 15.22(*b*)] is marked by a long furrow, the *olfactory sulcus*, which runs parallel to the longitudinal fissure. The olfactory bulb and tract lie in this sulcus.

Each cerebral hemisphere consists of a cortex, medulla, and cerebral nuclei, which are shown in a coronal section in Fig. 15.23.

The cerebral cortex consists of a superficial layer of gray matter. The formation of gyri, sulci, and fissures serves to increase the surface area of the cortex without increasing the size of the cerebral hemispheres.

The cerebral medulla consists of the white matter lying beneath the cortex. It is composed of bundles of myelinated fibers, which are classified as follows:

1. *Projection Fibers.* Fibers that run between the cerebral cortex and the brain stem. They transmit information from one level of the brain to another.

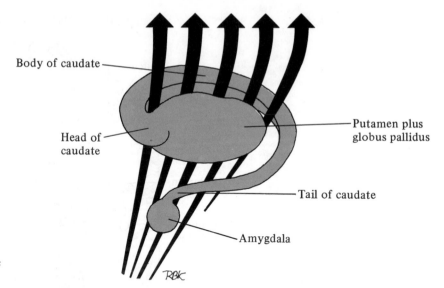

Body of caudate

Head of
caudate

Putamen plus
globus pallidus

Tail of caudate

Amygdala

Figure 15.24 Basal Ganglia. Arrows
represent fibers of internal capsule.

2. *Commissural Fibers.* Fibers that run between the right and left hemispheres. They transmit information from one hemisphere to the other.

3. *Association Fibers.* Fibers that run between gyri within the same hemisphere. They transmit information from one gyrus to another.

Four nuclear masses, or *basal ganglia,* are embedded deep in the white matter of each cerebral hemisphere. They are the *caudate, putamen, globus pallidus,* and *amygdala* (Fig. 15.24). The first three nuclei form the *corpus striatum,* an important part of the extrapyramidal motor system. The internal capsule (a broad band of myelinated projection fibers) separates the caudate nucleus from the globus pallidus and putamen. The fourth nucleus, the amygdala, is considered to be part of the limbic system, which is concerned with emotions and memory.

The cerebral cortex is the center of the highest functions of the nervous system. Neurologists (doctors who deal with the nervous system) have been able to map cortical function areas, which are described next and illustrated in Fig. 15.25.

1. *Frontal Cortex.* The frontal cortex initiates motor activities and is associated with the attributes of judgment and foresight.

   a. *Primary Somatomotor Area.* The primary somatomotor area is located within the precentral gyrus; large neurons give rise to tracts that constitute the *pyramidal motor system,* a system that initiates precise voluntary motor movements. There is a projection of the body image on the precentral gyrus [Fig. 15.26(*a*)] in

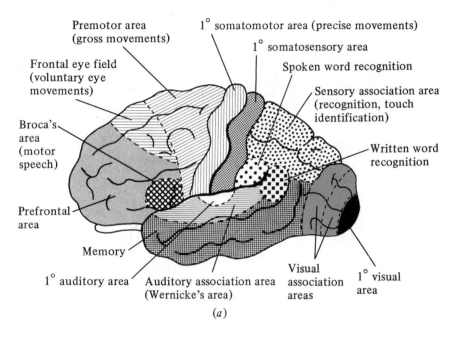

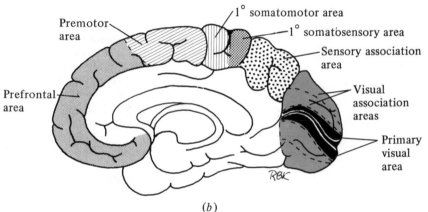

Figure 15.25 *Cortical Function Map.*
(a) Lateral aspect of cerebral hemisphere.
(b) Medial aspect of cerebral hemisphere.

which a body part is represented by a cortical area proportional to its use rather than its size.

b. *Premotor Area.* The premotor area is located in front of the precentral gyrus. Nerve fibers from this area plus the corpus striatum, subthalamic nucleus, red nucleus, and substantia nigra constitute the *extrapyramidal motor system.* This system initiates gross stereotyped movements that are almost unconscious in nature.

c. *Motor Speech Area.* The motor speech area (Broca's area) lies below and slightly anterior to the precentral gyrus. This area initiates vocalization and is almost always located on the *domi-*

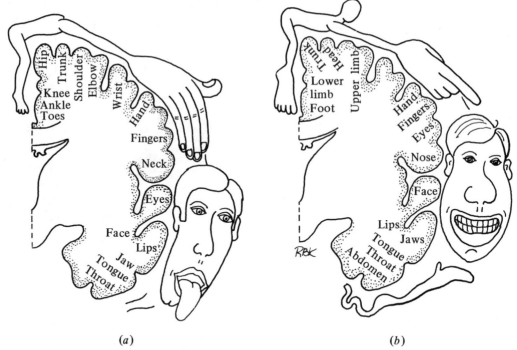

Figure 15.26 *Somatotopic Localization.* (a) For motor activity on precentral gyrus. Figure, called a motor homunculus, indicates representation of body parts on primary motor cortex as established by electrical stimulation. Figure is distorted because some body parts have a disproportionately large motor representation. (b) For sensory activity on postcentral gyrus. Sensory homunculus is also distorted because some body parts have a disproportionately large sensory representation.

nant hemisphere (the hemisphere that controls the dominant hand).

    d. Prefrontal Area. The prefrontal area is located in front of the premotor area. It is often considered the seat of human intelligence. The prefrontal area is associated with judgment, foresight, and choice of appropriate behavior. Persons who have had prefrontal lobotomies (removal or sectioning of the prefrontal area) show a marked deficit of these qualities.

2. *Parietooccipitotemporal Cortex.* The remainder of the cerebral cortex is devoted to sensory functions.

    a. Primary Somatosensory Area. The primary somatosensory area is located within the postcentral gyrus. It receives exteroceptive and proprioceptive information. There is also a projection of the body image on the postcentral gyrus [Fig. 15.26(b)] in which a body part is represented by a cortical area proportional to its number of sensory receptors rather than its size.

    b. Somatosensory Association Area. The somatosensory association area is located on the parietal lobe just posterior to the postcentral gyrus. This cortical area deals with one's awareness of body parts and their relationship with one another. It also deals with one's ability to identify objects by size, shape, and texture.

c. *Primary Auditory Area.* The primary auditory area is located on the superomedial part of the temporal lobe. It receives auditory information from both ears. There is a spatial localization of sounds ranging from high to low pitch on this cortical area.

d. *Auditory Association Area.* The auditory association area (Wernicke's area) surrounds the primary auditory area. It is concerned with the understanding of speech and the association of sounds with past experiences.

e. *Primary Visual Area.* The primary visual area is located on the tip of the occipital lobe and the lips of the calcarine fissure. Due to crossing of part of the optic nerve fibers, the right visual area receives information from the right half of each retina and vice versa.

f. *Visual Association Area.* The visual association area surrounds the primary visual area. It is concerned with visual speech (reading) and the association of sights with past experiences.

Ascending and Descending Pathways in the Central Nervous System

The ascending and descending pathways in the central nervous system are routes by which sensory information is transmitted from the spinal cord and brain stem to the cerebral cortex and by which information to initiate motor responses is transmitted in the opposite direction. Though there are many pathways in the central nervous system, the ones described here are those most frequently involved in body functions and are of clinical significance.

SOMATIC SENSORY SYSTEMS (ASCENDING TRACTS)

Somatic sensory systems generally consist of two or three neuron chains carrying sensory information from exteroceptors and proprioceptors to the primary somatosensory area.

Spinothalamic System

The spinothalamic system consists of small nerve fibers that transmit sensory information slowly and without precise localization. It includes the *lateral* and *anterior spinothalamic tracts* (Fig. 15.27).

Lateral Spinothalamic Tract

The lateral spinothalamic tract is a four-neuron chain that transmits pain and temperature information from the body to the cerebral cortex. The 1° (first-order) neuron has its sensory receptor in the skin and its cell body in the sensory ganglion of a spinal nerve. It enters the spinal cord and synapses on a 2° (second-order) neuron in the *substantia gelatinosa*, a sensory relay nucleus in the dorsal (posterior)

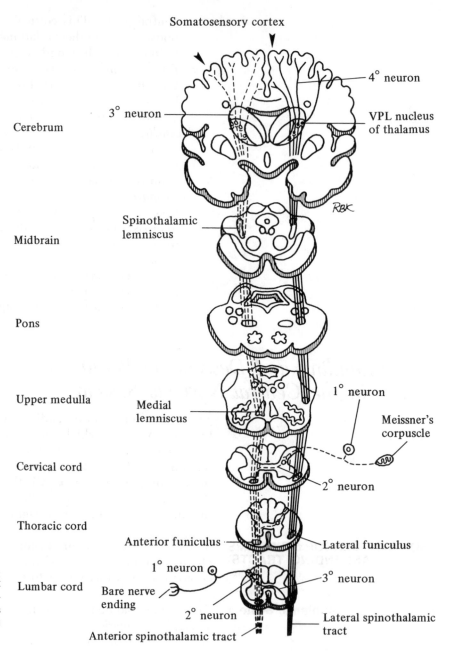

Somatosensory cortex

4° neuron

3° neuron

VPL nucleus
of thalamus

Cerebrum

Spinothalamic
lemniscus

Midbrain

Pons

1° neuron

Upper medulla

Meissner's
corpuscle

Medial
lemniscus

Cervical cord

2° neuron

Thoracic cord

Anterior funiculus

Lateral funiculus

1° neuron

Lumbar cord

3° neuron

Bare nerve
ending

2° neuron

Lateral spinothalamic
tract

Anterior spinothalamic tract

Figure 15.27 Spinothalamic Tracts.
Solid lines, lateral spinothalamic tract
fibers; *dashed lines,* anterior spinothalamic tract fibers. Note that both tracts cross
to contralateral (opposite) side in spinal
cord.

gray horn. The 2° neuron synapses with a 3° (third-order) neuron in
the base of the dorsal horn, which crosses to the opposite side of the
spinal cord and enters the lateral funiculus. It then ascends to the
thalamus, where it synapses with a 4° (fourth-order) neuron in the
ventral posterolateral nucleus (VPL), a thalamic relay nucleus. The

4° neuron ascends to the cerebral cortex, where it terminates in the postcentral gyrus. Pain and temperature information from the head is transmitted to the cerebral cortex by the *trigeminothalamic tract.*

Anterior Spinothalamic Tract

The anterior spinothalamic tract is a three-neuron chain that transmits crude (poorly localized) touch and pressure information from the body to the cerebral cortex. The 1° neuron has its sensory receptor in the skin and its cell body in the sensory ganglion of a spinal nerve. It enters the spinal cord and synapses on a 2° neuron in the base of the dorsal (posterior) horn. The 2° neuron crosses to the opposite side of the spinal cord and enters the anterior funiculus. It then ascends to the thalamus and synapses on a 3° neuron in the VPL nucleus. The 3° neuron ascends to the cerebral cortex, where it terminates in the postcentral gyrus. Crude touch information from the head is transmitted to the cerebral cortex by the *trigeminothalamic tract.*

Posterior Column System

The posterior column system consists of large nerve fibers that transmit sensory information with great speed and precise localization. It includes the *fasciculus gracilis* and *cuneatus* (Fig. 15.28).

Fasciculus Gracilis

The fasciculus gracilis is a three-neuron chain that transmits fine (precisely localized) touch, pressure, vibratory, and proprioceptive information from the lower half of the body to the cerebral cortex. The 1° neuron has its sensory receptor in the skin, a muscle, a joint, or a tendon and its cell body in the sensory ganglion of a spinal nerve. It enters the spinal cord and joins the dorsal (posterior) funiculus on the same side. It then ascends to the medulla where it synapses with a 2° neuron in the *nucleus gracilis,* a medullary relay nucleus. The 2° neuron crosses to the opposite side of the brain stem and ascends to the thalamus, where it synapses with a 3° neuron in the VPL nucleus. The 3° neuron ascends to the cerebral cortex, where it terminates in the postcentral gyrus.

Fasciculus Cuneatus

The fasciculus cuneatus is a three-neuron chain that transmits fine touch, pressure, vibratory, and proprioceptive information from the upper half of the body to the cerebral cortex. This tract is similar to the fasciculus gracilis except that the 1° neuron uses the fasciculus cuneatus to ascend to the medulla and synapses with the 2° neuron in the *nucleus cuneatus,* another medullary relay nucleus. Fine touch and proprioceptive information from the head is transmitted to the cerebral cortex by the *trigeminothalamic tract.*

Spinocerebellar System

The spinocerebellar system transmits proprioceptive information from the body to the cerebellar cortex (Fig. 15.29). This information is used in the coordination of muscle movements to perform skilled tasks and to regulate posture. It includes the posterior and anterior spinocerebellar tracts.

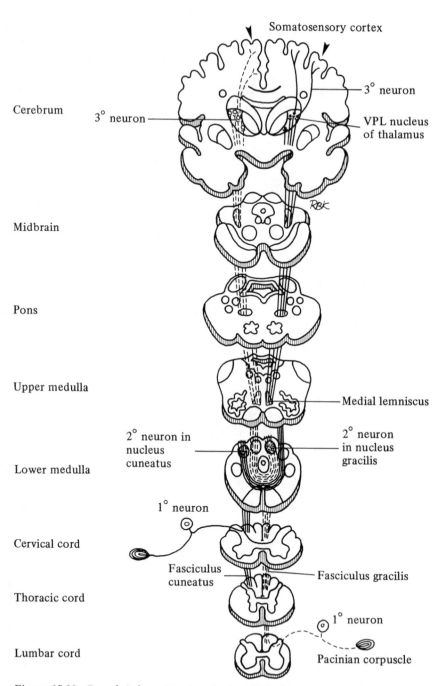

Figure 15.28 *Dorsal Column Tracts. Dashed lines,* fasciculus gracilis fibers; *solid lines,* fasciculus cuneatus fibers. Note that both tracts cross to contralateral (opposite) side in lower medulla.

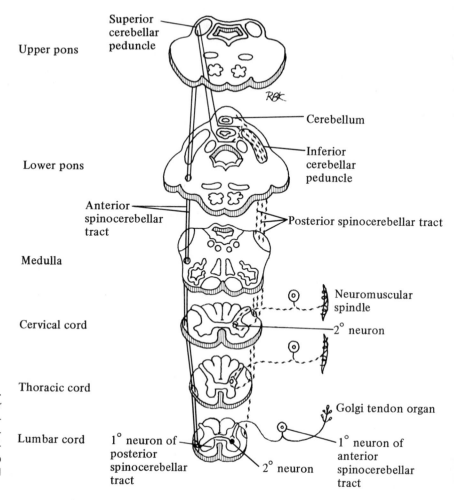

Upper pons

Superior cerebellar peduncle

Cerebellum

Lower pons

Inferior cerebellar peduncle

Anterior spinocerebellar tract

Posterior spinocerebellar tract

Medulla

Neuromuscular spindle

Cervical cord

2° neuron

Thoracic cord

Golgi tendon organ

Lumbar cord

1° neuron of posterior spinocerebellar tract

2° neuron

1° neuron of anterior spinocerebellar tract

Figure 15.29 Spinocerebellar Tracts. *Zigzag lines,* posterior spinocerebellar tract fibers; *solid lines,* anterior spinocerebellar tract fibers. Note that posterior spinocerebellar tract is uncrossed, while anterior spinocerebellar tract crosses to contralateral (opposite) side in spinal cord and recrosses to side of origin in pons.

Posterior Spinocerebellar Tract

The posterior (uncrossed) spinocerebellar tract is a two-neuron chain that transmits proprioceptive information from the skeletal muscles of the lower body to the cerebellar cortex. The 2° neuron originates in the base of the dorsal (posterior) horn of the spinal cord and ascends to the brain stem and terminates in the cerebellar cortex on the same side.

Anterior Spinocerebellar Tract

The anterior (crossed) spinocerebellar tract is a two-neuron chain that transmits proprioceptive information from the joints and tendons of the lower body to the cerebellar cortex. The 2° neuron originates in the base of the dorsal (posterior) horn of the spinal cord, crosses the cord, ascends to the brain stem and recrosses to terminate in the cerebellar cortex on the side of origin.

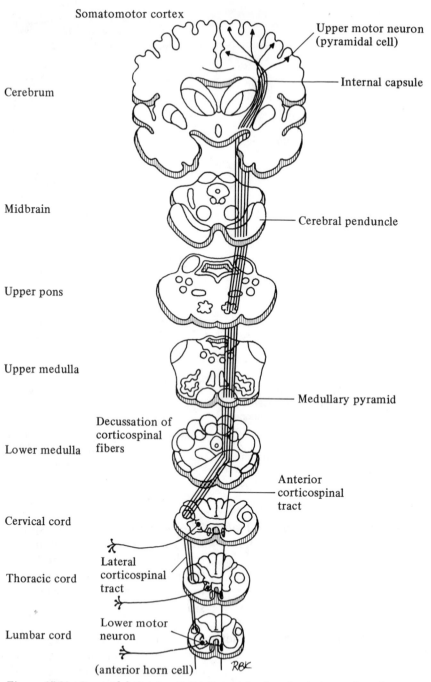

Somatomotor cortex

Cerebrum

Midbrain

Upper pons

Upper medulla

Lower medulla

Cervical cord

Thoracic cord

Lumbar cord

Upper motor neuron
(pyramidal cell)

Internal capsule

Cerebral penduncle

Medullary pyramid

Decussation of
corticospinal
fibers

Anterior
corticospinal
tract

Lateral
corticospinal
tract

Lower motor
neuron

(anterior horn cell)

RBK

Figure 15.30 Pyramidal Tracts. Solid lines, lateral and anterior corticospinal tract fibers. Note that majority of corticospinal fibers cross to contralateral (opposite) side in lower medulla to form lateral corticospinal tract.

SOMATIC MOTOR SYSTEMS (DESCENDING TRACTS)

Somatic motor systems consist of two or more neuron chains transmitting motor commands from the brain to the skeletal muscles.

Pyramidal System

The pyramidal system is responsible for initiating fine, precise muscle movements. It consists of all nerve fibers that (a) arise from large neurons in the primary somatomotor area, (b) pass through the medullary pyramids, and (c) descend in the pyramidal tracts of the spinal cord. The pyramidal system includes the *lateral* and *anterior corticospinal tracts* (Fig. 15.30).

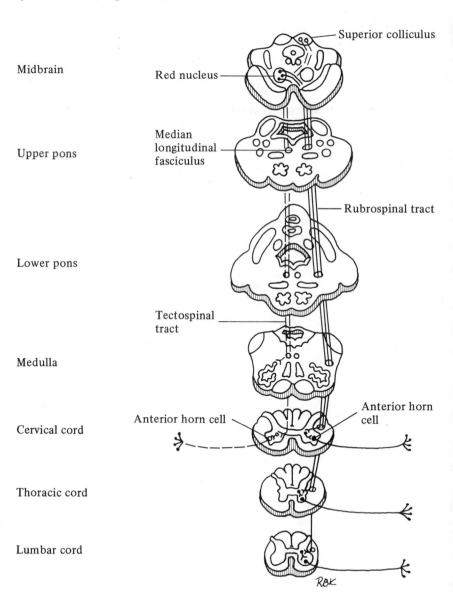

Figure 15.31   Extrapyramidal Tracts. Two of many extrapyramidal tracts are illustrated. *Solid lines,* rubrospinal tract fibers; *dashed lines,* tectospinal tract fibers. Rubrospinal tract originates in red nucleus, crosses to contralateral (opposite) side in midbrain, and descends to all levels of cord, where its fibers terminate by way of interneurons on anterior horn cells. The tectospinal tract originates in superior colliculus (part of roof or tectum of midbrain), crosses to contralateral side of midbrain, and descends primarily to cervical level of cord, where its fibers terminate by way of interneurons on anterior horn cells.

Midbrain

Superior colliculus

Red nucleus

Upper pons

Median longitudinal fasciculus

Lower pons

Rubrospinal tract

Tectospinal tract

Medulla

Anterior horn cell

Anterior horn cell

Cervical cord

Thoracic cord

Lumbar cord

Lateral Corticospinal Tract   The lateral corticospinal tract is a two-neuron chain that transmits motor commands from the primary somatomotor area to the skeletal muscles. It contains 75 to 90 percent of all corticospinal fibers. The upper motor neuron originates in the precentral gyrus and descends to the *medullary pyramid,* where it crosses to the opposite side of the brain stem. It enters the lateral funiculus of the spinal cord and descends to the appropriate level, where it synapses with a lower motor neuron in the *ventral (anterior) horn.* The lower motor neuron terminates on skeletal muscle fibers.

Anterior Corticospinal Tract   The anterior corticospinal tract is similar to the lateral corticospinal tract except that (a) the upper motor neuron crosses to the opposite side in the spinal cord rather than the medulla and (b) it descends in the ventral (anterior) funiculus of the spinal cord.

Extrapyramidal System   The extrapyramidal system is responsible for initiating gross, stereotyped muscle movements. It consists of all centers and tracts, exclusive of the corticospinal tracts, that influence lower motor neurons. The best example of an extrapyramidal tract is the *rubrospinal tract* (Fig. 15.31). The rubrospinal tract originates in the red nucleus of the midbrain, receives fibers from the cerebellar nuclei, crosses in the brain stem, and enters the lateral funiculus of the spinal cord. Rubrospinal fibers descend to the appropriate levels of the spinal cord, where they synapse on lower motor neurons.

The Meninges

The meninges consist of three connective tissue membranes that enclose and protect the brain and spinal cord. From outside in they include the *dura mater, arachnoid,* and *pia mater* (Fig. 15.32).

DURA MATER   The dura mater ("tough mother") is the thick fibrous connective tissue membrane that forms the outermost covering of the brain and spinal cord. The cranial portion of the dura consists of two layers that separate to enclose the *dural sinuses,* venous vascular channels that drain blood from the brain. The outer layer lines the cranial cavity, while the inner layer covers the surface of the brain. The inner layer also forms several folds that separate parts of the brain from one another. Some important dural folds (Fig. 15.33) are listed here.

1. *Falx Cerebri.*   A vertical fold that separates the cerebral hemispheres.
2. *Tentorium Cerebelli.*   A horizontal fold that separates the cerebrum from the cerebellum.
3. *Falx Cerebelli.*   A vertical fold that separates the cerebellar hemispheres inferiorly.

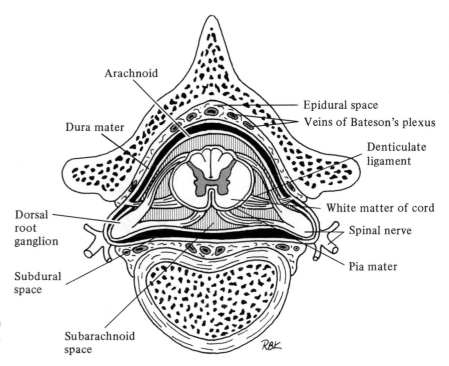

Figure 15.32 Meninges. Cross section shows meninges in relation to spinal cord, spinal nerves, and vertebra.

The spinal cord is covered by a single layer of dura that is continuous with the inner layer of cerebral dura. It is separated from the vertebral canal by the *epidural space,* a fat-filled space containing the vertebral venous plexus (of Bateson).

Subdural Space

The subdural space is a narrow potential space containing a thin film of serous fluid that separates the dura from the underlying arachnoid.

ARACHNOID

The arachnoid is a cobweblike connective tissue membrane situated between the dura and the pia mater. Slender arachnoid trabeculae extend across the subarachnoid space to attach the arachnoid to the pia. In the brain the arachnoid forms fingerlike projections, called *arachnoid villi,* that penetrate the dural sinuses. Cerebrospinal fluid is reabsorbed into the dural sinuses through these structures.

Subarachnoid Space

The subarachnoid space is a relatively wide space between the arachnoid and pia that contains cerebrospinal fluid. Cerebrospinal fluid tends to pool in large subarachnoid spaces called cisterns. The largest of these, the *cisterna magna,* is located between the cerebellum and medulla.

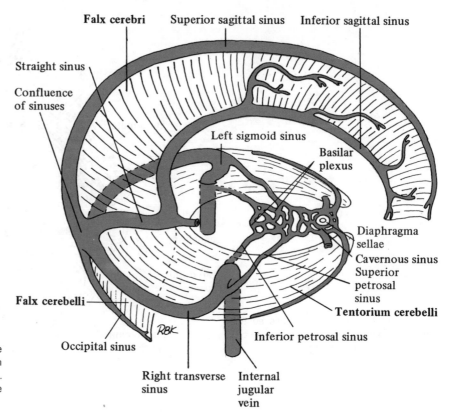

Figure 15.33 Dural Folds. These are reflexions of dura mater that partition cranial cavity and support brain parts. Venous sinuses they accommodate are also shown.

PIA MATER The pia mater ("tender mother") is the thin fibrous connective tissue membrane that forms the innermost covering of the brain and spinal cord. This highly vascular membrane adheres closely to the neural tissue beneath it. Extending laterally from the spinal portion of the pia to the dura are 21 pairs of triangular ligaments, the *denticulate ligaments*, that anchor the spinal cord to the dura.

Ventricular System

The ventricular system of the central nervous system is derived from the lumen of the neural tube. It consists of a series of cavities and canals that contain cerebrospinal fluid (Fig. 15.34).

The lateral (first and second) ventricles are large C-shaped cavities located in the cerebral hemispheres. Each lateral ventricle has a body plus anterior, inferior, and posterior horns that project into the frontal, temporal, and occipital lobes, respectively. A large choroid plexus projects into each lateral ventricle.

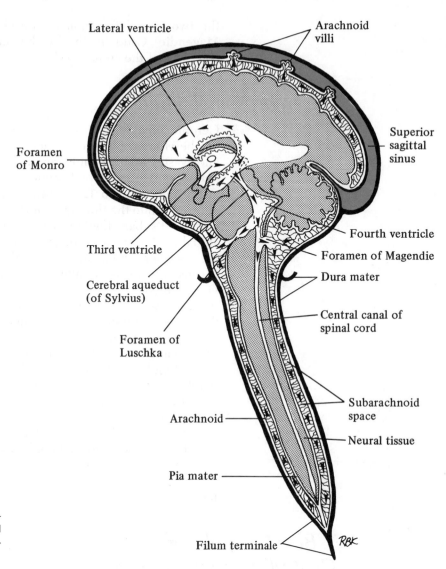

Lateral ventricle

Arachnoid villi

Foramen of Monro

Superior sagittal sinus

Third ventricle

Cerebral aqueduct (of Sylvius)

Fourth ventricle

Foramen of Magendie

Dura mater

Central canal of spinal cord

Foramen of Luschka

Arachnoid

Subarachnoid space

Neural tissue

Pia mater

Filum terminale

RBK

Figure 15.34 Ventricular System. Arrows show circulation of cerebrospinal fluid through ventricular system and subarachnoid space.

    The third ventricle is a narrow slitlike cavity located in the diencephalon. The lateral ventricles communicate with the third ventricle through the *interventricular foramina* (of Monro). The third ventricle also contains a choroid plexus.

    The *cerebral aqueduct* (of Sylvius) is a long narrow canal located in the midbrain. It connects the third and fourth ventricles.

    The fourth ventricle is a large diamond-shaped cavity located in the pons and medulla. It also contains a large choroid plexus. There are three exits from the fourth ventricle into the subarachnoid space:

the two *lateral foramina* (of Luschka) and the *medial foramen* (of Magendie). Caudally, the fourth ventricle is continuous with the central canal of the spinal cord.

Cerebrospinal Fluid

Cerebrospinal fluid is a clear colorless fluid formed from blood plasma by the choroid plexuses. It acts as a transport medium for nutrients and waste products and provides a protective fluid cushion for the central nervous system. Cerebrospinal fluid is produced continuously at the rate of 45 to 125 milliliters per day. It flows from the lateral ventricles into the third and fourth ventricles and then into the spinal canal and subarachnoid space. After circulating around the brain and spinal cord, it is reabsorbed into the dural venous sinuses, particularly the superior sagittal sinus, through the arachnoid villi. Reabsorption is caused by pressure differences between the cerebrospinal fluid and venous blood.

Excessive production or blocked drainage of cerebrospinal fluid produces a serious disorder called *hydrocephalus*. The increase in fluid volume in the ventricular system causes the brain to expand. If the fontanelles are still open, the skull enlarges to accommodate the expanding brain. If the fontanelles have closed, the brain cannot expand; neural tissue is destroyed unless the excess fluid is shunted into the cardiovascular system.

A spinal tap is a procedure for withdrawing cerebrospinal fluid from the subarachnoid space for diagnostic purposes. A large needle is inserted into the subarachnoid space between lumbar vertebrae *L4* and *L5*, which is below the termination of the spinal cord. The tip of the needle pushes aside any lumbar or sacral nerve roots it should happen to encounter.

Blood Supply of the Central Nervous System

BLOOD SUPPLY OF THE BRAIN
Arterial Supply

The brain receives blood from two sources: the internal carotid system and the vertebral system. These two systems unite at the base of the brain to form the arterial circle (of Willis). The circle of Willis provides alternative routes for blood to take when one of the major cerebral arteries is blocked. Components of the arterial systems of the brain are listed here and illustrated in Fig. 15.35.

1. *Internal Carotid System*
   a. *Anterior Cerebral Arteries.* Pair of arteries that supply the medial portions of the frontal and parietal lobes plus the corpus callosum.

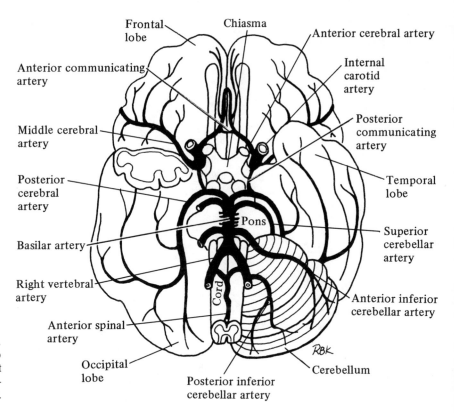

Figure 15.35 Arterial Supply of Brain.
Superficial arteries are shown in relation to
basal surface of brain. Portions of right
temporal lobe and right cerebellar hemisphere are removed to aid in visualization.

    b. Middle Cerebral Arteries. Pair of arteries that supply the lateral
      portions of the frontal, parietal, and temporal lobes.
    c. Hypophyseal Arteries. Two pairs of arteries that supply the
      pituitary and hypothalamus.
2. *Vertebral System*
    a. Posterior Cerebral Arteries. Pair of arteries that supply the oc-
      cipital lobe and most of the medial portion of the temporal lobe.
    b. Basilar Artery. An artery that supplies the pons.
    c. Cerebellar Arteries. Three pairs of arteries that supply the
      cerebellum.

Venous Drainage   The brain is drained by the cerebral and cerebellar veins, which are
tributaries of the dural sinuses. The major dural sinuses (Fig. 15.36)
are as follows:

1. *Superior Sagittal Sinus.* A single vessel that runs in the superior
    margin of the falx cerebri and ends in the right transverse sinus.
2. *Inferior Sagittal Sinus.* A single vessel that runs in the inferior
    margin of the falx cerebri and ends in the straight sinus.
3. *Straight Sinus.* A single vessel that runs along the attachment of

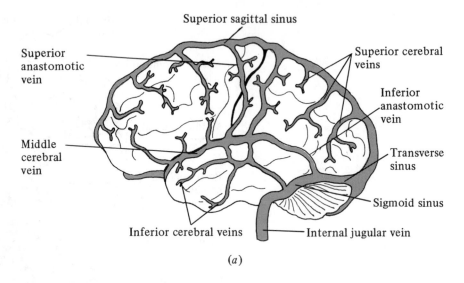

Superior sagittal sinus

Superior anastomotic vein

Superior cerebral veins

Inferior anastomotic vein

Middle cerebral vein

Transverse sinus

Sigmoid sinus

Inferior cerebral veins

Internal jugular vein

(a)

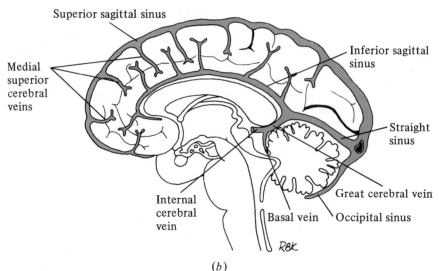

Superior sagittal sinus

Inferior sagittal sinus

Medial superior cerebral veins

Straight sinus

Internal cerebral vein

Great cerebral vein

Basal vein

Occipital sinus

RBK

(b)

Figure 15.36 Venous Drainage of Brain. Superficial and deep cerebral veins are shown. (*a*) Lateral surface. (*b*) Medial surface. These veins drain into dural venous sinuses.

the falx cerebri to the tentorium cerebelli and ends in the left transverse sinus.

4. *Transverse Sinuses.* A pair of vessels that lie in the lateral margins of the tentorium cerebelli.
5. *Confluence of Sinuses.* Point at which the superior sagittal, straight, and transverse sinuses meet.
6. *Sigmoid Sinuses.* A pair of S-shaped vessels that are continuous with the transverse sinuses and drain into the internal jugular veins.

BLOOD SUPPLY OF THE SPINAL CORD

The spinal cord receives blood from the anterior and posterior spinal arteries, which are branches of the vertebral arteries, plus the anterior and posterior radicular arteries, which are branches of the segmental arteries. It is drained by the anterior and posterior spinal veins, which are tributaries of the vertebral veins, and by the anterior and posterior radicular veins, which are tributaries of the segmental veins.

Diseases and Disorders of the Central Nervous System

PARALYSIS

Paralysis refers to complete loss of motor function. Paralysis of both lower extremities is known as *paraplegia*, while paralysis of all four extremities is known as *quadriplegia*. Paralysis of the extremities on one side of the body is known as *hemiplegia*, a condition often seen in stroke victims. Functionally, paralysis is divided into spastic and flaccid types.

Spastic Paralysis. Spastic paralysis (upper motor neuron disease) follows interruption of pyramidal and extrapyramidal tracts. Removal of cortical control over lower motor neurons results in spasticity. Spasticity is characterized by increased tone in affected muscles, exaggerated deep tendon reflexes, such as knee and ankle jerk, and absence of muscle atrophy.

Flaccid Paralysis. Flaccid paralysis (lower motor neuron disease) follows damage to lower motor neurons. It is characterized by loss of tone in affected muscles, loss of deep tendon reflexes, and muscle atrophy.

DISEASES AFFECTING THE SPINAL CORD

Poliomyelitis. Poliomyelitis is a viral disease that selectively attacks and kills lower motor neurons. It is characterized by flaccid paralysis without sensory disturbances. Once a major crippler of children and young adults, poliomyelitis has been virtually eliminated through immunization with the Salk or Sabin vaccines.

Tabes Dorsalis. Tabes dorsalis is a late neural manifestation of syphilis characterized by degeneration of the dorsal (posterior) white columns and the dorsal (posterior) or sensory nerve roots. It usually involves the lower half of the spinal cord. Knee and ankle jerk reflexes are lost in the tabetic patient. Fine touch, vibratory sense, and proprioception are lost in the lower extremities as the result of fasciculus gracilis interruption. When the tabetic patient is asked to stand with eyes closed and feet together, he or she loses balance and falls to the floor (Romberg test). He or she may also have intermittent attacks of sharp abdominal pains caused by irritation of dorsal nerve roots. There are no direct motor disturbances in tabes dorsalis.

Syringomyelia. Syringomyelia is a disease characterized by softening and cavitation of the gray matter around the central canal of the spinal cord. Degeneration often extends into the ventral (anterior) horns, destroying lower motor neurons. Syringomyelia is most commonly seen in the region of the cervical enlargement, so that the upper extremities are involved. Motor deficits are usually expressed as muscular weakness, while sensory deficits are expressed as loss of pain and temperature perception in the upper extremities.

SPINAL CORD SECTION

Complete or partial transection of the spinal cord may result from bullet wounds, stab wounds, or fracture-dislocation of the vertebral column.

Complete Transection. Complete transection of the spinal cord is followed by loss of all sensation below the level of the lesion because all ascending tracts have been interrupted. There is also complete loss of all voluntary movement below the level of the lesion because all descending tracts have been interrupted.

Partial Transection. Transection of half the spinal cord (hemisection) presents a different clinical picture. There are (1) spastic paralysis below the level of the lesion on the same side caused by interruption of the pyramidal and extrapyramidal tracts, (2) flaccid paralysis at the level of the lesion on the same side caused by injury to lower motor neurons, (3) complete loss of sensation at the level of the lesion on both sides caused by interruption of crossed and uncrossed sensory tracts, (4) loss of fine touch and proprioception below the level of the lesion on the same side caused by interruption of the posterior columns on that side, and (5) loss of pain and temperature sensation below the level of the lesion on the opposite side caused by interruption of the lateral spinothalamic tract.

DISEASES AFFECTING THE BRAIN

Parkinson's Disease. Parkinson's disease (paralysis agitans) is a common neurological disease of older persons. It is characterized by muscle rigidity, fine muscle tremors that are worse at rest, a slow and shuffling gait, lack of spontaneous arm movements while walking, and a masklike facial expression. Parkinson's disease is associated with depigmentation of melanin-containing neurons and lack of the neurotransmitter *dopamine* in various parts of the extrapyramidal motor system, especially in the substantia nigra and the corpus striatum. Relief and new hope have been given to the victims of Parkinson's disease through the use of a drug called L-dopa, a precursor of dopamine.

Cerebrovascular Accident. The brain has a great demand for oxygen. Any interruption in the oxygenated blood supply to the brain for more than a few minutes causes death of neural tissue. Interruption of cerebral blood flow may be caused by a blood clot (cerebral thrombosis) or rupture of a cerebral blood vessel (cerebral hemorrhage). These pathologies are referred to as cerebrovascular accidents (CVA). A relatively small CVA in the internal capsule, a bundle of projection fibers running between the brain stem and cerebrum, can cause widespread paralysis and sensory defects on the side opposite the lesion. This is due to the fact that most of the motor fibers going to and sensory fibers coming from one side of the body pass through the opposite internal capsule.

DISORDERS OF CORTICAL FUNCTION

Many brain lesions interfere with the cortical processing of sensory information (perception) and the initiation of appropriate motor responses. The different types of cortical dysfunction are termed *agnosia*, *aphasia*, and *apraxia*.

Agnosia. The term "agnosia" is used to indicate an inability to recognize the importance of sensory impressions. This may be expressed as inability of an affected individual to identify familiar objects by touch, sight, sound, taste, or smell.

Aphasia. The term "aphasia" is used to indicate an inability to understand or produce speech. This may be expressed as inability of an affected individual to understand spoken words, to read written words, to remember words, or to express himself or herself in speech or writing. For example, a patient with motor (Broca's) aphasia understands spoken words and knows what he or she wants to say but can only speak in a garbled fashion.

Apraxia. The term "apraxia" is used to indicate an inability to develop the concept and carry out a sequence of actions to perform a given task. This may be expressed as inability of an affected individual to perform a task on command because he or she either does not understand the command, cannot perform the individual actions, or is unable to put the actions in the correct sequence to perform the task.

Part C. **The Peripheral Nervous System**

The peripheral nervous system consists of the spinal and cranial nerves plus their associated ganglia and plexuses. Although the autonomic nerves qualify as peripheral nerves, they will be described separately.

Peripheral Nerve Structure

A peripheral nerve consists of bundles of nerve fibers enclosed in protective connective tissue wrappings. The nerve fiber consists of an axon and its sheath [Fig. 15.37(*a*)]. All peripheral axons are enclosed in a Schwann cell sheath. Axons of small diameter (up to 1 micron) are enclosed by a single fold of Schwann cell cytoplasm. Such axons are called *unmyelinated nerve fibers* because their sheaths cannot be seen under the light microscope. Axons of progressively larger diameter (from 1 to 20 microns) are enclosed by progressively thicker myelin sheaths and are called *myelinated axons*. The myelin sheath consists of Schwann cell membranes wrapped around the axon in jelly

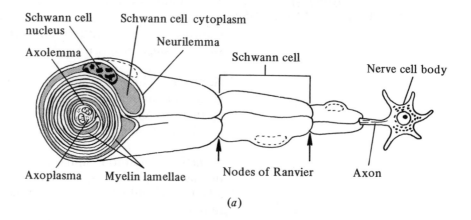

(*a*)

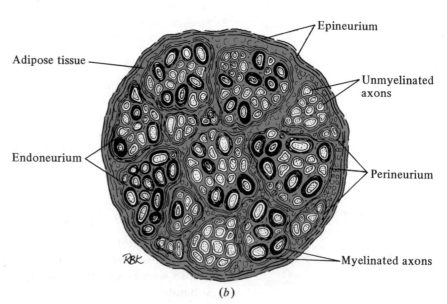

Figure 15.37 Peripheral Nerve Fiber. (*a*) Myelinated fiber (*b*) Histological cross section. Drawing based on light microscopic view.

(*b*)

roll fashion. During the wrapping process the cell membranes fuse with one another to form a lipoprotein complex called *myelin.* The myelin sheath is interrupted at regular intervals along the length of the axon by *nodes of Ranvier.* These are gaps between adjacent Schwann cells.

Surrounding each peripheral nerve fiber is a thin layer of reticular connective tissue called *endoneurium.* A bundle of nerve fibers with their endoneurial wrappings constitutes a *fascicle.* Each nerve fascicle is covered with a wrapping of areolar connective tissue called *perineurium.* An entire nerve consists of many fascicles, which are enclosed by a thick wrapping of fibrous connective tissue called *epineurium* [Fig. 15.37(*b*)].

Spinal Nerves

The spinal nerves are the peripheral nerves of the spinal cord (Fig. 15.38).

STRUCTURE OF THE SPINAL NERVE

A typical spinal nerve is illustrated in Fig. 15.39. Each spinal nerve originates from the spinal cord by two roots: a *dorsal (posterior) root* and a *ventral (anterior) root.* The dorsal root carries sensory fibers to the spinal cord. Each dorsal root bears a swelling called a dorsal root (spinal) ganglion. The dorsal root ganglion contains the cell bodies of 1° sensory neurons. The ventral root carries motor fibers away from the spinal cord.

The dorsal and ventral roots of the spinal nerve unite as the nerve enters its corresponding intervertebral foramen. The nerve that emerges from the vertebral column is thus a *mixed nerve,* containing both sensory and motor fibers. Just after it emerges from the vertebral column, the spinal nerve gives rise to two communicating branches, the *white* and *gray rami.* These rami connect the spinal nerve to the sympathetic trunk, a part of the autonomic nervous system. The spinal nerve also gives rise to two primary branches, the *posterior* and *anterior primary rami.* Posterior primary rami provide sensory and motor innervation for the posterior body wall, while the anterior primary rami provide sensory and motor innervation for the anterolateral body wall plus the extremities.

FUNCTIONAL COMPONENTS OF THE SPINAL NERVES

There are four functionally distinct types of nerve fibers that can be found in a spinal nerve. These types are as follows:

1. *General Somatic Efferent Fibers (GSE).* Fibers that transmit motor impulses from the spinal cord to skeletal muscle.
2. *General Visceral Efferent Fibers (GVE).* Fibers that transmit motor impulses from the spinal cord to smooth muscle, cardiac muscle, and glands.

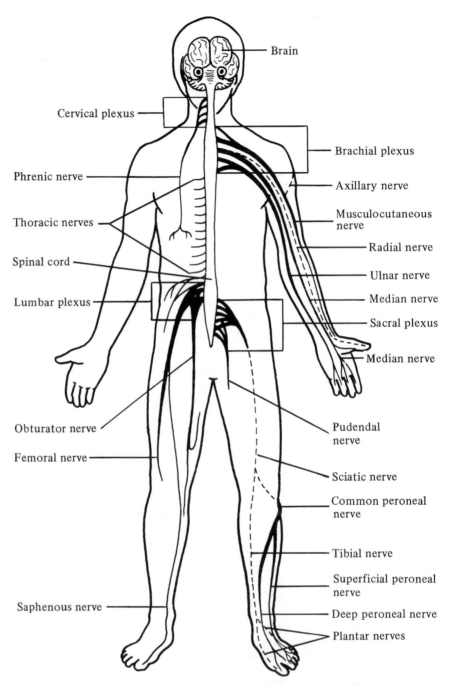

Figure 15.38 *Spinal Nerve Distribution.* Distribution of major peripheral nerves derived from anterior primary rami of spinal nerves. Nerves of anterior limb compartments are indicated by *solid lines*; nerves of posterior limb compartments are indicated by *dotted lines*.

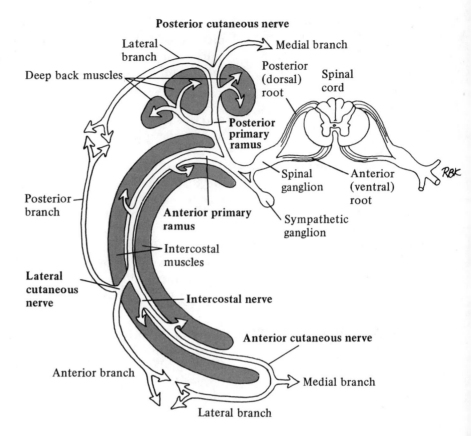

Posterior cutaneous nerve

Medial branch

Lateral branch

Posterior (dorsal) root

Spinal cord

Deep back muscles

Posterior primary ramus

Spinal ganglion

Anterior (ventral) root

Posterior branch

Anterior primary ramus

Sympathetic ganglion

Intercostal muscles

Lateral cutaneous nerve

Intercostal nerve

Anterior cutaneous nerve

Anterior branch

Medial branch

Lateral branch

Figure 15.39   Typical Spinal Nerve.

3. *General Somatic Afferent Fibers (GSA).* Fibers that transmit exteroceptive and proprioceptive impulses from the body to the spinal cord.

4. *General Visceral Afferent Fibers (GVA).* Fibers that transmit interoceptive impulses from the viscera to the spinal cord.

DISTRIBUTION OF THE SPINAL NERVES

There are 31 pairs of spinal nerves that correspond to the 31 spinal cord segments.

Cervical Nerves

There are eight pairs of cervical nerves. The first cervical nerve, the *suboccipital nerve,* emerges from the cerebral canal between the skull and the atlas. The second through seventh cervical nerves emerge above their corresponding vertebrae, while the eighth emerges below. The posterior primary rami of cervical nerves are distributed to the skin and deep muscles of the back of the head and neck, while the anterior primary rami take part in forming the cervical and brachial plexuses.

Thoracic Nerves   There are 12 pairs of thoracic nerves. Their posterior primary rami are distributed to the skin and deep muscles of the back. The anterior primary rami of nerves $T1$ to $T11$ lie between the ribs and are known as *intercostal nerves;* that of $T12$ lies below the twelfth rib and is known as the *subcostal nerve.* The intercostal and subcostal nerves are distributed to the skin and muscles of the thorax and abdomen. Also, the anterior primary rami of $T1$ and $T12$ take part in forming the brachial and lumbar plexuses, respectively.

Lumbar Nerves   There are five pairs of lumbar nerves. Their posterior primary rami are distributed to the skin of the buttocks (as far as the greater trochanter) and the deep muscles of the back. Their anterior primary rami take part in forming the lumbar and sacral plexuses.

Sacral and Coccygeal Nerves   There are five pairs of sacral nerves and one pair of coccygeal nerves. Their posterior primary rami are distributed to the skin of the medial part of the buttocks and the deep muscles of the back. The anterior primary rami of the sacral nerves take part in forming the sacral plexus.

SPINAL NERVE PLEXUSES   The anterior primary rami of cervical, lumbar, and sacral spinal nerves form complex networks called *plexuses.* Within each plexus the nerve fibers undergo considerable rearrangement, so that the peripheral nerves that emerge from the plexus may contain nerve fibers from more than one spinal cord segment. The major nerve plexuses are described here.

Cervical Plexus   The cervical plexus (Fig. 15.40) is composed of the anterior primary rami of spinal nerves $C1$ to $C4$. Cutaneous branches innervate the skin of the jaw, the back of the ear, the anterolateral aspect of the neck, and the upper anterior thorax. Muscular branches innervate the superficial muscles of the neck and upper back plus the infrahyoid muscles. The most important nerve of the cervical plexus is the *phrenic nerve,* which innervates the thoracic diaphragm. Injury to the phrenic nerve causes diaphragmatic paralysis.

Brachial Plexus   The brachial plexus (Fig. 15.41) is composed of the anterior primary rami of spinal nerves $C5$ to $T1$. It innervates the shoulder and upper extremity. Important nerves of the brachial plexus are listed here.

1. *Axillary Nerve.* Provides motor innervation for the teres minor and deltoid muscles plus sensory innervation for the skin over the deltoid muscle.
2. *Musculocutaneous Nerve.* Provides motor innervation for the flexor muscles of the arm plus sensory innervation for the skin on the anterolateral aspect of the forearm.
3. *Median Nerve.*   Provides motor innervation for the flexor-pronator muscles of the forearm (except flexor carpi ulnaris and the ulnar

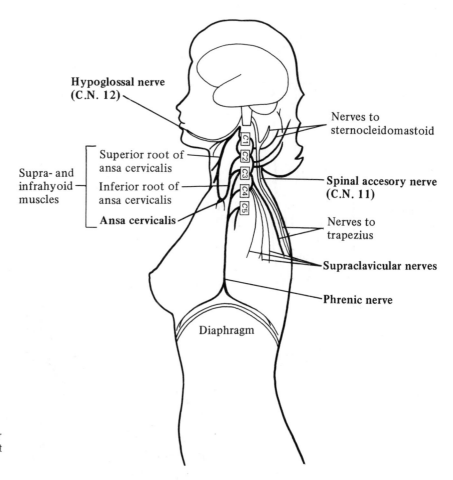

Hypoglossal nerve
(C.N. 12)

Nerves to
sternocleidomastoid

Supra- and
infrahyoid
muscles

Superior root of
ansa cervicalis

Inferior root of
ansa cervicalis

Ansa cervicalis

**Spinal accesory nerve
(C.N. 11)**

Nerves to
trapezius

Supraclavicular nerves

Phrenic nerve

Diaphragm

Figure 15.40 *Cervical Plexus and Major Nerves.* Branches innervating important muscles are indicated.

half of flexor digitorum profundus) plus the thenar muscles and the two lateral lumbricales. It also provides sensory innervation for the skin on the lateral half of the palm. Injury to the median nerve causes *apehand*.

4. *Radial Nerve.* Provides motor innervation for the extensor muscles of the arm plus the extensor-supinator muscles of the forearm. It also provides sensory innervation for the posterior aspect of the arm, forearm, and hand. Injury to the radial nerve causes *wristdrop*.

5. *Ulnar Nerve.* Provides motor innervation for the flexor muscles on the ulnar side of the forearm plus the hypothenar muscles, interossei, and the two medial lumbricals. It also provides sensory innervation for the skin on the anteromedial aspect of the forearm and medial half of the palm. Injury to the ulnar nerve causes *claw-hand*.

Lumbar Plexus   The lumbar plexus (Fig. 15.42) is composed of the anterior primary rami of spinal nerves *L1* to *L4*. It innervates anterior and medial thigh

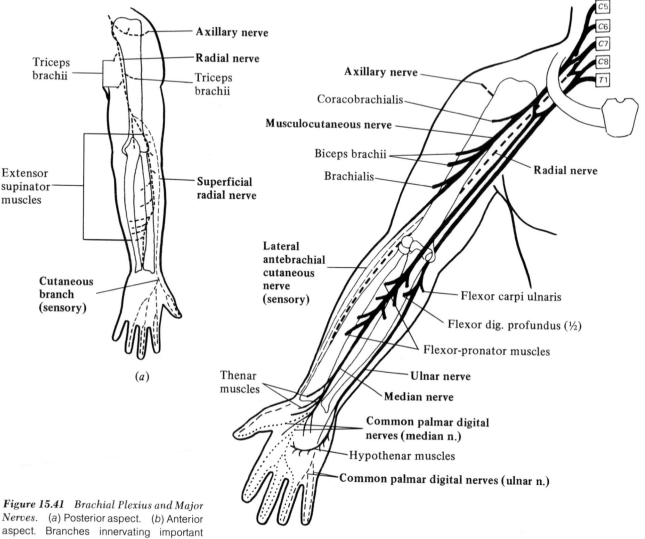

Axillary nerve

Radial nerve

Triceps brachii

Triceps brachii

Extensor supinator muscles

Superficial radial nerve

Cutaneous branch (sensory)

(a)

C5
C6
C7
C8
T1

Axillary nerve

Coracobrachialis

Musculocutaneous nerve

Biceps brachii

Brachialis

Radial nerve

Lateral antebrachial cutaneous nerve (sensory)

Flexor carpi ulnaris

Flexor dig. profundus (½)

Flexor-pronator muscles

Thenar muscles

Ulnar nerve

Median nerve

Common palmar digital nerves (median n.)

Hypothenar muscles

Common palmar digital nerves (ulnar n.)

(b)

Figure 15.41 *Brachial Plexius and Major Nerves.* (a) Posterior aspect. (b) Anterior aspect. Branches innervating important muscles are indicated.

muscles plus the skin of the lower abdominal wall, external genitalia, and the anteromedial aspects of the thigh and leg. Important nerves of the lumbar plexus are as follows:

1. *Femoral Nerve.* Provides motor innervation for the thigh flexor-leg extensor muscles of the thigh plus sensory innervation for the skin on the anteromedial aspect of the thigh and leg. Injury to this nerve causes *quadriceps paralysis.*
2. *Obturator Nerve.* Provides motor innervation for the adductor muscles of the thigh. Injury to this nerve causes *adductor paralysis.*

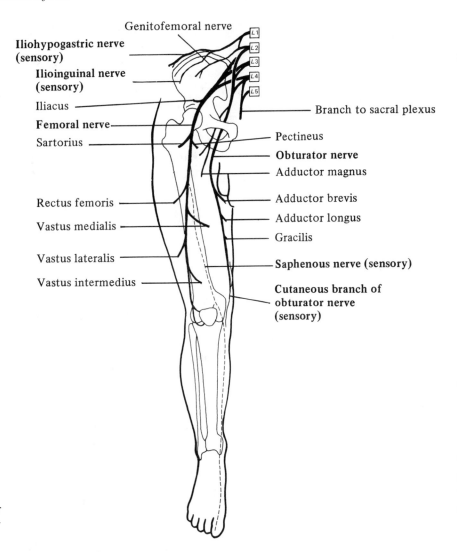

Genitofemoral nerve

Iliohypogastric nerve (sensory)

Ilioinguinal nerve (sensory)

Iliacus

Femoral nerve

Sartorius

Rectus femoris

Vastus medialis

Vastus lateralis

Vastus intermedius

Branch to sacral plexus

Pectineus

Obturator nerve

Adductor magnus

Adductor brevis

Adductor longus

Gracilis

Saphenous nerve (sensory)

Cutaneous branch of obturator nerve (sensory)

Figure 15.42 Lumbar Plexus and Major Nerves. Anterior aspect. Branches innervating important muscles are indicated.

Sacral Plexus   The sacral plexus (Fig. 15.43) is composed of the anterior primary rami of spinal nerves *L4* to *S4*. It innervates the muscles and skin of the buttocks, perineum, external genitalia, posterior thigh, leg, and foot. Important nerves of the sacral plexus are as follows:

1. *Sciatic Nerve (L4 to S3).* The longest and largest nerve in the human body. It provides motor innervation for the thigh extensor-leg flexor muscles of the thigh plus sensory innervation for the skin on the posterior aspect of the thigh. Injury to this nerve causes *hamstring paralysis.* The sciatic nerve in turn gives rise to two branches:

    a. Common Peroneal (Medial Popliteal) Nerve. Provides motor innervation for the dorsiflexor and peroneal muscles of the leg plus sensory innervation for the skin on the dorsum of the foot. Injury to this nerve causes *footdrop.*

   b. Tibial (Lateral Popliteal) Nerve. Provides motor innervation for the plantarflexor muscles of the leg and intrinsic muscles of the foot plus sensory innervation for the skin on the posterolateral aspect of the leg and sole of the foot. Injury to this nerve causes *clawfoot.*

2. *Pudendal Nerve (S2 to S4).* Provides motor innervation for the perineal muscles plus the external sphincters of the bladder and anus. It also provides sensory innervation for the skin of the perineum and external genitalia. Injury to this nerve causes loss of bladder and bowel control plus sexual dysfunction.

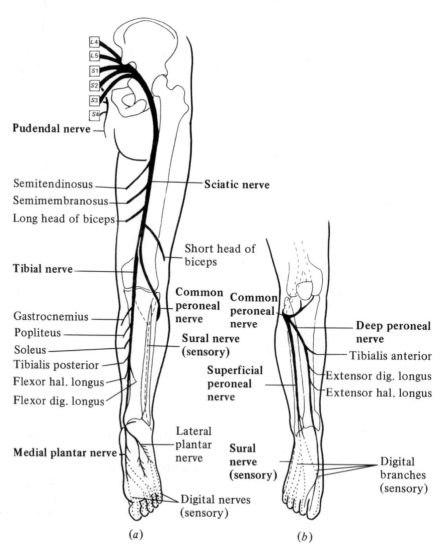

Figure 15.43 Sacral Plexus and Major Nerves. (*a*) Posterior aspect. (*b*) Anterior aspect. Branches innervating important muscles are indicated.

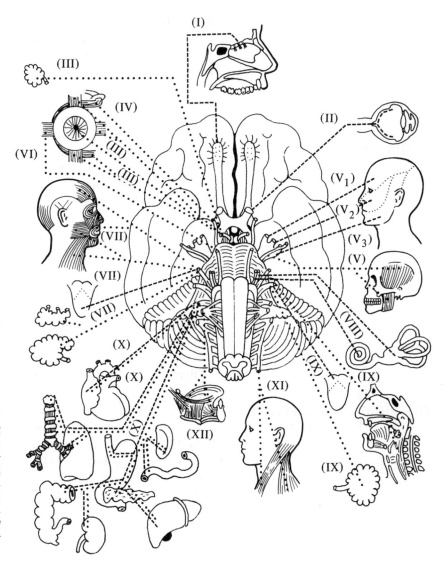

Figure 15.44 Distribution of Cranial Nerves. (I) Olfactory nerve; (II) optic nerve; (III) oculomotor nerve; (IV) trochlear nerve; (V) trigeminal nerve; (VI) abducens nerve; (VII) facial nerve; (VIII) vestibulo-cochlear nerve; (IX) glossopharyngeal nerve; (X) vagus nerve; (XI) spinal accessory nerve; (XII) hypoglossal nerve, *dashed lines,* sensory fibers; *dotted lines,* motor fibers.

Cranial Nerves

The cranial nerves are the peripheral nerves of the brain. The 12 pairs of cranial nerves have their superficial origin from the inferior surface of the brain (Fig. 15.44) and leave the skull by way of the foramina listed in Chapter 5.

STRUCTURE OF THE CRANIAL NERVES

The cranial nerves are more variable in structure than the peripheral nerves, especially those that innervate special sensory organs. Some carry only sensory fibers, some only motor fibers, and some are mixed.

A number of the mixed cranial nerves have separate sensory and motor roots. The sensory roots bear sensory ganglia, called cranial ganglia, that are similar to dorsal root ganglia. The sensory fibers of most cranial nerves terminate in sensory nuclei that are functionally equivalent to those in the dorsal (posterior) horns of the spinal cord; their motor fibers originate in motor nuclei that are functionally equivalent to those in the ventral (anterior) horns of the spinal cord.

FUNCTIONAL COMPONENTS

In addition to the general somatic and visceral afferent fibers and efferent fibers carried by the spinal nerves, the cranial nerves may carry the functional components listed next. It should be noted that no single cranial nerve carries all seven functional components.

1. *Special Somatic Afferent Fibers (SSA).* Fibers that transmit sensory impulses from the special sense organs of vision, equilibrium, and hearing to the brain.
2. *Special Visceral Afferent Fibers (SVA).* Fibers that transmit sensory impulses from the special sense organs of smell and taste to the brain.
3. *Special Visceral Efferent Fibers (SVE).* Fibers that transmit motor impulses from the brain to skeletal muscles derived from the brachial (gill) arches of the embryo. These include the muscles of mastication, facial expression, and swallowing.

DISTRIBUTION AND FUNCTION OF THE CRANIAL NERVES

The distribution and function of the cranial nerves are given next and are summarized in Table 15.1. The cranial nerves are numbered from 1 to 12 according to the anterior to posterior order in which they emerge from the brain. Ordinal numbers are used here to designate the cranial nerves rather than the more traditional but unfamiliar roman numerals, which are given in parentheses.

First Cranial Nerve: Olfactory (C.N. I)

The olfactory nerve is concerned with the sense of smell and carries sensory fibers (SVA). It originates from the olfactory mucosa and consists of 20 small bundles of nerve fibers that pass into the cranial cavity through the cribriform plate of the ethmoid bone. These nerve bundles enter the olfactory bulb, which is connected to the olfactory areas of the brain by the olfactory tract. Injury to the olfactory nerve causes anosmia (loss of ability to identify odors).

Second Cranial Nerve: Optic (C.N. II)

The optic nerve is concerned with the sense of vision and carries sensory fibers (SSA). It originates from the retina of the eye and passes from the orbit into the cranial cavity through the optic foramen. At the base of the brain, both optic nerves come together to form the optic chiasma (crossing) and separate again as the optic tracts. The optic tracts terminate in the thalamus. Injury to an optic nerve anterior to the chiasma causes blindness in the corresponding eye.

Third Cranial Nerve: Oculomotor (C.N. III)

The oculomotor nerve is concerned with eye movements. It emerges from the midbrain and passes through the superior orbital fissure into the orbit, where it innervates both extrinsic and intrinsic eye muscles. The oculomotor nerve carries motor fibers (GSE) to all the extrinsic eye muscles except the superior oblique and lateral rectus. It also carries parasympathetic motor fibers (GVE) destined for the sphincter pupillae muscle of the iris and the ciliary muscle of the lens. Injury to the oculomotor nerve causes ptosis (drooping) of the upper eyelid, diplopia (double vision), and dilation of the pupil.

Fourth Cranial Nerve: Trochlear (C.N. IV)

The trochlear nerve is also concerned with eye movements. It emerges from the midbrain and passes through the superior orbital fissure to the orbit, where it innervates the superior oblique muscle. The trochlear nerve carries motor fibers (GSE) to the superior oblique. Injury to the trochlear nerve causes diplopia plus inability to depress or rotate the affected eyeball downward.

Fifth Cranial Nerve: Trigeminal (C.N. V)

The trigeminal nerve, largest of the cranial nerves, is mainly concerned with facial sensation and jaw movements. It emerges from the pons and divides into three branches. The *ophthalmic branch* (V_1) leaves the cranial cavity through the superior orbital fissure and innervates the upper part of the face. It carries sensory fibers (GSA) from the skin of the forehead, the nasal bridge, the upper eyelid, the lacrimal gland, the cornea and conjunctiva, and the mucosa of frontal and ethmoid sinuses. The *maxillary branch* (V_2) leaves the cranial cavity through the foramen rotundum and innervates the middle part of the face. It carries sensory fibers (GSA) from the skin of the cheeks and nose, the lower eyelid, the upper lip, the upper teeth and gums, the nasal mucosa, the mucosa of the sphenoid and maxillary sinuses, and the meninges. The *mandibular branch* (V_3) leaves the cranial cavity through the foramen ovale and innervates the lower part of the face. It carries sensory fibers (GSA) from the skin of the jaw, the external ear, the lower lip, the lower teeth and gums, the oral mucosa, and the anterior two-thirds of the tongue. It also carries motor fibers (SVE) to and proprioceptive fibers (GSA) from the mylohyoid, digastric (anterior belly), tensor tympani, tensor palatini, and the muscles of mastication. Injury to the trigeminal nerve causes loss of sensation of the face, cornea, and anterior two-thirds of the tongue plus paralysis of the muscles of mastication on the side of injury. *Tic douloureux* (trigeminal neuralgia) is caused by irritation of the trigeminal nerve. Excruciating pain, which is difficult to relieve, follows the distribution of its sensory fibers.

Sixth Cranial Nerve: Abducens (C.N. VI)

The abducens nerve is concerned with eye movements. It emerges from the brain at the junction between the pons and medulla and passes through the superior orbital fissure to the orbit, where it in-

Table 15.1 Synopsis of Cranial Nerves

Name	Components*	Function	Location of cell bodies	Distribution
1. Olfactory (sensory)	SVA	Smell	Olfactory mucosa	Upper part of nasal cavity
2. Optic (sensory)	SSA	Vision	Retina	Retina
3. Oculomotor (motor)	GSE	Eye movements	Oculomotor nuclear complex (midbrain)	Extrinsic eye muscles except superior oblique and lateral rectus, levator palpebrae superioris
	GVE	Pupillary constriction and accommodation	Oculomotor nuclear complex (midbrain)	Sphincter pupillae and ciliary muscles
4. Trochlear (motor)	GSE	Eye movements	Trochlear nucleus (midbrain)	Superior oblique muscle
5. Trigeminal (mixed)	GSA	General sensation	Gassarian ganglion of trigeminal nerve	Skin and mucous membranes of face and head
	SVE	Mastication	Motor nucleus of trigeminal nerve (pons)	Muscles of mastication, digastric (anterior belly), mylohyoid, tensors tympani and palatini
	GSA	Proprioception	Mesencephalic nucleus of trigeminal nerve (midbrain)	Muscles of mastication
6. Abducens (motor)	GSE	Eye movements	Abducens nucleus (pons)	Lateral rectus muscle
7. Facial (mixed)	SVE	Facial expression	Motor nucleus of facial nerve (pons)	Muscles of facial expression, stylohyoid, digastric (posterior belly), stapedius
	GVE	Secretion	Superior salivatory nucleus (pons)	Lacrimal, submandibular, and sublingual glands
	GVA	Visceral sensation	Geniculate ganglion of facial nerve	Same as above
	SVA	Taste	Geniculate ganglion of facial nerve	Taste buds (anterior two-thirds of tongue)
8. Vestibulocochlear (sensory)				
Vestibular division	SSA	Equilibrium	Vestibular ganglion of vestibulocochlear nerve	Semicircular ducts, utricle, saccule
Cochlear division	SSA	Hearing	Spiral ganglion of vestibulocochlear nerve	Organ of Corti
9. Glossopharyngeal (mixed)	SVE	Swallowing	Nucleus ambiguus (medulla)	Stylopharyngeal muscle
	GVE	Secretion	Inferior salivatory nucleus (medulla)	Parotid gland
	GVA	Visceral sensation	Inferior ganglion of glossopharyngeal nerve	Same as above
	SVA	Taste	Inferior ganglion of glossopharyngeal nerve	Taste buds (posterior one-third of tongue)

Table 15.1 (Continued)

Name	Components*	Function	Location of cell bodies	Distribution
10. Vagus (mixed)	SVE	Swallowing and phonation	Nucleus ambiguus (medulla)	Muscles of pharynx and larynx
	GVE	Visceral muscle movements and secretion	Dorsal motor nucleus of vagus (medulla)	Heart, muscles and glands of respiratory tract, muscles and glands of digestive tract (down to descending colon)
	GVA	Visceral sensation	Inferior ganglion of vagus nerve	Same as above
	SVA	Taste	Inferior ganglion of vagus nerve	Taste buds (epiglottis)
11. Accessory (motor) Cranial portion	SVE	Swallowing and phonation	Nucleus ambiguus (medulla)	Muscles of pharynx and larynx
Spinal portion	GSE	Movements of head and shoulder	Cervical spinal cord to C5	Trapezius and sternocleidomastoid muscles
	GSA	Proprioception	Spinal ganglia C1 to C5	Same as above
12. Hypoglossal (motor)	GSE	Tongue movements	Nucleus of hypoglossal nerve (medulla)	Extrinsic and intrinsic muscles of tongue
	GSA	Proprioception	Unknown	Same as above

* GSE = general somatic efferent, GVE = general visceral efferent, GSA = general somatic afferent, GVA = general visceral afferent, SSA = special somatic afferent, SVA = special visceral afferent, SVE = special visceral efferent.

nervates the lateral rectus muscle. The abducens nerve carries motor fibers (GSE) to the lateral rectus. Injury to the abducens nerve causes diplopia and inward deviation of the affected eyeball.

Seventh Cranial Nerve: Facial (C.N. VII)

The facial nerve is concerned mainly with facial expression and the sense of taste. It emerges from the lower border of the pons and leaves the cranial cavity through the internal auditory meatus and facial canal to innervate facial structures. The facial nerve carries motor fibers (SVE) to the stylohyoid, the digastric (posterior belly), and the muscles of facial expression. It also carries motor fibers (GVE) destined for and sensory fibers (GVA) from the lacrimal, submandibular, and sublingual glands and receives sensory (SVA, taste) fibers from the anterior two-thirds of the tongue. Injury to the facial nerve as it leaves the pons causes facial paralysis plus loss of taste sensation on the anterior two-thirds of the tongue. Injury to the facial nerve within or after it emerges from the facial canal, as in Bell's palsy, causes facial paralysis without loss of taste sensation.

Eighth Cranial Nerve: Vestibulocochlear (C.N. VIII)

The vestibulocochlear nerve is concerned with the senses of equilibrium and hearing. It emerges from the lower border of the pons lateral to the facial nerve and leaves the cranial cavity through the internal auditory meatus, where it gives off vestibular and cochlear divisions to innervate structures of the inner ear. The vestibular division carries sensory fibers (SSA, equilibrium) from the utricle, saccule, and semicircular ducts, while the cochlear nerve carries sensory fibers (SSA, hearing) from the organ of Corti. Injury to the vestibular division causes vertigo (dizziness) and nystagmus (rapid involuntary eye movements). Injury to the cochlear division causes a partial hearing loss.

Ninth Cranial Nerve: Glossopharyngeal (C.N. IX)

The glossopharyngeal nerve is concerned mainly with swallowing and the sense of taste. It emerges from the side of the medulla and leaves the cranial cavity through the jugular foramen to innervate structures in the back of the oral cavity and in the pharynx. The glossopharyngeal nerve carries motor fibers (SVE) to the stylopharyngeal muscle plus motor fibers (GVE) destined for and sensory fibers (GVA) from the parotid gland. It also receives sensory fibers (GVA) from the posterior one-third of the tongue, the pharyngeal mucosa, and the middle ear plus sensory fibers (SVA, taste) from the posterior one-third of the tongue. The glossopharyngeal nerve also receives sensory fibers from the carotid body and carotid sinus that participate in the reflex regulation of blood pressure. Injury to the glossopharyngeal nerve causes loss of taste sensation on the posterior one-third of the tongue, difficulty in swallowing, loss of sensation in the throat, loss of the gag reflex, and blood pressure disturbances.

Tenth Cranial Nerve: Vagus (C.N. X)

The vagus nerve is concerned mainly with swallowing, speaking, and visceral function. It emerges from the side of the medulla below the glossopharyngeal nerve and leaves the cranial cavity through the jugular foramen along with the glossopharyngeal nerve. This nerve was named vagus ("wanderer") because it is distributed from the head down to the abdomen. The vagus nerve carries motor fibers (SVE) to the muscles of the pharynx (along with the accessory nerve) and larynx. It also carries motor fibers (GVE) to and sensory fibers (GVA) from the trachea, bronchi, heart, esophagus, stomach, small intestine, large intestine (down to the descending colon), liver, kidneys, and spleen. The vagus nerve also receives sensory fibers (SVA, taste) from the epiglottis. Injury to the vagus nerve causes difficulty in swallowing, vocal cord paralysis, and various visceral dysfunctions such as cardiac acceleration.

Eleventh Cranial Nerve: Accessory (C.N. XI)

The accessory nerve is concerned with swallowing, speaking, and movements of the head and neck. It is unusual in that it has both cranial and spinal portions. The cranial portion emerges from the medulla below the vagus and leaves the cranial cavity through the jugular foramen along with the glossopharyngeal and vagus nerves. It joins the vagus nerve in carrying motor fibers (SVE) to the muscles of the pharynx and larynx. It also has some overlapping GVE functions along with the vagus nerve. The spinal portion emerges from the upper four or five cervical cord segments and extends upward through the foramen magnum to join the cranial portion. It carries motor fibers (GSE) to and sensory fibers (GSA, proprioceptive) from the trapezius and sternocleidomastoid muscles. Injury to the spinal portion of the accessory nerve causes weakness in the muscles it innervates. Paralysis does not occur since the trapezius and sternocleidomastoid muscles are also innervated by cervical spinal nerves.

Twelfth Cranial Nerve: Hypoglossal (C.N. XII)

The hypoglossal nerve is concerned with tongue movements. It emerges from the ventral aspect of the medulla below the accessory nerve and leaves the cranial cavity through the hypoglossal canal to innervate the tongue muscles. The hypoglossal nerve carries motor fibers (GSE) to and sensory fibers (GSA, proprioceptive) to the intrinsic and extrinsic muscles of the tongue. Injury to the hypoglossal nerve affects speech, interferes with chewing and swallowing, and causes the tongue to deviate to the injured side when protruded.

Diseases and Disorders of the Peripheral Nervous System

Peripheral Nerve Injuries. Peripheral nerve injuries range from nerve bruises, which cause temporary loss of function, to complete nerve section, which may cause permanent loss of function. Most peripheral nerve injuries occur as a result of cuts, bone fractures, crush injuries, or penetrating wounds. Whenever a peripheral nerve is severed, an attempt is usually made to rejoin the severed ends so that regeneration of nerve fibers and restoration of function can occur. Regeneration can occur in the peripheral nervous system but not in the central nervous system.

Neuritis. Neuritis is the term generally applied to any disease of peripheral nerves. The cause is seldom inflammation, as the name suggests, but rather some form of trauma, infection, or metabolic disorder. When many nerves are involved, the condition is called *polyneuritis*. Polyneuritis is characterized by slowly progressive muscular weakness, muscular atrophy, tingling, and loss of sensation. Causes include chronic alcoholism, infectious disease, diabetes mellitus, or vitamin deficiencies.

Shingles. Shingles (herpes zoster) is an inflammation of peripheral nerves caused by *Herpesviris varicella*, the same virus that causes chickenpox in children. This disease is characterized by inflammation of spinal or cranial ganglia and by the eruption of numerous vesicles on one side of the body along the area of distribution of the sensory fibers of the affected ganglion. Recovery usually occurs in a few days to a few weeks, but the condition may be accompanied (and followed) by constant itching and severe pain. Treatment is aimed at relieving the itching and the pain while the infection runs its course.

Neurofibromatosis. Neurofibromatosis is a hereditary disease characterized by multiple benign tumors arising from peripheral nerve sheaths. There is no cure for this disease, but large unsightly tumors can be removed surgically.

Part D. The Autonomic Nervous System

The autonomic nervous system is a functional rather than a structural entity. It is the portion of the peripheral nervous system that innervates smooth muscle, cardiac muscle, and glands. The autonomic nervous system is thus the regulator, adjustor, and coordinator of vital visceral activities. Its principal function is to maintain the body's internal environment in a steady state.

The autonomic nervous system consists of the *visceral afferent system,* which transmits sensory information to the central nervous system from the viscera, and the *visceral efferent system,* which transmits motor commands from the central nervous system to the viscera.

Visceral Afferent System

The visceral afferent system consists of neurons whose receptors are located in visceral organs (interoceptors). Visceral afferent fibers travel along with somatic afferent fibers in cranial or spinal nerves to reach the central nervous system. Once in the central nervous system, these neurons either make synaptic contact with visceral efferent neurons to form reflex arcs or ascend the brain stem to terminate in autonomic centers in the thalamus or hypothalamus. Visceral afferent fibers are divided into two categories: physiological afferents and pain afferents. Physiological afferents are those that form the sensory limb of autonomic reflexes such as the micturition or defecation reflex. Pain afferents are those that respond to stimulation by disease or disturbed visceral function.

Visceral Efferent System

STRUCTURE
Neurons
The visceral efferent system differs from the somatic efferent system in that each end organ is innervated by a chain of two lower motor neurons (Fig. 15.45):

1. *Preganglionic Neuron.* The cell body of the preganglionic neuron is located in the central nervous system. It has a lightly myeli-

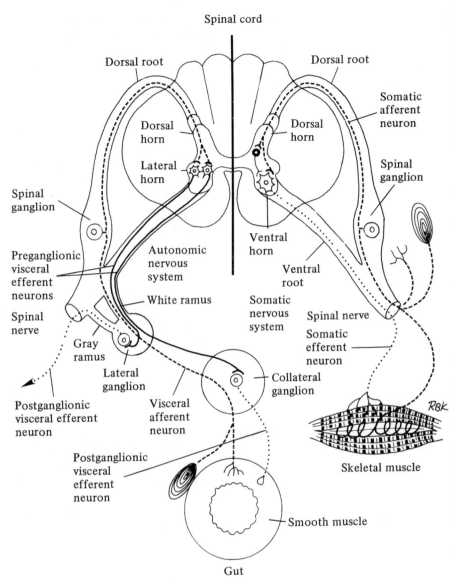

Spinal cord

Dorsal root

Dorsal root

Somatic
afferent
neuron

Dorsal
horn

Dorsal
horn

Spinal
ganglion

Lateral
horn

Spinal
ganglion

Ventral
horn

Preganglionic
visceral
efferent
neurons

Autonomic
nervous
system

Ventral
root

Spinal
nerve

White ramus

Somatic
nervous
system

Spinal nerve

Somatic
efferent
neuron

Gray
ramus

Lateral
ganglion

Visceral
afferent
neuron

Collateral
ganglion

Postganglionic
visceral efferent
neuron

Skeletal muscle

Postganglionic
visceral
efferent
neuron

Smooth muscle

Gut

Figure 15.45 Comparison of Autonomic and Somatic Nervous Systems. Autonomic nervous system, *left*; somatic nervous system, *right*. Note that visceral efferent system consists of two lower motor neurons: preganglionic neuron, which arises in lateral horn of cord and terminates in autonomic ganglion, and postganglionic neuron, which arises in autonomic ganglion and terminates in effector organ.

nated axon that travels to an autonomic ganglion, where it synapses with a postganglionic neuron. The preganglionic neuron releases acetylcholine as a neurotransmitter.

2. *Postganglionic Neuron.* The cell body of the postganglion neuron is located in an autonomic ganglion. It has an unmyelinated axon that terminates on or within the end organ it innervates. The postganglionic neuron releases acetylcholine as a neurotransmitter if it belongs to the parasympathetic division and norepinephrine if it belongs to the sympathetic division.

Ganglia   The ganglia of the autonomic nervous system are divided into lateral, collateral, and terminal groups.

Lateral Ganglia.   The lateral, or paravertebral, ganglia consist of 22 pairs of oval-shaped ganglia that lie lateral to the vertebral bodies: 3 cervical, 12 thoracic, 4 lumbar, and 4 sacral. Except for the cervical ganglia, each ganglion is connected to the nearest spinal nerve by a white communicating ramus, which carries myelinated preganglionic fibers to the ganglion, and by a gray communicating ramus, which carries unmyelinated postganglionic fibers back to the spinal nerve. The lateral ganglia are connected to one another by nerve fibers to form a longitudinal chain called the *sympathetic trunk.* The paired sympathetic trunks run parallel to the vertebral column from the base of the skull to the coccyx.

Collateral Ganglia.   The collateral, or prevertebral, ganglia consist of ganglia not associated with the vertebral column. The collateral ganglia in the head include the *ciliary, pterygopalatine, submandibular,* and *otic ganglia.* The collateral ganglia in the body cavities lie anterior to the vertebral column. The largest of these include the *celiac, superior mesenteric,* and *inferior mesenteric ganglia,* which lie near the arteries for which they are named.

Terminal Ganglia.   The terminal ganglia consist of many tiny ganglia lying close to or within visceral end organs.

Plexuses   The autonomic plexuses consist of complex networks of autonomic nerve fibers in the thoracic, abdominal, and pelvic cavities. They are derived from autonomic nerve fibers passing from the lateral and collateral ganglia to end organs other than the spleen. The fibers usually form their networks around blood vessels passing to the visceral end organs. The major autonomic plexuses are described next and illustrated in Fig. 15.46. Minor plexuses have been mentioned in previous chapters in connection with the end organs they supply.

Cardiac Plexus.   The cardiac plexus overlies the arch and ascending portion of the aorta. From this plexus and the secondary coronary, pulmonary, and esophageal plexuses, autonomic nerve fibers are distributed to the thoracic viscera.

Celiac Plexus.   The celiac plexus, also known as the *solar plexus,* overlies the aorta in the vicinity of the celiac trunk. From the main plexus a number of secondary plexuses are given off that follow the branches of the abdominal aorta. These include the phrenic, gastric, hepatic, splenic, gonadal, superior mesenteric, inferior mesenteric,

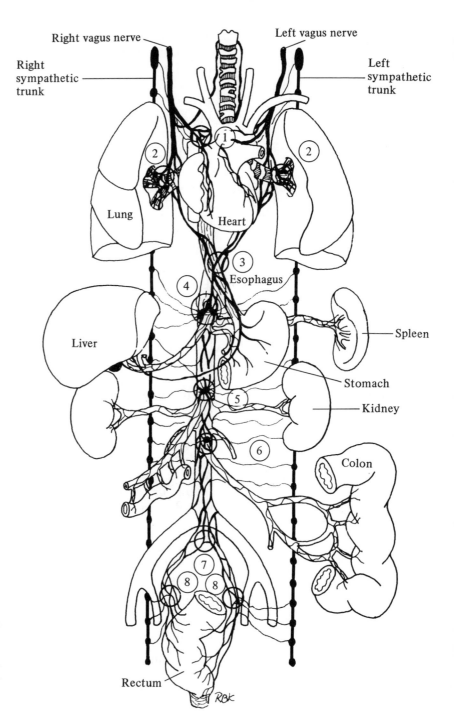

Right vagus nerve

Left vagus nerve

Right sympathetic trunk

Left sympathetic trunk

Lung

Heart

② ②

①

③ Esophagus

④

Liver

Spleen

Stomach

⑤

Kidney

⑥

Colon

⑦

⑧ ⑧

Rectum

Figure 15.46 Autonomic Plexuses. (1) Cardiac plexus; (2) pulmonary plexuses; (3) esophageal plexus; (4) celiac plexus; (5) superior mesenteric plexus; (6) inferior mesenteric plexus; (7) superior hypogastric plexus; (8) inferior hypogastric (pelvic) plexuses.

and hypogastric plexuses. From these plexuses autonomic fibers are distributed to the abdominal viscera and pelvic viscera of abdominal origin.

Hypogastric Plexuses. The superior hypogastric plexus overlies the upper part of the sacrum while the inferior hypogastric (pelvic) plexuses lie on each side of the lower part of the sacrum. From these plexuses plus the nearby inferior mesenteric plexus, autonomic nerve fibers are distributed to the pelvic viscera.

FUNCTION The visceral efferent system consists of *sympathetic* and *parasympathetic divisions*. These divisions usually have opposite effects on the organs they innervate; the one serves to stimulate activity, while the other serves to inhibit it. Innervation of the viscera by these two divisions is illustrated in Fig. 15.47 and summarized in Table 15.2.

Sympathetic Division The sympathetic division of the autonomic nervous system has been called the "fight or flight" division because its activities become more apparent under emergency or stress situations. The reactions of the sympathetic division include acceleration of the heart rate, increase in blood pressure, inhibition of digestive processes, retention of metabolic waste products, and the diversion of blood flow from the skin and viscera to the skeletal muscles.

The sympathetic division of the autonomic nervous system is also known as the *thoracolumbar outflow* because the cell bodies of sympathetic preganglionic neurons are located in the lateral gray horns of spinal cord segments *T1* to *L2*. The preganglionic fibers emerge from the spinal cord with the motor roots of corresponding spinal nerves. They leave the spinal nerves by way of the white communicating rami and enter the sympathetic trunks where they may do one of the following:

1. Synapse in lateral ganglia at the level of entry.
2. Ascend or descend the sympathetic trunk to synapse in lateral ganglia above or below the level of entry.
3. Pass through the sympathetic trunk by way of the splanchnic nerves to synapse in collateral ganglia.

Sympathetic postganglionic neurons are distributed to their end organs in the following manner:

1. Postganglion fibers from all lateral ganglia rejoin their corresponding spinal nerves by way of gray communicating rami. They are distributed to the skin, where they innervate blood vessels, hair follicles, and sweat glands.
2. Postganglionic fibers from the superior cervical ganglion are distributed to the dilator pupillae muscle of the iris by way of the internal carotid plexus.

3. Postganglionic fibers from the superior, middle, and inferior cervical ganglia are distributed to the heart by way of the cardiac plexus.
4. Postganglionic fibers from the upper four thoracic ganglia are distributed to the bronchi and lungs by way of the pulmonary plexuses and to the esophagus by way of the esophageal plexus.
5. Postganglion fibers from the celiac ganglion are distributed to the stomach, liver, spleen, kidneys, gonads, small intestine, and proximal colon (to the splenic flexure) by way of the celiac and superior mesenteric plexuses.
6. Postganglionic fibers from the superior and inferior mesenteric ganglia are distributed to the bladder, distal colon, rectum, anus, genital ducts, and external genitalia by way of the inferior mesenteric, hypogastric, and pelvic plexuses.

Parasympathetic Division   The parasympathetic division of the autonomic nervous system can be considered the "rest and relaxation" division because its activities, such as digestion of a meal, are not performed under stress. The reactions of the parasympathetic division include deceleration of the heart rate, decrease in blood pressure, stimulation of the digestive processes and accumulation of energy-rich nutrients, elimination of metabolic waste products, and promotion of blood flow to the skin and viscera.

The parasympathetic division is also known as the *craniosacral outflow* because the cell bodies of the parasympathetic preganglionic neurons are located in brain stem nuclei and in the lateral gray horns of spinal cord segments S2 to S4. The preganglionic fibers leave the central nervous system with the motor roots of cranial or spinal nerves and are distributed to their respective ganglia in the following manner:

1. Preganglionic fibers from the oculomotor nuclear complex (midbrain) travel by way of the oculomotor nerve to the ciliary ganglion.
2. Preganglionic fibers from the lacrimal nucleus (pons) travel by way of the facial nerve to the pterygopalatine ganglion.
3. Preganglionic fibers from the superior salivatory nucleus (pons) travel by way of the facial nerve to the submandibular ganglion.
4. Preganglionic fibers from the inferior salivatory nucleus (medulla) travel by way of the glossopharyngeal nerve to the otic ganglion.
5. Preganglionic fibers from the dorsal vagal nucleus (medulla) travel by way of the vagus nerve to terminal ganglia in the thoracic and abdominal viscera.
6. Preganglionic fibers from the sacral outflow leave the spinal cord with the motor roots of spinal nerves S2 to S4 and travel by way of the pelvic nerves to terminal ganglia in the pelvic viscera.

Table 15.2 Function of Autonomic Nervous System

System	Organ	Sympathetic stimulation	Parasympathetic stimulation
Visual	Iris muscles	Dilation of pupil (mydriasis)	Constriction of pupil (miosis)
	Ciliary muscle	No direct effect	Rounding up of lens (accommodation for near vision)
	Lacrimal gland	No direct effect	Tear secretion stimulated
Integumentary	Sweat glands	Copious secretion of sweat	No direct effect
	Pilomotor muscles	Contraction (goose bumps)	No direct effect
	Arteries	Vasoconstriction	No direct effect
Cardiovascular	Heart muscle	Acceleration of heart rate	Deceleration of heart rate
	Coronary arteries	Vasodilation	Vasoconstriction
	Skeletal muscle arteries	Vasodilation	No direct effect
	Visceral arteries	Vasoconstriction	No direct effect
Respiratory	Bronchioles	Bronchodilation	Bronchoconstriction
Digestive	Salivary glands	Scanty secretion of thick viscous saliva	Profuse secretion of thin watery saliva
	Pancreas	Secretion of thick viscous fluid	Secretion of thin watery fluid
	Gastric and intestinal glands	No direct effect	Secretion of digestive enzymes stimulated
	Digestive tract wall	Peristalsis decreased	Peristalsis increased
	Pyloric and anal sphincters	Contraction	Relaxation
Endocrine	Adrenal medulla	Secretion of catecholamines	No known effect
Genitourinary	Kidney	Urine output decreased	No known effect
	Ureters	Peristalsis decreased	Peristalsis increased
	Bladder wall (detrussor muscle)	Relaxation	Contraction
	Urethral sphincter	Contraction	Relaxation
	Penis	Ejaculation	Erection

Parasympathetic postganglionic neurons are distributed to their end organs in the following manner:

1. Postganglionic fibers from the ciliary ganglion terminate in the sphincter pupillae and ciliary muscles of the eye.
2. Postganglionic fibers from the pterygopalatine ganglion terminate in the lacrimal gland.
3. Postganglionic fibers from the submandibular ganglion terminate in the submandibular and sublingual glands.

Figure 15.47 Autonomic Nervous System. Diagram shows autonomic ganglia, nerves, and effector organs. Major autonomic ganglia are as follows: (1) ciliary ganglion; (2) pterygopalatine ganglion; (3) submandibular ganglion; (4) otic ganglion; (5) superior cervical ganglion; (6) middle cervical ganglion; (7) stellate ganglion; (8) unnamed sympathetic chain ganglia; (9) celiac ganglion; (10) superior mesenteric ganglion; (11) inferior mesenteric ganglion. Major autonomic nerves are as follows: *A*, greater splanchnic nerve; *B*, lesser splanchnic nerve; *C*, least splanchnic nerve; *D*, pelvic splanchnic nerve.

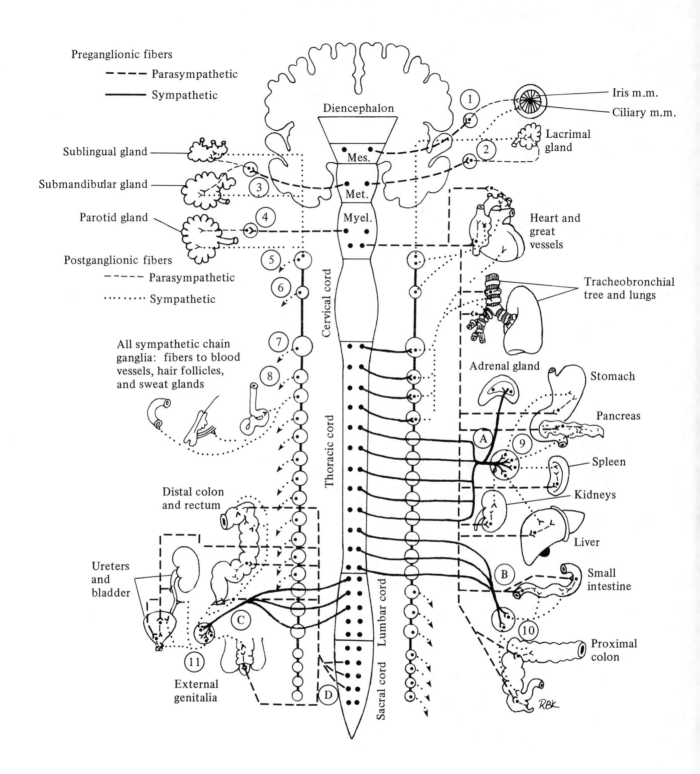

Preganglionic fibers

– – – Parasympathetic

——— Sympathetic

Diencephalon

Mes.

Met.

Myel.

Sublingual gland

Submandibular gland

Parotid gland

Postganglionic fibers

– – – Parasympathetic

·········· Sympathetic

All sympathetic chain
ganglia: fibers to blood
vessels, hair follicles,
and sweat glands

Cervical cord

Thoracic cord

Lumbar cord

Sacral cord

Distal colon
and rectum

Ureters
and
bladder

External
genitalia

Iris m.m.

Ciliary m.m.

Lacrimal
gland

Heart and
great
vessels

Tracheobronchial
tree and lungs

Adrenal gland

Stomach

Pancreas

Spleen

Kidneys

Liver

Small
intestine

Proximal
colon

RBK

4. Postganglionic fibers from the otic ganglion terminate in the parotid gland.
5. Postganglionic fibers from terminal ganglia in the thoracic and abdominal cavities terminate in the heart, lungs, esophagus, stomach, liver, spleen, kidneys, gonads, small intestine, and proximal colon.
6. Postganglionic fibers from terminal ganglia in the pelvic cavity terminate in the bladder, distal colon, rectum, anus, genital ducts, and external genitalia.

Diseases and Disorders of the Autonomic Nervous System

Horner's Syndrome. Horner's syndrome indicates a lesion of the sympathetic nervous system that involves the cervical portion of the sympathetic trunks or upper thoracic region of the spinal cord. It is characterized by bulging of the eyeball, constriction of the pupil, drooping of the upper eyelid, and flushing of the face on the side of the lesion.

Hirschsprung's Disease. Hirschsprung's disease, or megacolon, is a congenital disorder characterized by constriction of the distal portion of the colon and rectum. There is also tremendous compensatory dilation of the proximal colon above the level of constriction. This disease is caused by congenital absence or reduction in number of parasympathetic postganglionic neurons in the constricted region. The resulting chronic contraction of the smooth muscle and loss of peristaltic activity in the affected portion of the colon and rectum are due to unopposed activity of the sympathetic nervous system.

Raynaud's Disease. Raynaud's disease is a disease of the sympathetic nervous system characterized by intermittent attacks of vasoconstriction in the fingers, toes, ears, and/or tip of the nose. These attacks are usually precipitated by a temperature drop or emotional upsets and occur almost exclusively in young women. Raynaud's disease is sometimes treated by sectioning the sympathetic trunks.

Part E. The Special Senses

Smell

Receptors for the sense of smell, or olfaction, consist of modified bipolar neurons called *olfactory cells*. The olfactory cells are located in the olfactory mucosa in the upper part of the nasal cavity (Chapter 11). Each olfactory cell is a flask-shaped cell [Fig. 15.48(a)] whose apical

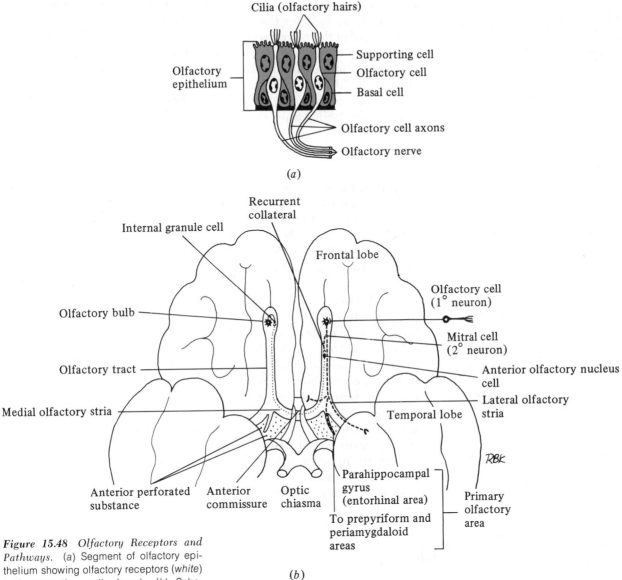

Cilia (olfactory hairs)

Olfactory epithelium

Supporting cell
Olfactory cell
Basal cell

Olfactory cell axons
Olfactory nerve

(a)

Recurrent collateral

Internal granule cell

Frontal lobe

Olfactory cell (1° neuron)

Olfactory bulb

Mitral cell (2° neuron)

Olfactory tract

Anterior olfactory nucleus cell

Medial olfactory stria

Lateral olfactory stria

Temporal lobe

Anterior perforated substance

Anterior commissure

Optic chiasma

Parahippocampal gyrus (entorhinal area)

To prepyriform and periamygdaloid areas

Primary olfactory area

RBK

Figure 15.48 *Olfactory Receptors and Pathways.* (a) Segment of olfactory epithelium showing olfactory receptors (*white*) and supporting cells (*gray*). (b) Schematic diagram of base of brain showing major olfactory structures and fibers, which terminate primarily in prepyriform, periamygdaloid, and entorhinal areas of temporal lobe.

(b)

end bears a cluster of specialized cilia called olfactory hairs. The olfactory hairs are stimulated by odorous substances dissolved in the watery nasal secretions. It is thought that odorous substances combine with receptor sites on the olfactory hairs, which causes the olfactory cell to fire. These hypothetical receptors can distinguish between camphoraceous, musky, floral, peppermint, ethereal, pungent, and putrid odors. An axon emerges from the basal end of the olfactory cell. It joins the axons of neighboring cells to form an olfactory nerve.

About twenty olfactory nerves pass through the cribriform plate of the ethmoid bone to enter the cranial cavity and join the *olfactory bulb* on the base of the corresponding cerebral hemisphere. Within the olfactory bulb the olfactory cells synapse with 2° neurons called mitral cells. Axons of the mitral cells travel to the brain in the *olfactory tract*. Most of the fibers in each olfactory tract terminate in the primary olfactory area of the corresponding cerebral hemisphere, which is located on the medial aspect of the temporal lobe [Fig. 15.48(*b*)].

Taste

Receptors for the sense of taste, or gustation, are located in ovoid bodies called taste buds. Taste buds are embedded in the epithelium of the tongue and epiglottis. Within each taste bud are two types of cells: *supporting cells* and *taste cells*. Each taste cell is a spindle-shaped epithelial cell [Fig. 15.49(*a*)] whose apical end bears a cluster of specialized microvilli called *taste hairs*. The taste hairs are stimulated by taste substances dissolved in saliva. It is thought that the taste substances combine with receptor sites on the microvilli to activate the taste cell. The activated taste cell then causes its neuron to fire. Taste receptors can distinguish between sweet, sour, bitter, and salty tastes. The tip of the tongue is more sensitive to sweet, the sides to sour, the back to bitter, and the center to salt. The basal end of each taste cell is in contact with the axon of a sensory nerve fiber that travels to the brain stem as part of the facial nerve (anterior two-thirds of the tongue), glossopharyngeal nerve (posterior one-third of the tongue), or vagus nerve (epiglottis). These fibers synapse with 2° neurons in taste nuclei in the brain stem. The axons of the 2° neurons cross to the opposite side of the brain stem and ascend to the thalamus, where they synapse with 3° neurons in a relay nucleus. The 3° neurons terminate in the primary taste area, which is located on the parietal lobe [Fig. 15.49(*b*)].

Vision

The eye is the special sense organ for vision. The receptors for vision are located in the neural coat, or tunica nervosa. The tunica nervosa is an outgrowth of the diencephalon and remains connected to the diencephalon by the optic nerve and tract. During the course of development, head mesenchyme condenses around the tunica nervosa to form the outer coats of the eyeball.

EXTRAOCULAR STRUCTURES Extraocular structures include the orbital cavity, the extrinsic eye muscles, the eyelids, and the lacrimal apparatus.

Orbital Cavity The orbital cavity, or orbit, is a bony cone-shaped socket in the skull

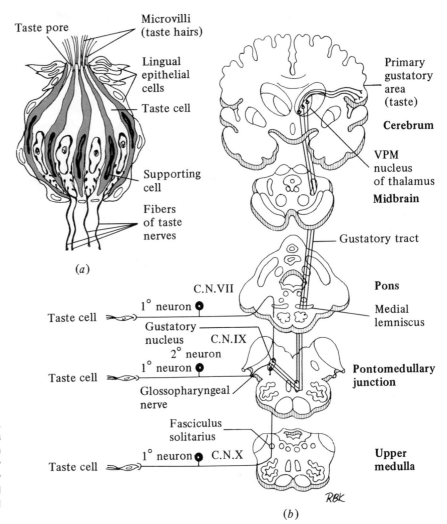

Figure 15.49 Taste Bud and Pathway. (a) Taste bud showing taste receptors (*white*) and supporting cells (*gray*). (b) Schematic diagram showing route taken by taste fibers to reach primary gustatory (taste) cortex, which is located at distal end of primary somatosensory strip in parietal lobe.

(Fig. 15.50) that contains the eyeball, extrinsic eye muscles, blood vessels, lacrimal apparatus, and fat. The eyeball is contained in Tenon's capsule, a protective fascial socket that permits the eyeball to move in the orbit without friction. The fat pad that lies between Tenon's capsule and the orbit acts as a protective cushion and shock absorber for the eyeball.

Extrinsic Eye Muscles   Each eyeball is connected to the orbit by six small extrinsic muscles: the superior, medial, inferior, and lateral rectus muscles plus the superior and inferior obliques. These muscles, described in Chapter 7, permit rotary movements of the eyeball in the orbit. A seventh ex-

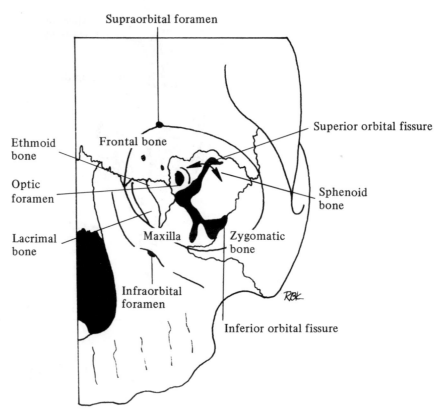

Supraorbital foramen

Superior orbital fissure

Ethmoid bone

Frontal bone

Sphenoid bone

Optic foramen

Lacrimal bone

Maxilla

Zygomatic bone

Infraorbital foramen

Inferior orbital fissure

Figure 15.50 Orbit. Apex contains optic foramen; floor contains orbital parts of maxilla and zygomatic bone; roof contains orbital plate of frontal bone and small wing of sphenoid; lateral wall contains orbital process of zygomatic bone and orbital part of great wing of sphenoid; medial wall contains frontal process of maxilla, lacrimal bone, ethmoid, and body of sphenoid; base above contains supraorbital part of frontal bone; laterally base contains zygomatic bone and zygomatic process of frontal bone; base below contains zygomatic bone and maxilla; medially base contains frontal bone and frontal process of maxilla.

trinsic muscle, the levator palpebrae superioris, inserts into and elevates the upper eyelid.

Eyelids
The palpebrae, or eyelids, form a pair of protective curtains in front of the eyeball. Their free margins contain a number of short stiff hairs, the eyelashes, which help keep particulate matter out of the eyes. Each eyelid is covered with thin skin externally and lined with a mucous membrane called the *conjunctiva* internally. The conjunctiva forms a protective sac for the eyeball. The *tarsus*, a plate of dense connective tissue that gives shape and substance to the eyelid, is located in its free margin. The tarsus contains *meibomian glands*, modified sweat glands whose oily secretion normally prevents the overflow of lacrimal fluid.

Lacrimal Apparatus
The *lacrimal apparatus* (Fig. 15.51) includes the lacrimal gland and its ducts, the lacrimal canals, plus the lacrimal sac and its duct. The almond-shaped lacrimal gland, which is located in the upper outer angle of the orbit, secretes *lacrimal fluid,* a watery substance more commonly known as tears. Tears keep the cornea moist and wash foreign particulate matter out of the conjunctival sac. They are secreted

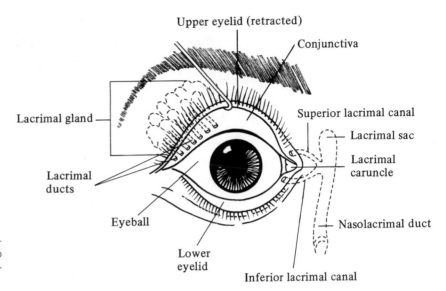

Upper eyelid (retracted)

Conjunctiva

Lacrimal gland

Superior lacrimal canal

Lacrimal sac

Lacrimal caruncle

Lacrimal ducts

Eyeball

Nasolacrimal duct

Lower eyelid

Inferior lacrimal canal

Figure 15.51 Lacrimal Apparatus. Apparatus (*dotted lines*) shown in relation to eye. Nasolacrimal duct terminates in inferior meatus of nasal cavity.

into ducts that open onto the upper eyelid and are drained off by a pair of lacrimal canals that extend from the inner angle of the eye to the lacrimal sac. The lacrimal sac is located in the lacrimal bone. It is connected to the corresponding nasal fossa by the nasolacrimal duct. When the nasolacrimal duct is blocked by a cold or when excess lacrimal fluid is secreted, tears overflow the conjunctival sac and stream down the cheeks.

THE EYEBALL
Coats of the Eyeball

The wall of the eyeball (Fig. 15.52) consists of three coats: the *tunica fibrosa* (fibrous coat), the *tunica vasculosa* (vascular coat), and the *tunica nervosa* (nervous coat).

Tunica Fibrosa

The tunica fibrosa is the outer coat of the eyeball. It corresponds to and is continuous with the dura mater. The tunica fibrosa consists of two unequal portions: the *sclera* and the *cornea*.

The sclera, or white of the eye (opaque portion), covers the posterior five-sixths of the eyeball. It consists of fibrous connective tissue that forms a tough protective covering for the eyeball.

The cornea (transparent portion) covers the anterior one-sixth of the eyeball. This structure, which permits light rays to enter the eyeball, consists of a layer of stratified squamous nonkeratinized epithelium, a thick layer (stroma) of collagen fibers, and a single layer of endothelial cells. Transparency of the cornea is due, in part, to the regular arrangement of collagen fibers in the stroma. Injuries to or infections of the cornea can cause corneal opacity and blindness. The limbus (zone between the cornea and sclera), contains a circular venous channel called *Schlemm's canal*. Schlemm's canal drains aqueous

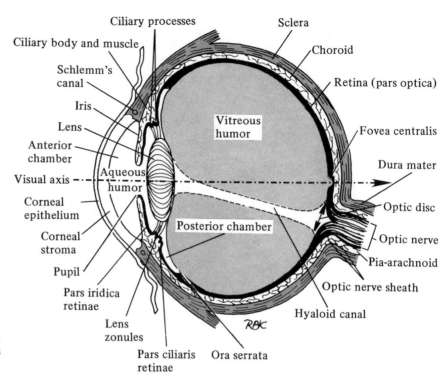

Figure 15.52 *Eyeball.* Sagittal section through human eyeball shows various coats and refractive media.

humor, a fluid similar to cerebrospinal fluid, from the interior of the eyeball. *Glaucoma,* a serious eye disease that can cause blindness, results from blockage of Schlemm's canal and buildup of aqueous humor in the eyeball.

Tunica Vasculosa   The tunica vasculosa, or *uvea,* is the middle coat of the eyeball. It corresponds to and is continuous with the arachnoid and pia mater. The uvea consists of loose, highly vascular, pigmented connective tissue. It is divided into the *choroid, ciliary body,* and *iris.* The choroid is the thin vascular membrane that lies beneath the sclera. The ciliary body is a thickened ring of vascular connective tissue that lies between the choroid and iris. It gives rise to the ciliary muscle and ciliary processes. The ciliary muscle regulates the shape of the lens, which is suspended from the ciliary body. The ciliary processes are fingerlike projections from the ciliary body that secrete aqueous humor into the eyeball. The iris is a thin pigmented diaphragm located between the lens and the cornea. Through its central hole, or *pupil,* the iris controls the amount of light admitted into the interior of the eyeball. Within the connective tissue stroma of the iris are two smooth muscle masses that regulate the diameter of the pupil. The sphincter pupillae causes *miosis* (constriction of the pupil), while the dilator pupillae causes *mydriasis* (dilation of the pupil).

Tunica Nervosa

The tunica nervosa, or *retina* (Fig. 15.53), is the inner coat of the eyeball. The retina consists of an outer pigmented layer and an inner neural layer. The pars optica (functional portion) of the retina lines the choroid, while the pars caecum (nonfunctional portion) lines the ciliary body and iris.

The neural layer of the pars optica contains the first three neurons in the visual pathway. The 1° neurons in the visual pathway are photoreceptors called *rods* and *cones*. The rods are dark-light receptors that function best at night and in dim light, while the cones are color receptors that function best during the day. There are three different

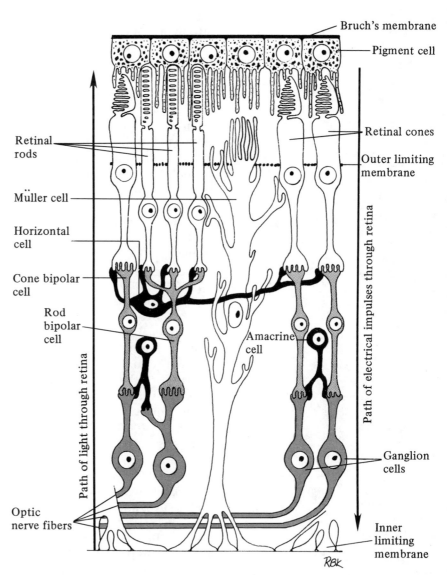

Figure 15.53 *Retina*. Note that retina is inverted, that is, light must pass through bulk of retina before reaching photoreceptors.

Bruch's membrane

Pigment cell

Retinal cones

Outer limiting membrane

Path of electrical impulses through retina

Ganglion cells

Inner limiting membrane

Retinal rods

Müller cell

Horizontal cell

Cone bipolar cell

Rod bipolar cell

Amacrine cell

Path of light through retina

Optic nerve fibers

RBK

kinds of cones: cones that respond best to red light rays, cones that respond best to blue light rays, and cones that respond best to green light rays. The colors that we see result from a mixture of impulses from all three cone types. The photoreceptors synapse with 2° neurons called *bipolar cells,* which in turn synapse with 3° neurons called *ganglion cells.* The axons of the ganglion cells leave the eyeball by way of the optic nerve. A serious eye disorder is *retinal detachment.* In this condition the neural retina separates from the protective pigment cell layer and the photoreceptors degenerate.

Not all regions of the retina are alike. Near the center of the neural retina is a small yellow spot called the *macula lutea.* There is a depression in the macula, the *fovea centralis,* that contains only cones and their fibers. Because of its high concentration of cones and low tissue density, the fovea is the center of visual acuity. As the distance from the fovea increases, the number of cones decreases and the number of rods increases. The *optic disc,* the retinal aspect of the optic nerve, is located a short distance from the nasal side of the fovea. It is the point where the optic nerve fibers leave the retina. There are no photoreceptors in the optic disc, which is why the optic disc is the blind spot of the retina. If intraocular pressure is high, as in glaucoma, the optic disc appears to be pushed backwards (cupped disc); if intracranial pressure is high, as in brain tumors, the optic disc appears to be pushed forward (choked disc).

Refracting Media
In addition to the cornea, the eyeball contains three other substances or structures that bend entering light rays and help focus them on the retina.

Aqueous Humor. Aqueous humor is a clear watery fluid that fills the anterior and posterior chambers of the eyeball. It is secreted by the ciliary processes into the posterior chamber of the eye, passes into the anterior chamber through the pupil, and is reabsorbed into the venous system through the canal of Schlemm. The rate of reabsorption is normally equal to the rate of secretion, which helps maintain constant pressure in the eyeball. Aqueous humor provides nutrients for and removes wastes from ocular structures that do not have their own blood supply.

The Lens. The lens is a transparent ovoid body located immediately behind the iris in the posterior chamber of the eye. It consists of epithelial cells filled with crystalline proteins that are responsible for the transparency of the lens. The lens is surrounded by an elastic capsule and is suspended from the ciliary body by tiny ligaments. Changes in tension exerted on the lens by the ciliary muscle through the suspensory ligaments alter its shape in order to keep objects in continuous focus on the retina. This is called *accommodation.* For near vision the lens rounds up, while for distant vision the lens flattens out. An opacity of the lens is called a *cataract*. Cataracts cause a gradual loss of vision, which is best treated by removal of the diseased lens. Following cataract surgery, corrective eyeglasses must be worn to compensate for the loss of the lens.

Vitreous Humor. The vitreous humor is a transparent gelatinous body that occupies the cavity in the eyeball between the lens and the retina. It helps maintain the shape of the eyeball and supports the retina.

BLOOD SUPPLY OF THE EYEBALL The retina receives blood from the central artery of the retina, a branch of the ophthalmic artery, and is drained by the central vein. The remainder of the eyeball receives blood from the anterior and posterior ciliary arteries and is drained by the superior and inferior ophthalmic veins. Examination of the central artery and its retinal branches provides a doctor with much valuable information about the cardiovascular system, for pathological changes in the system as a whole are reflected in changes in the retinal vessels.

NEURAL PATHWAY FOR VISION The 1°, 2°, and 3° neurons in the visual pathway are located in the retina. The optic nerve fibers (axons of the ganglion cells) leave the retina at the optic disc and travel in the optic nerve and tract. The two optic nerves meet at the base of the brain to form the optic chiasma, separate as the optic tracts, and enter the brain. Optic nerve fibers from the nasal half of each retina, which transmit images from the temporal half of the visual field, cross in the optic chiasma and enter the opposite optic tract. Fibers from the temporal half of each retina, which transmit images from the nasal half of the visual field, remain uncrossed and enter the optic tract on the side of origin (Fig. 15.54). Nerve fibers in each optic tract enter the brain stem and synapse with 4° neurons in the corresponding lateral geniculate nucleus, a sensory relay nucleus in the thalamus. The 4° neurons ascend to the occipital lobe of the corresponding cerebral hemisphere, where they terminate in the primary visual cortex.

Injury to an optic nerve anterior to the chiasma causes total loss of vision in the corresponding eye because fibers transmitting images from both halves of the visual field are damaged. Injury to the optic chiasma, such as by a pituitary tumor, causes *tunnel vision* (loss of the temporal visual fields) because the crossing fibers that transmit images from the temporal visual fields are damaged.

Equilibrium and Hearing

The ear is the special sense organ for equilibrium and hearing. Each ear consists of three anatomically distinct portions: the *external ear,* the *middle ear,* and the *inner ear.*

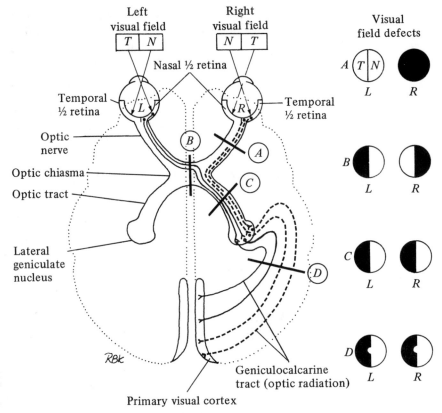

Figure 15.54 Visual Pathways and Defects Caused by Lesions. Schematic diagram shows effects of lesions at various points in right visual pathway. *A,* right optic nerve lesion—blindness of right eye; *B,* midline optic chiasma lesion—bitemporal hemianopia (blindness in temporal halves of visual fields); *C,* right optic tract lesion —left homonymous hemianopsia (blindness in left halves of visual fields); *D,* right geniculocalcarine tract lesion—left homonymous hemianopsia with sparing of central vision (for reasons as yet unknown).

EXTERNAL EAR The external ear (Fig. 15.55) is composed of the *pinna* (earflap), *external auditory meatus* (ear canal), and *tympanic membrane* (eardrum).

Pinna The pinna is a flaplike structure consisting of an elastic cartilage framework covered with skin. It functions as a trap to collect sound waves and funnel them through the ear canal to the eardrum. The pinna is less important in humans than in some of the lower mammals. To appreciate how these structures are used as sound traps, watch a dog move its pinnas when someone calls its name.

External Auditory Meatus The external auditory meatus is an S-shaped canal in the temporal bone that extends from the pinna to the eardrum, a distance of about $1\frac{1}{2}$ inches. The thin skin lining this canal contains numerous ceruminous glands whose waxy secretion lubricates and protects the canal. Overproduction of earwax can block the canal and impair hearing.

Tympanic Membrane The tympanic membrane (eardrum) is a delicate membrane that separates the ear canal from the middle ear. It has a central core of

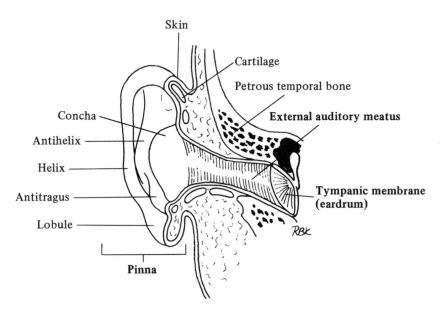

Skin

Cartilage

Petrous temporal bone

External auditory meatus

Concha

Antihelix

Helix

Antitragus

Lobule

Tympanic membrane (eardrum)

RBK

Pinna

Figure 15.55 *External Ear.* Frontal section.

connective tissue covered externally by very thin skin and internally by mucous membrane. Sound waves that enter the ear canal cause the eardrum to vibrate, which, in turn, moves the ossicles in the middle ear.

MIDDLE EAR The middle ear or tympanic cavity (Fig. 15.56) is an air-filled, mucous-membrane-lined cavity deep in the temporal bone. It communicates with the nasopharynx by way of the auditory tube. The auditory tube serves to equalize air pressure on both sides of the eardrum. If the air pressure is not equalized, the eardrum is unable to vibrate in response to sound waves and may even rupture.

Within the tympanic cavity are three small bones, or ossicles. The ossicles include the *malleus* (hammer), *incus* (anvil), and *stapes* (stirrup). The malleus is attached to the eardrum and articulates with the incus. The incus, in turn, articulates with the stapes. The footplate of the stapes is in contact with the oval window, a membrane-covered structure between the inner ear and middle ear. The chain of ossicles transmits vibrations received from the eardrum to the inner ear, where they are translated into fluid waves.

Protection against loud noises is provided by two tiny muscles in the middle ear. The *tensor tympani,* which inserts into the eardrum, tightens this structure to reduce its vibrations. The *stapedius,* which inserts into the stapes, pulls the stapes away from the oval window to reduce the intensity of the fluid waves.

INNER EAR The inner ear (Fig. 15.57) consists of a system of canals and cavities in the temporal bone known collectively as the *bony labyrinth*. The

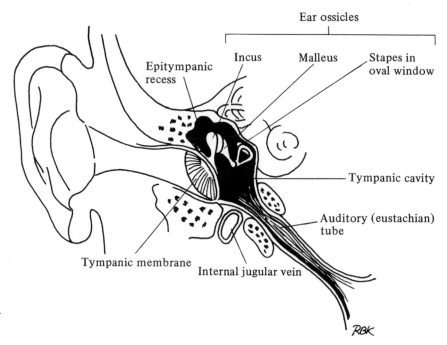

Ear ossicles

Epitympanic recess

Incus

Malleus

Stapes in oval window

Tympanic cavity

Auditory (eustachian) tube

Tympanic membrane

Internal jugular vein

Figure 15.56 Middle Ear. Frontal section.

three major divisions of the bony labyrinth are the *cochlea* (snail shell), *vestibule*, and *semicircular canals*. Several epithelial sacs and ducts, known collectively as the *membranous labyrinth*, are suspended in the fluid of the bony labyrinth. The four major divisions of the membranous labyrinth are the *cochlear duct*, which is suspended in the cochlea, the *utricle* and *saccule*, which are suspended in the vestibule, and the *semicircular ducts*, which are suspended in the semicircular canals. The spaces between the bony and membranous labyrinths are filled with a watery fluid called *perilymph*. The membranous labyrinth, in turn, is filled with a watery fluid called *endolymph*. The sensory receptors for equilibrium and hearing are located in the membranous labyrinth.

THE VESTIBULAR SYSTEM   The vestibular system is concerned with equilibrium (position sense). Its receptors detect both the position of the head and body in relation to the pull of gravity and sudden changes in movement of the body. Vestibular information is used for two purposes: to keep the eyes fixed on a given point despite changes in the position of the head and to maintain the body in an upright position.

The sensory receptors that detect static body balance and linear body movements are located in the utricle and saccule. These raised patches of sensory epithelium are known as the *macula utriculi* and *macula sacculi* [Fig. 15.58(*a*)]. The macula utriculi is oriented in the

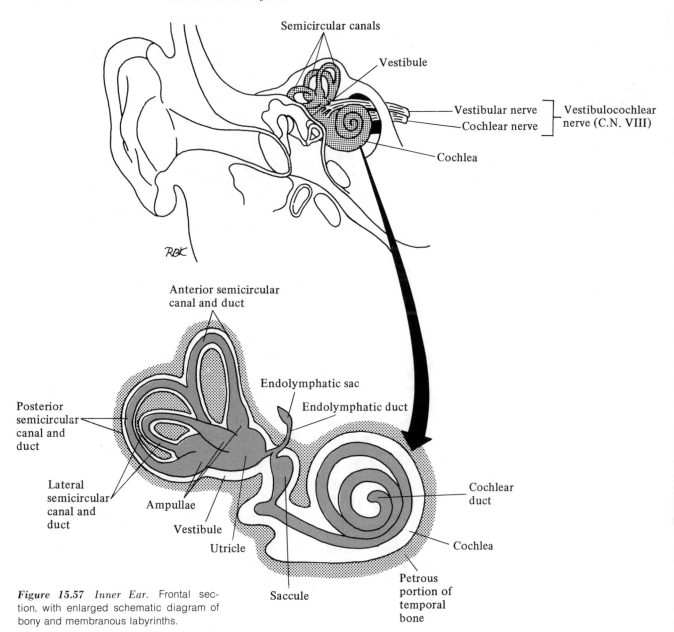

Figure 15.57 Inner Ear. Frontal section, with enlarged schematic diagram of bony and membranous labyrinths.

horizontal plane of the head, while the macula sacculi is oriented in the vertical plane of the head. Each macula [Fig. 15.58(*b*)] contains equilibrium receptors called *hair cells*. The microvilli (hairs) of the hair cells are embedded in a gelatinous membrane called the *otolithic membrane*. The otolithic membrane contains tiny concretions of pro-

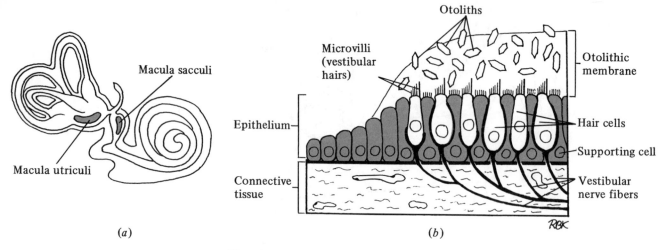

(a)

(b)

Figure 15.58 *Maculae.* (*a*) Gross anatomy. Locations of macula sacculi and macula utriculi. Note that maculae are oriented at right angles to one another. (*b*) Microscopic anatomy. Segment shows static equilibrium receptors *(white)*, supporting cells *(gray)*, and otolithic membrane.

tein and calcium carbonate called *otoliths* (ear stones). As the otoliths respond to the pull of gravity, they either pull or press on the hairs, which stimulates the hair cells. The stimulated hair cells then initiate impulses in afferent vestibular fibers that tell the brain about the position of the head and body in space.

The sensory receptors that detect dynamic body balance and angular movement are located in the semicircular ducts. The three semicircular ducts are located at right angles to one another in the transverse, frontal, and sagittal planes of the head. Each semicircular duct has a dilation at one end, called the ampulla, that contains a crista [Fig. 15.59(*a*)]. These raised ridges of sensory epithelium are known as *cristae ampullares*. The crista is similar in composition and appearance to a macula. The microvilli of its hair cells are embedded in a gelantinous membrane, called the *cupula* (dome), which projects into the ampula [Fig. 15.59(*b*)]. The cupula acts like a valve flap. Whenever movements of the body stop, start, accelerate, decelerate, or change direction, the endolymph moving through the semicircular canal deflects the cupula. As the cupula is deflected, the hair cells in the crista are stimulated. Stimulation is maximal when movement of endolymph is in the plane of the duct. The stimulated hair cells then initiate impulses in afferent vestibular fibers that tell the brain how fast and in what direction the body is moving.

NEURAL PATHWAYS FOR EQUILIBRIUM

The 1° neuron in the vestibular pathway is a bipolar neuron whose dendrite is in contact with a hair cell and whose axon enters the brain stem to synapse with a 2° neuron in one of the vestibular nuclei. The

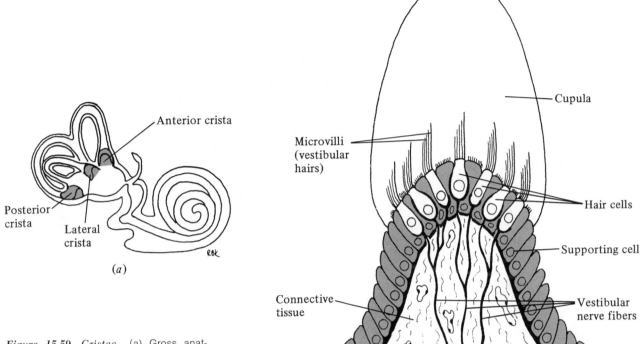

Figure 15.59 Cristae. (a) Gross anatomy. (b) Microscopic anatomy. Crista ampullaris showing dynamic equilibrium receptors (*white*), supporting cells (*gray*), and cupula.

2° vestibular neurons either (a) ascend the brain stem to terminate in the motor nuclei of cranial nerves innervating extrinsic eye muscles, (b) descend the spinal cord to terminate on lower motor neurons innervating neck muscles, or (c) project to the cerebellum (Fig. 15.60). The vestibular system thus coordinates movements of the eyes and head as well as postural muscles of the body. An unpleasant and poorly understood disease of the vestibular system is *Meniere's syndrome.* The person with Meniere's syndrome suffers from an irritation of the vestibular system that causes tinnitus (ringing in the ears), vertigo (dizziness), nausea, and vomiting.

THE AUDITORY SYSTEM   The auditory system is concerned with hearing. Its receptors detect sounds of different pitch and intensity.

The sole component of the auditory system is the *cochlear duct,* an epithelial tube that is wrapped around the *modiolus* (the bony core of the cochlea). The cochlear duct is triangular in shape: its roof is formed by the *vestibular membrane,* its lateral wall is formed by the *stria vascularis,* and its floor is formed by the *basilar membrane.* The

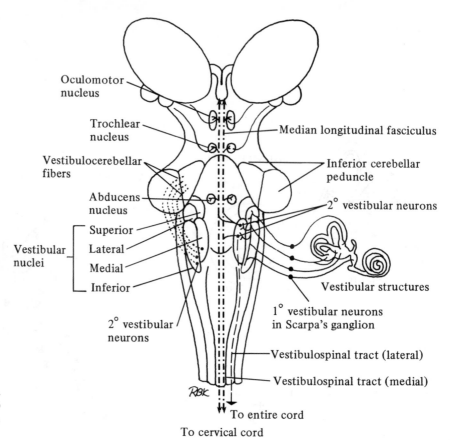

Figure 15.60 *Pathways for Equilibrium.*
Pathways taken by 2° vestibular fibers to
cerebellum, nuclei of extrinsic eye muscles,
and spinal cord.

Labels on figure:

Oculomotor nucleus

Trochlear nucleus

Vestibulocerebellar fibers

Abducens nucleus

Vestibular nuclei — Superior, Lateral, Medial, Inferior

2° vestibular neurons

Median longitudinal fasciculus

Inferior cerebellar peduncle

2° vestibular neurons

Vestibular structures

1° vestibular neurons in Scarpa's ganglion

Vestibulospinal tract (lateral)

Vestibulospinal tract (medial)

To entire cord

To cervical cord

cochlear duct stretches across the bony canal of the cochlea, dividing
it into two parts. The part above the cochlear duct is called the *scala
vestibuli*, while the part below is called the *scala tympani*. The
cochlear duct is also called the *scala media*. The scala vestibuli and
scala tympani contain perilymph, while the scala media contains
endolymph [Fig. 15-61(*a*)].

Sitting on the basilar membrane is a complex sensory receptor
called the *organ of Corti*. The spiral organ of Corti consists of a ribbon
of hair cells and supporting cells that run the entire length of the
cochlear duct. There are three rows of outer hair cells and one row of
inner hair cells separated by a tiny tunnel. The microvilli of the hair
cells are embedded in a gelatinous membrane called the *tectorial
membrane* [Fig. 15.61(*b*)]. Vibrations transmitted from the eardrum
are converted to pressure waves in the perilymph at the oval window,
a membrane-covered structure in contact with the scala vestibuli. As
the pressure waves travel up the scala vestibuli to the apex of the
cochlea, they cause the basilar membrane to vibrate. High-pitched

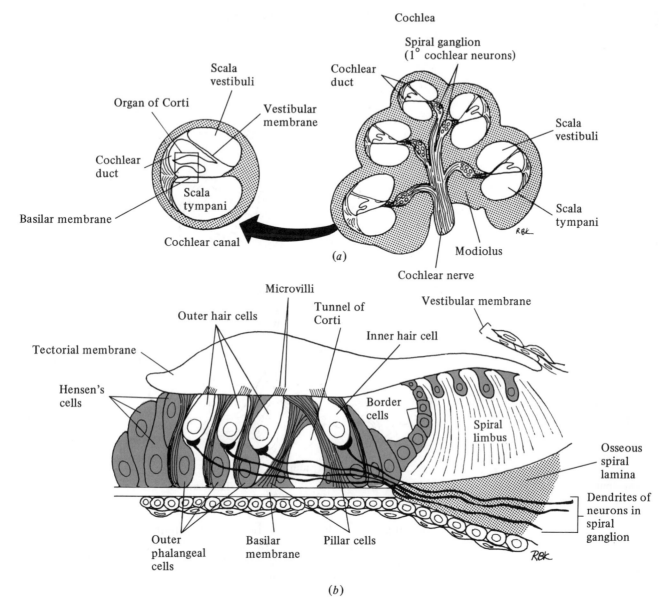

Cochlea

Spiral ganglion (1° cochlear neurons)

Cochlear duct

Scala vestibuli

Organ of Corti

Vestibular membrane

Scala vestibuli

Cochlear duct

Scala tympani

Basilar membrane

Scala tympani

Cochlear canal

Modiolus

Cochlear nerve

(a)

Microvilli

Tunnel of Corti

Outer hair cells

Vestibular membrane

Inner hair cell

Tectorial membrane

Border cells

Hensen's cells

Spiral limbus

Osseous spiral lamina

Dendrites of neurons in spiral ganglion

Outer phalangeal cells

Basilar membrane

Pillar cells

(b)

Figure 15.61 *Cochlea.* (a) Gross anatomy. *Inset* of enlarged cross section of cochlear canal shows location of organ of Corti in cochlear duct.

Organ of Corti. (b) Microscopic anatomy. Segment shows auditory receptors (*white*) and supporting cells (*gray*).

sounds cause the basilar portion to vibrate maximally, while low-pitched sounds cause the apical portion to vibrate maximally. When the flexible basilar membrane vibrates and is displaced into the scala tympani, the rigid tectorial membrane does not move; this causes a

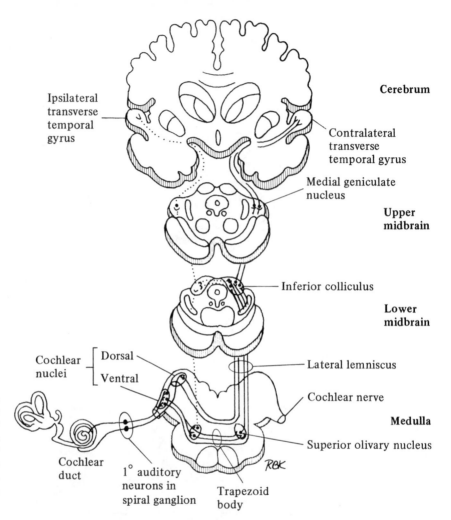

Cerebrum

Ipsilateral transverse temporal gyrus

Contralateral transverse temporal gyrus

Medial geniculate nucleus

Upper midbrain

Inferior colliculus

Lower midbrain

Cochlear nuclei — [Dorsal
Ventral]

Lateral lemniscus

Cochlear nerve

Medulla

Superior olivary nucleus

Cochlear duct

1° auditory neurons in spiral ganglion

Trapezoid body

Figure 15.62 Auditory Pathways. Schematic diagram showing the various pathways taken by 2°, 3°, 4° and 5° auditory fibers. Note that most 2° auditory fibers cross to the contralateral (opposite) side of the brain stem, but some remain on the ipsilateral (same) side. Thus there is a bilateral cortical projection from each organ of Corti.

shearing motion to develop between the hair cells and the tectorial membrane, which deflects the hairs and stimulates the hair cells. The stimulated hair cells initiate impulses in afferent auditory fibers that transmit sounds of different pitch and intensity to the brain.

NEURAL PATHWAYS FOR HEARING

The 1° neuron in the auditory pathway is a bipolar neuron whose dendrite is in contact with a hair cell and whose axon enters the brain stem to synapse with a 2° neuron in one of the cochlear nuclei. Many 2° neurons cross to the opposite side of the brain stem and synapse with 3° neurons in the superior olivary complex. The 3° neurons ascend to the inferior colliculus, where some may synapse with 4° neurons. The 3° and 4° neurons then ascend to the medial geniculate nucleus of the thalamus, a sensory relay nucleus where they synapse with 4° or 5° neurons. These neurons in turn ascend to the corresponding cerebral hemisphere, where they terminate in the primary auditory cortex (Fig. 15.62). The uncrossed neurons in the auditory pathway take a similar route but on the side of entry.

Injury to or disease of the auditory system causes either *conduction, nerve,* or *central deafness.* Conduction deafness is caused by a ruptured eardrum or improper function of the ossicles. Nerve deafness follows degeneration of hair cells in the organ of Corti or damage to the cochlear nerve. Central deafness is caused by damage to central auditory pathways or to the primary auditory cortex.

Suggested Readings

AMOORE, JOHN E., JAMES W. JOHNSTON, and MARTIN RUBIN. "The stereochemical theory of odor," *Sci. Am., 210,* no. 2 (February 1964), pp. 42–49.

AXELROD, JULIUS. "Neurotransmitters," *Sci. Am., 230,* no. 6 (June 1974), pp. 58–71.

BENZINGER, T. H. "The human thermostat," *Sci. Am., 204,* no. 1 (January 1961), pp. 134–147.

BOTELHO, STELLA Y. "Tears and the lacrimal gland," *Sci. Am., 211,* no. 4 (October 1964), pp. 78–86.

COTZIAS, GEORGE C. "L-Dopa in Parkinson's disease," *Hosp. Pract., 4,* no. 9 (September 1969), pp. 35–41.

DeARMOND, STEPHEN J., MADELINE M. FUSCO, and MAYNARD M. DEWEY. *Structure of the Human Brain: A Photographic Atlas.* New York: Oxford University Press, 1976.

ECCLES, SIR JOHN. "The synapse," *Sci. Am., 212,* no. 1 (January 1965), pp. 56–66.

FENDER, DEREK H. "Control mechanisms of the human eye," *Sci. Am., 211,* no. 1 (July 1964), pp. 24–33.

FRENCH, J. D. "The reticular formation," *Sci. Am., 196,* no. 5 (May 1957), pp. 54–60.

GESCHWIND, NORMAN. "Language and the brain," *Sci. Am. 226,* no. 4 (April 1972), pp. 76–83.

GORDON, BARBARA. "The superior colliculus of the brain," *Sci. Am., 227,* no. 6 (December 1972), pp. 72–82.

HEIMER, LENNART. "Pathways in the brain," *Sci. Am., 225,* no. 1 (July 1971), pp. 48–60.

HUBEL, DAVID H. "The visual cortex of the brain," *Sci. Am., 209,* no. 5 (November 1963), pp. 54–62.

KIMURA, DOREEN. "The asymmetry of the human brain," *Sci. Am., 228,* no. 3 (March 1972), pp. 70–78.

LERIMAN, SIDNEY. "Cataracts," *Sci. Am., 206,* no. 3 (March 1962), pp. 106–114.

LLINÁS, RODOLFO R. "The cortex of the cerebellum," *Sci. Am., 232,* no. 1 (January 1975), pp. 56–71.

LOEWENSTEIN, WERNER R. "Biological transducers," *Sci. Am., 203,* no. 2 (August 1960), pp. 98–108.

LURIA, A. R. "The functional organization of the brain," *Sci. Am.*, *222*, no. 1 (March 1970), pp. 66–78.

MacNICHOL, EDWARD F., JR. "Three-pigment color vision," *Sci. Am.*, *211*, no. 6 (December 1964), pp. 48–56.

MICHAEL, CHARLES R. "Retinal processing of visual images," *Sci. Am.*, *220*, no. 5 (May 1969), pp. 104–114.

NETTER, FRANK H. *The Ciba Collection of Medical Illustrations.* Vol. 1: *The Nervous System.* Summit, N.J.: Ciba, 1962.

° PETERS, ALAN, SANFORD L. PALAY, and HENRY DEF. WEBSTER. *The Fine Structure of the Nervous System: The Neurons and Supporting Cells.* Philadelphia: Saunders, 1976.

ROSENZWEIG, MARK R. "Auditory localization," *Sci. Am.*, *205*, no. 4 (October 1961), pp. 132–142.

RUSHTON, W. A. H. "Visual pigments and color blindness," *Sci. Am.*, *232*, no. 3 (March 1975), pp. 64–74.

——— . "Visual pigments in man," *Sci. Am.*, *207*, no. 5 (November 1962), pp. 120–132.

° SHEPARD, GORDEN M. *The Synaptic Organization of the Brain: An Introduction.* New York: Oxford University Press, 1974 (paperback).

Sperry, R. W. "The great cerebral commissure," *Sci. Am. 210*, no. 1 (January 1964), pp. 42–52.

° STEEGMANN, A. T., and H. H. WHITE. *Examination of the Nervous System.* Chicago: Year Book, 1970 (paperback).

VAN HEYNINGEN, RUTH. "What happens to the human lens in cataract," *Sci. Am.*, *233*, no. 6 (December 1975), pp. 70–81.

WERBLIN, FRANK S. "The control of sensitivity in the retina," *Sci. Am.*, *228*, no. 1 (January 1973), pp. 70–79.

WHITTAKER, VICTOR P. "Membranes in synaptic function," *Hosp. Pract.*, *9*, no. 4 (April 1974), pp. 111–119.

WILSON, VICTOR J. "Inhibition in the central nervous system," *Sci. Am.*, *214*, no. 5 (May 1966), pp. 102–110.

YOUNG, RICHARD W. "Visual cells," *Sci. Am.*, *223*, no. 4 (October 1970), pp. 80–91.

° Advanced level.

Glossary

abdomen (ab-DO-men). The portion of the body located between the thorax and the pelvis; belly.

abduct (ab-DUKT). To move away from the midline of the body.

abortion (ah-BOR-shun). Termination of pregnancy before the product of conception is able to survive outside the uterus.

acetabulum (as′e-TAB-u-lum). Cup-shaped cavity in the hipbone that receives the head of the femur.

acinus (AS-ĭ-nus). Grapelike structure; flask-shaped glandular secretory unit.

acromegaly (ak′ro-MEG-ah-lē). Abnormal enlargement of the bones of the face, hands, and feet caused by overproduction of growth hormone in an adult.

acromion (ah-KRO-mē-on). Tip of the scapular spine.

active transport (AK-tiv TRANS-port). Movement of a molecule or ion across the cell membrane against a chemical or electrical gradient. The movement involves the expenditure of energy by the cell.

adduct (ah-DUKT). To move towards the midline of the body.

adenine (AD-ē-nīn). A purine present in nucleic acids.

adenohypophysis (ad′ĕ-nō-hī-POF-ĭ-sis). Glandular part of the hypophysis (pituitary).

adenoid (AD-e-noid). Glandlike; referring to the nasopharyngeal tonsil.

adenosine triphosphate (ah-DEN-o-sēn trī-FOS-fāte). A nucleotide containing adenine, a five-carbon sugar, and three phosphoric acids; an important energy source for cellular activities.

adipose (AD-ĭ-pōs). Fatty.

adrenal (ah-DRE-nal). Near the kidney; a small endocrine gland that rests on top of the kidney.

adrenalin (ah-DREN-ah-lin). Trade name for epinephrine.

adrenergic (ad′ren-ER-jik). Activated or transmitted by epinephrine or norepinephrine; pertaining to nerve fibers that release norepinephrine at their synapses.

adventitia (ad′ven-TISH-yah). Fibrous outer coat of a hollow muscular organ, tube, or vessel.

afferent (AF-er-ent). Conducting toward the center of the body.

afferent nerve (AF-er-ent nerv). A nerve that conducts electrical impulses from the periphery of the body toward the central nervous system.

aldosterone (al-DOS-ter-ōn). The principal mineral-regulating hormone of the adrenal cortex.

alveolus (al-VĒ-o-lus). A small sac or cavity; one of the air sacs of the lung.

amenorrhea (a-men′o-RĒ-ah). Absence or abnormal stoppage of the menstrual periods.

amino acid (ah-MĒ-nō AS-id). An organic compound containing an NH_3 and a COOH group in its molecular structure. Amino acids are the structural units of proteins.

amnion (AM-nē-on). The protective fluid-filled sac formed by the inner fetal membrane.

amoeboid (ah-MĒ-boid). Resembling an amoeba.

ampulla (am-PUL-ah). A flask-shaped dilation of a tube or duct.

anastomosis (ah-nas′to-MŌ-sis). A natural or surgical connection between two tubular structures, such as blood vessels.

androgen (AN-drō-jen). Any substance that stimulates development of male secondary sex characteristics, especially the hormone testosterone.

anemia (ah-NĒ-mē-ah). A blood condition in which hemoglobin or red blood cells are deficient.

aneurysm (AN-u-rizm). A saclike dilation of the wall of an artery or vein.

angstrom (ANG-strom). A unit of length equivalent to 10^{-7} of a millimeter; abbreviated Å.

antagonist (an-TAG-ō-nist). An agent, such as a muscle, that exerts an action opposite to that of another.

anterior (an-TĒR-ē-or). Toward the front part of the body or body structure.

antibody (AN-ti-bod′ē). A type of protein produced by immunologically competent cells in response to contact with an antigen.

antigen (AN-tĭ-jen). Any substance not recognized as a self-component that stimulates immunologically competent cells to produce a specific antibody.

antrum (AN-trum). A cavity that is almost completely closed, especially one within a bone.

anus (Ā-nus). Distal opening of the digestive tract.

aorta (ā-OR-tah). The great arterial vessel arising from the left ventricle of the heart.

aperture (AP-er-chur). An opening.

apex (Ā-peks). The top or pointed end of a cone-shaped structure.

aphasia (ā-FA-ze-ah). Loss of ability to

understand or transmit language by reading, writing, or speaking.

aponeurosis (ap′o-nū-RŌ-sis). A flat tendon.

appendage (ah-PEN-dij). A less important part attached to a main body structure.

aqueduct (AK-wĕ-dukt′). A canal within a body structure that conducts a fluid.

arachnoid (ah-RAK-noid). Like a spider's web; specifically, the membrane located between the outermost and innermost of the membranes that cover the brain and spinal cord.

areola (ah-RĒ-ō-lah). A small area; the pigmented ring of skin surrounding the nipple of the mammary gland.

arteriole (ar-TĒ-rē-ōl). A minute arterial branch with a muscular wall.

arteriosclerosis (ar-tē′rē-ō-sklĕ-RŌ-sis). Hardening of the arteries.

artery (AR-ter-ē). A vessel that conducts blood away from the heart.

arthritis (ar-THRĪ-tis). Inflammation of one or more joints.

articular (ar-TIK-ū-lar). Pertaining to a joint.

ascites (ah-SĪ-tēz). Accumulation of an abnormal amount of fluid in the abdominal cavity.

aspirate (AS-pī-rāte). To withdraw fluids by suction; to draw pathogenic material into the respiratory tract.

asthma (AZ-mah). A bronchial disease characterized by shortness of breath, wheezing, and coughing.

astrocyte (AS-trō-sīt). A star-shaped neuroglial cell that provides physical and metabolic support for neurons in the central nervous system.

ataxia (ah-TAK-sē-ah). Lack or loss of muscular coordination.

ATP. (See *adenosine triphosphate.*)

atrium (Ā-trē-um). A chamber; one of the upper chambers of the heart.

atrophy (AT-rō-fē). Decrease in size of a normally developed cell, tissue, organ, or body part.

auricle (AW-rĭ-kl). The flap of the ear; the flap-shaped appendage of either atrium of the heart.

autograft (AW-tō-graft). A graft transferred from one part of a person's body to another.

autonomic (aw′to-NŌM-ik). Not under voluntary control.

axial (AK-sē-al). Pertaining to the axis of a structure or body part.

axilla (ak-SIL-ah). The armpit.

axon (AK-son). Process that conducts electrical impulses away from the nerve cell body.

basal (BĀ-sal). Pertaining to a base.

base (bās). The lower part of an object; the widest part of a pyramidal or cone-shaped structure.

basophil (BĀ-sō-fil). Any substance that stains with basic dyes; a white blood cell whose granules stain deep blue with basic dyes.

benign (bē-NĪN). Of a mild nature; a tumor that is nonmalignant.

biceps (BĪ-seps). A muscle having two heads.

bicuspid (bī-KUS-pid). Having two cusps; the left atrioventricular valve of the heart; a premolar tooth.

bifurcate (bī-FUR-kāte). To divide into two branches.

bilateral (bī-LAT-er-al). Having two sides; pertaining to both sides of the body.

bilirubin (bil-ĭ-ROO-bin). A reddish orange bile pigment produced from the breakdown of hemoglobin.

biopsy (BĪ-op-sē). Examination of tissue removed from the living body.

blast (blast). An immature form of a cell.

blastocyst (BLAS-tō-sist). Thin-walled cystic structure produced by cleavage of a fertilized ovum in the course of mammalian embryonic development.

blood (blud). The fluid that circulates through the cardiovascular system.

bolus (BŌ-lus). A rounded mass; a mass of food entering the esophagus at one swallow.

Bowman's capsule (BŌ-manz KAP-sul). The cup-shaped blind end of a kidney tubule that surrounds a tuft of capillaries.

brachial (BRĀ-kē-al). Pertaining to the brachium or arm.

breast (brest). Mammary gland; a female secondary sex organ that produces milk.

bronchial (BRONG-kē-al). Pertaining to the bronchi.

bronchiole (BRONG-kē-ōl). Minute branch of a bronchus.

bronchitis (brong-KĪ-tis). Inflammation of the bronchi.

bronchus (BRONG-kus). One of a series of tubes that conduct air from the trachea into the substance of the lungs.

buccal (BUK-al). Pertaining to the inner surface of the cheeks.

bursa (BUR-sah). Purselike sac; a small fluid-filled sac that permits body structures to move over one another without friction.

bursitis (bur-SĪ-tis). Inflammation of a bursa.

buttock (BUT-ok). One of two fleshy masses formed by the gluteal muscles.

cadaver (kah-DAV-er). A dead human body.

calculus (KAL-kū-lus). A stone formed in a hollow muscular organ or tube.

canal (kah-NAL). A narrow tubular passage.

canaliculus (kan′ah-LIK-ū-lus). A minute canal.

capillary (KAP-ĭ-ler′ē). One of many minute blood vessels that connect arterioles to venules; a minute lymph vessel.

carbohydrate (kar′bō-HĪ-drāte). An organic compound containing carbon, hydrogen, and oxygen in certain proportions. The chief carbohydrates are sugars and starches.

carcinoma (kar′sĭ-NŌ-mah). A malignant tumor arising from cells in an epithelial sheet.

caries (KĀR-ēz). Decay, especially of the teeth.

carotid (kah-ROT-id). Pertaining to the principal artery of the neck.

carpal (KAR-pal). Pertaining to the carpus, or wrist.

castration (kas-TRĀ-shun). Removal of the gonads.

cataract (KAT-ah-rakt). Loss of transparency of the lens of the eye.

caudal (KAW-dal). Pertaining to the tail or distal end of the body.

cavity (KAV-ĭ-tē). A hollow space.

cecum (sē-kum). A blind pouch; the first part of the large intestine.

celiac (SĒ-lē-ak). Pertaining to the abdomen.

cementum (se-MEN-tum). Modified bone covering the root of a tooth.

centimeter (SEN-ti-mě′ter). One-hundredth of a meter or 0.3937 of an inch, abbreviated cm.

cephalic (se-FAL-ik). Pertaining to the head or cranial end of the body.

cerebellum (ser′ě-BEL-um). Portion of the brain located behind the cerebrum and above the pons and fourth ventricle.

cerebral cortex (SER-ě-bral KOR-teks). Gray matter of the cerebral hemispheres.

cerebrum (SER-ě-brum). Portion of the brain that consists of the two cerebral hemispheres and the diencephalon.

cerumen (sě-ROO-men). Earwax.

cervix (SER-viks). The neck of an organ, especially the neck of the uterus.

chiasma (kī-AZ-ma). A crossing, such as the crossing of the optic nerves.

cholesterol (kō-LES-ter-ol). A fatty organic alcohol found in all animal tissues and in blood, bile, and gallstones.

cholinergic (kō′lin-ER-jik). Pertaining to nerve fibers that liberate acetylcholine at their synapses.

cholinesterase (kō′lin-ES-ter-ās). An enzyme that breaks down acetylcholine.

chorion (KŌ-rē-ōn). The outer fetal membrane; forms the fetal portion of the placenta.

choroid (KO-roid). Membranelike structure, especially the middle coat of the eyeball and the lining of the ventricular system of the brain.

chromatid (KRŌ-mah-tid). One of two longitudinal strands making up a chromosome that has undergone DNA replication.

chromatin (KRO-mah-tin). The substance in the nucleus of the nondividing cell that contains DNA and basic proteins.

chromosome (KRŌ-mo-sōm). One of many dark-staining bodies containing DNA and basic proteins that appear in the nucleus at the time of cell division. Chromosomes transmit genetic information from the parent cell to the daughter cells.

chyle (kīl). A milky fluid, consisting of lymph and finely emulsified fat, taken up by the lacteals of the small intestine during digestion.

chyme (kīm). The semifluid material produced in the stomach by action of gastric juice and muscular churning on food masses received from the esophagus.

cilia (SIL-ē-ah). Tiny motile hairlike processes that project from free surfaces of certain cells.

circadian rhythms (ser′kāh-DE-an RIthmz). Rhythmic repetition of body phenomena in 24-hour cycles.

circumcision (ser′kum-SIZH-un). Surgical removal of the foreskin of the penis.

clavicle (KLAV-ĭ-kl). Curved bone lying at the root of the neck; collarbone.

cleavage (KLĚV-ij). Splitting of a fertilized ovum into smaller cells by successive mitotic divisions.

clot (klot). A semisolid mass, such as a blood clot.

coccyx (KOK-siks). Small bone located below the sacrum in humans; the tailbone.

cochlea (KŌK-lē-ah). Shaped like a snail shell; the spiral canal of the inner ear.

colitis (kō-LĪT-is). Inflammation of the colon.

collagen (KOL-ah-jen). Type of protein found in the white fibers of connective tissue, cartilage, and bone.

collateral (kō-LAT-er-al). Secondary or accessory branch of a nerve or blood vessel.

colloid (KOL-oid). Particles of matter evenly dispersed in a gaseous, liquid, or solid medium that do not precipitate or sediment under the influence of gravity.

colon (KŌ-lon). The portion of the large intestine that extends from the cecum to the rectum.

column (KOL-um). A supporting part or pillarlike structure.

comedo (KOM-ě-dō). A mass of keratin and sebum that plugs up the opening of a hair follicle.

commissure (KOM-i-shūr). A band of nerve fibers connecting opposite parts in the brain and spinal cord.

conception (kon-SEP-shun). Union of male and female sex cells to form a new individual.

concha (KONG-kah). A shell-shaped structure; one of the bony projections from the lateral walls of the nasal cavity.

condom (KON-dum). A protective sheath worn over the penis during sexual intercourse to prevent conception and infection.

condyle (KON-dīl). A rounded prominence on the articular end of a bone.

congenital (kon-JEN-ĭ-tal). Present at and usually before the time of birth.

contraception (kon′trah-SEP-shun). Prevention of conception.

contraction (kon-TRAK-shun). A shortening or increase in tension, as in the contraction of muscular tissue.

coracoid (KOR-ah-koid). Like a crow's beak; hooked process projecting anteriorly from the superior border of the scapula.

cornea (KOR-nē-ah). The transparent anterior portion of the outermost coat of the eyeball.

coronal (kō-RŌ-nal). Pertaining to the crown of the head.

coronary (KOR-o-na′rē). Vessels, nerves, or ligaments that encircle anatomical structures like a crown.

corpus (KOR-pus). The body as a whole or the body of an organ.

corpuscle (KOR-pusl). A small body or mass.

corpus luteum (KOR-pus LŪ-tē-um). Yellow body; endocrine tissue formed in the ovary in the site of a ruptured ovarian follicle.

cortex (KOR-teks). An outer layer; the outer layer of many solid glandular organs.

cortisol (KOR-tĭ-sol). The principal carbohydrate-regulating hormone of the adrenal cortex.

cortisone (KOR-tĭ-sōn). Hormone produced by the adrenal cortex that is used in the treatment of arthritis and certain allergic conditions.

costal (KOS-tal). Pertaining to the ribs.

coxa (KOK-sah). The hip or hip joint.

cranium (KRĀ-nē-um). The bones of the head exclusive of the facial bones.

cretinism (KRĚ-ti-nizm). A state of physical and mental retardation caused by congenital lack of thyroxin.

cribriform (KRIB-ri-form). Having small perforations like a sieve.

cricoid (KRĪ-koid). Shaped like a signet ring.

crista (KRIS-tah). A crest.

cruciate (KROO-shē-āte). Shaped like a cross.

crypt (kript). A pit or tubelike depression.

cryptorchism (krip-TOR-kism). Failure of one or both testes to descend into the scrotum.

cubital (KYŪ-bi-tal). Pertaining to the elbow or ulna.

cutaneous (kyū-TĀ-nē-us). Pertaining to the cutis (skin).

cyanosis (sī'ah-NŌ-sis). A bluish discoloration of the skin or nails due to an excessive amount of reduced hemoglobin in the blood.

cystic (SIS-tik). Pertaining to a bladder, especially the urinary bladder.

cytology (sī-TOL-ŏ-jē). The study of cells.

cytoplasm (SĪ-tō-plazm). The protoplasm of the cell outside the nucleus.

deafness (DEF-nes). Impairment of hearing.

deciduous (de-SID-ū-us). A structure that is shed at maturity.

decubitus (de-KYŪ-bĭ-tus). Lying down; position assumed in bed.

decussation (dĕ'kus-SĀ-shun). The point where one band of nerve fibers crosses over another in the central nervous system.

defecation (def'ĕ-KĀ-shun). Elimination of fecal material from the digestive tract.

deferent (DEF-er-ent). Conducting away from or down from a specific part of the body.

degeneration (dē-jen'ĕ-RĀ-shun). Regressive changes in cells or tissues in which structure is altered and certain functions are inhibited or lost.

deglutition (deg'loo-TISH-un). The act of swallowing.

deltoid (DEL-toid). Triangular in outline.

dendrite (DEN-drīt). Process that conducts electrical impulses toward the nerve cell body.

dentate (DEN-tāte). Tooth-shaped.

deoxyribonucleic acid (dē-OK-sĭ-rī'bō-nu-klā'ic AS-id). DNA; a large organic molecule composed of phosphoric acid, deoxyribose, two kinds of purine bases (adenine and guanine), and two kinds of pyrimidine bases (thymine and cytosine). DNA is present in chromosomes and serves as the carrier of genetic information.

dermis (DER-mis). The connective tissue layer of the skin that lies beneath the epidermis.

diabetes (dī'ah-BĒ-tez). A disease characterized by formation of excessive amounts of urine; two major types are diabetes mellitus and diabetes incipidus.

dialysis (dī-AL-ĭ-sis). The diffusion of molecules in solution through a semi-permeable membrane from the side of higher concentration to the side of lower concentration.

diaphragm (DĪ-ah-fram). A partition; the dome-shaped muscle that separates the thoracic and abdominal cavities.

diaphysis (dī-AF-ĭ-sis). The shaft of a long bone.

diastole (dī-AS-tō-lē). The phase of the cardiac cycle in which the heart relaxes and its chambers fill with blood.

diencephalon (dī'en-SEF-ah-lon). Posterior part of the forebrain; the part of the brain located between the two cerebral hemispheres.

diffusion (dī-FŪ-zhun). The random movement of ions, molecules, or small particles in a solution or suspension to achieve uniform distribution.

digit (DIJ-it). A finger or toe.

dilate (DĪ-lāte). To enlarge or expand.

diploid (DIP-loid). The number of chromosomes characteristic of the somatic cells of an organism. In humans the diploid chromosome number is 2 × 23, or 46.

diplopia (di-PLŌ-pē-ah). Double vision; seeing a single object as two objects.

disease (dĭ-ZĒZ). Disorder of body functions, organs, or systems.

dislocation (dis'lō-KĀ-shun). Displacement of one or more bones from a joint.

dissect (dĭ-SEKT). To cut, expose, and separate body structures for anatomical study.

distal (DIS-tal). Farthest from a given point of reference, such as the trunk or point of attachment.

distention (dis-TEN-shun). The state of being stretched out.

diverticulum (dī'ver-TIK-ū-lum). A blind pouch.

dorsal (DOR-sal). Toward the back or posterior part of the human body.

duct (dukt). A canal or tubule for the passage of secretions or excretions.

duodenum (du'o-DE-num). Proximal part of the small intestine.

dura mater (DU-rah MA-ter). Outermost of the three connective tissue membranes covering the brain and spinal cord.

ectoderm (EK-tō-derm). Outermost of the three primary embryonic germ layers.

ectopic (ek-TOP-ik). Situated outside of the normal place.

edema (ĕ-DĒ-mah). Accumulation of an excessive amount of fluid in the intercellular spaces of the body.

effector (ĕ-FEK-tor). A peripheral organ that responds to nerve impulse stimulation by contraction (muscle) or secretion (gland).

efferent (EF-er-ent). Conducting away from the center of the body.

electrolyte (ē-LEK-trō-līt). Any compound capable of conducting an electrical current when dissolved in water and separated into charged particles called ions.

embolus (EM-bō-lus). A mass of material transported by the bloodstream from a larger vessel to a smaller one, where it obstructs circulation.

embryo (EM-brē-ō). The human organism from the time of conception until the end of the second month of development.

emphysema (em'fi-SĒ-mah). Abnormal presence of air within body tissues; a chronic lung disease characterized by dilation of the alveoli.

encephalon (en-SEF-ah-lon). The brain.

endocarditis (en'dō-kar-DĪ-tis). Inflammation of the lining of the heart.

endochondral (en'dō-KON-dral). Formed within cartilage.

endocrine (EN-dō-krin). Pertaining to internal secretion; ductless glands that secrete into the bloodstream.

endoderm (EN-dō-derm). Innermost of the three primary embryonic germ layers.

endometrium (en'dō-MĒ-trē-um). The uterine mucosa.

endoplasmic reticulum (en-dō-PLAZ-mik re-TIK-ū-lum). Microscopic membrane-lined channels in the cytoplasm.

endosteum (en-DOS-tē-um). Membrane lining the marrow cavity of a hollow bone.

enzyme (EN-zīm). A protein substance that acts as a catalyst to accelerate biochemical reactions.

eosinophil (ē'ō-SIN-ō-fil). A white blood cell whose large granules stain bright red with eosin-containing dyes.

epidermis (ep'ĕ-DER-mis). The epithelial layer of the skin.

epimysium (ep'ĭ-MIS-ē-um). The fibrous connective tissue sheath that surrounds a skeletal muscle

epinephrine (ep'ĭ-NEF-rin). Vasopressor

hormone produced by the adrenal medulla.

epineurium (ep′i-NŪ-rē-um). The fibrous connective tissue sheath that surrounds a peripheral nerve fiber.

epiphysis (e-PIF-ĕ-sis). Part of a bone separated from the main portion by cartilage during the period of bone growth; generally refers to the end of a long bone.

epithelium (ep′ĭ-THĒ-lē-um). Type of tissue that covers body surfaces and lines body cavities and tubes.

erectile (ē-REK-tĭl). Tissue or organ capable of erection.

erection (ē-REK-shun). The process of becoming swollen and rigid when filled with blood.

erythrocyte (e-RITH-rō-sīt). Red blood cell.

estrogen (ES-trō-jen). Any of several female sex hormones secreted by the ovarian follicle that are responsible for female secondary sex characteristics.

ethmoid (ETH-moid). Sievelike.

evagination (ē-vaj′ĭ-NĀ-shun). An outpouching.

eversion (ē-VER-zhun). The process of turning outward.

excretion (ek-SKRĒ-shun). The elimination of waste products.

exocrine (EK-sō-krin). Pertaining to external secretion; glands that secrete through ducts.

extrapyramidal (eks′trah-pi-RAM-ĭ-dal). Outside the pyramidal tracts.

extremity (ek-STREM-ĭ-tē). A limb of the body.

extrinsic (ek-STRIN-sik). Originating outside of a given part or structure.

fascia (FASH-ē-ah). A sheet of fibrous connective tissue that covers, binds, or separates body structures.

fascicle (FAS-ĭ-kl). A small bundle, especially of muscle or nerve fibers.

feces (FĒ-sēz). Waste products expelled from the large intestine; stools.

femoral (FEM-ō-ral). Pertaining to the femur or to the thigh.

fertilization (fer′tĭ-lĭ-ZĀ-shun). The union of a sperm with an ovum; conception.

fetus (FĒ-tus). The human organism from the beginning of the third month of development until birth.

fiber (FĪ-ber). An elongated threadlike structure; may refer to a cell, a cell process, or connective tissue elements.

fibrin (FĪ-brin). A protein essential in blood clot formation.

fibrinogen (fī-BRIN-o-jen). A high-molecular-weight plasma protein that is converted into fibrin by the action of thrombin.

fibroblast (FĪ-brō-blast). An immature fiber-producing connective tissue cell; the mature form is called a fibrocyte.

filiform (FIL-ĭ-form). Threadlike.

filtration (fil-TRĀ-shun). Passage of a substance, generally a fluid, through a structure that prevents passage of certain molecules.

fimbria (FIM-brē-ah). One of the components of a fringed structure.

fissure (FISH-er). A narrow slit or cleft.

flagella (flah-JEL-ah). Long motile cytoplasmic projections from certain cells.

follicle (FOL-ĭ-kl). A small saclike structure.

foramen (fō-RĀ-men). An opening or passage, especially into a bone.

fossa (FOS-ah). A pit or depression.

fovea (FŌ-vē-ah). A small pit or depression.

frenulum (FREN-ū-lum). A narrow fold of skin or mucous membrane that limits the movement of an organ or body part.

fundus (FUN-dus). The bottom or base of a structure; the part of a hollow muscular organ that is farthest from its exit.

funiculus (fu-NIK-ū-lus). A cordlike structure; especially one of the large bundles of nerve tracts making up the white matter of the spinal cord.

fusiform (FŪ-zi-form). Spindle-shaped.

gamma globulin (GAM-ah GLOB-ū-lin). A class of plasma proteins produced by plasma cells and B-lymphocytes in response to certain antigens.

ganglion (GANG-glē-on). A group of nerve cell bodies located outside the central nervous system.

gastric (GAS-trik). Pertaining to the stomach.

gene (jēn). One of the units of heredity; a segment of a DNA molecule capable of directing the formation of an enzyme or structural protein.

genitalia (jen′ĭ-TA-lē-ah). The reproductive organs.

genu (JE-nu). The knee or any angular structure that resembles the flexed knee.

gestation (jes-TĀ-shun). The development of a new individual from conception to birth.

gingiva (jin-JI-vah). The gum.

gland (gland). An organ that secretes a specific product or products.

glaucoma (glaw-KŌ-mah). An eye disease characterized by increased intraocular pressure, which causes optic nerve damage and blindness if not treated properly.

glenoid (GLE-noid). Resembling a pit or socket.

glia (GLĒ-ah). The supporting cells of the nervous system.

globin (GLŌ-bin). The protein portion of hemoglobin.

glomerulus (glō-MER-ū-lus). A small cluster or tuft of capillaries; the tuft of capillaries associated with the beginning of each uriniferous tubule.

glossal (GLOS-al). Pertaining to the tongue.

glucagon (GLOO-kah-gon). A pancreatic hormone that increases the concentration of glucose in the blood.

glucocorticoid (gloo′kō-KOR-tĭ-koid). Class of hormones of the adrenal cortex that promotes the formation of glucose by the liver from noncarbohydrate molecules.

glucose (GLOO-kōs). A simple sugar; the principal sugar in human blood and body fluids.

gluteal (GLOO-tē-al). Pertaining to the buttocks.

glycogen (GLĪ-kō-jen). Animal starch; a polysaccharide that is the chief storage form of carbohydrate material in animals.

goiter (GOI-ter). A chronic enlargement of the thyroid gland that causes a bulge in the front of the neck.

gonad (GŌ-nad). A sex gland; the ovary in the female and the testis in the male.

groin (groin). The inguinal region; the junction of the abdomen and thigh.

gustatory (GUS-tah-to′rē). Pertaining to the sense of taste.

gyrus (JĪ-rus). A convolution; one of the

rounded elevations on the surface of the cerebral hemispheres.

hallux (HAL-uks). The big toe.

haploid (HAP-loid). The number of chromosomes characteristic of the germ cells of an organism. In humans the haploid chromosome number is 23.

haustrum (HAWS-trum). One of the pouches of the colon.

haversian canal (ha-VER-shan kah-NAL). One of a series of longitudinal canals in compact bone that transmit vessels and nerves.

helix (HĒ-liks). A coiled structure.

hematocrit (hē-MAT-ō-krit). Percentage of formed elements per volume of whole blood.

hematopoiesis (hem′ah-tō-poi-Ē-sis). Production of the formed elements of the blood.

hemiplegia (hem′i-PLĒ-jē-ah). Paralysis of the extremities of one half of the body.

hemoglobin (hē′mō-GLŌ-bin). The oxygen-binding pigment produced in red blood cells.

hemophilia (hē′mō-FIL-ē-ah). A disease characterized by impaired blood clotting.

hemorrhage (HEM-ō-rij). An outpouring of blood from a ruptured blood vessel.

hemorrhoids (HEM-ō-roidz). Piles; varicosites of the external hemorrhoidal veins.

heparin (HEP-ah-rin). An anticoagulant occurring naturally in the tissues; it is most abundant in the liver.

hepatic (he-PAT-ik). Pertaining to the liver.

hernia (HER-nē-ah). The abnormal protrusion of an organ or part of an organ through the wall of the cavity containing it.

hilus (HĪ-lus). A depression where ducts, vessels, and nerves enter or leave an organ.

Hodgkin's disease (HOJ-kinz di-ZĒZ). A progressive and fatal condition characterized by generalized enlargement of lymphoid tissue.

homograft (HŌ-mō-graft). A tissue graft between animals of the same species but having different genotypes.

homologous (hō-MOL-ō-gus). Corresponding in structure, position, and origin to another organ.

hormone (HŌR-mōn). A secretion produced by a ductless gland and released into the bloodstream to reach its target organ(s).

humerus (HŪ-mer-us). The bone of the upper arm.

hyaline (HĪ-ah-līn). Glassy; translucent.

hymen (HĪ-men). Membranous fold that partly closes the external vaginal opening.

hyoid (HĪ-oid). U-shaped, pertaining to the tongue bone.

hyperglycemia (hī′per-glī-SĒ-mē-ah). Excess of glucose in the blood.

hyperplasia (hī′per-PLĀ-zē-ah). Increase in size of a tissue or organ caused by an increase in the number of cells.

hypertension (hī′per-TEN-shun). High blood pressure; a condition in which the resting systolic pressure is greater than 160 millimeters of mercury and the diastolic pressure is greater than 90 millimeters of mercury.

hypertrophy (hī-PER-tro-fe). Increase in size of a tissue or organ caused by an increase in size of existing cells.

hypochondriac (hī′pō-KON-drē-ak). Pertaining to the abdominal region just below the ribs.

hypoglycemia (hī′pō-glī-SĒ-mē-ah). An abnormally low level of glucose in the blood.

hypophysis (hī-POF-i-sis). An outgrowth; the pituitary gland.

ileum (IL-ē-um). The distal portion of the small intestine.

ilium (IL-ē-um). The flaring lateral portion of the hipbone.

immunity (i-MŪ-ni-tē). Resistance of the body to infectious agents or their products.

immunosuppressive (im′ū-nō-su-PRES-iv). Pertaining to an agent that inhibits the formation of antibodies, especially to transplant antigens.

impermeable (im-PER-mē-ah-bl). Not permitting passage of a substance, especially a fluid.

implantation (im′plan-TĀ-shun). Embedding of the blastocyst in the lining of the uterus.

impulse (IM-puls). A sudden pushing force; electrical activity propagated along nerve fibers.

inclusion (in-KLOO-zhun). A nonliving, often temporary substance in the cytoplasm of a cell.

incus (ING-kus). Anvil; the middle of the three ear ossicles.

infant (IN-fant). A child from birth to 2 years of age.

infarct (IN-farkt). A localized area of necrosis caused by interruption of the blood supply to an organ or body part.

inferior (in-FĒR-ē-or). Situated below a specified structure; directed toward the soles of the feet.

inflammation (in′flah-MĀ-shun). A tissue response to injury characterized by pain, vascular congestion, and the outpouring of serum and leukocytes.

infundibulum (in′fun-DIB-ū-lum). A funnel-shaped structure; the stalk connecting the posterior lobe of the pituitary to the brain.

ingestion (in-JES-chun). The introduction of food, drink, or medication into the digestive tract.

inguinal (ING-gwī-nal). Pertaining to the groin.

inhalation (in′hah-LĀ-shun). Drawing air or medicated vapor into the lungs.

inhibition (in′hi-BISH-un). Arrest or restraint of a process.

in situ (in SĪ-tū). In the normal place or confined to the site of origin.

insulin (IN-su-lin). A pancreatic hormone that regulates carbohydrate metabolism and lowers the concentration of glucose in the blood.

intercellular (in′ter-SEL-ū-lar). Between cells.

intercostal (in′ter-KOS-tal). Between the ribs.

interneuron (in′ter-NŪ-ron). A neuron situated between the primary sensory neuron and the final motor neuron.

interstitial (in-ter-STISH-al). Pertaining to spaces or gaps in a tissue.

intima (IN-ti-mah). Innermost coat of a blood vessel.

intracellular (in′tra-SEL-ū-lar). Inside a cell or cells.

intrinsic (in-TRIN-sik). Originating inside a given part or structure.

invaginate (in-VAJ-i-nāt). To insert part of a structure into another part of the same structure.

inversion (in-VER-zhun). The process of turning inward.

involuntary (in-VOL-un-tār′ē). Independent of the will.

involution (in′vō-LU-shun). A regressive or degenerative change.

ion (Ī-on). An atom or group of atoms bearing a positive or negative charge of electricity.

ischemia (is-KĒ-me-ah). Deficiency of blood supply to a given organ or body part due to vascular constriction or obstruction.

jaundice (JAWN-dis). Yellow coloration of the skin and eyes caused by an excess of bile pigment in the blood.

jejunum (je-JOO-num). The middle portion of the small intestine.

joint (joint). The junction of two or more bones.

jugular (JUG-ū-lar). Pertaining to the neck.

juice (jōōs). Any fluid secretion.

keratin (KER-ah-tin). A sulfur-containing protein found mainly in integumentary structures such as hair and nails.

kidney (KID-nē). One of two bean-shaped glandular organs that secrete urine.

kyphosis (kī-FŌ-sis). Abnormal increase in the convexity of the thoracic spine.

labium (LĀ-bē-um). A lip or liplike structure.

labor (LĀ-bor). The act of expelling the fetus from the uterus through the birth canal.

labrum (LĀ-brum). Lip-shaped structure.

lacrimal (LAK-rĭ-mal). Pertaining to tears.

lactation (lak-TĀ-shun). The secretion of milk.

lacteal (LAK-tē-al). Pertaining to milk; a lymph capillary in the small intestine.

lactose (LAK-tōs). Milk sugar.

lacuna (lah-KŪ-nah). A small pit, cavity, or depression.

lamella (lah-MEL-ah). A thin flat plate, especially of bone.

larynx (LAR-ingks). The voice box; the portion of the respiratory tract located between the pharynx and the trachea.

lateral (LAT-er-al). Pertaining to the side; farther away from the midline of the body.

lesion (LĒ-zhun). Any form of tissue damage.

leukemia (lū-KĒ-me-ah). Malignant disorder of blood-forming tissues characterized by proliferation of abnormal white blood cells.

ligament (LIG-ah-ment). A tough band of fibrous connective tissue that connects bone to bone or supports an organ.

lingual (LING-gwahl). Pertaining to the tongue.

lipid (LIP-id). Fat or fatlike substance.

liter (LĒ-ter). Unit of liquid measurement equivalent to 1.0567 quarts.

liver (LIV-er). Large accessory digestive gland located in the upper right quadrant of the abdomen.

lobe (lōb). Large subdivision of a gland or organ.

lobotomy (lo-BOT-o-mē). Cutting of nerve fibers to disconnect a lobe of the brain from the thalamus.

loin (loin). The part of the back between the ribs and the pelvis.

lumbar (LUM-bar). Pertaining to the lower back.

lumen (LŪ-men). The cavity inside a hollow muscular organ or tube.

lung (lung). One of the two main organs of respiration that lie within the thoracic cavity.

lymph (limf). The colorless fluid circulating in the lymphatic system.

lymphocyte (LIM-fō-sīt). A type of white blood cell formed in lymphoid tissue from a bone marrow precursor.

lysosome (LĪ-sō-sōm). Membrane-bound sac of digestive enzymes found in the cytoplasm of a cell.

macrophage (MAK-rō-fāj). A large phagocytic cell.

macroscopic (mak′ro-SKOP-ik). Something that is visible to the naked eye.

macula (MAK-u-lah). A spot.

malignant (mah-LIG-nant). Something that goes from bad to worse; a tumor that has the properties of uncontrolled growth and distant spread.

malleus (MAL-ē-us). Hammer; the ossicle attached to the eardrum.

mammary (MAM-er-ē). Pertaining to the breast.

mandible (MAN-di-bl). The lower jawbone.

manubrium (mah-NŪ-brē-um). Handle; the uppermost portion of the sternum.

mastication (mas′ti-KĀH-shun). The act of chewing.

mastoid (MAS-toid). Breast-shaped; rounded process of the temporal bone.

matrix (MĀ-triks). The intercellular substance of a tissue.

meatus (mē-ā-tus). A passage or channel; the external opening of a canal.

medial (MĒ-dē-al). Pertaining to the center; closer to the midline of the body.

mediastinum (mē′dē-ah-STI-num). A medium septum or partition; the mass of tissues and organs situated in the middle of the thoracic cavity.

medulla (mē-DUL-ah). The central or inner portion of a solid glandular organ.

megakaryocyte (meg′ah-KAR-ē-ō-sīt). The giant bone marrow cell that gives rise to blood platelets.

meiosis (mī-Ō-sis). A special type of cell division that occurs in the formation of sex cells; during meiosis the number of chromosomes per cell is reduced from the diploid to the haploid number.

membrane (MEM-brān). A thin layer of tissue that covers a surface, lines a cavity, or partitions an organ or a space.

meninges (me-NIN-jēz). The connective tissue membranes that cover the brain and the spinal cord.

menopause (MEN-ō-pawz). The period when the ovaries stop functioning and the menstrual cycles cease.

menstrual cycle (MEN-stroo-al SĪ-kl). The cyclic changes in the uterine lining that culminate in menstruation.

menstruation (men′stroo-ā-shun). The discharge of blood and tissue from the uterus at the end of each menstrual cycle.

mesencephalon (mes′en-SEF-ah-lon). The midbrain; the middle of the three primary brain vesicles.

mesenchyme (MES-eng-kīm). A loose meshwork of embryonic connective tissue derived from mesoderm.

mesenteric (mes′en-TER-ik). Pertaining to a mesentery.

mesentery (MES-en-ter′ē). A peritoneal fold; the fold that suspends the small intestine from the posterior abdominal wall.

mesoderm (MEZ-ō-derm). The middle of the three primary embryonic germ layers.

mesothelium (mez′ō-THĒ-lē-um). A type

of simple squamous epithelium derived from mesoderm that forms the surface layer of all true serous membranes.

metabolism (me-TAB-ō-lizm). The total of all the physical and chemical reactions taking place in the body.

metastasis (me-TAS-tah-sis). The spread of disease, especially cancer, from one body part or organ to another.

microglia (mī-CROG-lē-ah). Glial cells of mesodermal origin that act as phagocytes in the central nervous system.

micron (MĪ-kron). Unit of linear measurement equivalent to 0.001 millimeter; also called a micrometer.

microscopic (mī′krō-SKOP-ik). Something that can be seen only with the aid of a microscope.

microtubules (mī′krō-TŪ-byūls). Straight hollow tubules found in the cytoplasm of a cell.

microvilli (mī′krō-VIL-lē). Minute cytoplasmic protrusions from the free surface of certain cells.

micturition (mik′tu-RISH-un). The act of passing urine.

milliliter (MIL-ī-lē′ter). Unit of volume equivalent to 0.001 liter.

millimeter (MIL-ī-mē′ter). Unit of linear measurement equal to 0.025 inch, 0.001 meter, or 0.1 centimeter.

mineralocorticoid (min′er-al-ō-KOR-ti-koid). Class of hormones of the adrenal cortex that promotes the retention of sodium and the loss of potassium.

mitochondria (mi′tō-KON-drē-ah). Rod-shaped cytoplasmic organelles that serve as the principal energy source of the cell.

mitosis (mī-TŌ-sis). Type of division in which a parent cell gives rise to two new daughter cells, each having the same kind and number of chromosomes as the parent cell.

mitral (MĪ-tral). Shaped like a bishop's miter; pertaining to the mitral valve.

modiolus (mo-DĪ-ō-lus). The central core of spongy bone around which the cochlear duct is wrapped.

monocyte (MON-ō-sīt). The largest of the white blood cells.

morphology (mor′FOL-ō-jē). The science that deals with the forms and structures of plants and animals.

morula (MOR-u-lah). Shaped like a mulberry; the solid mass of cells (blastomeres)

formed by the early cleavage divisions of the fertilized ovum.

motile (MŌ-til). Something that is able to move spontaneously.

motor (MŌ-tor). Pertaining to motion.

motor neuron (MŌ-tor NŪ-ron). A neuron whose axon carries electrical impulses to or toward muscles and glands.

mucosa (mū-KŌ-sah). Mucous membrane; a lining membrane that consists of a layer of mucus-secreting epithelium supported by a layer of connective tissue.

mucus (MŪ-kus). The thick slimy secretion of goblet cells and mucous glands; mucus consists of a protein (mucin) and inorganic salts suspended in water.

muscle (MUS-l). An organ composed of contractile cells called muscle fibers.

myelin (MĪ-ē-lin). A lipoprotein material that forms a protective sheath around axons of myelinated nerve fibers.

myeloid (MĪ-ē-loid). Pertaining to the bone marrow.

myocardium (mī′ō-KAR-dē-um). The muscular coat of the heart.

myotome (MĪ-o-tōm). The part of the somite that gives rise to skeletal muscle.

navicular (nah-VIK-ū-lar). Boat-shaped; adjective applied to certain bones.

neoplasia (nē′ō-PLA-zē-ah). A pathological process characterized by the formation of new growths.

neoplasm (NĒ-ō-plazm). A new growth or tumor.

nephron (NEF-ron). The functional unit of the kidney.

neurilemma (nū′ri-LEM-ah). The outer covering of a nerve fiber.

neuroglia (nū-ROG-lē-ah). The supporting cells of the nervous system.

neurohypophysis (nū′rō-hī-POF-ī-sis). The neural part of the hypophysis (pituitary).

neuron (NŪ-ron). The nerve cell body and all its processes.

neutrophil (NŪ-trō-fil). A granular white blood cell with a three-to-five-lobed nucleus and small, lavender-staining cytoplasmic granules.

node (nōd). A swelling or protuberance; an encapsulated mass of tissue.

nodule (NOD-ūl). A small node.

nuchal (NŪ-kal). Pertaining to the back of the neck.

nucleoside (NŪ-klē-ō-sīd). A compound that consists of a sugar (pentose or hexose) and a purine or pyrimidine base.

nucleotide (NŪ-klē-ō-tīd). A compound consisting of a purine or pyrimidine base, a pentose or hexose sugar, and phosphoric acid.

nucleus (NŪ-klē-us). A spherical body in a cell that serves as its control center; a collection of neurons in the central nervous system.

occipital (ok-SIP-ī-tal). Pertaining to the back of the head.

odontoid (ō-DON-toid). Toothlike.

olecranon (o-LEK-rah-non). The point of the elbow; a large process on the proximal end of the ulna.

oligodendrocytes (ol′i-gō-DEN-drō-sītz). Glial cells that myelinate nerve fibers in the central nervous system.

oocyte (Ō-o-sīt). An immature ovum.

oogenesis (ō′o-JEN-e-sis). The development of mature ova from oogonia.

oogonium (ō′o-GO-nē-um). The female germ cell.

oophorectomy (ō′o-fo-REK-tō-mē). Surgical removal of an ovary.

ophthalmic (of-THAL-mik). Pertaining to the eye.

orbit (OR-bit). A cavity in the skull containing the eyeball and its muscles, nerves, and blood vessels.

orchiectomy (or′kē-EK-tō-mē). Surgical removal of a testis.

orgasm (OR-gazm). The climax of sexual excitement.

os (os). A bone; also, an opening or mouth.

osmosis (oz-MŌ-sis). The diffusion of water through a semipermeable membrane from the side of lesser solute concentration to the side of greater solute concentration.

osseous (OS-ē-us). Bony.

ossicle (OS-i-kl). A small bone; one of the three small bones of the middle ear.

ossification (os′i-fi-KĀ-shun). Formation of or conversion to bone.

ovary (Ō-var-ē). Female sex gland.

ovulation (o′vu-LĀ-shun). The process by which an ovum is released from an ovary.

ovum (Ō-vum). Egg; the female sex cell.

oxyhemoglobin (ok′sē-hē′mō-GLŌ-bin). Oxygen combined with hemoglobin.

oxytocin (ok′sē-TO-sin). A hormone produced in the hypothalamus that stimulates uterine contraction during labor and causes release of milk during breast-feeding.

palate (PAL-at). The roof of the oral cavity.

pancreas (PAN-krē-as). Digestive and endocrine gland located in the curve of the duodenum.

papilla (pa-PIL-ah). A small nipple-shaped projection.

paralysis (pah-RAL-ĭ-sis). Loss of ability to move body parts.

parasympathetic (par′ah-sim′pah-THET-ik). Pertaining to the division of the autonomic nervous system that is concerned with restorative processes.

parathyroid glands (par′ah-THĪ-roid glands). Two pairs of endocrine glands located on or in the thyroid gland whose secretion regulates calcium and phosphate metabolism.

parenchyma (pah-RENG-ki-mah). The soft or functional tissue of an organ.

parietal (pah-RĪ-ĕ-tal). Pertaining to the walls of a cavity or hollow structure.

parotid glands (pah-ROT-id glands). Pair of large salivary glands located near the ears.

patella (pah-TEL-ah). Kneecap.

patent (PA-tent). Open, unobstructed.

pathology (pah-THOL-ō-jē). The study of disease, especially the structural and functional changes produced by disease.

pectineal (pek-TIN-ē-al). Pertaining to the pubis.

pectoral (PEK-tor-al). Pertaining to the chest.

peduncle (pe-DUNG-kl). A stemlike part; name applied to bundles of nerve fibers connecting different portions of the central nervous system.

pelvis (PEL-vis). Basinlike structure; the bony basin formed by the hipbones and the lower portion of the vertebral column.

penis (PĒ-nis). The external male genital organ used for urination and sexual intercourse.

pennate (PEN-āt). Shaped like a feather.

pericardium (per′ĭ-KAR-dē-um). Fibroserous membrane enclosing the heart.

perineum (per′ĭ-NĒ-um). Diamond-shaped region between the thighs that forms the floor of the pelvis.

periosteum (per′ē-OS-tē-um). Connective tissue membrane that covers bone.

peripheral (pĕ-RIF-er-al). Pertaining to or situated at the periphery.

peristalsis (per′ĭ-STAL-sis). Wormlike contraction waves that propel solids or liquids down hollow muscular tubes.

peroneal (per′o-NĒ-al). Pertaining to the fibula.

petrous (PET-rus). Resembling a rock.

phagocyte (FAG-ō-sīt). A cell that ingests microorganisms, particulate matter, or other cells.

phalanges (fa-LAN-jēz). The bones of the fingers or toes.

pharynx (FAR-ingks). The throat.

phrenic (FREN-ik). Pertaining to the diaphragm.

pia mater (PĒ-ah MA-ter). The delicate innermost meningeal membrane covering the brain and spinal cord.

pilomotor (pi′lō-MŌ-ter). Causing hairs to move.

pineal (PIN-ē-al). Like a pine cone; a small endocrine gland arising from the roof of the diencephalon.

pinocytosis (pi′nō-si-TŌ-sis). The uptake of fluid by minute invaginations of the cell membrane, which close to form fluid-filled vacuoles.

pisiform (PI-sĭ-form). Like a pea in size and shape.

plantar (PLAN-tar). Pertaining to the sole of the foot.

plasma (PLAZ-mah). The fluid portion of the blood.

platelet (PLĀT-let). A disc-shaped, non-nucleated blood element involved in the process of blood clotting.

pleura (PLOO-rah). The serous membrane that covers a lung and lines the adjacent thoracic wall.

plexus (PLEK-sus). A network of nerves or veins.

plicae (PLĪ-sē). Folds or ridges; the circular folds of mucosa in the small intestine.

pons (ponz). Bridge; part of the brain stem situated between the medulla and metencephalon.

popliteal (pop-LIT-ē-al). Pertaining to the back of the knee.

portal (POR-tal). Pertaining to an entrance, especially the porta hepatis.

posterior (pos-TĒR-ē-or). Toward the back of the body.

progesterone (prō-JES-tĕ-rōn). Female sex hormone secreted by the corpus luteum that prepares the uterine lining for implantation, protects the embryo, and promotes the development of the placenta.

progestins (prō-JES-tinz). Hormones with progestational activity.

prolapse (PRŌ-laps). The downward displacement of an organ or body part or its protrusion through a body opening.

proliferation (prō-lif′ĕ-RĀ-shun). The multiplication of cells.

pronation (prō-NĀ-shun). The act of moving the forearm so that the palm of the hand faces downward.

prostate (PROS-tāt). An accessory reproductive gland that surrounds the proximal portion of the male urethra.

protoplasm (PRŌ-tō-plazm). The essential living matter of all cells.

protuberance (prō-TŪ-ber-ans). A projection or swelling.

psoas (SŌ-as). One of the two muscles of the loins.

pterygoid (TER-ĭ-goid). Wing-shaped; applied to two pairs of jaw muscles.

ptosis (TŌ-sis). Abnormal downward displacement; drooping of the upper eyelid.

puberty (PYŪ-ber-tē). Stage of growth at which the reproductive organs become functional.

pudenda (pyū-DEN-dah). External genitalia, especially the female genitalia.

quadrate (KWOD-rāt). Square or squared.

quadriceps (KWOD-rĭ-seps). A muscle having four heads.

quickening (KWIK-en-ing). The time when the fetal movements are first felt by a pregnant woman.

radius (RĀ-dē-us). The bone on the lateral or thumb side of the forearm.

ramus (RĀ-mus). A branch, especially of a nerve, artery, or vein.

receptor (rē-SEP-tor). A sensory nerve

ending that responds to stimulation by transmitting nerve impulse to the central nervous system.

recipient (rē-SIP-ē-ent). A person who receives a blood transfusion or a tissue or organ graft.

rectus (REK-tus). Straight.

reflex (RĒ-fleks). An automatic response by the nervous system to a given stimulus.

rejection (rē-JEK-shun). The attack upon and ultimate destruction of a graft by the recipient's antibodies.

renal (RĒ-nal). Pertaining to the kidney.

renin (RĒ-nin). An enzyme produced by the kidneys in response to decreased renal blood flow or decreased reabsorption of sodium from the kidney tubules.

rennin (REN-in). A milk-clotting enzyme found in the gastric juice; may not be present in humans.

reproduction (rē'prō-DUK-shun). The process by which a living organism produces a new individual of its own kind.

respiration (res'pi-RĀ-shun). The exchange of oxygen and carbon between the lungs and the blood (external respiration), between the cell and its environment (internal respiration), and within the cell (cell respiration).

reticular (re-TIK-ū-lar). Resembling a network.

retina (RET-i-nah). Innermost coat of the eyeball; the coat that contains the photoreceptors (rods and cones).

retinaculum (ret'i-NAK-ū-lum). A restraining band or ligament.

ribonucleic acid (rī'bō-nū-KLĀ-ik AS-id). RNA; an organic molecule composed of phosphoric acid, ribose, two kinds of purine bases (adenine and guanine), and two kinds of pyrimidine bases (uracil and cytosine). The three principal types of RNA found in the cell are messenger RNA, transfer RNA, and ribosomal RNA.

ribosome (RĪ-bo-sōm). A cytoplasmic organelle composed of RNA and protein that is involved in protein synthesis.

rugae (ROO-gē). Wrinkles or folds in the mucous membrane of a hollow muscular organ.

sac (sak). A pouchlike organ or structure.

saccule (SAK-ūl). A little sac.

sacrum (SĀ-krum). The wedge-shaped bone that forms the base of the spine.

sagittal (SAJ-ĭ-tal). A plane coinciding with or parallel to the long axis of the body.

saliva (sah-LĪ-vah). The secretion of the salivary glands.

salpinx (SAL-pinks). A trumpet-shaped tube; refers to the uterine or auditory tube.

sarcoma (sar-KŌ-mah). A malignant tumor derived from tissue of mesodermal origin.

Schwann cells (shwahn selz). Glial cells that myelinate peripheral nerve fibers.

sciatic (sī-AT-ik). Pertaining to the ischium; a large peripheral nerve that emerges from the greater sciatic notch of the hipbone and runs down the back of the thigh.

sebum (SĒ-bum). The oily secretion of sebaceous glands.

semen (SĒ-men). Fluid discharged from the male reproductive tract during ejaculation.

semilunar (sem'-ĭ-LŪ-nar). Shaped like a half-moon.

sensory neuron (SEN-so-rē NŪ-ron). A neuron whose axon carries electrical impulses from peripheral structures to the central nervous system.

septum (SEP-tum). A partition.

serosa (se-RŌ-sah). A serous membrane.

serous (SĒ-rus). Producing a thin watery secretion.

serum (SĒ-rum). Blood plasma minus the clotting substances.

sesamoid (SES-ah-moid). Shaped like a sesame seed.

sigmoid (SIG-moid). S-shaped.

sinus (SĪ-nus). A cavity; an atypical venous channel.

skeleton (SKEL-e-ton). The bony framework of the body.

somatic (sō-MAT-ik). Pertaining to the body (soma).

spastic (SPAS-tik). Characterized by spasms (involuntary contraction of a muscle or group of muscles).

sperm (sperm). The male sex cell.

spermatogenesis (sper'mah-tō-JEN-ē-sis). The entire process of sperm formation.

spermatogonium (sper'mah-tō-GŌ-nē-um). The male germ cell.

spermiogenesis (sper'mē-ō-JEN-ē-sis). The transformation of a spermatid into a mature sperm.

sphenoid (SFĒ-noid). Wedge-shaped; the wedge-shaped bone at the base of the skull.

sphincter (SFINGK-ter). A circular muscle that constricts a duct or closes a natural body opening.

spleen (splēn). Large lymphoid organ located in the upper left quadrant of the abdomen.

squamous (SKWA-mus). Scalelike.

stapes (STĀ-pēz). Stirrup; ear ossicle in contact with the oval window of the inner ear.

stenosis (ste-NŌ-sis). A narrowing of a vessel, duct, or body opening.

steroid (STE-roid). A large family of organic compounds related to cholesterol that comprises many hormones, vitamins, body constituents, and drugs.

stimulus (STIM-ū-lus). Any agent that produces a reaction in a muscle, nerve, gland, or other excitable tissue.

stomach (STUM-ak). A J-shaped saclike enlargement of the digestive tract that is situated between the esophagus and duodenum.

stratum (STRA-tum). A sheetlike layer.

striated (STRĪ-āt-ed). Having streaks or stria (lines).

stroke (strōk). Condition caused by the rupture or blockage of a blood vessel in the brain.

stroma (STRŌ-mah). The connective tissue framework of an organ.

subcutaneous (sub'kyū-TĀ-nē-us). Beneath the dermal layer of the skin.

sulcus (SUL-kus). A groove or furrow.

superior (soo-PĒR-ē-or). Situated above a specified structure; directed toward the crown of the head.

supination (soo'pi-NĀ-shun). The act of moving the forearm so that the palm of the hand faces upward.

suprarenal (soo'prah-RĒ-nal). Above the kidney; adrenal.

suture (SOO-tūr). A type of fibrous joint uniting adjacent skull bones.

sympathetic (sym'pah-THET-ik). Pertaining to the division of the autonomic nervous system that is concerned with the expenditure of energy.

symphysis (SIM-fi-sis). A type of joint in which the ends of the participating bones are united by fibrocartilage.

synapse (SIN-aps). The place where a nerve impulse is transmitted from one neuron to another.

syndesmosis (sin'-des-MŌ-sis). A type of fibrous joint in which the participating bones are united by an interosseous membrane or a ligament.

syndrome (SIN-drōm). A combination of symptoms resulting from a single cause.

synovial fluid (sī-NŌ-vē-al FLOO-id). The lubricating fluid secreted by synovial membranes.

syphilis (SIF-ĭ-lis). A contagious venereal disease caused by a microorganism, *Treponema pallidum.*

system (SIS-tem). A collection of organs that perform a common function or group of related functions.

systemic (sis-TEM-ik). Pertaining to or affecting the entire body.

systole (SIS-to-lē). Contraction of the heart, particularly contraction of the ventricles.

talus (TA-lus). Ankle bone that articulates with the tibia and fibula above and the calcaneus below.

tectum (TEK-tum). A rooflike structure; roof of the brain stem.

tegmentum (teg-MEN-tum). A covering; central portion of the brain stem.

tela (TĒ-lah). A thin weblike tissue or structure.

temporal (TEM-po-ral). Pertaining to the temple (lateral region of the head above the zygomatic arch).

tendon (TEN-don). A cord or band of fibrous connective tissue that connects muscle to muscle or muscle to bone.

testosterone (tes-TOS-te-rōn). Male sex hormone secreted by the interstitial cells in the testis that is responsible for male secondary sex characteristics.

thorax (THŌ-raks). The part of the body between the neck and abdomen; chest.

thrombus (THROM-bus). A clot formed in a blood vessel or one of the chambers of the heart.

toxic (TOK-sik). Poisonous, harmful to the body.

tract (trakt). A longitudinal collection of tissues or organs; a bundle of nerve fibers having a common origin, function, and termination.

trauma (TRAW-mah). A wound or injury, usually produced by external force.

trigeminal (trī-JEM-ĭ-nal). Triple; pertaining to the fifth cranial nerve, which has three sensory roots.

trigone (TRĪ-gōn). Triangle; smooth triangular region on the posterior wall of the urinary bladder.

trochlear (TROK-lē-ar). Pertaining to a trochlea or pulley; the fourth cranial nerve.

tube (tūb). A hollow cylindrical organ.

tubercle (TŪ-ber-kl). A small rounded projection.

tumor (TŪ-mor). Swelling; a mass of new and actively growing tissue.

tunic (TŪ-nik). Coat; a membrane or tissue layer forming part of the wall of a hollow organ.

turbinate (TUR-bĭ-nāte). Cone-shaped; a nasal concha.

tympanic (tim-PAN-ik). Pertaining to the tympanum (middle ear).

ulcer (UL-ser). A localized defect on a mucous or cutaneous surface.

ulna (UL-nah). The bone on the medial or little finger side of the forearm.

umbilicus (um-BIL-i-kus). Navel; the scar marking the site of the umbilical cord in the fetus.

urea (ū-RĒ-ah). The end product of protein metabolism in the body and the chief nitrogenous component of urine.

uremia (ū-RĒ-mē-ah). Condition caused by an excessive amount of urea and other nitrogenous wastes in the blood.

utricle (Ū-tre-kl). Little sac; the larger of the two membranous sacs in the vestibule of the inner ear.

vagina (vah-JĪ-nah). A sheath; the canal extending from the vaginal vestibule to the uterus.

vagus (VĀ-gus). Wandering; the tenth cranial nerve.

valve (valv). A membranous fold in a canal or vessel that usually prevents backflow of material passing through it.

varicose (VAR-ĭ-kos). Unnaturally swollen or distended.

vas (vas). A vessel, duct, or canal.

vascular (VAS-kyū-lar). Pertaining to or full of blood vessels.

vasectomy (vah-SEK-tō-mē). Excision of a portion of the vas deferens as a means of insuring male sterility.

vasoconstriction (vas'ō-kon-STRIK-shun). Contraction of blood vessels.

vasodilation (vas'ō-dī-LĀ-shun). Dilation of blood vessels.

vastus (VAS-tus). Wide, great.

vein (vān). A blood vessel that conducts blood to or toward the heart.

ventral (VEN-tral). Directed to or toward the belly surface; interchangeable with anterior in humans.

ventricle (VEN-tri-kl). A small cavity or chamber, especially in the brain or heart.

vesicle (VES-ĭ-kl). A small saclike structure.

vestibule (VES-tĭ-būl). A space or cavity at the entrance to another structure.

vestibulocochlear (ves-tib'ū-lō-KOK-lē-ar). Pertaining to the eighth cranial nerve.

viable (VĪ-ah-bl). Capable of living, especially a fetus able to live outside the uterus.

villus (VIL-us). A projection, especially from the surface of a mucous membrane.

viscera (VIS-er-ah). The internal organs.

visceral (VIS-er-al). Pertaining to the viscera.

viscous (VIS-kus). Sticky.

vitamin (VĪ-tah-min). One of a group of organic compounds, present in minute quantities in natural foodstuffs, that are essential to normal body metabolism.

volar (VŌ-lar). Pertaining to the palm of the hand or the sole of the foot.

wart (wort). A small, hard, abnormal growth on the skin or adjacent mucous membrane that is caused by a virus.

xiphoid (ZIF-oid). Sword-shaped; distal portion of the sternum.

X rays (EKS-rāz). Electromagnetic radiation of very short wavelength used for diagnostic purposes and for treating various diseases.

zone (zōn). A restricted area.

zygomatic (zī'gō-MAT-ik). Pertaining to something that is yoke-shaped.

zygote (ZĪ-gōt). The fertilized ovum.

References and Supplemental Readings

Because students in the allied health professions carry a heavy course load, they scarcely have time to complete their assigned readings, let alone do any supplemental readings in their basic science courses. However, there are times when the student of anatomy may want to get more information on a given topic than is provided by this book, either to satisfy his or her own curiosity or to write a report or term paper. Accordingly, I have provided a list of sources of additional information on anatomy and related subjects. The books and articles listed here and at the end of each chapter—sources with which I am personally familiar—represent only a small fraction of the total amount of references and supplemental reading material available. However, because most of these sources have their own bibliographies, the student is provided with an additional entry into the scientific literature if he or she wishes to pursue a given topic.

Although anatomists publish extensively and in a wide variety of journals, I have not included original research papers in my lists. This is a deliberate omission because I feel that much of present-day anatomical research is too technical for the beginning student and deals mainly with nonhuman subject matter. Instead, I have chosen review articles written specifically for nonexperts. These articles, taken from the magazines *Scientific American* and *Hospital Practice*, present the structural and functional aspects of new and often highly sophisticated research and clinical materials in an easily readable form. Moreover, they are accompanied by simplified, vividly colored drawings and charts that facilitate comprehension of the text.

The references and supplemental readings listed here vary in difficulty of comprehension. No asterisk indicates that a beginning anatomy student should have little or no difficulty in comprehending the material. One asterisk (°) indicates that the material is of moderate difficulty for a beginning anatomy student, especially if he or she has a limited background in biology and chemistry, and may require some extra effort for comprehension. Two asterisks (°°) indicate that the material is highly detailed or technical and more suited for an advanced rather than a beginning anatomy student; I have included this material for those students who wish to pursue a topic in depth and are willing to expend a great deal of effort for comprehension.

Books Covering Anatomical and Related Subjects

On occasion the student may wish to do some in-depth reading in anatomical or related subjects or wish to see how anatomical material is presented in other texts. A copy of *Gray's Anatomy* (of any edition), which many beginning anatomy students receive from friends or relatives, is an excellent reference. Another useful book is a medical dictionary, which is a reference not only for new or unfamiliar terms but also for clinical procedures. The group of books listed here, mainly textbooks and review books, are categorized according to field and are not mentioned again in the individual chapters.

Anatomy (General and Gross Anatomy)

BASMAJIAN, J. V. *Grant's Method of Anatomy.* 8th ed. Baltimore: Williams & Wilkins, 1971.

———. *Primary Anatomy.* 7th ed. Baltimore: Williams & Wilkins, 1976.

CHRISTENSEN, JOHN B., and IRA R. TELFORD. *Synopsis of Gross Anatomy.* 2d. ed. New York: Harper & Row, 1972 (paperback).

CLEMENTE, CARMINE D. *Anatomy: A Regional Atlas of the Human Body.* Philadelphia: Lea & Febiger, 1975.

°CRAFTS, RODGER C. *A Textbook of Human Anatomy.* New York: Ronald Press, 1966.

CROUCH, J. L. *Functional Human Anatomy.* 3d ed. Philadelphia: Lea & Febiger, 1978.

DAWSON, HELEN L. *Basic Human Anatomy.* 2d ed. New York: Appleton-Century-Crofts, 1974.

FRANCIS, CARL C. *Introduction to Human Anatomy.* St. Louis: Mosby, 1968.

GARDNER, WESTON D., and WILLIAM A. OSBURN. *Structure of the Human Body.* 3d ed. Philadelphia: Saunders, 1978.

°GOSS, CHARLES MAYO, ed. *Gray's Anatomy of the Human Body.* 29th ed. Philadelphia: Lea & Febiger, 1973.

GRANT, J. C. B. *An Atlas of Anatomy.* 6th ed. Baltimore: Williams & Wilkins, 1972.

°HOLLINGSHEAD, W. HENRY. *Textbook of Anatomy.* 3d ed. Hagerstown, Md.: Harper & Row, 1974.

NILLSON, LENNART, with JAN LINDBERG. *Behold Man.* Boston: Little, Brown, 1973.

PANSKY, BEN, and EARL LAWRENCE HOUSE. *Review of Gross Anatomy.* 3d ed. New York: Macmillan, 1975 (paperback).

PERROTT, J. W. *Anatomy for Students and Teachers of Physical Education.* 2d ed. London: Arnold, 1970 (paperback).

ROYCE, JOSEPH. *Surface Anatomy.* Philadelphia: Davis, 1965 (paperback).

°SCHNEIDER, LAWRENCE. *Anatomical Case Histories.* Chicago: Year Book, 1976.

°SNELL, RICHARD S. *Clinical Anatomy for Medical Students.* Boston: Little, Brown, 1973.

°THOREK, PHILIP. *Anatomy in Surgery.* Philadelphia: Lippincott, 1962.

°WOODBURNE, RUSSELL T. *Essentials of Human Anatomy.* 5th ed. New York: Oxford University Press, 1963.

YOKUCHI, C. *Photographic Anatomy of the Human Body.* Baltimore: University Park Press, 1971.

Anatomy and Physiology

ANDERSON, PAUL D. *Clinical Anatomy and Physiology for Allied Health Sciences.* Philadelphia: Saunders, 1976.

ANTHONY, CATHERINE PARKER, and NORMA JANE KOLTHOFF. *Textbook of Anatomy and Physiology.* 9th ed. St. Louis: Mosby, 1975.

CHAFFEE, ELLEN E., and ESTHER M. GREISHEIMER. *Basic Anatomy and Physiology.* 3d ed. Philadelphia: Lippincott, 1974.

DECOURSEY, RUSSELL MYLES. *The Human Organism.* 4th ed. New York: McGraw-Hill, 1974.

GREISHEIMER, ESTHER M., and MARY P. WIEDEMAN. *Physiology and Anatomy.* 9th ed. Philadelphia: Lippincott, 1972.

JACOB, STANLEY W., and CLARICE ASHWORTH FRANCONE. *Structure and Function in Man.* 3d ed. Philadelphia: Saunders, 1974.

KING, BARRY G., and MARY JANE SHOWERS. *Human Anatomy and Physiology.* 6th ed. Philadelphia: Saunders, 1969.

LANGLEY, L. L., IRA R. TELFORD, and JOHN B. CHRISTENSEN. *Dynamic Anatomy and Physiology.* 4th ed. New York: McGraw-Hill, 1974.

MEMMLER, RUTH LUNDEEN, and RUTH BYERS RADA. *The Human Body in Health and Disease.* Philadelphia: Lippincott, 1970.

PANSKY, BEN. *Dynamic Anatomy and Physiology.* New York: Macmillan, 1975.

PITAL, MARTHA, and MILDRED SCHELLIG. *Clinical Coordination of Anatomy and Physiology.* New York: Springer, 1965.

Cellular Biology

°AVERS, CHARLOTTE J. *Cell Biology.* New York: Van Nostrand, 1976.

°BITTAR, E. EDWARD, ed. *Cell Biology in Medicine.* New York: Wiley, 1973.

°DEROBERTIS, E. D. P., FRANCISCO A. SAEZ, and E. M. F. DEROBERTIS, JR. *Cell Biology.* 6th ed. Philadelphia: Saunders, 1975.

°DUPRAW, ERNEST J. *Cell and Molecular Biology.* New York: Academic Press, 1968.

Readings from *Scientific American. The Molecular Basis of Life: An Introduction to Molecular Biology.* San Francisco: Freeman, 1968.

°WATSON, JAMES D. *Molecular Biology of the Gene.* 2d ed. New York: Benjamin, 1970.

Embryology (Developmental Anatomy)

GASSER, RAYMOND F. *Atlas of Human Embryos.* Hagerstown, Md.: Harper & Row, 1975.

°LANGMAN, J. *Medical Embryology.* 3d ed. Baltimore: Williams & Wilkins, 1975.

°MOORE, KEITH L. *The Developing Human.* Philadelphia: Saunders, 1973.

°SNELL, RICHARD S. *Clinical Embryology for Medical Students.* 2d ed. Boston: Little, Brown, 1975.

Genetics

MCKUSICK, VICTOR A. *Human Genetics.* 2d ed. Englewood Cliffs, N.J.: Prentice-Hall, 1969.

° ———, and ROBERT CLAIBORNE, eds. *Medical Genetics.* New York: HP Publishing Co., 1973.

°THOMPSON, JAMES S., and MARGARET W. THOMPSON. *Genetics in Medicine.* 2d ed. Philadelphia: Saunders, 1973.

Histology (Microscopic Anatomy)

°BEVELANDER, GARRIT, and JUDITH A. RAMALEY. *Essentials of Histology.* 7th ed. St. Louis: Mosby, 1974.

°°BLOOM, WILLIAM, and DON W. FAWCETT. *A Textbook of Histology.* 10th ed. Philadelphia: Saunders, 1975.

FAWCETT, DON W. *The Cell: An Atlas of Fine Structure.* Philadelphia: Saunders, 1968.

JUNQUEIRA, LUIS C., JOSÉ CARNEIRO, and ALEXANDER N. CONTOPOULOS. *Basic Histology.* 2d ed. Los Altos, Calif.: Lange, 1977.

LEESON, C. ROLAND, and THOMAS S. LEESON. *Histology.* 3d ed. Philadelphia: Saunders, 1976.

LENTZ, THOMAS L. *Cell Fine Structure.* Philadelphia: Saunders, 1971.

PORTER, KEITH R., and MARY A. BONNEVILLE. *Fine Structure of Cells and Tissues.* 4th ed. Philadelphia: Lea & Febiger, 1973.

REITH, EDWARD J., and MICHAEL H. ROSS. *Atlas of Descriptive Histology.* 2d ed. New York: Harper & Row, 1970.

RHODIN, JOHANNES A. G. *An Atlas of Histology.* New York: Oxford University Press, 1975.

STILES, KARL A. *Handbook of Histology.* 5th ed. New York: McGraw-Hill, 1968.

Neuroanatomy

°BARR, MURRAY L. *The Human Nervous System.* 2d ed. Hagerstown, Md.: Harper & Row, 1974 (paperback).

°CARPENTER, MALCOLM B. *Core Text of Neuroanatomy.* Baltimore: Williams & Wilkins, 1972 (paperback).

°CHUSID, J. G. *Correlative Neuroanatomy and Functional Neurology.* 16th ed. Los Altos, Calif.: Lange, 1976.

CLARK, RONALD G. *Manter and Gatz's Essentials of Clinical Neuroanatomy and Neurophysiology.* 5th ed. Philadelphia: Davis, 1975.

°°GUYTON, ARTHUR C. *Structure and Function of the Nervous System.* 2d ed. Philadelphia: Saunders, 1976.

°NOBACK, CHARLES R., and ROBERT J. DEMAREST. *The Human Nervous System.* 2d ed. New York: McGraw-Hill, 1977 (paperback).

Pathology

°ANDERSON, W. A. D., and THOMAS M. SCOTTI. *Synopsis of Pathology.* 9th ed. St. Louis: Mosby, 1972.

°BOLANDE, ROBERT P. *Cellular Aspects of Developmental Pathology.* Philadelphia: Lea & Febiger, 1967.

BOYD, WILLIAM, and HUNTINGTON SHELDON. *An Introduction to the Study of Disease.* 7th ed. Philadelphia: Lea & Febiger, 1977.

HOPPS, HOWARD C. *Principles of Pathology.* 2d ed. New York: Appleton-Century-Crofts, 1964.

KING, DONALD WEST, ed. *Ultrastructural Aspects of Disease.* New York: Harper & Row, 1966.

°°ROBBINS, STANLEY L. *Pathologic Basis of Disease.* Philadelphia: Saunders, 1974.

° ———, and MARCIA ANGELL. *Basic Pathology.* 2d ed. Philadelphia: Saunders, 1976.

°SUCHESTON, MARTHA E., and M. SAMUEL CANNON. *Congenital Malformations.* Philadelphia: Davis, 1973.

Physiology

°GANONG, W. F. *Review of Medical Physiology.* 7th ed. Los Altos, Calif.: Lange, 1975.

GUYTON, ARTHUR C. *Function of the Human Body.* 4th ed. Philadelphia: Saunders, 1974.

° ° _____. *Textbook of Medical Physiology.* 5th ed. Philadelphia: Saunders, 1976.

RATCLIFF, J. D. *Your Body and How It Works.* New York: Dell (Reader's Digest/Delacorte Press), 1975.

VANDER, ARTHUR J., JAMES H. SHERMAN, and DOROTHY S. LUCIANO. *Human Physiology.* 2d ed. New York: McGraw-Hill, 1975.

Terminology

Dorland's Illustrated Medical Dictionary. 25th ed. Philadelphia: Saunders, 1974.

MILLER, B. F., and C. B. KEANE. *Encyclopedia and Dictionary of Medicine, Nursing, and Allied Health.* 2d ed. Philadelphia: Saunders, 1978.

Stedman's Medical Dictionary. 22d ed. Baltimore: Williams & Wilkins, 1972.

THOMAS, CLAYTON L., ed. *Tabor's Cyclopedic Medical Dictionary.* 12th ed. Philadelphia: Davis, 1973.

Tests and Treatments

°BENSON, R. C. *Handbook of Obstetrics and Gynecology.* 6th ed. Los Altos, Calif: Lange, 1977 (paperback).

CHATTON, M. J. *Handbook of Medical Treatment.* 15th ed. Los Altos, Calif: Lange, 1977 (paperback).

°KRUPP, M. A., and M. J. CHATTON, eds. *Current Medical Diagnosis and Treatment.* Los Altos, Calif: Lange, 1977 (paperback).

°_____, N. J. SWEET, E. JAWETZ, E. G. BIGLIERI, and R. L. ROE. *Physician's Handbook.* 18th ed. Los Altos, Calif.: Lange, 1976.

°WILSON, J. L., ed. *Handbook of Surgery.* 5th ed. Los Altos, Calif: Lange, 1973.

Index

79 80 81 82 9 8 7 6 5 4 3 2 1